Bolyai Society Mathematical Studies

Volume 32

This series will typically consist of survey volumes. It may also include conference proceedings and monographs.

Gyula O. H. Katona · Balázs Patkós ·
Casey Tompkins
Editors

Sum(m)it280

Surveys in Extremal Combinatorics and Combinatorial Geometry

BOLYAI SOCIETY

Springer

Editors
Gyula O. H. Katona
Alfréd Rényi Institute of Mathematics
Budapest, Hungary

Balázs Patkós
Alfréd Rényi Institute of Mathematics
Budapest, Hungary

Casey Tompkins
Alfréd Rényi Institute of Mathematics
Budapest, Hungary

ISSN 1217-4696 ISSN 2947-9460 (electronic)
Bolyai Society Mathematical Studies
ISBN 978-3-032-18809-0 ISBN 978-3-032-18810-6 (eBook)
https://doi.org/10.1007/978-3-032-18810-6

This Springer imprint is published by the registered company Springer Nature Switzerland AG
The registered company address is: Gewerbestrasse 11, 6330 Cham, Switzerland

Preface

The years 1953 and 1954 were exceptionally good years for Hungarian mathematics. Péter Frankl, Zoltán Füredi, Ervin Győri, and János Pach were born in these years. I started a seminar on Extremal Set Theory for students at the beginnings of the 1970s. These four students joined the seminar in the years 1973–1975. They were supposed to read and tell us the contents of new papers in the area. But soon they started to tell their solutions for the open problems found in these papers. I was extremely lucky and happy.

János simultaneously was also working in Geometry, he was connecting Combinatorics and Geometry. The other three kept working mainly in Extremal Combinatorics. Their life-works (until now) are huge. According to MathSciNet, the 4 of them have written 1178 papers, 16,336 papers cite them. Altogether they have 47 scientific descendants. They really deserve a nice celebration. In July, 2024, the Rényi Institute and the János Bolyai Mathematical Society organized a week-long conference Sum(m)it280 in Budapest to celebrate their 70th birthdays. This event mobilized the people working in the area, there were about 250 participants, among them many strong mathematicians. The present volume is the reflection of this conference. It contains mainly surveys related to the works of the celebrated friends. But there are some new results, too.

Budapest, Hungary
December, 2025

Gyula O. H. Katona

Contents

Some Extremal Problems of Zero-Sum Theory in Additive Combinatorics

Sukumar Das Adhikari

Abstract Consider a finite abelian group G (written additively). A sequence $S = (g_1, \ldots, g_l)$ over G is called a *zero-sum sequence* if $g_1 + \cdots + g_l = 0$, where 0 is the identity element of the group. Inspired by a well known result of Erdős, Ginzburg and Ziv, and some questions of K. Rogers and Davenport related to factorization in number fields, the area of zero-sum theorems in combinatorial number theory has seen a rapid growth in recent years. The area of zero-sum theorems has many interesting results and several unanswered questions. Several authors have introduced interesting elementary algebraic techniques to deal with these problems. We describe some experiments with these elementary algebraic methods and some combinatorial ones, in a weighted generalization of some extremal problems in the area of Zero-sum Combinatorics.

1 Introduction

Let G be a finite abelian group (written additively). A sequence $S = (g_1, g_2, \ldots, g_l)$ of elements (not necessarily distinct) of G, is called a *zero-sum sequence* if $g_1 + \cdots + g_l = 0$, where 0 is the identity element of the group.

One is interested in the study of various zero-sum constants associated to G, which are the lengths of the smallest sequence of elements of G guaranteeing a non-empty zero-sum subsequence with or without restriction on its size.

A weighted version of some zero-sum constants was initiated in some papers of Adhikari et al. [5], Adhikari and Rath [10], Thangadurai [46], Adhikari and Chen [4] and Adhikari et al. [3].

The purpose of this expository article is to go through some of the zero-sum constants with some interesting weights and have a glimpse of some elementary algebraic methods and some combinatorial ones, while dealing with these problems.

S. Das Adhikari (✉)
Department of Mathematics, Ramakrishna Mission Vivekananda Educational and Research Institute, Howrah, West Bengal, India
e-mail: adhikarisukumar@gmail.com

G. O. H. Katona et al. (eds.), *Sum(m)it280*, Bolyai Society Mathematical Studies 32,
https://doi.org/10.1007/978-3-032-18810-6_1

There are already some nice expository articles [20, 25] and books [26, 31, 39] dealing with the classical zero-sum constants. Regarding the weighted generalizations, we shall discuss the problems of finding some weighted constants and we shall mainly consider them with weight sets different from those in the existing expository articles such as [2] on the weighted version. Also we shall consider only two of the weighted constants here. We shall start with a short account of the classical case.

We shall write $\mathbb{Z}_n$ for the group $\mathbb{Z}/n\mathbb{Z}$ and for a finite set S, $|S|$ will denote the number of elements of S.

2 Classical Case: The Constants $D(G)$ and $E(G)$

Given a positive integer n, let us try to see whether a sequence of integers of a certain length must have a subsequence of length n whose elements add up to a multiple of n.

One observes that the following sequence of length $2n-2$ does not have any subsequence of length n such that the sum of the elements of the subsequence is zero modulo n.

$$(\underbrace{0, 0, \ldots, 0}_{(n-1)\text{ times}}, \underbrace{1, 1, \ldots, 1}_{(n-1)\text{ times}}) \tag{1}$$

We now increase the length of the sequence by 1.

If $n = 2$, in a sequence of $2n - 1 = 3$ integers, two of them will have the same parity and hence we get a subsequence of 2 elements whose sum is 0 modulo 2.

If $n = 3$, in a sequence of $2n - 1 = 5$ integers, if there are elements $\equiv i \pmod 3$, for each $i = 0, 1, 2$, then clearly we get a subsequence of 3 elements whose sum is 0 modulo 3. Otherwise, there are three elements congruent to each other modulo 3 and their sum would be 0 modulo 3.

The 1961 result of Erdős, Ginzburg and Ziv [23] (usually called the *EGZ-theorem*) says that *for any positive integer n, any sequence $a_1, a_2, \ldots, a_{2n-1}$ of $2n-1$ integers has a subsequence of n elements whose sum is 0 modulo n.*

A prototype of zero-sum theorems, the EGZ-theorem has several proofs (see [13] for instance); some among the most interesting proofs involve elementary algebraic and combinatorial techniques. We would like to mention that the essence of the EGZ theorem lies in the case when n is a prime.

For a finite abelian group G, the constant $E(G)$ is the smallest natural number k such that any sequence of k elements in G has a zero-sum subsequence of length $|G|$. The EGZ theorem, together with the example (1), implies that $E(\mathbb{Z}_n) = 2n - 1$. Again, by a simple inductive argument, from the EGZ-theorem, appealing to the structure theorem for finite abelian groups, it follows that *any sequence of length $2n - 1$ of elements in an abelian group G of order n contains a zero-sum subsequence of length n and hence $E(G) \leq 2|G| - 1$.*

We now proceed to define the constant $D(G)$.

Given any sequence of n integers $(a_1, a_2, \ldots, a_n)$, considering the sums

$$s_1 = a_1, s_2 = a_1 + a_2, \ldots, s_n = a_1 + \cdots + a_n,$$

by the pigeonhole principle, it follows that either an s_i is $0 \pmod n$ or at least two of the s_i's are equal modulo n.

This gives us an elementary combinatorial result which says that given any sequence of n integers, the sum of the elements of a non-empty subsequence is $0 \pmod n$. One can ask for a natural generalization of the above problem for an arbitrary finite abelian group.

Indeed, in the 1960s, for a finite abelian group G, the following problem was posed.

Determine the smallest natural number k (known as *the Davenport constant* and denoted by $D(G)$) such that any sequence of k elements in G has a non-empty zero-sum subsequence. Though the problem of finding $D(G)$ had been proposed by H. Davenport in the 'Midwestern Conference on Group Theory and Number Theory' at Ohio State University held in April 1966, Rogers [43] had studied it in 1962 (attributing the question to Sudler).

Since the paper of Rogers was somehow missed out by the early authors in this area, being attributed to Davenport, it is now known as the Davenport constant.

Davenport posed the question in the connection that if G is the class group of an algebraic number field K, then $D(G)$ is the maximal number of prime ideals (counting multiplicity) required in the decomposition of an irreducible integer in K. Rogers too had studied it in connection with the same question. The problem of finding $D(G)$ for a finite abelian group G, also has applications in graph theory and in some other problems which include the proof of the infinitude of Carmichael numbers by Alford, Granville and Pomerance [12]. By the argument we have seen earlier in this section, for a finite abelian group G of order n,

$$D(G) \leq n.$$

When $G = \mathbb{Z}_n$, observing that the sequence

$$\underbrace{(1, 1, \ldots, 1)}_{(n-1) \text{ times}}$$

does not have any non-empty subsequence whose sum is zero, one has the equality $D(\mathbb{Z}_n) = n$.

$$\text{Given} \quad G = \mathbb{Z}_{n_1} \times \mathbb{Z}_{n_2} \times \cdots \times \mathbb{Z}_{n_r},$$

with $n_1 | n_2 | \cdots | n_r$, writing $M(G) = 1 + \sum_{i=1}^{r} (n_i - 1)$, one observes that

$$M(G) \leq D(G) \leq |G|. \tag{2}$$

The equality $D(G) = |G|$ holds if and only if $G = \mathbb{Z}_n$, the cyclic group of order n. Olson [40, 41] proved that $D(G) = M(G)$ for all finite abelian groups of rank 2 and for all finite abelian p-groups (via an identity in the group-ring of the group G over the rational integers).

First Baayen and then van Emde Boas and Kruyswijk [18] gave examples of groups G with $D(G) > M(G)$. Geroldinger and Schneider [28] showed that there are infinitely many G with rank 4 such that $D(G) > M(G)$. Interesting results on the upper bound of $D(G)$ for a non-cyclic abelian group G, have been obtained by Alon et al. [11], Caro [19], Ordaz and Quiroz [42], Balasubramanian and Bhowmik [17] and others.

Emde Boas and Kruyswijk [18], Baker and Schmidt [16] and Meshulam [38] gave upper bounds involving the exponent m of the group and the cardinality of the group G. The best among them (proved again by Alford et al. [12]) is:

$$D(G) \leq m\left(1 + \log\frac{|G|}{m}\right).$$

Remark 1 We would like to mention that the proof of the above bound by Alford, Granville, Pomerance is a clever generalization of Olson's [40] idea for obtaining the exact value of the Davenport constant for a finite abelian p-group G via an identity in the group-ring of the group G over the rational integers.

Conjecture 1 (Gao and Geroldinger) *With notations as in (2),*
$D(G) = M(G)$*, if* G *is of rank* 3 *or* $G = \mathbb{Z}_n^r$.

Remark 2 The above conjecture has been established asymptotically for $G = \mathbb{Z}_n^r$ as $n \to \infty$ by Girard [29].

Conjecture 2 (Narkiewicz and Śliwa) *With notations as in (2),*

$$D(G) \leq \sum_{i=1}^{r} n_i.$$

For a finite abelian group G of order n, Gao [24] (see also [1], [26, Proposition 5.7.9]) established the following relation between the constants $D(G)$ and $E(G)$:

$$E(G) = D(G) + n - 1. \tag{3}$$

It is easy to observe that by appending a sequence of $n - 1$ zeroes to a sequence of length $D(G) - 1$ with no non-empty zero-sum subsequence (which exists by the definition of $D(G)$), the resulting sequence of length $D(G) + n - 2$ will have no zero-sum subsequence of length n and hence to establish (3), one has to prove

$$E(G) \leq D(G) + n - 1. \tag{4}$$

Remark 3 Since for a finite abelian group of order n, $D(G) \leq n$, from the result (3) of Gao, it follows that $E(G) \leq 2n - 1$ and hence the EGZ theorem follows from (3). The constant $E(G)$ is called the *Gao constant* by some authors.

3 Weighted Versions of the Constants $D(G)$ and $E(G)$

Given a finite abelian group G with exponent $\exp(G) = m$ and a non-empty subset A of $\{1, 2, \ldots, m-1\}$, the *Davenport constant of G with weight A*, denoted by $D_A(G)$, is defined to be the least positive integer k such that for any sequence $(x_1, \ldots, x_k)$ of length k, with $x_i \in G$, there exists a non-empty subsequence $(x_{j_1}, \ldots, x_{j_l})$ and $a_1, \ldots, a_l \in A$ such that $\sum_{i=1}^{l} a_i x_{j_i} = 0$.

Similarly, for any such weight set A, if the abelian group G has order n, *the constant* $E_A(G)$ is defined to be the least positive integer t such that for any sequence $(x_1, \ldots, x_t)$ of t elements with $x_i \in G$, there exists an *A-weighted zero-sum subsequence* of length n, that is, there exist indices $j_1, \ldots, j_n \in \mathbb{N}$, $1 \le j_1 < \cdots < j_n \le t$, and $\vartheta_1, \ldots, \vartheta_n \in A$ with $\sum_{i=1}^{n} \vartheta_i x_{j_i} = 0$.

For $A = \{1\}$, $D_A(G)$ and $E_A(G)$ correspond respectively to the classical *Davenport constant* $D(G)$ and the *Gao constant* $E(G)$.

The generalizations $D_A(G)$ and $E_A(G)$ were initially considered (see [5, 10]) for the particular group $\mathbb{Z}_n$. Later in [4, 46], they were introduced for an arbitrary finite abelian group G.

Even though the weighted version was taken up mainly as some curious questions, and also with some hope that this could shed some light on the conjectures on bounds of the classical ones, there are now arithmetical interpretations of the weighted Davenport constant and its further generalizations by Halter-Koch [34] and others [27]. Also there are interactions with coding theory (see Marchan et al. [37]) and several other interesting developments prompted by these weighted considerations.

While obtaining some precise results for certain weights were interesting, research on these problems gave rise to some new ideas and applications of some new results on set addition.

We start with some early results on $D_A(G)$ and $E_A(G)$, with $G = \mathbb{Z}_n$, for some weight sets A.

(I) The case $A = \{1, -1\}$**.**
In the very first paper [5] on this theme, the weight set $A = \{1, -1\}$ was considered. Here, one has $D_A(\mathbb{Z}_n) \le [\log_2 n] + 1$ by the pigeonhole principle; and considering the sequence $(1, 2, \ldots, 2^r)$, where r is defined by $2^{r+1} \le n < 2^{r+2}$, it follows that $D_A(\mathbb{Z}_n) \ge [\log_2 n] + 1$. Thus

$$D_A(\mathbb{Z}_n) = [\log_2 n] + 1. \tag{5}$$

It was shown by Adhikari et al. [5] that

$$E_A(\mathbb{Z}_n) = n + [\log_2 n]. \tag{6}$$

(II) The case when $A = \{a \in \{1, 2, \ldots, n-1\} | (a, n) = 1\}$**.**
Observing that writing $n = p_1 \cdots p_s$ where the p_i's are primes, the sequence $(1, p_1, p_1 p_2, \ldots, p_1 p_2 \cdots p_{s-1})$ does not have any non-empty A-weighted zero-sum

subsequence and hence $D_A(n) \geq 1 + \Omega(n)$, it was conjectured in [5] that

$$E_A(n) = n + \Omega(n), \tag{7}$$

where $\Omega(n)$ denotes the number of prime factors of n, multiplicity included.

It was first proved by Luca [36] and later by Griffiths [30] by a different method. From

$$n + \Omega(n) = E_A(n) \geq n - 1 + D_A(n) \geq n + \Omega(n),$$

it follows that

$$D_A(n) = 1 + \Omega(n). \tag{8}$$

(III) The case where $n = p$, a prime, and A is the set of quadratic residues (mod p).

In this case, it was shown by Adhikari and Rath [10] that

$$D_A(\mathbb{Z}_p) = 3, \quad E_A(\mathbb{Z}_p) = p + 2. \tag{9}$$

Proof Here an elementary algebraic method, which had been used in [13] for the proof of the EGZ theorem, worked very well. Given any sequence $(s_1, \ldots, s_{p+2})$ of elements of $\mathbb{Z}_p$, one considers the following system of equations over the finite field $\mathbb{F}_p$ of p elements.

$$\sum_{i=1}^{p+2} s_i x_i^2 = 0, \quad \sum_{i=1}^{p+2} x_i^{p-1} = 0,$$

By Chevalley-Warning Theorem (see [39, 45], for instance), there is a nontrivial solution $(y_1, \ldots, y_{p+2})$ of the above system. Writing $I = \{i : y_i \neq 0\}$, from the first equation it follows that $\sum_{i \in I} a_i s_i = 0$, where a_i's belong to the set of squares in $(\mathbb{Z}_p)^*$. By Fermat's little theorem, from the second equation we have $|I| = p$ and hence one obtains

$$E_A(\mathbb{Z}_p) \leq p + 2.$$

Then considering a sequence $v_1, -v_2$, where v_1 is a quadratic residue and v_2 a quadratic non-residue (mod p), one has

$$3 \leq D_A(\mathbb{Z}_p) \leq E_A(\mathbb{Z}_p) - p + 1 \leq 3,$$

and hence (9) is established. □

(IV) The case when $n = p$, a prime, and $A = \{1, 2, \ldots, r\}$, where r is an integer such that $1 < r < p$.

In this case, it was proved by Adhikari and Rath [10] that

$$D_A(\mathbb{Z}_p) = \left\lceil \frac{p}{r} \right\rceil, \quad E_A(p) = p - 1 + \left\lceil \frac{p}{r} \right\rceil. \tag{10}$$

To establish the result on $D_A(\mathbb{Z}_p)$, a simple combinatorial argument was required. Consider any sequence $S = (s_1, \ldots, s_{\lceil \frac{p}{r} \rceil})$ of elements of $\mathbb{Z}/p\mathbb{Z}$ of length $\lceil \frac{p}{r} \rceil$. Considering the sequence

$$S' = (\overbrace{s_1, s_1, \ldots, s_1}^{r \text{ times}}, \overbrace{s_2, s_2, \ldots, s_2}^{r \text{ times}}, \ldots, \overbrace{s_{\lceil \frac{p}{r} \rceil}, \ldots, s_{\lceil \frac{p}{r} \rceil}}^{r \text{ times}}),$$

obtained from S by repeating each element r times, and observing that the length of this sequence is $\geq p$, it follows that

$$D_A(p) \leq \left\lceil \frac{p}{r} \right\rceil. \tag{11}$$

We shall come to the proof of the result on $E_A(p)$ later.

The above observations lead people (Adhikari and Rath [10], Adhikari and Chen [4], Thangadurai [46], Griffiths [30]) to the conjecture that for a finite abelian group G and any non-empty subset A of $\{1, \ldots, \exp(G) - 1\}$, the following *weighted generalization of Gao's theorem* holds.

$$E_A(G) = D_A(G) + n - 1. \tag{12}$$

For $G = \mathbb{Z}_p$, p prime, for the results in (10), the method of deriving $E_A(\mathbb{Z}_p) = p - 1 + \lceil \frac{p}{r} \rceil$ from $D_A(\mathbb{Z}_p) = \lceil \frac{p}{r} \rceil$, by using the Cauchy-Davenport theorem, employed by Adhikari and Rath [10], works for establishing (12) for any weight set $A \subset [1, p - 1]$. Therefore, we get a proof of (12) for $G = \mathbb{Z}_p$. We shall here see a sketch of a later proof of this result by Adhikari et al. [7] using the permanent lemma.

For the weight $A = \{1, -1\}$, one needed some new ideas to establish the result $E_A(\mathbb{Z}_n) = n + [\log_2 n]$. The ideas here were suitably generalized by Adhikari and Chen [4], to prove the following.

Theorem 1 *Let G be a finite abelian group of order n, and $A = \{a_1, a_2, \ldots, a_r\}$ be a finite subset of $\mathbb{Z}$ with $|A| = r \geq 2$ and* $\gcd(a_2 - a_1, a_3 - a_1, \ldots, a_r - a_1, n) = 1$. *Then we have $E_A(G) = D_A(G) + n - 1$.*

Here, one needs to consider two cases: when the given sequence is complete in G or not. A sequence $(x_1, \ldots, x_N)$ with elements $x_i \in G$ is called *complete in G* if for any subgroup H of G we have

$$|\{j | x_j \notin H\}| \geq \frac{|G|}{|H|} - 1.$$

The conjecture (12) was established for cyclic groups by Yuan and Zeng [49] and for general finite abelian groups, by Grynkiewicz et al. [33]. An important ingredient in both these proofs is a result of Devos et al. [22].

We now proceed to give a sketch of the proof of (12) for $G = \mathbb{Z}_p$ by Adhikari et al. [7] which uses the permanent lemma.

Since the result (3) of Gao takes care of the case with $A = \{1\}$, it was enough to prove the following.

Theorem 2 *Whenever the weight set $A \subset [1, p-1]$ satisfies $|A| \geq 2$, we have*

$$E_A(\mathbb{Z}_p) = D_A(\mathbb{Z}_p) + p - 1.$$

Let us recall that given an n by n matrix $A = [a_{i,j}]$ over a field F, its *permanent*, denoted by $per\ A$, is defined by

$$per\ A = \sum a_{1,\alpha(1)} a_{2,\alpha(2)} \cdots a_{n,\alpha(n)},$$

where the summation is taken over all permutations α of $[n]$.

We shall require a special case of the *permanent lemma*, first proved in [15] (see also [14]):

Lemma 1 *For a prime p, let $A = (a_{i,j})$ be a $(p-1) \times (p-1)$ matrix over the finite field $\mathbb{F}_p$. Suppose that its permanent $Per(A)$ is non-zero. Then given any sequence of elements $b_1, b_2, \ldots, b_{p-1}$ in $\mathbb{F}_p$, there exists $\epsilon_1, \epsilon_2, \ldots, \epsilon_{p-1} \in \{0, 1\}$ such that*

$$\sum_{j=1}^{p-1} \epsilon_j a_{ij} \neq b_i, \ \forall i.$$

Sketch of a proof of Theorem 2. Let A contain a, b with $1 \leq a < b \leq p-1$. Let $c_1, \ldots, c_m$ be any sequence of m elements of $\mathbb{F}_p$ where $m = D_A(p) + p - 1$.

First we assume that the number of non-zero elements in the above sequence is at least $p-1$. Without loss of generality, we assume that $c_i \neq 0$ for $1 \leq i \leq p-1$.

Consider the matrix $A = (a_{ij})_{(p-1)\times(p-1)}$ where for all i, $a_{ij} = (b-a)c_j$, $1 \leq j \leq p-1$.

Now, consider the element $-a(c_1 + \cdots + c_p)$ in $\mathbb{F}_p$ and let $b_1, \ldots, b_{p-1}$ be the remaining elements of $\mathbb{F}_p$.

Since $per(A) = (p-1)!(b-a)^{p-1} c_1 c_2 \cdots c_{p-1} \neq 0$, by using the permanent lemma (Lemma 1), there exists $\epsilon_1, \epsilon_2, \ldots, \epsilon_{p-1} \in \{0, 1\}$ such that

$$\sum_{j=1}^{p-1} \epsilon_j (b-a) c_j = -a(c_1 + \cdots + c_p),$$

which can be re-written as

$$\sum_{j=1}^{p-1} \left(\epsilon_j (b-a) c_j + a c_j\right) + a c_p = 0.$$

Clearly, this shows that a sum of the subsequence $(c_1, \ldots, c_p)$ with coefficients in $\{a, b\}$ is zero.

In the case when the given sequence has less than $p-1$ non-zero elements, it is taken care of by a combinatorial argument. □

Remark 4 Clearly, Theorem 2 follows from Theorem 1.

Many people worked on questions and conjectures related to these generalizations. We record here some results on $D_A(\mathbb{Z}_n)$ for some weight sets A.

For general n, if A consists of the squares in the group of units in the cyclic group $\mathbb{Z}_n$, it was proved by Adhikari et al. [6] that if n is square-free and coprime to 6, then

$$D_A(\mathbb{Z}_n) = 2\Omega(n) + 1. \tag{13}$$

Chintamani and Moriya [21] proved that (13) holds when n is coprime to $30 = 2 \times 3 \times 5$; removing the requirement that n is square-free. They also proved that (13) holds when n is a power of 3. In this direction, Grynkiewicz and Hennecart [32] proved that, for an odd integer n if A is the set of squares in the group of units in the cyclic group $\mathbb{Z}_n$, then for a prime p writing $v_p(n)$ to denote the p-valuation of n, one has

$$D_A(\mathbb{Z}_n) = 2\Omega(n) + \min\{v_5(n), v_3(n)\} + 1, \tag{14}$$

if either $v_3(n) = 0$ or $v_3(n) \geq v_5(n)$. The proof uses some combinatorial tools in design theory among other things.

Generalizing (9), in a different direction, Adhikari et al. [3] showed that if $W \subseteq (\mathbb{Z}_p)^*$ satisfies $|W| \geq \frac{p}{3}$ and there exists α in $(\mathbb{Z}_p)^*$ such that $\frac{a}{b} \neq \alpha$ for any a, b in W, then $D_W(\mathbb{Z}_p) = 3$.

Taking A to be a subset of the set of quadratic residues $\pmod p$, satisfying $|A| \geq \frac{p}{3}$, since for any quadratic non-residue α, we have $\frac{a}{b} \neq \alpha$ for any a, b in A, the above result can be applied in this case to obtain the result $D_A(\mathbb{Z}_p) = 3$.

Now we come to a result of Griffiths [30] which says that for an odd integer $n = p_1^{a_1} \cdots p_k^{a_k}$, where the p_i's are distinct primes, so that $\mathbb{Z}_n$ is isomorphic to $\mathbb{Z}_{p_1^{a_1}} \times \cdots \times \mathbb{Z}_{p_k^{a_k}}$, if for each s, $A_s \subset \mathbb{Z}_{p_s^{a_s}}$ is a subset with its size $|A_s| > p_s^{a_s}/2$, then with $A = A_1 \times \cdots \times A_k$,

$$D_A(\mathbb{Z}_n) \leq \Omega(n) + 1.$$

Griffiths [30] also had a similar result when n is even.

With suitable modifications of the method of Griffiths [30], it was shown by Adhikari and Mazumdar [9] that if with the above notation, $|A_s| > (4/9)p_s^{a_s}$ for each s, then

$$D_A(\mathbb{Z}_n) \leq 2\Omega(n) + 1.$$

Coming to generalizations of (10), for a general positive integer n, with $A = \{1, 2, \ldots, r\}$, where r is an integer such that $1 < r < n$, Xia and Li [48] showed that by the argument in [10], which was used to prove (10), one can establish the following:

$$D_A(\mathbb{Z}_n) = \left\lceil \frac{n}{r} \right\rceil. \tag{15}$$

While for a positive integer n, with $A = \mathbb{Z}_n \setminus \{0\}$, it was observed in [5] that $D_A(\mathbb{Z}_n) = 2$, Xia [47] proved that if A is an arithmetic progression with length $l = \lceil \frac{n}{2} \rceil$ and common difference 1, then $D_A(\mathbb{Z}_n) = 2$.

Since for weight sets A, B, with $A \subset B$, clearly $D_B(\mathbb{Z}_n) \leq D_A(\mathbb{Z}_n)$, for any C with

$$\{1, 2, \ldots, \left\lceil \frac{n}{2} \right\rceil\} \subset C \subset \{1, 2, \ldots, n-1\}, \tag{16}$$

we have $D_C(\mathbb{Z}_n) = 2$.

Clearly, in (16), $\{1, 2, \ldots, \lceil \frac{n}{2} \rceil\}$ can be replaced by any arithmetic progression contained in $\{1, 2, \ldots, n-1\}$ with length $l = \lceil \frac{n}{2} \rceil$ and common difference 1.

Another result of Xia [47] says that if n is a positive integer and p a prime satisfying $p^k || n$ and $A = \{a : 1 \leq a \leq n, \ (a, p) = 1\}$, then $D_A(\mathbb{Z}_n) = k + 1$.

These results of Xia have been further generalized by Adhikari and Hegde [8]; where apart from some elementary combinatorial arguments, Scherk's lemma [44] (see also [35]) has been used.

Acknowledgements The author would like to thank the referee for pointing out several typos and suggesting some small corrections.

References

1. Adhikari, S.D.: The relation between two combinatorial group invariants: history and ramifications. Math. Stud. **77**(1–4), 131–143 (2008)
2. Adhikari, S.D.: Plus-minus weighted zero-sum constants: a survey, In: Analytic Number Theory, Modular Forms and q-Hypergeometric Series. Springer Proceedings in Mathematics & Statistics, vol. 221, pp. 1–9. Springer (2017)
3. Adhikari, S.D., Balasubramanian, R., Pappalardi, F., Rath, P.: Some zero-sum constants with weights. Proc. Indian Acad. Sci. (Math. Sci.) **118**, 183–188 (2008)
4. Adhikari, S.D., Chen, Y.G.: Davenport constant with weights and some related questions-II. J. Comb. Theory Ser. A **115**, 178–184 (2008)
5. Adhikari, S.D., Chen, Y.G., Friedlander, J.B., Konyagin, S.V., Pappalardi, F.: Contributions to zero-sum problems. Discrete Math. **306**, 1–10 (2008)
6. Adhikari, S.D., David, C., Urroz, J.J.: Generalizations of some zero-sum theorems. Integers **8**, (2008). Paper A 52
7. Adhikari, S.D., Gun, S., Rath, P.: Remarks on some zero-sum theorems. Proc. Indian Acad. Sci. (Math. Sci.) **119**(3), 275–281 (2009)
8. Adhikari, S.D., Hegde, S.: Zero-sum constants involving weights. Proc. Indian Acad. Sci. Math. Sci.**131**(2), 9 (2021). Paper No. 37
9. Adhikari, S.D., Mazumdar, E.: Modifications of some methods in the study of zero-sum constants. Integers **14**, (2014). Paper A 25
10. Adhikari, S.D., Rath, P.: Davenport constant with weights and some related questions. Integers **6**, A30 (2006)
11. Alon, N., Bialostocki, A., Caro, Y.: Extremal zero-sum problems. Manuscript
12. Alford, W.R., Granville, A., Pomerance, C.: There are infinitely many Carmichael numbers. Ann. Math. **139**(3), 703–722 (1994)
13. Alon, N., Dubiner, M.: Zero-sum sets of prescribed size. Combinatorics, Paul Erdős is eighty (Volume 1), Keszthely (Hungary) 30–50 (1993)

14. Alon, N., Linial, N., Meshulam, R.: Additive bases of vector spaces over prime fields. J. Comb. Theory Ser. A **57**, 203–210 (1991)
15. Alon, N., Tarsi, M.: A nowhere-zero point in linear mappings. Combinatorica **9**, 393–395 (1989)
16. Baker, R.C., Schmidt, W.: Diophantine problems in variables restricted to the values of 0 and 1. J. Number Theory **12**, 460–486 (1980)
17. Balasubramanian, R., Bhowmik, G.: Upper bounds for the Davenport constant. Integers **7**(2), A03 (2007)
18. van Emde Boas, P., Kruswjik, D.: A combinatorial problem on finite abelian group III. Z. W. 1969-008 (Math. Centrum, Amsterdam)
19. Caro, Y.: Zero-sum subsequences in abelian non-cyclic groups. Israel J. Math. **92**, 221–233 (1995)
20. Caro, Y.: Zero-sum problems—a survey. Discrete Math. **152**, 93–113 (1996)
21. Chintamani, M.N., Moriya, B.K.: Generalizations of some zero sum theorems. Proc. Indian Acad. Sci. (Math. Sci.) **122**(1), 15–21 (2012)
22. DeVos, M., Goddyn, L., Mohar, B.: Representation of finite abelian group elements by subsequence sums. J. Théor. Nombres Bordeaux **21**(3), 559–587 (2009)
23. Erdős, P., Ginzburg, A., Ziv, A.: Theorem in the additive number theory. Bull. Res. Counc. Isr. **10F**, 41–43 (1961)
24. Gao, W.D.: A combinatorial problem on finite abelian groups. J. Number Theory **58**, 100–103 (1996)
25. Gao, W.D., Geroldinger, A.: Zero-sum problems in finite abelian groups; a survey. Expo. Math. **24**(4), 337–369 (2006)
26. Geroldinger, A., Halter-Koch, F.: Non-Unique Factorizations: Algebraic, Combinatorial and Analytic Theory. Pure and Applied Mathematics (Boca Raton), vol. 278. Chapman & Hall/CRC, Boca Raton, FL (2006)
27. Geroldinger, A., Halter-Koch, F., Zhong, Q.: On monoids of weighted zero-sum sequences and applications to norm monoids in Galois number fields and binary quadratic forms. Acta Math. Hungar. **168**(1), 144–185 (2022)
28. Geroldinger, A., Schneider, R.: On Davenport's constant. J. Combin. Theory Ser. A **61**(1), 147–152 (1992)
29. Girard, B.: An asymptotically tight bound for the Davenport constant. J. Éc. polutech. Math. **5**, 605–611 (2018)
30. Griffiths, S.: The Erdős-Ginzburg-Ziv theorem with units. Discrete Math. **308**(23), 5473–5484 (2008)
31. Grynkiewicz, D.J.: Structural Additive Theory. Developments in Mathematics. Springer (2013)
32. Grynkiewicz, D.J., Hennecart, F.: A weighted zero-sum problem with quadratic residues. Unif. Distrib. Theory **10**(1), 69–105 (2015)
33. Grynkiewicz, D.J., Marchan, L.E., Ordaz, O.: A weighted generalization of two theorems of Gao. Ramanujan J. **28**(3), 323–340 (2012)
34. Halter-Koch, F.: Arithmetical interpretation of weighted Davenport constants. Arch. Math. (Basel) **103**(2), 125–131 (2014)
35. Kemperman, J.H.B., Scherk, P.: Complexes in abelian groups. Can. J. Math. **6**, 230–237 (1954)
36. Luca, F.: A generalization of a classical zero-sum problem. Discret. Math. **307**(13), 1672–1678 (2007)
37. Marchan, L.E., Ordaz, O., Santos, I., Schmid, W.A.: Multi-wise and constrained fully weighted Davenport constants and interactions with coding theory. J. Combin. Theory Ser. A **135**, 237–267 (2015)
38. Meshulam, R.: An uncertainty inequality and zero subsums. Discrete Math. **84**(2), 197–200 (1990)
39. Nathanson, M.B.: Additive Number Theory: Inverse Problems and the Geometry of Sumsets. Springer (1996)
40. Olson, J.E.: On a combinatorial problem of finite Abelian groups, I. J. Number Theory **1**, 8–10 (1969)

41. Olson, J.E.: On a combinatorial problem of finite Abelian groups, II. J. Number Theory **1**, 195–199 (1969)
42. Ordaz, O., Quiroz, D.: The Erdős-Ginzburg-Ziv theorem in abelian non-cyclic groups. Divulg. Mat. **8**(2), 113–119 (2000)
43. Rogers, K.: A combinatorial problem in Abelian groups. Proc. Cambridge Phil. Soc. **59**, 559–562 (1963)
44. Scherk, P.: Solution to problem 4466. Am. Math. Monthly **62**, 46–47 (1955)
45. Serre, J.-P.: A Course in Arithmetic. Springer (1973)
46. Thangadurai, R.: A variant of Davenport's constant. Proc. Indian Acad. Sci. (Math. Sci.) **117**(2), 147 –158 (2007)
47. Xia, X.: Two generalized constants related to zero-sum problems for two special sets. Integers **7**, (2007). Paper A 52
48. Xia, X., Li, Z.: Some davenport constants with weights and Adhikari & Rath's conjecture. Ars Combin. **88**, 83–95 (2008)
49. Yuan, P., Zeng, X.: Davenport constant with weights. Eur. J. Comb. **31**, 677–680 (2010)

Problems and Results in Extremal Combinatorics–V

Noga Alon

Dedicated to Péter Frankl, Zoltán Füredi, Ervin Győri and János Pach, for their 70th birthday

Abstract Frankl, Füredi, Győri and Pach obtained fundamental results in extremal combinatorics, graph theory and discrete geometry. In this paper we describe solutions or partial solutions of several problems in these areas. The problems considered deal with bipartite coverings of graphs, triangle-free graphs of diameter 2, and geometric and combinatorial designs and partial designs. This is a (short) sequel of several previous papers of the same flavour. Each section here is essentially self contained, and can be read separately.

1 Introduction

The conference Summit280, celebrating the 70th birthday of Péter Frankl, Zoltán Füredi, Ervin Győri and János Pach, took place in Budapest in July 2024. The remarkable contributions of these four researchers to Extremal Combinatorics, Graph Theory and Discrete Geometry had a profound impact on the development of modern discrete mathematics. Their results, problems, papers and lectures form an ideal representation of the influential Hungarian school of Combinatorics. The present short article contains several recent solutions or partial solutions of several problems in extremal combinatorics and geometry that originated in this school. These are described in the following three sections, each of which can be read separately. As

Research supported in part by NSF grant DMS-2154082.

N. Alon (✉)
Princeton University, Princeton, NJ, USA

Tel Aviv University, Tel Aviv, Israel
e-mail: nalon@math.princeton.edu

G. O. H. Katona et al. (eds.), *Sum(m)it280*, Bolyai Society Mathematical Studies 32,
https://doi.org/10.1007/978-3-032-18810-6_2

the title of the paper suggests, this is a (short) sequel of four previous papers [1–4] of the same flavour.

In Sect. 2 we consider bipartite coverings of graphs, extending old results of Hansel and of Katona and Szemerédi, and slightly strengthening recent work of Kim and Lee.

In Sect. 3 we prove that for any $0 < \varepsilon < 1/6$ and $n > n_0(\varepsilon)$, one can add to any triangle-free graph on n vertices with maximum degree at most $n^{1/2-\varepsilon}$ less than $3n^{2-\varepsilon}$ edges, transforming it to a triangle-free graph with diameter 2. This settles an open problem of Erdős and Gyárfás and is closely related to a problem of Erdős and Pach and to a result of Füredi and Seress.

In Sect. 4 we prove that there is a partial design with n vertices in which each block is of size at least $\Omega(\sqrt{n})$, so that every set that intersects all blocks contains at least $\Omega(\log n)$ points of one of them. We also show that the number of sequences $n \geq x_1 \geq x_2 \geq x_m \geq 2$ so that there is a block design on n elements with blocks of sizes $x_1, x_2, \ldots, x_m$ is at least $2^{\Omega(n^{1/2} \log n)}$, significantly larger than the number of such sequences that correspond to geometric designs that can be realized by points and lines in the Euclidean plane. This settles two problems of Erdős.

2 Biparite Coverings of Graphs and Multigraphs

2.1 Results

A bipartite covering $\mathcal{H} = \{H_1, \ldots, H_m\}$ of a graph G on the set of vertices $[n] = \{1, 2, \ldots, n\}$ is a collection of bipartite graphs H_i on $[n]$, so that each edge of G belongs to at least one of them. Note that each H_i is not necessarily a subgraph of G, the only assumption is that it is a bipartite subgraph of the complete graph on $[n]$. The capacity cap($\mathcal{H}$) of the cover is the sum $\sum_i |V(H_i)|$ of the numbers of vertices of these bipartite graphs. A known result of Hansel [15] is that the capacity of any bipartite covering of the complete graph K_n on n vertices is at least $n \log_2 n$. This bound is tight when n is power of 2.

In [20] Kim and Lee consider the analogous problem, where the complete graph K_n is replaced by the complete multigraph K_n^λ in which every pair of distinct vertices is connected by λ parallel edges. A bipartite covering here is a collection of bipartite graphs so that each edge belongs to at least λ of them. They prove that the capacity of each bipartite covering of K_n^λ is at least

$$\max\left\{2\lambda(n-1), n\left[\log n + \lfloor(\lambda-1)/2\rfloor \log\left(\frac{\log n}{\lambda}\right) - \lambda - 1\right]\right\},$$

where all logarithms here and in the rest of this section are in base 2. They also establish an upper bound: there exists a bipartite covering of K_n^λ of capacity at most

$$n(\log(n-1) + (1+o(1))\lambda \log\log n).$$

This shows that for $\lambda = (\log n)^{1-\Omega(1)}$ the smallest possible capacity is $n\log n + \Theta(n\lambda \log\log n)$ but leaves an additive gap of $\Omega(\lambda n \log\log n)$ between the upper and lower bounds.

The proofs in [20] proceed by studying a more general problem regarding graphons. Our first contribution in this section is a shorter combinatorial proof of the results above, slightly improving the bounds. Let $\mathrm{cap}(n, \lambda)$ denote the minimum possible capacity of a bipartite covering of K_n^λ.

Theorem 2.1 (Lower bound) *For positive integers $n \geq 2$ and λ,*

$$cap(n,\lambda) \geq \max\left\{2\lambda(n-1), n\left[\log n + \lfloor(\lambda-1)/2\rfloor \log\left(\frac{2\log n}{\lambda-1}\right)\right]\right\}$$

Theorem 2.2 (Upper bound) *Let $k(n, \lambda)$ denote the minimum length of a binary error correcting code with distance at least λ which has at least n codewords. Then $cap(n, \lambda) \leq n \cdot k(n, \lambda)$. Therefore*

1. *For any $n \geq 2$*
$$cap(n,2) \leq n(\lceil \log n\rceil + 1) < n(\log n + 2).$$

2. *For any n and $\lambda \leq 0.5 \log n$*
$$cap(n,\lambda) \leq n\left[\log n + (\lambda-1)\left(\log\left(\frac{\log n}{\lambda-1}\right) + 4\right)\right]$$

3. *For any $0 < c < 1/2$, and for $\lambda \geq c\frac{\log n}{1-H(c)}$ where $H(x) = -x\log x - (1-x)\log(1-x)$ is the binary entropy function,*
$$cap(n,\lambda) \leq \frac{\lambda}{c} n.$$

4. *For any fixed λ there are infinitely many values of n so that*
$$cap(n,\lambda) \leq n[\log n + \lfloor(\lambda-1)/2)\rfloor \log\log n + 2].$$

Katona and Szemerédi [21] proved the following generalization of the result of Hansel, dealing with the capacity of bipartite coverings of general graphs.

Theorem 2.3 ([21]) *Let G be a graph on the set of vertices* $[n]$ *and let* $d_1, d_2, \ldots, d_n$ *denote the degrees of its vertices. Then the capacity of any bipartite covering of G is at least*

$$\sum_{i=1}^{n} \log\left(\frac{n}{n-d_i}\right).$$

Our second contribution here is the following strengthening of this result.

Theorem 2.4 *Let G be a graph on the set of vertices* $[n]$. *For each vertex i let* α_i *denote the maximum size of an independent set of G that contains the vertex i. Then the capacity of any bipartite covering of G is at least*

$$\sum_{i=1}^{n} \log\left(\frac{n}{\alpha_i}\right).$$

Since it is clear that $\alpha_i \leq n - d_i$ for every i, this is indeed a strengthening of the Katona-Szemerédi result (Theorem 2.3). The binomial random graph $G = G(n, 0.5)$ is one example for which Theorem 2.4 is strictly stronger than Theorem 2.3. Indeed, with high probability for $G = G(n, 0.5)$, $d_i = (1/2 + o(1))n$ for every i and $\alpha_i = (2 + o(1)) \log n$ for every i. Therefore the lower bound of Theorem 2.3 for this G is typically $(1 + o(1))n$, whereas the lower bound provided by Theorem 2.4 is $n \log n - (1 + o(1))n \log\log n$. This is tight since the chromatic number of $G = G(n, 0.5)$ is, with high probability, $\chi_n = (1 + o(1))\frac{n}{2\log n}$ implying that G admits a bipartite covering consisting of $\lceil \log \chi_n \rceil$ spanning bipartite graphs, and the corresponding capacity is $n \log n - (1 + o(1))n \log\log n$.

The rest of this section contains the (short) proofs of the results above.

2.2 Complete Multigraphs: The Lower Bound

Let $\mathcal{H} = \{H_1, \ldots, H_m\}$ be a bipartite covering of the complete multigraph K_n^λ on the set of vertices $[n] = \{1, 2, \ldots, n\}$. We first prove that $\text{cap}(\mathcal{H}) \geq 2\lambda(n-1)$. Let n_i denote the number of vertices of H_i. Since it is bipartite the number of its edges is at most $n_i^2/4$. As the edges of all these graphs cover each of the $n(n-1)/2$ edges of the complete graph on $[n]$ at least λ times, and since $n_i \leq n$ for all i, it follows that

$$\frac{n}{4}\sum_{i=1}^{m} n_i \geq \sum_{i=1}^{m} \frac{n_i^2}{4} \geq \lambda n(n-1)/2.$$

This implies that $\text{cap}(\mathcal{H}) = \sum_{i=1}^{m} n_i \geq 2\lambda(n-1)$, as needed.

Note that for any even n this inequality is tight for infinitely many (large) values of λ. In particular it is tight for $\lambda = \frac{n}{4n-4}\binom{n}{n/2}$, and if there is a Hadamard matrix of order n then it is tight for $\lambda = n/2$ as well. In addition, if for some fixed n it is tight for λ_1 and λ_2 then it is also tight for their sum $\lambda = \lambda_1 + \lambda_2$.

We next prove the second inequality, that

$$\text{cap}(\mathcal{H}) \geq n\left[\log n + \lfloor(\lambda - 1)/2\rfloor \log\left(\frac{2\log n}{\lambda - 1}\right)\right].$$

Without loss of generality assume that each of the bipartite graphs H_i in $\mathcal{H}$ is a complete bipartite graph, and let $L_i, R_i \subset [n]$ denote its two color classes. For each vertex $j \in [n]$ let A_j denote the set of indices i for which the vertex j belongs to the vertex class L_i of H_i and let B_j be the set of indices i for which $j \in R_i$. Let $x_j = |A_j| + |B_j|$ be the total number of bipartite graphs H_i that contain the vertex j. Note that $x_j \geq \lambda$ for each j, as any edge incident with j must be covered at least λ times.

Put $r = \lfloor(\lambda - 1)/2\rfloor$ and let $v = (v_1, v_2, \ldots, v_m)$ be a uniform random binary vector of length m. For each j, $1 \leq j \leq n$, let E_j denote the event that the number of indices i that belong to A_j for which $v_i = 1$ plus the number of indices i that belong to B_j for which $v_i = 0$ is at most r. It is clear that the probability of E_j is exactly the probability that the binomial random variable $B(x_j, 1/2)$ is at most r, which is

$$p(x_j, 1/2) = \frac{\sum_{q=0}^{r}\binom{x_j}{q}}{2^{x_j}}.$$

Note, crucially, that the events E_j are pairwise disjoint. This is because for every two distinct vertices j and j' there are at least $\lambda > 2r$ indices i for which j and j' belong to the two distinct vertex classes of H_i. Therefore

$$\sum_{j=1}^{n} p(x_j, 1/2) \leq 1.$$

The desired lower bound for $\text{cap}(\mathcal{H}) = \sum_{j=1}^{n} x_j$ can be deduced from the last inequality by a convexity argument. We proceed with the details. Note, first, that for every x, $p(x, r) \geq (x/r)^r 2^{-x}$. Therefore

$$\sum_{j=1}^{n}(x_j/r)^r 2^{-x_j} \leq 1. \tag{1}$$

Recall also that for each j, $x_j \geq \lambda \geq 2r + 1$. Consider the function $f(x) = (x/r)^r 2^{-x}$. A simple computation shows that for $r = 1$ its second derivative is $(\ln 2)2^{-x}[x\ln 2 - 2]$ which is positive for all $x \geq 2r + 1 = 3$. For $r \geq 2$ the second derivative of $f(x)$ is

$$(x/r)^{r-2}2^{-x}[((x/r)\ln 2 - 1)^2 - 1/r].$$

It is not difficult to check that this is positive for all $x \geq 2r+1$. This shows that $f(x)$ is convex in the relevant range. Therefore, by (1) together with Jensen's Inequality, if we denote $x = \text{cap}(\mathcal{H}) = \sum x_j$ we get

$$n\left(\frac{x}{nr}\right)^r 2^{-x/n} \leq 1$$

implying that

$$x \geq n\left[\log n + r\log\left(\frac{x}{nr}\right)\right].$$

Since $x/n \geq \log n$ this shows that

$$x \geq n\left[\log n + r\log\left(\frac{\log n}{r}\right)\right] \geq n\left[\log n + \lfloor(\lambda-1)/2\rfloor \log\left(\frac{2\log n}{\lambda-1}\right)\right].$$

This completes the proof of Theorem 2.1. □

2.3 Complete Multigraphs, The Upper Bound

In this subsection we prove Theorem 2.2. Put $k = k(n, \lambda)$ and let $A = (a_{ij})$ be the k by n binary matrix whose columns are n of the codewords of a binary code of length k with minimum distance (at least) λ. For each i, $1 \leq i \leq k$, let H_i be the complete bipartite graph on the classes of vertices $L_i = \{j : a_{ij} = 0\}$ and $R_i = \{j : a_{ij} = 1\}$. It is easy to see that these bipartite graphs cover every edge of the complete graph on $[n]$ at least λ times. The capacity of this covering is kn, establishing the first part of the theorem. The subsequent items in the theorem follow by considering appropriate known error correcting codes, see, e.g. [22].

For the first item simply take the code consisting of all 2^{k-1} codewords with even Hamming weight. Since $2^{k-1} \geq n$ for $k = \lceil \log n \rceil + 1$ the claimed result follows. The second and third items follow from the Gilbert-Varshamov bound which gives that the maximum cardinality of a binary code with length k and distance λ is at least

$$\frac{2^k}{\sum_{i=0}^{\lambda-1}\binom{k}{i}}.$$

This quantity is at least $2^k(\frac{ek}{\lambda-1})^{-(\lambda-1)}$, implying the second item. For any $\lambda = ck \leq k/2$ this quantity is also at least $2^{(1-H(c))k}$, where $H(x)$ is the binary entropy function. This yields the third item.

The fourth follows by considering an appropriate augmented BCH code. For any k which is a power of 2 and for any d this is a (linear) binary code of length k with

$$n = \frac{2^k}{2k^{d-1}}$$

codewords and minimum distance $2d$. For $d - 1 = \lfloor(\lambda - 1)/2\rfloor$, $2d \geq \lambda$ and

$$k = \log n + 1 + (d - 1)\log k \leq \log n + \lfloor(\lambda - 1)/2\rfloor \log\log n + 2.$$

This completes the proof of Theorem 2.2. □

2.4 General Graphs

In this subsection we prove Theorem 2.4. We need the following simple lemma.

Lemma 1 *Let $E_i, i \in I$ be a finite collection of events in a (discrete) probability space. Suppose that for every point x in the space, if $x \in E_i$ then the total number of events E_j in the collection that contain x is at most a_i. Then*

$$\sum_{i \in I} \frac{Prob(E_i)}{a_i} \leq 1. \tag{2}$$

It is worth noting that the above holds (with the same proof) for any probability space, the assumption that it is discrete here is merely because this is the case we need, and it slightly simplifies the notation in the proof.

Proof Let x be an arbitrary point of the space, and let $p(x)$ denote its probability. Suppose it belongs to r of the events E_i, let these be $E_{i_1}, \ldots E_{i_r}$. By the definition of the numbers a_i it follows that $a_{i_j} \geq r$ for all $1 \leq j \leq r$. Therefore the total contribution of the point x to the sum in the left-hand-side of (2) is

$$\sum_{j=1}^{r} \frac{p(x)}{a_{i_j}} \leq \sum_{j=1}^{r} \frac{p(x)}{r} \leq p(x).$$

The desired result follows by summing over all points x in the space. □

Proof of Theorem 2.4: Let G be a graph on the set of vertices $[n]$, let α_i denote the maximum cardinality of an independent set of G containing the vertex i, and let $\mathcal{H} = \{H_1, H_2, \ldots, H_m\}$ be a bipartite covering of G.

As in the proof of Theorem 2.1 we may and will assume, without loss of generality, that each of the bipartite graphs H_i in $\mathcal{H}$ is a complete bipartite graph. Let $L_i, R_i \subset [n]$ denote its two color classes. For each vertex $j \in [n]$ let A_j denote the set of indices i for which the vertex j belongs to the vertex class L_i of H_i and let B_j be the set of indices i for which $j \in R_i$. Let $x_j = |A_j| + |B_j|$ be the total number of bipartite graphs H_i that contain the vertex j. Our objective is to prove a lower bound for the capacity of $\mathcal{H}$, which is exactly the sum $\sum_{j=1}^{n} x_j$.

Let $v = (v_1, v_2, \ldots, v_m\}$ be a uniform random binary vector of length m. For each j, $1 \leq j \leq n$, let E_j denote the event that $v_i = 0$ for every index i that belongs to A_j and $v_i = 1$ for every index i that belongs to B_j. Note that the probability of E_j is exactly 2^{-x_j}. Note also that if some point $v = (v_1, v_2, \ldots, v_m\}$ belongs to the events E_j, $j \in J$, then the set of vertices $J \subset [n]$ is an independent set of G. Indeed, if some two vertices in J are adjacent, then the edge connecting them belongs to at least one of the graphs H_i implying that one of these vertices belongs to L_i whereas the other lies in R_i and showing that they can't both satisfy the requirement given by v_i. It thus follows that any point v that lies in E_j belongs to at most α_j of the events $E_{j'}$. Therefore, by Lemma 1

$$\sum_{j=1}^{n} 2^{-x_j - \log \alpha_j} = \sum_{j=1}^{n} \frac{2^{-x_j}}{\alpha_j} \leq 1.$$

By the arithmetic-geometric means inequality this implies

$$n 2^{-(\sum_{j=1}^{n} x_j + \sum_{j=1}^{n} \log \alpha_j)/n} \leq 1,$$

giving

$$2^{\sum_{j=1}^{n} x_j} \geq n^n 2^{-\sum_{j=1}^{n} \log \alpha_j} = 2^{\sum_{j=1}^{n} (\log n - \log \alpha_j)}.$$

Therefore

$$\sum_{j=1}^{n} x_j \geq \sum_{j=1}^{n} \log(\frac{n}{\alpha_j}),$$

completing the proof. □

3 Triangle-Free Graphs of Diameter 2

The following problem was raised by Erdős and Gyárfás ([12], see also [9], Problem number 134).

Problem 1 Let $\varepsilon, \delta > 0$ be two fixed positive reals, and suppose n is large as a function of ε, δ. Let G be a triangle-free graph on n vertices with maximum degree smaller than $n^{1/2-\varepsilon}$. Can G be made into a triangle-free graph with diameter 2 by adding at most δn^2 edges?

They proved that the conclusion holds if the maximum degree is at most $\log n / \log\log n$. It is mentioned in [12] that Simonovits showed that this does not necessarily hold if the maximum degree is $C\sqrt{n}$ for some large fixed C. In fact, maximum degree $(1+o(1))\sqrt{n/2}$ suffices as shown by the incidence graph of the lines and points of a projective plane of order p. This is a bipartite $(p+1)$-regular (triangle-free, of course) graph on $n = 2(p^2+p+1)$ vertices, where p is a prime power. Any two vertices of the same vertex class in this graph have a common neighbor. Therefore one cannot add any edge connecting two vertices of the same vertex class without creating a triangle. It follows that in order to reduce the diameter to 2 one must add all missing edges between pairs of nonadjacent vertices that do not lie in the same vertex class. The number of these missing edges is $(1/4 - o(1))n^2 = \Omega(n^2)$.

We describe a short proof of the following, which settles the Erdős-Gyárfás problem in a strong form. Here and in what follows we do not make any effort to optimize the absolute constants.

Theorem 3.1 *Let $G = (V, E)$ be a triangle-free graph with n vertices and maximum degree $d \le c(n)\sqrt{n}$, where*

$$2\frac{(\log n)^{1/3}}{n^{1/6}} \le c = c(n) \le \frac{1}{10}$$

and n is sufficiently large. Then one can add to G at most $2.5cn^2$ edges and get a triangle-free graph of diameter 2.

Note that by taking $c(n) = n^{-\varepsilon}$ the above theorem implies that if in the Erdős-Gyárfás problem the maximum degree is at most $n^{1/2-\varepsilon}$ then it suffices to add at most $O(n^{2-\varepsilon})$ edges.

A related problem of Erdős and Pach dealing with triangle-free graphs appears right before the problem above in [12], see also [9], Problem 133.

Problem 2 Let $f(n)$ denote the smallest integer for which there is a triangle-free graph G on n vertices, diameter 2 and maximum degree $f(n)$. What is the order of growth of $f(n)$?

Erdős and Pach conjectured that $f(n)/\sqrt{n}$ tends to infinity as n tends to infinity.

A tight asymptotic expression for $f(n)$ is still open. Clearly $f(n) \ge (1-o(1))\sqrt{n}$ as any graph with maximum degree f and diameter 2 can have at most $1 + f + f(f-1) = f^2 + 1$ vertices. We suspect that $f(n) = (1+o(1))\sqrt{n}$. It is known that for $r \in \{1, 2, 3, 7\}$ and possibly also for $r = 57$ there is an r-regular triangle-free graph of diameter 2 with $r^2 + 1$ vertices. Although the Hoffman-Singleton Theorem [18] asserts that there are no such graphs for additional values of r, it is possible that there are triangle-free graphs of diameter 2, maximum degree r and $(1-o(1))r^2$ vertices for infinitely many values of r.

3.1 The Proof

Proof of Theorem 3.1. Let $G = (V, E)$, d and $c = c(n)$ be as in the statement of the theorem. Throughout the proof we assume, whenever this is needed, that $n > n_0$ where n_0 is a sufficiently large constant. We first apply (a variant of) the triangle-free process for $m = c^2 n^{3/2}$ steps as follows. Starting with $G = G_0$, in each step i for $1 \le i \le m$ let G_i be obtained from G_{i-1} by adding a random edge chosen uniformly among all pairs of nonadjacent vertices of G_{i-1} that are both of degree smaller than $2c\sqrt{n}$ and that do not have a common neighbor. Note that by construction the maximum degree of G_m (and hence of all the graphs during the process) is at most $2c\sqrt{n}$. In addition, by construction, G_m (and all graphs during the process) are triangle-free.

Claim With high probability G_m does not contain an independent set of size $5cn$.

Proof Fix an independent set U of $5cn$ vertices of $G = G_0$. We estimate the probability that it stays independent in G_m. Since the maximum degree in each G_i is at most $2c\sqrt{n}$, the number of pairs of vertices in U that have a common neighbor is at most

$$n\binom{2c\sqrt{n}}{2} < 2c^2n^2.$$

In addition, the total number of vertices whose degrees have been increased already to $2c\sqrt{n}$ is at most $2cn$ (since the total number of edges added is at most $c^2n^{3/2}$ so the graph consisting of all added edges can have at most $2cn$ vertices of degree at least $c\sqrt{n}$). It follows that in every G_i during the process there are at least

$$\binom{|U| - 2cn}{2} - 2c^2n^2 > 2c^2n^2$$

pairs of vertices of U that are of degree smaller than $2c\sqrt{n}$ and that do not have a common neighbor. Each such pair can be chosen as the selected random edge in each step, and the probability none of these edges have been chosen during the process is at most

$$(1 - 4c^2)^m \le e^{-4c^4n^{3/2}}.$$

There are at most

$$\binom{n}{5cn} \le 2^{H(5c)n} < 2^{10c\log(1/c)n}$$

possible sets U, where H is the binary entropy function. Our choice of c ensures that

$$2^{10c\log(1/c)n}e^{-4c^4n^{3/2}} = o(1).$$

The assertion of the claim thus follows by the union bound.

Returning to the proof of the theorem, fix a graph G_m satisfying the conclusion of the claim and add to it, repeatedly, edges to make it a maximal (with respect to containment) triangle-free graph. In other words, as long as there is a pair of nonadjacent vertices with no common neighbor, add the edge connecting them. This creates a triangle-free graph G' of diameter 2, and its independence number is at most that of G_m, which is smaller than $5cn$, by the claim. This implies that the maximum degree in G' is smaller than $5cn$, as the set of all neighbors of any vertex is an independent set. Therefore G' contains at most $2.5cn^2$ edges, completing the proof of Theorem 3.1. □

Problem 2 has been solved up to a constant factor already in the 80s by Hanson and Seyffarth [16] who proved that $f(n) \le 2\sqrt{n}$. Additional constructions appear in [10, 17]. The problem of deciding whether or not $f(n) = (1 + o(1))\sqrt{n}$ remains open. The constructions in all three papers above are Cayley graphs of Abelian groups, and it is easy to see that using such a construction cannot provide graphs with maximum degree smaller than $(\sqrt{2} + o(1))\sqrt{n}$. A better upper estimate of $(2/\sqrt{3} + o(1))\sqrt{n}$ is described by Füredi and Seress in [14]. See also [5] for some related constructions.

4 Blocking Partial Designs and Block-Compatible Sequences

4.1 Results

In this section we consider two related open problems of Erdős on block designs. Recall that a family of subsets $A_1, A_2 \ldots, A_m$ of a finite set X is a (pairwise balanced) block design if every pair of distinct elements of X is contained in exactly one of the subsets A_i. It is a partial design if every pair of distinct elements of X is contained in at most one of the subsets A_i (equivalently, if $|A_i \cap A_j| \le 1$ for all $1 \le i < j \le m$.)

The first problem deals with partial designs and appears in [13], see also [9], Problem number 664.

Problem 3 Is it true that for every fixed positive constant $c < 1$ there is a finite constant $C = C(c)$ so that the following holds. For every m and n and for every family of subsets $\{A_1, A_2, \ldots, A_m\}$ of $[n] = \{1, 2, \ldots, n\}$ that satisfies $|A_i| > c\sqrt{n}$ for all $1 \le i \le m$, and $|A_i \cap A_j| \le 1$ for all $1 \le i < j \le m$, there is a subset $B \subset [n]$ so that $0 < |B \cap A_i| \le C$ for all $1 \le i \le m$?

The second problem appears in [11], page 35, see also [9], Problem number 732.

Problem 4 Call a sequence $n \ge x_1 \ge x_2 \ge \ldots \ge x_m \ge 2$ *block-compatible for n* if there is a pairwise balanced block design $A_1, A_2, \ldots, A_m$ of m subsets of $[n]$ such that $|A_i| = x_i$ for $1 \le i \le m$. Is there an absolute constant $c > 0$ so that for all large n there are at least $e^{cn^{1/2} \log n}$ sequences that are block-compatible for n?

We show that the answer to the first problem is "no" and the answer to the second is "yes". The proofs are short, based on appropriate modifications of the family of lines of a projective plane which form a block design with $m = n = q^2 + q + 1$ subsets of cardinality $q + 1 = (1 + o(1))\sqrt{n}$ of a set of size $n = q^2 + q + 1$. It is well known that such a plane exists for any prime power q.

4.2 Proofs

Throughout the proofs we assume, whenever this is needed, that the parameter n is sufficiently large. All logarithms are in base 2, unless otherwise specified. To simplify the presentation we omit all floor and ceiling signs whenever these are not crucial. We make no serious attempt to optimize the absolute constants that appear in the proofs.

The following result settles Problem 3.

Theorem 4.1 *Let q be a (large) prime power and put $m = n = q^2 + q + 1$. Then there is a partial design consisting of m subsets $A_1, A_2, \ldots, A_m$ of an n element set P, so that $|A_i| > 0.4\sqrt{n}$ for all $1 \le i \le m$, $|A_i \cap A_j| \le 1$ for all $1 \le i < j \le m$, and for any subset B of P that has a nonempty intersection with all sets A_j, there is some $1 \le i \le m$ so that $|B \cap A_i| \ge 0.1 \log n$.*

Proof Let P be the set of $n = q^2 + q + 1$ points of a projective plane of order q, and let $L_1, L_1, \ldots, L_m \subset P$ be the sets of points of the m lines of this plane. Thus each point lies in $q + 1$ lines and each line is of size $q + 1$. For each $1 \le i \le m$ let A_i be a random subset of L_i obtained by picking every point of L_i, randomly and independently, to lie in A_i with probability $1/2$. Note that the choices of distinct subsets A_i here are independent. By the standard estimates for Binomial distributions (c.f., e.g., [6], Appendix A) together with the union bound, with high probability (that is, with probability tending to 1 as q (or n) tends to infinity) the following conditions hold:

1. Each set A_i is of cardinality $(1/2 + o(1))\sqrt{n}$.
2. Each point lies in $(1/2 + o(1))\sqrt{n}$ of the sets A_i.
3. $|A_i \cap A_j| \le 1$ for all $1 \le i < j \le m$.

Claim With high probability there is no subset $B \subset P$ of cardinality at most $0.3\sqrt{n} \log n$ that intersects all the sets A_i.

Proof : Let B be a fixed set of cardinality at most $0.3\sqrt{n} \log n$. Consider the set of pairs $S = \{(b, L_i) : b \in B, 1 \le i \le m, b \in L_i\}$. Clearly $|S| = |B|(q + 1)$. If $J = J(B)$ is the set of all indices i so that $|B \cap L_i| \ge 0.4 \log n$ then $|S| \ge |J| 0.4 \log n$. Thus

$$|J| \le \frac{|B|(q+1)}{0.4 \log n} \le (3/4 + o(1))n < 0.8n.$$

Therefore there are at least $0.2n$ lines L_i that contain less that $0.4 \log n$ points of B. For each such line L_i, the probability that A_i contains no point of B is at least $2^{-0.4 \log n} = n^{-0.4}$. Therefore, the probability that there is no such line (that is, that B intersects all these sets A_i) is at most

$$(1 - 1/n^{0.4})^{0.2n} \le e^{-0.2n^{0.6}}.$$

Since the total number of possible sets B as above is smaller than

$$n^{0.3\sqrt{n}\log n} = 2^{0.3\sqrt{n}(\log n)^2} = o(2^{0.2n^{0.6}}) \; (\; = o(e^{0.2n^{0.6}}) \;),$$

the union bound implies the assertion of the claim.

Returning to the proof of the theorem, fix a choice of the sets A_i that satisfy the conditions 1, 2, 3 above and the assertion of Claim 4.2. If a set B intersects all subsets A_i then it must satisfy $|B| > 0.3\sqrt{n} \log n$. Since each point is contained in $(1/2 + o(1))\sqrt{n}$ of the sets A_i this implies, by averaging, that the intersection of B with some set A_i is of size at least $|B|(1/2 + o(1))\sqrt{n}/n > 0.1 \log n$. This completes the proof. □

The next result settles Problem 4.

Theorem 4.2 *Let q be a large prime power and put $n = q^2 + q + 1$. Let $S = (x_1 \ge x_2 \ge x_3 \ge \ldots \ge x_m)$ be any sequence of integers satisfying*

$$q + 1 \ge x_1 \ge x_2 \ge x_3 \ldots \ge x_n \ge 3,$$

$$m = n + \sum_{i=1}^{n}[\binom{q+1}{2} - \binom{x_i}{2}],$$

and $x_i = 2$ for all $n < i \le m$. Then S is block-compatible for n. Therefore, the number of sequences that are block-compatible for n is at least

$$\binom{n+q-2}{q-2} = 2^{(0.5+o(1))n^{1/2} \log n}.$$

Proof Let P and $L_1, L_2, \ldots, L_n \subset P$ be, as in the proof of Theorem 4.1, the set of points of a projective plane of order q and the sets of points of the n lines of P. For each $1 \le i \le n$ let X_i be a subset of cardinality x_i of L_i. Consider the block design consisting of the n blocks X_i together with the following additional

$$\sum_{i=1}^{n}[\binom{q+1}{2} - \binom{x_i}{2}]$$

blocks of cardinality 2: for each $1 \le i \le n$, every pair of distinct elements of L_i which is not contained in X_i. This is clearly a block design and the ordered sequence of cardinalities of its blocks is the sequence S. This completes the first part of the proof. For the estimate note that the number of possibilities for the subsequence $x_1 \geq x_2 \geq \ldots \geq x_n \geq 3$ in the above construction is the number of ordered sequences of $q-1$ nonnegative integers whose sum is n, which is

$$\binom{n+q-2}{q-2}.$$

□

4.3 Remarks

- It is easy to see that the estimate in Theorem 4.1 is tight up to constant factors, for every partial design in which all blocks are of sizes $\Theta(\sqrt{n})$. Indeed the following simple fact can be proved by choosing the set B randomly and by applying the union bound and the standard estimates for Binomial distribution.

Proposition 4.1 *For any two positive constants $c_1 < c_2$ there are two positive constants $C_1(c_1, c_2) < C_2(c_1, c_2)$ so that the following holds. For every m and n and for every family of subsets $\{A_1, A_2, \ldots, A_m\}$ of $[n] = \{1, 2, \ldots, n\}$ that satisfies $c_1\sqrt{n} \leq |A_i| \leq c_2\sqrt{n}$ for all $1 \leq i \leq m$, and $|A_i \cap A_j| \leq 1$ for all $1 \leq i < j \leq m$, there is a subset $B \subset [n]$ so that $C_1 \log n < |B \cap A_i| \leq C_2 \log n$ for all $1 \leq i \leq m$.*

In fact the conclusion holds, of course, even without any assumption on the sizes of the intersections of pairs of blocks A_i.

- Problem 3 for block designs (and not for partial designs) was also asked by Erdős in [11]. This remains open although we suspect that the answer here is negative as well. We suggest the following conjecture which, if true, would establish this negative answer.

Conjecture 4.2 *Let q be a (large) prime power, put $n = q^2 + q + 1$, let P be the set of n points of a projective plane of order q and let $L_1, L_2, \ldots, L_n$ be the sets of points of its lines. Let R be a random subset of P obtained by picking each point of P randomly and independently to lie in R with probability $1/2$. Then with high probability the smallest cardinality of a subset B of R that intersects all the subsets $L_1 \cap R, L_2 \cap R, \ldots, L_n \cap R$ satisfies $|B|/q > f(q)$ for some function $f(q)$ tending to infinity as q tends to infinity. In fact, this may even be true with $f(q) = \Omega(\log q)$.*

This conjecture remains open, although related results have been proved in [7, 8] using the container method. The parameters in these papers are very different and it seems that a proof here, if true, would require additional ideas.

- As mentioned by Erdős in [11], the $e^{cn^{1/2}\log n}$ lower bound for the number of block-compatible sequences for n is tight up to the absolute constant c. For completeness we include a brief proof of the upper bound. It is clear that if $n \geq x_1 \geq \ldots \geq x_m \geq 2$ is block-compatible for n, then $m \leq \binom{n}{2} < n^2/2$. Therefore, the number of choices of all x_i which are, say, at most $2\sqrt{n}$ is smaller than

$$\binom{n^2/2 + 2\sqrt{n}}{2\sqrt{n}} < 2^{4\sqrt{n}\log n}.$$

Next, observe that for each block-compatible sequence x_i for n, the number of indices i with $x_i > 2\sqrt{n}$ is smaller than $\sqrt{n}$. Indeed, otherwise the size of the union of a set of blocks $A_i, i \in I$ with $|A_i| = x_i > 2\sqrt{n}$ and $|I| = \sqrt{n}$ in a block design realizing the sequence is at least

$$\sum_{i \in I} |A_i| - \sum_{i,j \in I, i<j} |A_i \cap A_j| > \sqrt{n} \cdot 2\sqrt{n} - \binom{\sqrt{n}}{2} > n,$$

which is impossible. Thus there are less than $\sqrt{n}$ such large x_i, and the number of choices for those is at most

$$\binom{n + \sqrt{n}}{\sqrt{n}} < 2^{\sqrt{n}\log n}.$$

This gives the required $2^{O(\sqrt{n}\log n)}$ upper bound for the total number of block-compatible sequences for n.

- Call a sequence $n \geq x_1 \geq x_2 \geq \ldots \geq x_m \geq 2$ *line-compatible for* n if there is a set P of n points in the Euclidean plane R^2 so that for the family $L_1, L_2, \ldots, L_m$ of all lines in R^2 determined by the points of P, $|L_i \cap P| = x_i$ for $1 \leq i \leq m$. Note that every line-compatible sequence for n is also block-compatible for n, but the converse is not true. Erdős conjectured in [11] (see also [9], Problem 733) that the number of sequences which are line compatible for n is only $2^{O(n^{1/2})}$. This upper bound was proved by Szemerédi and Trotter in [24]. Note that in view of Theorem 4.2 this is much smaller than the number of block-compatible sequences for n.
  Indeed, there are far more block designs on n points than designs that can be described by the lines determined by a set of points in the plane. This is demonstrated by the following result.

Proposition 4.3 *1. The number of hypergraphs on n labelled vertices whose edges form a block design is* $2^{\Theta(n^2 \log n)}$.

2. The number of hypergraphs whose vertices are n labelled points in R^2 *and whose edges are the sets of points contained in the lines determined by the points is only* $2^{\Theta(n \log n)}$.

Proof The lower bound in the first part follows from the known lower bound for the number of Steiner triple systems on n points, which, as proved in [19], is $2^{(1+o(1)n^2 \log n/6}$. (This is the case for n congruent to 1 or 3 modulo 6. For the other values of n we can include some blocks of size 2 as needed.) To prove the upper bound let $x_1, x_2, \ldots x_m$ be the sizes of the blocks. Then $\sum_i \binom{x_i}{2} = \binom{n}{2}$ and thus $\sum x_i \leq n^2$. The number of choices of the number of blocks m and their sizes x_i is at most $2^{O(\sqrt{n} \log n)}$. Given those, the number of ways to choose subsets of cardinalities $x_1, x_2, \ldots, x_m$ in $[n]$ is at most

$$\binom{n}{x_1}\binom{n}{x_2}\cdots\binom{n}{x_m} < n^{\sum_i x_i} \leq n^{n^2} = 2^{n^2 \log n}.$$

The lower bound in the second part is proved by considering all possible labelings of the hypergraph consisting of a matching of $\lfloor n/3 \rfloor$ pairwise disjoint edges of size 3 and additional $\binom{n}{2} - \lfloor n/3 \rfloor$ edges of size 2. It is easy to realize this hypergraph by points in the plane and the lines they determine by placing 3 points on each of $\lfloor n/3 \rfloor$ parallel lines where no other line contains more than two points.

We prove the upper bound in the second part by using the result of [23] on the number of zero patterns of a sequence of polynomials.

If $f = (f_1, \ldots, f_a)$ is a sequence of polynomials in b variables over a field K, then the *zero pattern* of f evaluated at the point $u \in K^b$ is the set

$$Z_f(u) = \{i \in [a] : f_i(u) = 0\}.$$

Let Z_f denote the total number of distinct zero patterns that appear as u ranges over K^b. We need the following result of Rónyai et al. [23].

Theorem 4.3 ([23]) *Let $f = (f_1, \ldots, f_a)$ be a sequence of polynomials in b variables over a field K, and let d_i denote the degree of f_i. Then*

$$Z_f \leq \binom{b + \sum_{i=1}^{a} d_i}{b}.$$

In our case the field is R, each point can be described by 2 real variables, so $b = 2n$, and for each set of 3 points there is a degree 2 polynomial that vanishes iff they lie on a line. The zero pattern of these $\binom{n}{3}$ real polynomials of degree 2 in $2n$ variables determines all the lines, and therefore, by Theorem 4.3, the number of possible hypergraphs here is at most

$$\binom{2n + (n^3/6)2}{2n} = 2^{(4+o(1))n \log n}.$$

This completes the proof. □

Acknowledgements I thank Ishay Haviv and Wojciech Samotij for helpful comments.

References

1. Alon, N.: Problems and results in extremal combinatorics I. Discret. Math. **273**, 31–53 (2003)
2. Alon, N.: Problems and results in extremal combinatorics II. Discret. Math. **308**, 4460–4472 (2008)
3. Alon, N.: Problems and results in Extremal Combinatorics III. J. Comb. **7**, 233–256 (2016)
4. Alon, N.: Problems and results in extremal combinatorics IV (2020). arXiv:2009.12692
5. Alon, N., Erdős, P., Holzman, R., Krivelevich, M.: On k-saturated graphs with restrictions on the degrees. J. Graph Theory **23**, 1–20 (1996)
6. Alon, N., Spencer, J.H.: The Probabilistic Method, 4th edn, pp xiv+375. Wiley (2016)
7. Balogh, J., Samotij, W.: An efficient container lemma. Discrete Anal. 56 (2016). Paper No. 17
8. Balogh, J., Solymosi, J.: On the number of points in general position in the plane. Discrete Anal. 20 (2018). Paper No. 16
9. Bloom, T.: Erdős' open problems. https://www.erdosproblems.com/all
10. Clark, W.E., Pedersen, J.: Sum-free sets in vector spaces over GF (2). J. Comb. Theory Ser. A **61**, 222–229 (1992)
11. Erdős, P.: On the combinatorial problems which I would most like to see solved. Combinatorica 25–42 (1981)
12. Erdős, P.: Some old and new problems in various branches of combinatorics. Discret. Math. **165**(166), 227–231 (1997)
13. Erdős, P.: Some unsolved problems. In: Combinatorics, Geometry and Probability (Cambridge, 1993), pp. 1–10 (1997)
14. Füredi, Z., Seress, Á.: Maximal triangle-free graphs with restrictions on the degrees. J. Graph Theory **18**(1), 11–24 (1994)
15. Hansel, G.: Nombre minimal de contacts de fermeture nécessaires pour réaliser une fonction booléenne symétrique de n variables. C. R. Acad. Sci. Paris **258**, 6037–6040 (1964)
16. Hanson, D., Seyffarth, K.: k-saturated graphs of prescribed maximum degree. Congr. Numer. **44**, 127–138 (1984)
17. Haviv, I., Levy, D.: Symmetric complete sum-free sets in cyclic groups. Israel J. Math. **227**, 931–956 (2018)
18. Hoffman, A.J., Singleton, R.R.: Moore graphs with diameter 2 and 3. IBM J. Res. Dev. **5**, 497–504 (1960)
19. Keevash, P.: Counting Steiner triple systems. In: European Congress of Mathematics. European Mathematical Society, Zürich, pp. 459–481 (2018)
20. Kim, J., Lee, H.: Covering multigraphs with bipartite graphs (2023). arXiv:2304.11691
21. Katona, G., Szemerédi, E.: On a problem of graph theory. Studia Sci. Math. Hungar. **2**, 23–28 (1967)
22. MacWilliams, F.J., Sloane, N.J.A.: The theory of error-correcting codes, I. North-Holland Mathematical Library, vol. 16. North-Holland Publishing Co., Amsterdam, New York, Oxford (1977)
23. Rónyai, L., Babai, L., Ganapathy, M.K.: On the number of zero-patterns of a sequence of polynomials. J. Am. Math. Soc. **14**, 717–735 (2001)
24. Szemerédi, E., Trotter, W.T.: Extremal problems in discrete geometry. Combinatorica **3**(3–4), 381–392 (1983)

Monotone Boolean Reconstruction

Levon Aslanyan and Hasmik Sahakyan

Abstract Monotone Boolean reconstruction refers to a process in the field of combinatorial optimization and information theory, where the goal is to recover a hidden binary function (a function that takes as input the values 0 or 1) based on limited information obtained from observations. The term "monotone" indicates that the function maintains the non-decreasing property: if one of the inputs is changed from 0 to 1, the output will not decrease. In applications such as data mining and network analysis, reconstructing the structure of monotone Boolean functions can help in understanding underlying relationships or dependencies within data. This reconstruction problem is significant in various fields including economics, bioinformatics, and machine learning. Key components of monotone Boolean reconstruction involve: *Observations*: Limited data or samples that provide insights into the behavior of a Boolean function. *Reconstruction*: The process of inferring the original function from the available observations, often requiring algorithms and techniques from combinatorial and optimization theory. *Applications*: Implications for various practical scenarios where understanding monotonic relationships between variables is crucial. The complexity of this problem can vary, and often involves algorithmic challenges, in particular, to determining the minimum set of observations required for accurate reconstruction. Researchers use different structural methods and mathematical frameworks to efficiently tackle such reconstruction tasks. Boolean functions form a descriptive model of compatibility of sets of independent constraints, but the class of all functions is unambiguously complex and it is of interest to find also relatively simple subclasses of monotone functions. In this survey, we summarise existing results in the area of monotone Boolean recognition, including those related to special classes of monotone Boolean functions. Results related to the multivalued domain are also briefly presented.

L. Aslanyan · H. Sahakyan (✉)
Institute for Informatics and Automation Problems, Yerevan, Armenia
e-mail: hsahakyan@sci.am

L. Aslanyan
e-mail: lasl@sci.am

G. O. H. Katona et al. (eds.), *Sum(m)it280*, Bolyai Society Mathematical Studies 32,
https://doi.org/10.1007/978-3-032-18810-6_3

1 Introduction

Monotone Boolean reconstruction refers to a process in the field of combinatorial optimization and information theory, where the goal is to recover a hidden binary function (a function that takes as input the values 0 or 1) based on limited information obtained from observations.

Many problems with monotone Boolean functions (MBFs) appear in logical and physical level design of systems [8], but also in artificial intelligence [6], data science and computational learning theory [4], hypergraph theory [67, 68] and other areas (e.g., [20, 24, 45, 55, 80]). MBFs are used to encode extremely important constructions in various combinatorial optimizations providing a natural way of describing compatible subsets of a finite group of constraints (see e.g., [5, 25, 75]).

Let us start with a simple interpretation. In an optimization problem a set of n independent constraints defines the feasibility domain. Consider a function $\nu(\alpha)$,– compatibility indicator for $\alpha \in P([n])$, where $[n] = \{1, 2, \ldots, n\}$, and $P([n])$ is the power set of $[n]$. $\nu(\alpha)$ encodes a subset of compatible constraints by value 1, otherwise, the value is 0. Sometimes checking the compatibility of subsets α becomes a separate intractable problem. The question naturally arises how to minimize the checks/tests in a global optimization procedure. Obviously, if a compatible subset α_1 is given and $\alpha_2 \subseteq \alpha_1$, then α_2 is compatible. The overall structure of compatibility is given by these inclusions and then by monotone Boolean functions, and the idea behind the recognition of these functions is to find the maximal compatible subsets of constraints by applying fewer checks and procedures in the optimality search processes.

The class of MBFs is one of the pre-complete classes of Boolean functions in Post's famous functional completeness Theorem of 1941 [65].

The problem of determining the number of elements of the free distributive lattice on n generators (equivalently–the number of MBFs of n variables) was posed by Dedekind [29] in 1897.

A number of applications (e.g., wireless sensor network design, dead-end tests of tables and data mining algorithms) [6, 8, 53, 55] use optimization with MBF, where MBFs are represented by hypercube constructions such as chains and antichains [35, 41]. Other similar applications with MBF can be added to this list [20, 23, 25, 75].

There are a number of known effective tools and methods for analyzing MBFs [36, 42, 50, 76] and new approaches are constantly being sought, investigated, and applied (e.g., [11, 12, 16, 20, 56, 69]. Several problems in this area include the reconstruction problem of bounded classes of Boolean functions with randomization of queries and functions, or the use of combinatorial cube-splitting and chain-splitting technique of the Boolean domain [7, 14, 15, 17, 43, 63].

A well-known approach concerning MBFs recognition is query-based identification–recognition of an unknown MBF of n variables using an oracle and membership queries to it.

In this survey, we summarise the existing results of applying this approach to both Boolean and multi-valued cases.

In order to obtain solutions for bounded classes of MBFs, it is necessary to find a way to the structural properties of these classes [19, 22, 57]. In this survey, we present results concerning some special classes of monotone Boolean functions, including:

- Class of spherical MBFs that has an interpretation in association rule mining applications [6]. For functions of this class, the function value is 1 on all vertices that are above a certain layer of the hypercube.
- Class of positive 2-monotonic Boolean functions [18].
- Class of KK-MBF functions related to well-known Kruskal-Katona theorem [34, 47, 48, 54, 70] that describes the exact optimal monotone constructions of shadow minimization (constraint minimization) [19, 22, 44, 58] and existence of Sperner systems for a given set of parameters. A practical value of KK-MBF recognition consists in the following. KK-MBFs appear when compatibility is not only controlled by the inclusion of subsets of constraints. Suppose we are given two constrained, k-subsets α_1 and α_2, such that α_2 alphabetically proceeds α_1. In some applications, constraint compatibility should have the following property: compatibility of α_2 is a simple consequence of the compatibility of α_1 because of the alphabetical priorities of constraints. In this case, the structure of constraint compatibility is given by KK-MBF functions.
- Class of k-reason MBFs [37].

The survey is organized as follows. Section 2 provides necessary definitions, preliminaries, and basic concepts. Existing results in the area of monotone Boolean recognition are summarised in Sect. 3. Section 4 addresses some special classes of monotone Boolean functions. Results related to the multivalued domain are presented in Sect. 5.

2 Preliminaries

2.1 Hypercube, Hypergrid

The hypercube, binary cube or unit cube, as we call it in different occasions, is a basic structure in mathematical logic and finite set theory. The extension of this structure to the multivalued case we often call a lattice (particular case), and also a multidimensional multivalued grid, a hypergrid.

The element set of a hypercube corresponds to the elements of the power set of a finite set, and for a hypergrid it represents a collection of finite multi-sets. Hypercube and Hypergrid (Lattice) graphs appear when neighbouring elements (vertices) are connected by edges.

Hypercube

Let $B^n = \{(x_1, \ldots, x_n) \mid x_i \in \{0, 1\},\ i = 1, \ldots, n\}$ denote the set of vertices of the n-dimensional hypercube. Let $\alpha = (\alpha_1, \ldots, \alpha_n)$ and $\beta = (\beta_1, \ldots, \beta_n)$ be two vertices of B^n. We say that α ***precedes*** β (by *component-wise order*), denoted as $\alpha \preccurlyeq \beta$, if and only if $\alpha_i \leq \beta_i$ for $1 \leq i \leq n$. α and β are *comparable* if $\alpha \preccurlyeq \beta$ or $\beta \preccurlyeq \alpha$, otherwise, they are *incomparable*. A set of incomparable vertices in B^n is also called a ***Sperner*** family. A set $\{\alpha^1, \ldots, \alpha^k\}$ of elements of B^n is a growing **chain** if $\alpha^i \prec \alpha^j$ for $1 \leq i < j \leq k$. A chain $\{\alpha^1, \ldots, \alpha^k\}$ is *simple* if for $\alpha^i \prec \alpha^j$ there is no α^r such that $\alpha^i \prec \alpha^r \prec \alpha^j$.

The **serial number** of the vertex $\alpha = (a_1, \ldots, a_n)$ is the natural number $a_1 2^{n-1} + a_2 2^{n-2} + \cdots + a_n 2^0$, whose binary representation is $a_1 a_2 \cdots a_n$.

We will also use the **lexicographic order** of vertices: α *precedes* lexicographically β ($\alpha \preccurlyeq_{\text{lex}} \beta$) if either there exists an integer k, $1 \leq k \leq n$, such that $a_k < b_k$ and $a_i = b_i$ for $i < k$, or $\alpha = \beta$.

In literature there are different, sometimes confusing definitions of lexicographic order [71]. Mathematical definition is set-theoretical that uses the value 1 to code the presence of an element in a subset: $x <_{\text{lex}} y$ if $\min(x \Delta y) \in x$. Appearance of elements in a subset is coded by a binary vector, and vector α precedes vector β when α precedes β alphabetically. Here $1 < 0$. When we apply to combinatorial settings, the hypercube is considered, and lexicographical order of vertices of B^n, defined alphabetically, uses relation $0 < 1$. We will use the standard set-theoretical definition indicated as *lex* in Fig. 1. *revlex* is the reverse of this, i.e. $(00000,\ 00001,\ \ldots,\ 11111)$. *colex* in set-theoretical settings means: $x <_{co-lex} y$ if $\max(x \Delta y) \in y$. This is convenient when we add a new dimension to the vectors intending to continue the ordered sequence when the initial segment remains unchanged. But as we see in Fig. 1, *colex* unlike the *lex* or *revlex* preserves neither the order of numerical values nor the order on the layers of the hypercube graphically, in the **Hasse diagram**. The *sim* (simplicial) order lists hypercube vertices layer after the layer (see below for the concept of a layer); in each layer, vertices are ordered according to their serial number. We will use the *revlex* basically, but when there is no confusion we will refer to it simply as *lex*.

Lexicographic, co-lexicographic and simplicial orders are shown in Fig. 1 on the set of vertices of B^5.

We define also **partition/splitting** of B^n into two $(n-1)$-dimensional sub-cubes according to the values of the binary variables; for arbitrary x_i:

$$B^{n-1}_{x_i=0} = \{(x_1, \ldots, x_n) \in B^n | x_i = 0\} \text{ and}$$
$$B^{n-1}_{x_i=1} = \{(x_1, \ldots, x_n) \in B^n | x_i = 1\}$$

Any subset $\mathcal{M} \subseteq B^n$ will be partitioned into

$$\mathcal{M}_{x_i=1} \subseteq B^{n-1}_{x_i=1},\ and\ \mathcal{M}_{x_i=0} \subseteq B^{n-1}_{x_i=0}.$$

lex	#	co-lex	#
11111	31	00000	00
11110	30	10000	16
11101	29	01000	08
11100	28	11000	24
11011	27	00100	04
11010	26	10100	20
11001	25	01100	12
11000	24	11100	28
10111	23	00010	02
10110	22	10010	18
10101	21	01010	10
10100	20	11010	26
10011	19	00110	06
10010	18	10110	22
10001	17	01110	14
10000	16	11110	30
01111	15	00001	01
01110	14	10001	17
01101	13	01001	09
01100	12	11001	25
01011	11	00101	05
01010	10	10101	21
01001	09	01101	13
01000	08	11101	29
00111	07	00011	03
00110	06	10011	19
00101	05	01011	11
00100	04	11011	27
00011	03	00111	07
00010	02	10111	23
00001	01	01111	15
00000	00	11111	31

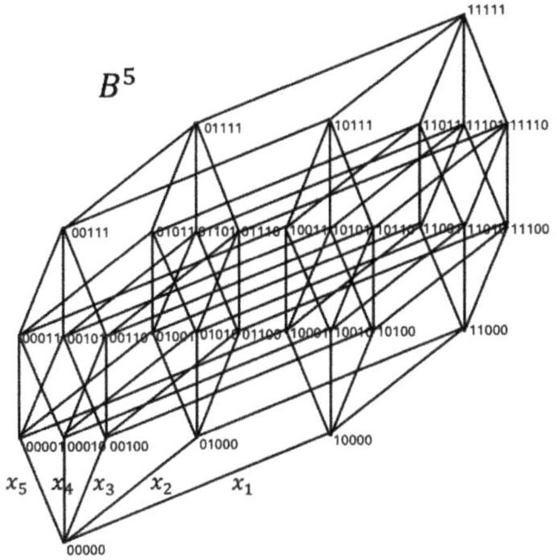

sim	#
11111	31
11110	30
11101	29
11011	27
10111	23
01111	15
11100	28
11010	26
11001	25
10110	22
10101	21
10011	19
01110	14
01101	13
01011	11
00111	07
11000	24
10100	20
10010	18
10001	17
01100	12
01010	10
01001	09
00110	06
00101	05
00011	03
10000	16
01000	08
00100	04
00010	02
00001	01
00000	00

Fig. 1 Lexicographic, co-lexicographic and simplicial orders of vertices of B^5. The sign # marks the natural order which coincides with the numerical value of binary representation of vertices

B^n can also be partitioned according to a set of variables. Partitioning according to variables $x_{i_1}, \ldots, x_{i_k}$, we get 2^k number of $(n-k)$-dimensional sub-cubes, where in each of them the values of $x_{i_1}, \ldots, x_{i_k}$ are fixed in appropriate way; for example,

$$B^{n-k}_{x_{i_1}=1,\ldots,x_{i_k}=1} = \left\{(x_1, \ldots, x_n) \in B^n \middle| x_{i_1} = 1, \ldots, x_{i_k} = 1\right\}.$$

Figure 2 illustrates the split of the B^6 according to two groups of variables. Let $L_k = \left\{(x_1, \ldots, x_n) \in B^n \middle| \sum_{i=1}^n x_i = k\right\}$. We call L_k the k-th **layer** of B^n. The **shadow** $\delta^i \mathcal{M}$ of $\mathcal{M} \subseteq B^n$ is the set of vertices of L_i, which are less than some vertex of $\mathcal{M}$.

In the case when all vertices of $\mathcal{M}$ are from the same **layer**, e.g., from the k-th layer, then the lower (respectively, upper) shadow of $\mathcal{M}$ is $\delta^{k-1}\mathcal{M}$ (respectively, $\delta^{k+1}\mathcal{M}$), i.e., the set of vertices from the $(k-1)$-th layer, which are less than some vertex of $\mathcal{M}$ (respectively, from the $(k+1)$-th layer, which are greater than some vertex of $\mathcal{M}$).

Boolean function $f : B^n \to \{0, 1\}$ is called **monotone** (it is called also positive) if for every two vertices $\alpha, \beta \in B^n$, if $\alpha \prec \beta$ then $f(\alpha) \leq f(\beta)$. Vertices of B^n, where f takes the value "1" are called *units* or *true points* of the function; vertices, where f takes the value "0" are called *zeros* or *false points* of the function. α^1 is a *lower unit* (or *minimal true point*) of the function if $f(\alpha^1) = 1$, and $f(\alpha) = 0$ for

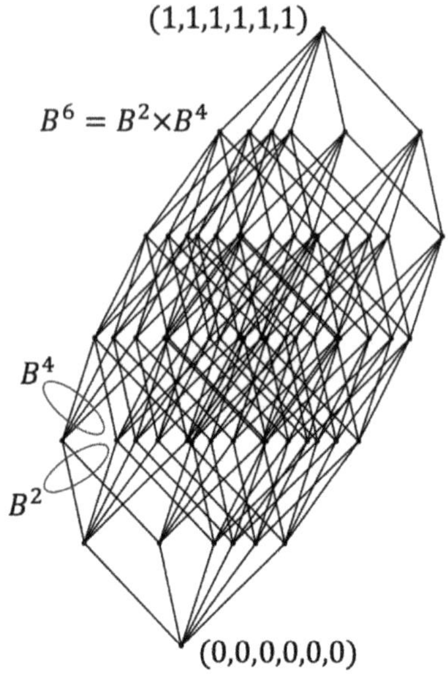

Fig. 2 Split of the B^6 into the $B^4 \times B^2$. This split generates 4 subcubes of type B^4 and 16 subcubes of type B^2. Two representatives of these groups are highlighted with ovals. If we consider the marked B^2, then a B^4 is incident to each of its vertices. The same structure holds for the marked B^4 and its vertices

every $\alpha \in B^n$, such that $\alpha \prec \alpha^1$. α^0 is an *upper zero* (or *maximal false point*) of the function if $f(\alpha^0) = 0$, and $f(\alpha) = 1$ for every $\alpha \in B^n$ such that $\alpha^0 \prec \alpha$. min $T(f)$ and max $F(f)$ denote the sets of minimal true points and maximal false points, respectively. Obviously, min $T(f)$ and max $F(f)$ are Sperner families in B^n. We denote by $\mathcal{M}_n$ the set of all monotone Boolean functions defined on B^n.

Formally, the early references to MBFs started in 1897, with the issue of counting their number [28]. The first algorithmic and complexity-related considerations belong to [50], where, in particular, the valuable concept of *resolving subsets* was introduced. The final asymptotic estimate about the number of MBFs of n variables was obtained in [51, 52]. The technique on how to introduce and analyze MBFs, is basically presented in [42, 51, 57].

Hypergrid

Let $\Xi_{m+1} = \{0, 1, \ldots, m\}$ and $\Xi^n_{m_1+1,m_2+1,\ldots,m_n+1}$ denote the set of vertices of the n-**dimensional multy-valued grid** defined as:

$$\Xi^n_{m_1+1,m_2+1,\ldots,m_n+1} = \{(a_1, \ldots, a_n) : a_i \in \Xi_{m_i+1} \text{ for all } i \in \overline{1,n} = \{1, 2, \ldots, n\}\}.$$

Factorization

As an example interpretation of this structure, we consider a natural number k with its canonical factorization $k = p_1^{m_1} p_2^{m_2} \cdots p_n^{m_n}$. The sum $m_1 + m_2 + \cdots + m_n$ is a degree of k. $\Xi^n_{m_1+1,m_2+1,\ldots,m_n+1}$ simply and effectively represents the structure of all

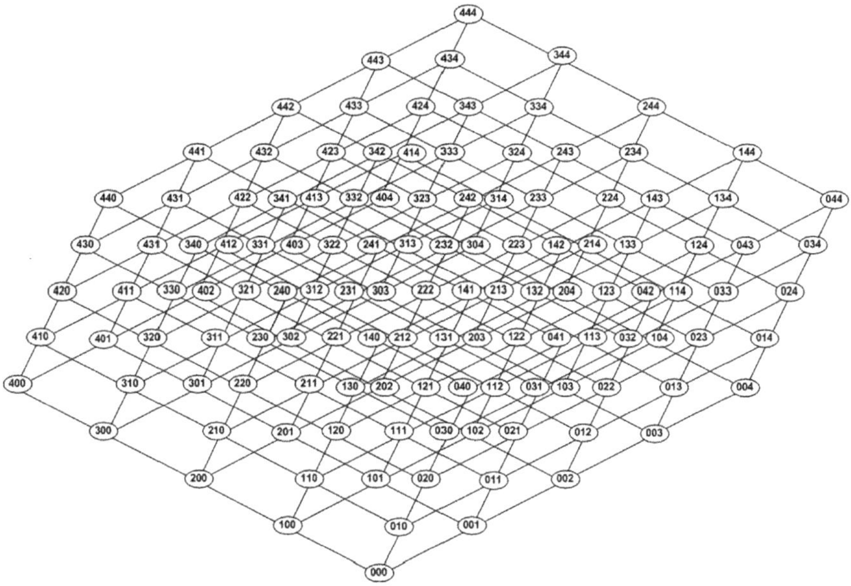

Fig. 3 Hasse diagram of Ξ_5^3, circles correspond to vertices

divisors of k. A traditional research topic here is the investigation of the parameters of collections of divisors of k that are pairwise indivisible, that is, are parameters of Sperner families over the $\Xi^n_{m_1+1,m_2+1,\ldots,m_n+1}$. A particular case of $\Xi^n_{m_1+1,m_2+1,\ldots,m_n+1}$ when all $m_i = m$ is also important, we denote it as Ξ^n_{m+1}.

Define a component-wise partial order on Ξ^n_{m+1}: $(a_1, \ldots, a_n) \le (b_1, \ldots, b_n)$ if and only if $a_i \le b_i$ for all $i \in \overline{1, n}$. Ξ^n_{m+1}, with this order is a partially ordered set. Its Hasse diagram has $m \cdot n + 1$ layers/levels, numbered from 0 (lower level) to $m \cdot n$; the k-th layer contains all vertices with $a_1 + \cdots + a_n = k$. $\frac{m \cdot n}{2}$-th is the middle layer of Ξ^n_{m+1}, when $m \cdot n$ is even, otherwise, Ξ^n_{m+1} has two middle layers: $\frac{m \cdot n+1}{2}$-th and $\frac{m \cdot n-1}{2}$-th. Edges connect those vertices in neighbour layers related by a cover relation. Figure 3 demonstrates the Hasse diagram of Ξ_5^3.

Chains (simple chains), Sperner families can be defined here similar to the case of hypercube.

Consider a binary function $f \colon \Xi^n_{m+1} \to \{0, 1\}$. We say that f is *monotone* if for any two vertices $a, b \in \Xi^n_{m+1}$, if $a > b$ then $f(a) \ge f(b)$. For $m = 1$ we get monotone Boolean functions defined on the B^n. $a^1 \in \Xi^n_{m+1}$ is a *lower unit* of some monotone function f if $f(a^1) = 1$, and $f(a) = 0$ for every $a \in \Xi^n_{m+1}$, which is less than a^1. $a^0 \in \Xi^n_{m+1}$ is an *upper zero* of monotone function f if $f(a^0) = 0$, and $f(a) = 1$ for every $a \in \Xi^n_{m+1}$, which is greater than a^0. We denote by $\mathcal{M}(n, m)$ the set of all monotone functions defined on Ξ^n_{m+1}.

2.2 Query-Based Reconstruction

Let f be an arbitrary binary function defined with the help of a certain operator (also called an oracle) such that receiving any vertex a of Ξ^n_{m+1}, the oracle gives $f(a)$, the value of the function. If there is no a-priori information about the function then $(m+1)^n$ accesses to the oracle are required for identification of the function. If some property of the function is known then it may not be necessary to access to the oracle for all vertices of the cube. For example, if $f \in \mathcal{M}(n,m)$, then $f(a) = 1$ for some vertex $a \in \Xi^n_{m+1}$ implies $f(b) = 1$ for all $b > a$. All algorithms of identification of monotone Boolean functions differ from each other by the set of vertices presented to the oracle, and by the method of selecting vertices and diffusing their values to other vertices. The problem is in identification of the function by as far as possible small number of accesses to the oracle. We differ exact reconstruction, adaptive algorithms, random queries and approximate decisions.

2.3 Chains and Sperner Families

Chains and Sperner families are largely studied [32, 49]. De Bruijn, Ebbenhorst Tengbergen, and Kruyswijk (BETK) in [27] introduced a symmetric chain split over the set of divisors of natural numbers, which, as we have already mentioned, coincides with the vertex set of a multidimensional multivalued grid, and chains are symmetric with respect to the middle layer/layers of the grid:

Theorem 1 *The set of divisors of a natural number m can be partitioned into disjoint symmetric chains.*

The special case of squarefree m, that is when m is the product of n different primes, is a theorem of Sperner [74].

The fundamental idea of splitting the grid of divisors by a set of disjoint chains, symmetric with respect to the middle layer of the grid, is applied further in different occasions.

The chain splitting in Theorem 1 is constructed by induction, by the number of primes in the factorization. Figure 4 represents the general step of the induction. Let $k = p_1^{m_1} p_2^{m_2} \cdots p_n^{m_n}$ be the canonical factorization of k. And let $q_1, q_2, \ldots, q_l$ be an arbitrary symmetric chain of the chain splitting of number k. We proceed to prove the induction statement for number kp^m (new prime is incorporated). We consider the divisors $q_i p^j$, $i = 1, \ldots, l$, $j = 1, \ldots, m$. Figure 4 exactly shows the allocation of these divisors into the new set of chains that are generated in the basis of the initial chain $q_1, q_2, \ldots, q_l$. The proof that chains are disjoint, symmetrical, and cover the space, can be found in [27].

The next step in the chain splitting is done by Hansel [42] in 1966, who observed a powerful property of these chains and this played one of the central roles in MBF-related algorithmic techniques after that.

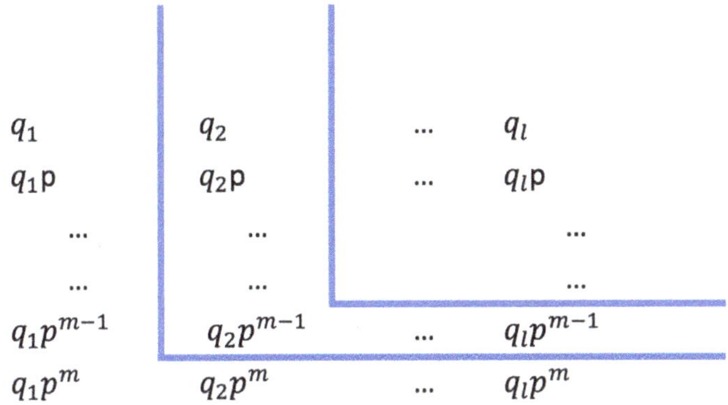

Fig. 4 General step of building chain split of $\Xi^n_{m_1+1,m_2+1,\dots,m_n+1}$ by [60]

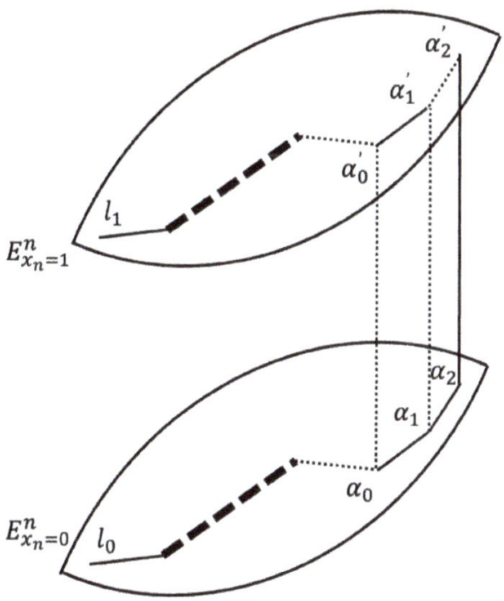

Fig. 5 Recurrent step of constructing Hansel chains in B^n

Theorem 2 *The hypercube B^n can be covered by $\binom{n}{[n/2]}$ disjoint chains of vertices, which have the following properties:*
(a) The number of chains of length $n-2p+1$ is $\binom{n}{p}-\binom{n}{p-1}$, $1 \le p \le [n/2]$
(b) The minimal vertex of any chain of length $n-2p+1$ has p 1s and $n-p$ 0s,
(c) For any three sequential elements $\alpha_1 \prec \alpha_2 \prec \alpha_3$ of a chain of length $n-2p+1$, the complement to α_2 in the interval α_1, α_3 belongs to a chain of length $n-2p-1$.

Here (b) corresponds to the property of chain symmetries. (c) is a new property on **"relative completion"** that interconnects chains and their elements together. The immediate consequence of this property is the Shannon complexity of MBF reconstruction, obtained by Hansel.

Hansel chains may be constructed by an induction, similar to the chains of divisors (see the last two lines at the picture Fig. 4). This is outlined in the picture of Fig. 5. During the inductive step the edge (α_1', α_2') is deleted from the chain (l_1, α_2'), and an additional edge (α_2, α_2') to the chain (l_0, α_2) is appended. Vertices $\alpha_1, \alpha_2, \alpha_2'$ form a new fragment of immediately consecutive vertices, and their relative complement α_1' belongs to the modified chain (l_1, α_1'), that is, to a 2 shorter chain. For other similar fragments point (c) of Theorem 2 is the supposition of induction.

To use the chain split data structure one have to keep this structure in the computer memory which is ineffective. The next principal step towards this technique was done by Tonoyan [76] in 1979, who introduced the "chain algebra",—a set of simple procedures that serve all the actual queries about Hansel chains, providing a technical solution to all the problems related to algorithms with Hansel chains, without constructing and keeping them in computer memory. A slightly modified and simplified version of chain algebra tools are given in [73].

Hypercube Chain Algebra

The technique achieved by Tonoyan [76] is related to the problem of recovering Boolean functions by queries, but it is more universal and is applicable to solve other similar problems. Therefore, we present "Tools" from [76], which are independent procedures that perform particular local operations on the chains (e.g., find the sequential number of a vertex on the chain, find a chain containing the relative complement of three consecutive vertices, etc.) without constructing and keeping them in computer memory. They can be applied in global recognition type problems. A specific application is the synthesis of association rules of data mining, based on the chain split algorithms.

A set of numbers $K(\alpha) = (K_1(\alpha), K_2(\alpha), \ldots, K_n(\alpha))$, associated with vertex $\alpha = (\alpha_1, \alpha_2, \ldots, \alpha_n)$ of B^n, is used in the Tools, where:

$$K_i(\alpha) = \begin{cases} K_{i-1}(\alpha) + 1 \; if \; \alpha_i = 0 \\ K_{i-1}(\alpha) - 1 \; if \; K_{i-1}(\alpha) \geq 2, \alpha_i = 1 \\ 1 \; if \; K_{i-1}(\alpha) = \alpha_i = 1 \end{cases}$$

$$K_1(\alpha) = \begin{cases} 2 \; if \; \alpha_1 = 0 \\ 1 \; if \; \alpha_1 = 1 \end{cases}$$

Tool 1 "VERTEX SEQUENTIAL NUMBER ON THE CHAIN"

For a given vertex α this tool finds the the length of the chain containing α, and the sequential number of α on the chain.

Tool 2 "NEIGHBOUR VERTEX SEQUENTIAL NUMBER ON THE CHAIN"
For a given vertex α and its neighbor β, Tool 2 finds the sequential number of β on the chain of β.

Tool 3 "LENGTH OF CHAIN ADJACENT TO THE VERTEX NEIGHBOUR TO THE GIVEN ONE"

For a given vertex α and its neighbor β, Tool 3 finds the length of the chain containing β.

Tool 4 "CLIMBING UP"

Tool 5 "CLIMBING DOWN"
For a given chain and a vertex α on it Tool 4 and Tool 5 find the upper and lower neighbors of α, correspondingly.

Tool 6 "ALL LOWER VERTICES"

Tool 7 "ALL UPPER VERTICES"
Tool 6 and Tool 7 find the lowest and highest vertices of the chains, correspondingly.

Tool 8 "THE CHAIN OF A RELATIVE COMPLEMENT VERTEX"
For given vertices $\alpha_1 \prec \alpha_2 \prec \alpha_3$, this tool finds the chain containing their relative complement.

Hypergrid Chain Properties

The chain split technique is less developed for multivalued grids. In [1] V. Alekseev represented a chain splitting of $\Xi^n_{m_1+1,m_2+1,\ldots,m_n+1}$, which actually matches to BETK.

Theorem 3 $\Xi^n_{m_1+1,m_2+1,\ldots,m_n+1}$ *can be split into chains of vertices with the following properties:*
(1) the chains are disjoint and contain all the vertices of $\Xi^n_{m_1+1,m_2+1,\ldots,m_n+1}$
(2) vertices in each chain immediately follow one another
(3) all chains are symmetric with respect to the middle layer/layers
(4) if α, β, γ *are any three non-collinear consecutive vertices of some chain of length* l*, then the set conjugate to them lays on a chain of length* $l - 2$.

Thus, the property "conjugate" or "relative complement" is also possible for chain splits on $\Xi^n_{m_1+1,m_2+1,\ldots,m_n+1}$. The same time, the means of chain algebra for these systems are not yet known.

3 Monotone Boolean Functions: Basic Results

Boolean functions in their diversity represent a wide class of objects that are actually grouped around certain properties such as linearity, compactness, threshold, monotonicity, self-duality, etc. Different classes occur in different types of applications and depending on the group of functions, problems of their representation, minimization, simplification or approximation, etc. appear. The class of monotone Boolean functions is characterized by problems of using the strong property of monotonicity to find the minimal kernel of these functions. This is similar to how a point of a numerical interval is recognized by successive dichotomies of logarithmic complexity. Structural, quantitative and qualitative features of monotone Boolean functions have been intensively studied by many authors.

3.1 Number of Functions

The problem posed by R. Dedekind in 1897 is about the number of monotone Boolean function of n variables. It is solved by the author for small numbers, until $n = 4$ [29]. Then for some of the next dimensions the problem was considered by different authors and the recent results are [64, 78], that obtained the numbers for $n = 9$:

$$286{,}386{,}577{,}668{,}298{,}411{,}128{,}469{,}151{,}667{,}598{,}498{,}812{,}366$$

and the number

$$789{,}204{,}635{,}842{,}035{,}040{,}527{,}740{,}846{,}300{,}252{,}680$$

for the case of inequivalent functions.

For n variable MBFs, since 1954, several formulas were found for their number $\psi(n)$:

$$2^{\binom{n}{[n/2]}} \leq \psi(n) \leq n^{\binom{n}{[n/2]}+2}$$

by Gilbert, then an improved upper bound

$$\psi(n) \leq 2^{B\binom{n}{[n/2]}(1+\epsilon_n)}$$

where $B = \frac{3log_2 3}{(\sqrt[3]{9}-1)^{3/2}}$, $\epsilon_n \to 0$ and $n \to \infty$, which imply

$$2^{4.23\binom{n}{[n/2]}}$$

is obtained by Korobkov.

Hansel's Theorem 2 reduced the upper bound to $3^{\binom{n}{[n/2]}}$.

Next improvement is by Kruskal [54]:

$$2^{(1+\alpha_n\binom{n}{[n/2]})} \leq \psi(n) \leq 2^{(1+\beta_n\binom{n}{[n/2]})}$$

with $\alpha_n = ce^{-n/4}$ and $\beta_n = c'(logn)n^{1/2}$.

In the paper "The number of monotone Boolean functions" A. Korshunov obtained the asymptotic formula for the number of monotone Boolean functions [52]:

Theorem 4 $\psi(n) = (1 + o(1)) \cdot 2^{\binom{n}{\lfloor n/2 \rfloor}} \exp a(n)$ *for even n, and*

$$\psi(n) = (1 + o(1)) \cdot 2^{\binom{n}{\lfloor n/2 \rfloor + 1}} \exp(b(n) + c(n)) \; for \; odd \; n, where$$

$a(n) = \binom{n}{n/2} - 1(2^{-n/2} + n^2 2^{-n-5} - n 2^{-n-4}),$
$b(n) = \binom{n}{(n-3)/2}(2^{-(n+3)/2} + n^2 2^{-n-6} - n 2^{-n-3})$, *and*
$c(n) = \binom{n}{(n-1)/2}(2^{-(n+1)/2} + n^2 2^{-n-4}),$

3.2 Reconstruction of Boolean Functions, Generale

Two types of recognition problems about the monotone Boolean functions are raising in different applications. One is to determine whether an unknown function is monotone, and the second is to identify an unknown function, given that it is monotone. For the former, it is to mention almost linear randomized testing algorithms for monotonicity [30]. Goldreich et al. [40] proposed a simple monotonicity tester:

Query the function value on a pair of vertices differing by exactly one coordinate (an edge) and reject the claim if monotonicity is violated.

This is a one-way algorithm (always accepts a monotone Boolean function), has a constant error rate (in rejecting truly monotone functions), and is bounded by $O(n/\epsilon)$ complexity (queries to the oracle).

All the basic results in this direction start with the known edge isoperimetric inequalities by Margulis and Talagrand. A series of improvements of initial results lead to [21] with a complexity estimate $npolylogn$. The recent paper by Khot, Minzer and Safra reduces this to the optimal (apart from powers of $logn$), $\sqrt{n}polylogn$. They use a new tester and a new isoperimetric inequality. Fischer et al. [33] provides the same level lower estimate of complexity.

We basically address the second topic of Boolean reconstruction, which in fact is correlated to the first problem, but has other applications, such as the problem of generating frequent itemsets in data mining [6].

Suppose that an arbitrary (unknown to us) function $f \in \mathcal{M}_n$ is given by an operator A_f, which receiving input $\alpha \in B^n$, returns the value $f(\alpha)$. Given the operator A_f it is required to fully restore the function. After each call to the operator which resumes the value $f(\alpha)$ for a point $\alpha \in B^n$, certain points of B^n become determined through the monotonicity property. Obviously, we should strive for optimality of these algorithms, i.e., minimising the number of steps of applying to A_f.

Consider the set R of all algorithms that solve this problem. That is, for a monotone Boolean function f an algorithm from R exploiting the operator A_f restores the complete table of values of f. The work of algorithms consists of several stages.

Selection: algorithm selects a point $\alpha \in B^n$ and with the help of operator A_f computes $f(\alpha)$, then inserts it into the table of computed values.

Extension: the table is extended by the monotonicity. For example, if $f(\alpha) = 1$ then $f(\delta)$ must be 1 for all points δ, $\delta \succ \alpha$, and these values are filled in the table, accordingly.

The next step is the rule of selecting inputs for the operator A_f.

This process is repeated until the table of values is filled completely.

It is natural to associate the pair ¡algorithm $r \in R$, monotone function f> with the number of calls to the operator A_f during filling the table of f by the algorithm r. Let $\varphi_r(f)$ denote the minimum number of calls sufficient to identify f by r, and $\varphi_r(n)$ is the minimum number of calls sufficient to identify an arbitrary monotone Boolean function of n variables by the algorithm r.

It is appropriate to evaluate the quality of the algorithms of R using the function $\varphi_R(n) = \min_r \varphi_r(n)$; and this characterise the Shannon complexity of recognition of the class of monotone Boolean functions of n variables. Further, we will refer it as $\varphi(n)$.

Resolving sets

Let us introduce some general terms on function recognition [50].

Suppose we are given a certain class F of Boolean functions and a function f, belonging to this class. A set of vertices of B^n (denoted as $G(f, F)$) is called resolving set for the pair (f, F), if from the fact that

(a) a function g belongs to F,
(b) the values of f and g are the same on the set $G(f, F)$

it follows that $f = g$.

To restore the table of values of a function it is sufficient to determine the values of the function on some of its resolving sets.

Resolving set $G(f, F)$ is called a deadlock resolving set for (f, F), if no subset of it is resolving for the pair (f, F).

Every monotone Boolean function has a unique deadlock resolving set that is included in all its resolving sets, and this is the set $G(f) = minT(f) \cup maxF(f)$ (recall that $maxF(f)$ and $minT(f)$ denote the sets of all upper zeros and lower units of a monotone function f correspondingly). We should note that this is not the case for other classes, for instance, for the class of symmetric Boolean functions.

Estimates for $\varphi(n)$

Kororbkov [50], in an early stage, has obtained the following result concerning the upper and lower bounds for $\varphi(n)$.

Theorem 5 (Korobkov's Theorem)

$$\binom{n}{[n/2]} + \binom{n}{[n/2]+1} \leq \varphi(n) \leq b \cdot \binom{n}{[n/2]}(1 + \epsilon_n),$$

where

$$b = 8/(\sqrt[3]{16} - 1)^{3/2}$$

and $\epsilon_n \to 0$ *when* $n \to \infty$.

The proof of this theorem uses the following monotone Boolean function $h(x_1, x_2, \ldots, x_n)$, defined as follows:

$$h(\alpha_1, \alpha_2, \ldots, \alpha_n) = \begin{cases} 1 & if\ [\frac{n}{2}] + 1 \leq \sum_{i=1}^{n} \alpha_i \\ 0 & if\ 0 \leq \sum_{i=1}^{n} \alpha_i \leq [n/2]. \end{cases}$$

It is obvious that in this case the resolving set $G(h)$ has exactly $\binom{n}{[n/2]} + \binom{n}{[n/2]+1}$ points. This provides a lower bound for $\varphi(n)$:

$$\varphi(n) \geq \binom{n}{[n/2]} + \binom{n}{[n/2]+1}.$$

Upper bound is obtained from the Theorem 2. According to the property (b) of the theorem, if the value of a monotone Boolean function f is known on all the vertices of chains of length $n - 2p - 1$, then on each chain of length $n - 2p + 1$ there are at most two (neighbouring) vertices, where the function value is unknown. Thus the function can be recognised by the following number of queries:

$$number\ of\ chains\ of\ the\ shortest\ length\ +\ 2(number\ of\ the\ remaining\ chains)$$

Simple calculations show that

$$\varphi(n) \leq \binom{n}{[n/2]} + \binom{n}{[n/2]+1}.$$

Thus,

Theorem 6 ([42]) *The minimum number of queries to the oracle for arbitrary monotone Boolean function's recognition is equal to:*

$$\binom{n}{[n/2]} + \binom{n}{[n/2]+1}.$$

Another recognition structure is used in [73] and in its extensions. For even n, B^n is split according to two variables and the recognition starts in the two middle layer sub-cubes of this construction. For odd n, firstly B^n is split according to one variable, then as each sub-cube now has an even size, the procedure for even sizes is applied. This provides optimal recognition of all MBFs in the sense of Shannon complexity.

Complexity by the resolving set size

Finally, it is to mention the work [36] that considers not the Shannon complexity but the individual complexity of MBF given by its resolving set size.

Ganinanov's Procedure. Given an arbitrary vector $\alpha \in B^n$ and a monotone Boolean function f of n variables. Gainanov [36] presented a simple algorithm that effectively enlarges the information of $f(\alpha)$ by the monotonicity. The goal is to find elements of the set $minT(f) \cup maxF(f)$. Let α belong to the l-layer L_l of B^n, and $f(\alpha) = 1$. Consider the vector $\beta = ON(\alpha)$, which consists of the indices of 1-valued coordinates in the vector α. For $i = 1, 2, \ldots, l$, define $\alpha^0 = \alpha$, and: $\alpha^i = \alpha^{i-1} - e_{\beta_i}\ if\ f(\alpha^{i-1} - e_{\beta_i}) = 1\ and\ \alpha^i = \alpha^{i-1}\ otherwise.$ e_{β_i} denotes the β_i'th unite vector. It is easy to see that the last α^i in this chain with $f(\alpha^i) = 1$ is an element of $minT(f)$. The case of $f(a) = 0$ is treated symmetrically, and the algorithm produces an element of $maxF(f)$. In either case, at most $n + 1$ vectors are tested by membership queries and the total time required for this task is $O(n)$ locally and $O(n(|minT(f)| + |maxF(f)|))$ globally. To give a more complete picture of the field, we present a number of established studies and results in various formats.

4 Constraint Monotone Boolean Function Recognition

In general, tasks related to the recognition of MBFs may have different formulations. One objective is to recognize a particular unknown function, knowing that it belongs to the class of MBFs or to one of its subclasses. Another goal is to start with partial knowledge about the unknown function, trying to complete the information. One more case is when the number of queries is restricted by some number k and the goal is to maximize the recognized part of the function [69]. Similar problems can be formulated for specific classes of Boolean functions. Examples of classes are as follows:

KK-MBF (Kruskal-Katona MBFs) arise as a result of the shadow minimization theorem [47, 48, 54]. $KK - MBF$s are monotone Boolean functions but they also intersect the cube layers along their initial segments of the lexicographic order. The complement of the $KK - MBF$ area in B^n has a similar property; it is related to the initial segments of the reverse-lexicographic order, and they are anti-monotone.

Symmetric MBF This is a trivial class of functions that takes a constant value on the cube layers. Trivial, but these functions are practically important. Examples are majority functions, parity functions, and others [61].

Threshold MBF Functions are defined by a linear inequality of weighted sums of variables.

Matroid MBF Monotone Boolean function f is called a matroid function if for each $\alpha, \beta \in min\ T(f)$ with $\alpha_i = 1,\ \beta_i = 0$, there exists a coordinate j with $\alpha_j = 0,\ \beta_j = 1$ such that vertex α', obtained from α by replacing α_i with 0 and α_j with 1, belongs to min $T(f)$.

The combinatorial complexity of reconstruction in these and other subclasses of MBFs is not well studied. For example, monotone Boolean functions, with zeros and units separated by two middle layers of the cube, are the most difficult functions for query-based reconstruction when only the monotonicity of the function is given. But if it is known that the function belongs to the class of symmetric functions, the reconstruction of this function can be done by n queries.

In this section, we present results concerning the reconstruction of some special classes of monotone Boolean functions.

4.1 Spherical Boolean Functions

Consider a class of partially defined monotone Boolean functions f, such that $f(\alpha) = 1$ for all α above some k-th layer of B^n, $0 < k < [n/2]$ (k-th layer is not included). We denote this class as $\mathcal{M}_n^k$. An application of these functions can be found in terms of the data mining theory, namely, in terms of association rule mining algorithms, where specifically, this class of monotone Boolean functions is actual [6].

Theorem 7 ([6]) $\varphi(n) = \binom{n}{k} + \binom{n}{k-1}$ *for the class of functions* $\mathcal{M}_n^k$.

Proof To prove the lower bound we consider the following monotone Boolean function of $\mathcal{M}_n^k$:

$$h(\alpha_1, \alpha_2, \ldots, \alpha_n) = \begin{cases} 0, & if\ 0 \le \sum_{i=1}^n \alpha_i \le k \\ 1, & if\ k < \sum_{i=1}^n \alpha_i \le n. \end{cases}.$$

$G(h)$, the deadlock resolving set of function $h(\alpha_1, \alpha_2, \ldots, \alpha_n)$, consists of vertices of the $(k-1)$-th and k-th layers, and thus $\varphi(n) \ge \binom{n}{k} + \binom{n}{k-1}$.

To prove the upper bound one may consider a chain starting at the k-th layer, we know the values of function at all points on these chains except the starting points, i.e. points located at the k-th layer. By making one request for each of these chains, we will fully recover the function on these chains. Having the values of the function at all points of the chains that start at the k-th layer, we can use the property (b) of Hansel's Theorem 2 and property of monotonicity to determine the function's values at the points of chains starting at the $k-1$ layer. As a result it turns out that for the chains starting at the $k-1$ layer the functions values remain unknown at no more than two points.

After performing this procedure for the remaining chains, we conclude that for the complete recognition of a function on the chains starting at the k-th layer it is required to determine the function's value only at one point, and at most at two points for the other chains.

In this way,
$\varphi(n) \le$ (the number of chains starting at the k-th layer) +
2*(number of other chains starting below the k-th layer).

Let us substitute values using the formula for calculating the number of chains.

$$\varphi(n) \leq \binom{n}{k} - \binom{n}{k-1} + 2\left(\binom{n}{k-1} - \binom{n}{k-2} + \cdots + \binom{n}{1} - \binom{n}{0} + 1\right).$$

After making all reductions we obtain: $\varphi(n) \leq \binom{n}{k} + \binom{n}{k-1}$.

By combining both results we obtain:

$$\varphi(n) = \binom{n}{k} + \binom{n}{k-1}.$$

4.2 Threshold Monotone Functions

Boolean function $f(x_1, x_2, \ldots, x_n)$ is called threshold function, if there $\exists a_1, a_2, \ldots, a_n, b \in \mathbb{R}$ such that

$$f(x_1, x_2, \ldots, x_n) = \begin{cases} 1 & if\ a_1x_1 + a_2x_2 + \cdots + a_nx_n \geq b \\ 0 & otherwise. \end{cases}$$

A subclass of this with $\forall a_i = 1$ and $b = n/2$ is known as majority functions. Boolean function is k-summable, if for some j, $2 \leq j \leq k$ there exist j not necessarily distinct vectors $x_1^0, \ldots, x_j^0 \in f_0^{-1}$ and $x_1^1, \ldots, x_j^1 \in f_1^{-1}$ so that $x_1^0 + \cdots + x_j^0 = x_1^1 + \cdots + x_j^1$. Otherwise f is called k-asummable. If f is k-asummable for all k, then it is called asummable.

Theorem 8 ([31]) *Function f is threshold, iff it is asummable.*

Assumability is a fine theoretical descriptor. In the general case, the asummability condition is difficult to verify, which corresponds to the complexity of the problem of recognizing the threshold functions. However, in some special cases, the conditions can be relaxed, clarified. Thus, for a monotone function f, its lower units can be taken as the set of units in the summability condition, or its upper zeros can be used as zeros. Further clarifications are also possible.

4.3 2-Monotonic Functions

There is a chain of investigations about the positive 2-monotonic functions.

We define k-assignment on B^n as selecting k variables $x_{i_1}, x_{i_2}, \ldots, x_{i_k}$ and assigning them values $\{0, 1\}$. n-assignment gives a vertex α of B^n, and the value of a function f, $f(\alpha)$. If $k < n$ we get a subcube where the values of the assigned k variables are fixed and the others vary. In this case, we get a restricted function.

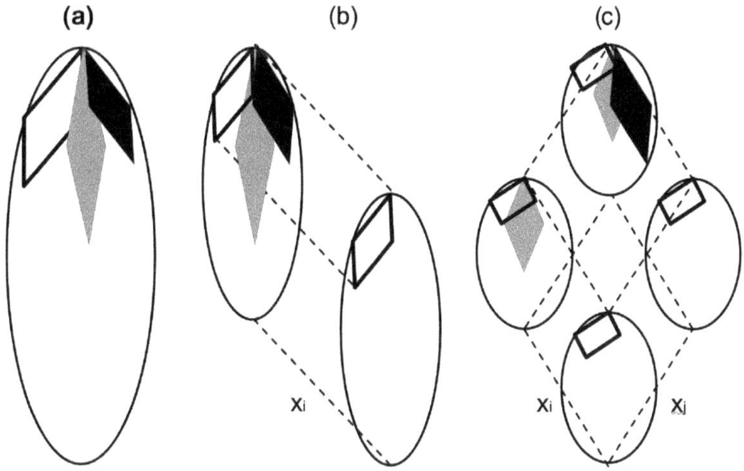

Fig. 6 In an example monotone function the area of units of the function is composed of three subcubes, white, gray, and black. **a** The black one has fixed values of partition variables x_i and x_j. The white subcube is split in both directions x_i and x_j; **b** the (1, 0) part of the function is $\geq$ of the part (0, 1). Similar subcubes can be many, but if there is a subcube with variable x_i and fixed x_j then the resulting function cannot be 2-monotonic; **c** and the gray subcube is partitioned only in direction x_j

We say that f is comparable at α if $f(\alpha) \leq f(\overline{\alpha})$ or $f(\overline{\alpha}) \leq f(\alpha)$, where $\overline{\alpha}$ is the complement of α. f is k-comparable if either $f(\alpha) \leq f(\overline{\alpha})$ or $f(\overline{\alpha}) \leq f(\alpha)$ holds for all k-assignments.

If a function f is i-comparable for every i such that $1 \leq i \leq k$, then f is said to be k-monotonic.

2-monotonic functions are an important special case of this. Let us consider the case when f is a positive (monotone) 2-monotonic function. First, 1-monotonic in this case means that $T(f)$, the set of units of the function, represents a bunch of some subcubes $B_1, B_2, \ldots, B_l$ around the vertex **1** (with all coordinates 1).

Let b^j be the set of indices of those subcubes B_i where the variable x_j is not fixed (assigned). In a 1-assignment by the variable x_j $f_{x_j=0}$ consists of partitions of all b^j subcubes that belong to $B^{n-1}_{x_j=0}$. So $f_{x_j=0} \leq f_{x_j=1}$.

For positive 2-monotonic functions, 2-assignments values to some variables x_i and x_j, together with their negations compose two pairs: 00, 11 and 01, 10. The first case is similar to the 1-assignment. Now, let $x_i = 0$ and $x_j = 1$. In accord to this let us consider subcubes B_k where neither i nor j are fixed. For the assignment 01 the generated subcubes are $B_k(x_i = 0, x_j = 1)$, and this must be comparable with $B_k(x_i = 1, x_j = 0)$. This requires that either b^i is a subset of b^j, or b^j is a subset of b^i. Further generalization of this shows that all b^i must be ordered by inclusion and this is the essence of 2-monotonicity. An example is given in Fig. 6.

Boros et al. [18] describes two algorithms that check whether an unknown function is 2-monotonic. In the following theorems the time complexities are given.

Theorem 9 ([18]) *For an unknown positive function* f *of* n *variables, the algorithm decides whether* f *is 2-monotonic, and if* f *is 2-monotonic, it outputs* $minT(f)$ *and* $maxF(f)$. *The time required is* $O(n^3m)$ *and the number of queries to the oracle is* $O(n^3m)$, *where* $m = |minT(f)| + |maxF(f)|$.

Theorem 10 ([18]) *For an unknown positive function* f *of* n *variables, the algorithm decides whether* f *is 2-monotonic, and if* f *is 2-monotonic, it outputs* $minT(f)$ *and* $maxF(f)$. *The time required is* $O(nm^2 + n^2m)$ *and the number of queries to the oracle is* $O(nm)$, *where* $m = |minT(f)| + |maxF(f)|$.

The algorithms are similar to Gainanov's procedure, but they are multi-block and consider more data structures. They take into account both the complexity of queries and the computational complexity of individual algorithmic blocks. Makino and Ibaraki [59] developed an algorithm with time and query requirement as n^2m.

The mentioned algorithms generalize the interval division method. Many interval problems can be solved with logarithmic complexity. This fully applies to symmetric Boolean functions. Monotonic functions are more complex, they are multidirectional, but some analogy is applicable here.

4.4 KK-MBF Recognition

In this section, we consider a special class of monotone Boolean functions, related to the well-known Kruskal-Katona theorem, which describes the exact optimal monotone constructions of shadow minimization (constraint minimization) problem, and the existence problem of Sperner systems for a given set of parameters.

Introduction to KK-MBF Type Functions

Let f be a monotone Boolean function on B^n. f is called a $KK - MBF$ type function if zeros of f on the layers of B^n compose initial segments of the reverse-lexicographic order (an example is given in Fig. 7, where unmarked vertices correspond to zeros of the function).

The initial formulations of the shadow theorem in [47, 48, 54] are given in terms of co-lexicographic order, but this framework was later simplified to the simple lexicographic ordering [9]. The basic result was obtained for two neighbor layers of the cube. The name KK now refers to an extension of the basic result of the shadow minimization theorem to many layers of B^n, as well as to results on the existence of Sperner families [23, 26]. Usually, a $KK - MBF$ function f is given through its characteristics, $\#minT(f) = \langle p_{i_1}, p_{i_2}, \dots, p_{i_r} \rangle$, where p_{i_j} is the number of lower units of f on the i_j-th layer. An example is given in Fig. 7.

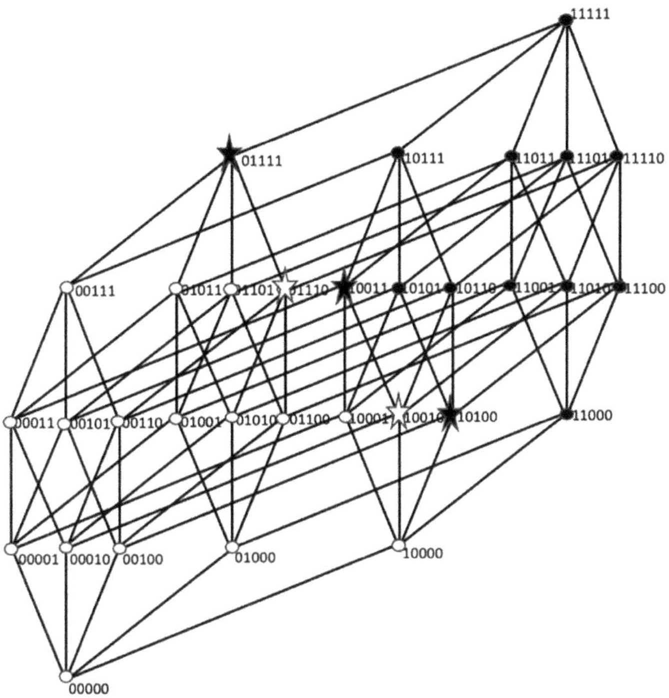

Fig. 7 $KK - MBF$ function on B^5 with $p_2 = 2$ (vertices 11000 and 10100), $p_3 = 1$ (vertex 10011), and $p_4 = 1$ (vertex 01111). Uncolored vertices show zeros, and the dark vertices–units of the function. Stars indicate the corner points

Resolving Sets for KK-MBF

We formulate two simple properties for $KK - MBF$ type functions f and call them *horizontal* and *vertical* conditions, denoting them as *Cond-h* and *Cond-v*.

Cond-h:

(1) if $f(\alpha) = 0$ for a vertex α of some layer L_k, then $f(\beta) = 0$ for all $\beta \in L_k$ reverse-lexicographically preceding α, ($\beta \preccurlyeq_{\text{revlex}} \alpha$),
(2) if $f(\alpha) = 1$ for a vertex α of some layer L_k, then $f(\beta) = 1$ for all β of L_k lexicographically preceding α, ($\beta \preccurlyeq_{\text{lex}} \alpha$).

Cond-v:

(1) if $f(\alpha) = 0$ for a vertex α, then $f(\beta) = 0$ for all β, $\beta \preccurlyeq \alpha$ (component-wise order),
(2) if $f(\alpha) = 1$ for a vertex α, then $f(\beta) = 1$ for all β, $\beta \succcurlyeq \alpha$ (component-wise order).

These conditions, applied recursively, define a *domain* for each vertex $\alpha \in f$; denote it by $d(f, \alpha)$.

We also define the notion of *corner points* for $KK - MBF$ type functions.

A zero vertex α of a function f is called a *zero-corner point* if:

(1) $f(\beta) = 1$ for all β from the same layer, such that $\beta \prec_{\text{lex}} \alpha$, and
(2) $f(\beta) = 1$ for all β, $\alpha \prec \beta$ (component-wise order).

Similarly, a unit vertex α of a $KK - MBF$ type function f is called *one-corner point* if:

(1) $f(\beta) = 0$ for all β from the same layer such that $\beta \succ_{\text{lex}} \alpha$, and
(2) $f(\beta) = 0$ for all β, $\beta \prec \alpha$ (component-wise order).

Let $z(f)$ denote the set of all zero-corner points, and $o(f)$ denote the set of all one-corner points of function f.

Summarizing all the above reasoning, we formulate the following statement.

Theorem 11 ([5]) *Every monotone Boolean function f of class $KK - MBF$ has a unique deadlock resolving set which is included in all its resolving sets, this is $G(f) = z(f) \cup o(f)$.*

The following proposition provides an upper bound for $|z(f) \cup o(f)|$. Due to the resolving property of the set $z(f) \cup o(f)$, the estimate will show the number of required tests for reconstruction of $KK - MBF$ functions.

Theorem 12 ([5]) *For monotone Boolean functions f of class $KK - MBF$*
$|z(f) \cup o(f)| \leq 2(n-1)$.

A series of $KK - MBF$ functions are introduced in [5], which provide the following lower bound for $|z(f) \cup o(f)|$, $|z(f) \cup o(f)| \geq n$.

Improved Bounds

In this section, we present an improved lower bound for $|z(f) \cup o(f)|$ [62].

We introduce/describe a special "compressed" function f in KK-MBF class, and prove that $|z(f) \cup o(f)| \geq 2(n-4)$ for this function.

Let us fix two layers in B^n, L_{k_1} and L_{k_2}, where $1 \leq k_1 \leq k_2 \leq n-1$, and define a function of KK-MBF class, whose lower units and upper zeros are situated between those layers (inclusive).
A starting vertex (lexicographically smallest lower unit of the function on L_{k_1}) can also be given, otherwise, we suppose that we start from the lexicographically smallest vertex of L_{k_1}, which can be either a lower unit of the function, denoted as $\alpha_1^{k_1}$ or upper zero, denoted as $\alpha_0^{k_1}$. Upper index shows the layer, and lower index indicates whether the vertex is unit or zero of the function.

"Compressed" function is defined layer by layer.

<u>Layers from L_0 to $L_{k_{1-1}}$</u>:
$f(\alpha) = 0$ for all α belonging to these layers.

Layer L_{k_1}:
If we start with the unit vertex $\alpha_1^{k_1}$ then:
$f(\alpha_1^{k_1}) = 1$ and we assign $f(\alpha) = 1$ for all α of the domain $d(f, \alpha_1^{k_1})$. Let $\alpha_0^{k_1}$ denote the lexicographically greatest neighbor of $\alpha_1^{k_1}$. We assign $f(\alpha_0^{k_1}) = 0$, and also $f(\alpha) = 0$ for all α in $d(f, \alpha_0^{k_1})$.
If we start with the zero vertex $\alpha_0^{k_1}$ then:
$f(\alpha_0^{k_1}) = 0$, and we assign $f(\alpha) = 0$ for all α of the domain $d(f, \alpha_0^{k_1})$.

Layer $L_{k_{1+1}}$:
Let $\alpha_1^{k_1+1}$ denote the lexicographically greatest vertex of the upper shadow (on the $L_{k_{i+1}}$-th layer) of $\alpha_0^{k_1}$. Assign $f(\alpha_1^{k_1+1}) = 1$ and $f(\alpha) = 1$ for all α of the domain of $\alpha_1^{k_1+1}$. Assign $f(\alpha) = 0$ for all α lexicographically succeeding $\alpha_1^{k_1+1}$ (if such vertices exist). Let $\alpha_0^{k_1+1}$ denote the lexicographically smallest of them, i.e., the vertex immediately succeeding $\alpha_1^{k_1+1}$ lexicographically.

We continue the process until:

1. we reach a layer L_k, $k < k_2$, such that the function value is known for all vertices of B^n, and we set $L_{k_2} = L_k$; or
2. we reach the layer L_{k_2}.

Layers from $L_{k_{2+1}}$ to L_n:
$f(\alpha) = 1$ for all α belonging to these layers.

Thus, L_{k_1} is the first layer where we take vertices $\alpha_1^{k_1}$ and $\alpha_0^{k_1}$ ($\alpha_1^{k_1}$ may not exist if we started with the zero vertex as initial), and L_{k_2} is the last layer, where we take vertices $\alpha_1^{k_2}$ and $\alpha_0^{k_2}$ ($\alpha_0^{k_2}$ may not exist). Note that α_0^k does not exist on some layer L_k only when the function is equal to 1 on layers above L_k, and thus the function becomes completely defined.

It is easy to check that the defined function belongs to the class KK-MBF, and $(\alpha_1^{k_1}, \alpha_0^{k_1})$, $(\alpha_1^{k_1+1}, \alpha_0^{k_1+1})$, $\ldots$, $(\alpha_1^{k_2}, \alpha_0^{k_2})$ ($\alpha_1^{k_1}$ and/or $\alpha_0^{k_2}$ may not exist), are pairs of corner points located on the layers from L_{k_1} to L_{k_2}, where the lower index indicates whether it is 1- or 0-corner point.

We call this function compressed KK-MBF function, and if the corner points are located on layers L_{k_1} to L_{k_2}, both inclusive, we denote the function as: $f^{k_1 \to k_2}$. Depending on the starting vertex we obtain different functions $f^{k_1 \to k_2}$ but for the same start the function $f^{k_1 \to k_2}$ is unique.

Theorem 13 *Let f be a compressed KK-MBF function defined on B^n with corner points located between k_1-th and k_2-th layers inclusive, where $0 < k_1 < k_2 < n$. Then there exists a compressed KK-MBF function $\overline{f}$ with corner points located on layers from $n - k_2$ to $n - k_1$ such that α_1^k is a 1-corner point of f if and only if $\overline{\alpha}_1^{n-k}$ is 0-corner point of $\overline{f}$, where $\overline{\alpha}_1^k$ is the complement (inversion) of α_1^k in B^n.*

Proof Obviously, if α_1^k belongs to L_k, then $\overline{\alpha}_1^{n-k}$ belongs to L_{n-k}.

Let $(\alpha_1^{k_1}, \alpha_0^{k_1})$, $(\alpha_1^{k_1+1}, \alpha_0^{k_1+1}), \ldots, (\alpha_1^{k_2}, \alpha_0^{k_2})$ be pairs of corner points of $f^{k_1 \to k_2}$ ($\alpha_1^{k_1}$ and/or $\alpha_0^{k_2}$ may not exist). We construct $\overline{f}^{n-k_2 \to n-k_1}$ in the following way:

$\overline{f}^{n-k_2 \to n-k_1}(\alpha) = 0$ for all α belonging to layers 0 to $n - k_2 - 1$.

$\overline{f}^{n-k_2 \to n-k_1}(\overline{\alpha}_0^{n-k_2}) = 1$ and $\overline{f}^{n-k_2 \to n-k_1}(\alpha) = 1$ for all α of the domain of $\overline{\alpha}_0^{n-k_2}$.

$\overline{f}^{n-k_2 \to n-k_1}(\overline{\alpha}_1^{n-k_2}) = 0$ and $\overline{f}^{n-k_2 \to n-k_1}(\alpha) = 0$ for all α of the domain of $\overline{\alpha}_1^{n-k_2}$.

We continue the process as in the construction of the function $f^{k_1 \to k_2}$.

Clearly, $\overline{f}^{n-k_2 \to n-k_1}$ is compressed KK-MBF, and $\overline{\alpha}_0^{n-k_2}, \overline{\alpha}_1^{n-k_2}, \ldots, \overline{\alpha}_0^{n-k_1}, \overline{\alpha}_1^{n-k_2}$ are its corner points, where it is 1-corner point, if the lower index is 0, and 0-corner point, otherwise.

Theorem 14 *Let $n \geq 5$. There exists a compressed KK-MBF function $f^{2 \to n-r}$ in B^n with two corner points on layers from 2 to $n - r$, where $r \leq 3$.*

The proof is by the induction on n.

Corollary 1 *There exists KK-MBF function in B^n with at least $2(n - 4)$ corner points.*

Identification of KK-MBF Type Functions: Main Procedures

Hansel chain based MBF recognition methods are global tools and can be applied to recognize any class of monotone Boolean functions, including $KK - MBF$. However, we aim to exploit specific properties of this class of functions to achieve more efficient recognition. To go beyond the chain-based analysis, in this section we consider the recognition of $KK - MBF$ functions at the level of analysis of particular layers of B^n.

If the vertices on each layer of B^n are ordered lexicographically, we can apply layer by layer bisections and find a corner vertex candidate α_k on the k-th layer with no more than $\log_2 (\binom{n}{k} + 1)$ queries.

In this manner, a $KK - MBF$ type function can be recognized by no more than $\sum_{k=1}^{n} \log_2 (\binom{n}{k} + 1)$ queries. A very rough estimate of this total value would be $O(n^2)$, as the largest layer is the middle k-th layer with $k = [n/2]$, where $\binom{n}{[n/2]} \sim \frac{2^{n+1}}{\sqrt{2\pi n}}$ with $n \to \infty$.

Theorem 15 ([5]) *Time complexity of reconstruction of monotone Boolean functions of class $KK - MBF$ by layer by layer bisections is $O(n^2)$.*

For $KK - MBF$ functions the time complexity of Hansel algorithm remains $\sim \binom{n}{[n/2]}$, so $O(n^2)$ is a valuable reduction of the complexity characteristics. But the techniques of [73, 76] for the reduction of memory used in Hansel algorithm can not be applied in the case of our layer bisection approach.

To find a way to reduce the memory complexity we continue to address algorithmic questions on layers from an additional perspective. In order to apply bisections on layers, we need to keep all 2^n vertices (ordered lexicographically, layer-by-layer) in computer memory. Otherwise, we need to find a vertex by its number in lexicographic order, or by a distance from a certain vertex (again, in lexicographic order). There are no explicit formulas for this. Thus, both cases will require appropriate time and memory resource.

One way to avoid this situation is using sub-cube structures on layers when partitioning the cube. This is easily applicable to find the 0–1 boarder on layers, although it will require more than logarithmic queries on layers.

We will use α_j^k for the j-th element of L_k in the lexicographic order. Then the smallest element of L_k in the lexicographic order is $\alpha_1^k = (\overbrace{11\cdots1}^{k}\overbrace{00\cdots0}^{n-k})$, and the largest one is the element $\alpha_{C_n^k}^k = (\overbrace{00\cdots0}^{n-k}\overbrace{11\cdots1}^{k})$.

Let $B^{n-1}_{x_1=1}$ and $B^{n-1}_{x_1=0}$ be the partitions of B^n according to x_1, and $L_{k,x_1=1}$ and $L_{k,x_1=0}$ denote the parts of L_k in $B^{n-1}_{x_1=1}$ and $B^{n-1}_{x_1=0}$, respectively. Then $\alpha_1 = (0\,\overbrace{11\cdots1}^{k}\overbrace{00\cdots0}^{n-k-1})$ is the lexicographically smallest element of $L_{k,x_1=0}$, and $\alpha_0 = (1\overbrace{00\cdots0}^{n-k}\overbrace{11\cdots1}^{k-1})$ is the lexicographically largest element of $L_{k,x_1=1}$.

Instead of taking the arithmetical middle vertex of L_k to ask/test the function value, we take either α_1 or α_0 (for certainty, we will take α_1). If $f(\alpha_1) = 1$, then $f(\alpha) = 1$ for all $\alpha \in L_{k,x_1=1}$; therefore, the next vertex that we will take to ask the function value, is the largest element of $L_{k,x_1=0,x_2=0}$ (the part of L_k in $B^{n-2}_{x_1=0,x_2=0}$), this is element $\alpha_2 = (00\,\overbrace{11\cdots1}^{k}\overbrace{00\cdots0}^{n-k-2})$.

If $f(\alpha_1) = 0$, then $f(\alpha) = 0$ for all $\alpha \in L_{k,x_1=0}$; therefore the next vertex that we will take is the largest element of $L_{k,x_1=1,x_2=0}$ (the part of L_k in $B^{n-2}_{x_1=1,x_2=0}$), this is the element $\alpha_2 = (10\,\overbrace{11\cdots1}^{k-1}\overbrace{00\cdots0}^{n-k-1})$.

In general, in $B^{n-i}_{x_1=\sigma_1,\cdots x_i=\sigma_i}$ the largest element of k-th layer is

$\sigma_1\cdots\sigma_i\overbrace{11\cdots1}^{x}\overbrace{00\cdots0}^{y}$, where x is k minus the number of 1s in $\sigma_1\cdots\sigma_i$, and y is $(n-k)$ minus the number of 0s in $\sigma_1\cdots\sigma_i$.

In this way, after each query, we continue in a smaller sub-cube, and hence, the number of queries in each layer can be at most $n-1$. For the most heavy layer, for $k = n/2$, we get the same estimate but without either keeping all vertices in computer memory or calculating the given j-th vertex in the lexicographic order. Indeed, the number of vertices of middle layer equals $\binom{n}{[n/2]} \sim \frac{2^{n+1}}{\sqrt{2\pi n}}$ when $n \to \infty$. Logarithm of this, the number of steps in dichotomy, is equivalent to $n-1$, and so, the partition by cube structures is simple and effective.

Theorem 16 ([5]) *Reconstruction of monotone Boolean functions of class $KK - MBF$ can be organized with complexity $O(n^2)$ without keeping the cube structure in computer memory.*

So far, when recognizing $KK - MBF$ functions layer by layer, we have only used the fact that the function satisfies the condition Cond-h and have not used the monotonicity of the function. Using the monotonicity will narrow the space for taking the next vertex for testing.

4.5 k-Reasons Functions

The variant of MBF f that has k-"reasons" to be 1, i.e., the functions with k lower units is investigated in [37]. An asymptotic formula on the number of queries needed for fixed k is found for the adaptive case. The basic result is for the non-adaptive version of the problem, where the queries are asked at the same time. The exact number of queries for most values of k and n is determined:

Theorem 17 ([37]) *Let $n \geq 3$. Then*

$$h(n,k) = \begin{cases} n, & \text{if } k = 1, \\ \sum_{i=n-k}^{n-1} \binom{n}{i} - 1, & \text{if } 1 < k < n, \\ 2^n - 4, & \text{if } n \leq k \leq \binom{n-2}{[n/2]-1} + 1. \end{cases}$$

An important part of this result is the structures of sets of minimal tests. In the first case, where $k = 1$, this is the set of all vertices of the $(n-1)$-th layer of B^n. In the second case, $1 < k < n$, the set of tests consists of all elements of layers from $n-1$ to $n-k$ except the one single element of layer $n-1$. In the third case, these are all vertices of B^n except four vertices: $\emptyset$, $[n]$, n, and $[n-1]$. Recall that for the class of all MBFs, in the non-adaptive case, reconstruction requires an appeal to an oracle over all the 2^n vertices of B^n. In our case, it turns out that in the presence of n lower units, it is necessary to ask almost the same number of questions,–only 4 questions less.

5 Monotone Binary Reconstruction

In this section, we consider identification of monotone binary functions defined on multi-valued grids.

We will use similar definitions and notations here as for the case of hypercube. Particularly, $\phi(n)$ denotes the Shannon function for the recognition in the hypergrid.

5.1 Estimates for Complexity

Several lower and upper bounds are obtained for $\phi(n)$ in [50]. The n-th Cartesian degree is considered of an arbitrary finite partially ordered set R, and monotone binary functions are defined on this structure. Let $\phi_R(n)$ denote the minimum number of accesses to the oracle for identification of these functions. The following estimate is formulated by V. Korobkov.

Theorem 18 ([50])

$$C_1(R)\frac{|R|^n}{\sqrt{n}} \leq \phi_R(n) \leq C_2(R)\frac{|R|^n}{\sqrt{n}},$$

where $C_1(R)$ and $C_2(R)$ are constants depending on R.

In [1] an algorithm U_0 (generalization of Hansel's algorithm) is constructed by V. Alekseev for identification of monotone binary functions defined on the grid $\Xi_{k_1 k_2 \ldots k_n} = \Xi_{k_1} \times \Xi_{k_2} \cdots \times \Xi_{k_n}$, where $\Xi_{k_i} = \{0, 1, \ldots, (k_i - 1)\}$, and proved that:

Theorem 19 ([1])

$$\frac{t(U_0)}{t(U_{opt})} \leq \frac{1}{2}\lceil log_2(k-1)\rceil,$$

where $t(U_0)$ is the complexity of the algorithm U_0 and $t(U_{opt})$ is the complexity of an optimal algorithm U_{opt}, and $k = max k_i$.

The complexities of the algorithms are also presented in terms of middle layers of the hypergrid.

Let M and N denote the sets of vertices of middle layers of the hypergrid:

$$M = \{(a_1, \ldots, a_n) \in \Xi_{k_1} \times \Xi_{k_2} \cdots \times \Xi_{k_n} \mid a_1 + \cdots + a_n = \lceil \frac{1}{2}\sum_{i=1}^{n}(k_i - 1)\rceil\},$$

$$N = \{(a_1, \ldots, a_n) \in \Xi_{k_1} \times \Xi_{k_2} \cdots \times \Xi_{k_n} \mid a_1 + \cdots + a_n = \lceil \frac{1}{2}\sum_{i=1}^{n}(k_i - 1)\rceil + 1\}$$

It is known also ([1]) that

$$t(U_{opt}) \geq |M| + |N|$$

and

$$t(U_0) \leq |M| + \lfloor log_2 k\rfloor \cdot |N|.$$

Applying this to the monotone functions defined on Ξ_{m+1}^n, gives:

$$t(U_0) \leq |M'| + \lfloor log_2(m+1)\rfloor \cdot |N'|,$$

where M' and N' are the middle layers of Ξ_{m+1}^n.

Formulas for the cardinalities of the middle levels of the grid can be brought out by the help of [2, 10, 46, 66], where a formula for the number of ordered partitions of the given integer into the given number of parts, each of size at least 0 but no larger than the given size, is introduced. Using also the technique of [50, 83] deepens this result for a particular case. Combining these analyses, at least in an asymptotic for large n, m yields a simple and closed-form expression for complexities of the recognition (cardinalities of M and N) [7]. Details will be given in Sect. 5.4.

5.2 On a Special Decomposition of the Multi-Valued Cube

The section introduces a split technique and a special decomposition of Ξ_{m+1}^n through the structures, isomorphic to binary cubes, which is called *cube-split* technique [7].

In this manner, the known effective constructions designed so far for the hypercube domain, can be transfered through the cube-split technique to the area of Ξ_{m+1}^n. In fact, for structuring Ξ_{m+1}^n there are two perspectives:
(1) to develop G. Hansel type chain-splitting algorithms for Ξ_{m+1}^n, and
(2) to reduce constructions for Ξ_{m+1}^n to the binary cube domain and structures, and then use existing approaches for this domain.

The main achievement in (1) is by Alekseev [1] who introduced a proper Hansel type chain-split but these chains are specific, and they use *long collinear segments* (the case of sequential increase of the same component) that lead to the increase of algorithmic complexities. If to construct such chains but with an additional "zigzag" property, then the minimal asymptotic algorithmic complexity of monotone recognition will be achieved. In [72] chains with no collinear paths were constructed for one particular sub-case.

The cube-split in (2), on a theoretical level provides the splitting of Ξ_{m+1}^n recursively into smaller domains of the same type objects–hypergrids (*grid type partition*), and then into binary cubes and chains of smaller sizes. On a practical level it is important that the cube-split makes the transfer of effective tools of chain computations [3, 76] into the domain of grid computations possible.

Before introducing the special decomposition of grids we first bring some necessary concepts, and distinguish several classes of vertices in Ξ_{m+1}^n [7].

Middle layer vertices. $m_{mid+} = (\lceil m/2\rceil, \dots, \lceil m/2\rceil)$ and $m_{mid-} = (\lfloor m/2\rfloor, \dots, \lfloor m/2\rfloor)$ are *middle vertices* of Ξ_{m+1}^n.

Upper vertices, lower vertices. A vertex $(a_1, \dots, a_n)$ of Ξ_{m+1}^n is called *upper vertex* if $(a_1, \dots, a_n) \geq m_{mid+}$. A vertex $(a_1, \dots, a_n)$ of Ξ_{m+1}^n is called *lower vertex* if $(a_1, \dots, a_n) \leq m_{mid-}$. $\hat{H}$ and $\check{H}$ denote the sets of all upper and lower vertices, respectively.

$$|\hat{H}| = |\check{H}| = ((m+1)/2)^n \ for\ odd\ m.$$

$$|\hat{H}| = |\check{H}| = (m/2+1)^n \ for\ even\ m.$$

V-equivalence. Vertices $(a_1, \ldots, a_n)$ and $(b_1, \ldots, b_n)$ of Ξ^n_{m+1} are *vertically equivalent* if $a_i \in \{b_i, m - b_i\}$ for $1 \leq i \leq n$. $V(a)$ denotes the V-equivalence class of a vertex a. In $V(a)$ we distinguish two vertices $\hat{a}$ and $\check{a}$, with components defined as follows:

$$\hat{a}_i = \begin{cases} a_i, & \text{if } a_i \geq m_{mid+} \\ m - a_i, & \text{if } a_i < m_{mid-} \end{cases}$$

$$\check{a}_i = \begin{cases} m - a_i, & \text{if } a_i \geq m_{mid+} \\ a_i, & \text{if } a_i \leq m_{mid-} \end{cases}$$

These are the only vertices of $V(a)$ that belong to $\hat{H}$ and $\check{H}$, respectively. Thus all vertices of $V(a)$ can be constructed from the upper and/or lower elements (similarly, from an arbitrary vertex) of $V(a)$ inverting (with respect to m) all groups of components. It is evident that the V-equivalence classes of different vertices of $\hat{H}$ ($\check{H}$) are disjoint.

This provides a complete partition of Ξ^n_{m+1} into $|\hat{H}|$ equivalence classes, which are uniquely defined by elements of $\hat{H}$. For a given $a \in \Xi^n_{m+1}$ consider an integer defined as $k = |\{a_i | a_i \neq (m - a_i)\}|$ ($k = n$ for odd m). Then $|V(a)| = 2^k$. We identify each vertex of $V(a)$ with α, a binary sequence of length n, such that $\alpha_i = 1$ if and only if $a_i = \hat{a}_i$. In this manner, $V(a)$ becomes isomorphic to the k-dimensional binary cube B^k: the 0-th level of B^k contains the lower vertex of $V(a)$ belonging to $\check{H}$; the i-th level consists of all vertices of $V(a)$ obtained from the lower vertex by applying i number of component inversions. Thus Ξ^n_{m+1} is partitioned into $|\hat{H}|$ disjoint classes—that are identical by structure to binary cubes. This partitioning of Ξ^n_{m+1} we will refer as *cube type* partition. Figure 8 illustrates the equivalence classes of points (3, 4, 3), (2, 3, 4) and (4, 2, 2) in Ξ^3_5.

It is worth mentioning that in the usual chain-split (as is the partition of the binary cube in [42]), chains are simple, which is not the case in the cube-split.

In case of odd m we obtain $((m+1)/2)^n$ V-equivalence classes, and every class/binary-cube has dimension n. The following formula relates sizes and partitions for odd m

$$((m+1)/2)^n \cdot 2^n = (m+1)^n.$$

The case of even m is not so homogeneous. For vertices of $\hat{H}$ let k be the number of components, not equal to $m/2$, $0 \leq k \leq n$. Thus $n - k$ components are fixed to $m/2$ and k components are actual, creating $(m/2)^k$-vertex sets in $\hat{H}$. Each of these vertices has its V-equivalence class of size 2^k. The following formula relates sizes and partitions for even m

$$\sum_{k=0}^{n} (\binom{n}{k} \cdot 2^k \cdot (m/2)^k) = \sum_{k=0}^{n} (\binom{n}{k} \cdot m^k) = (m+1)^n.$$

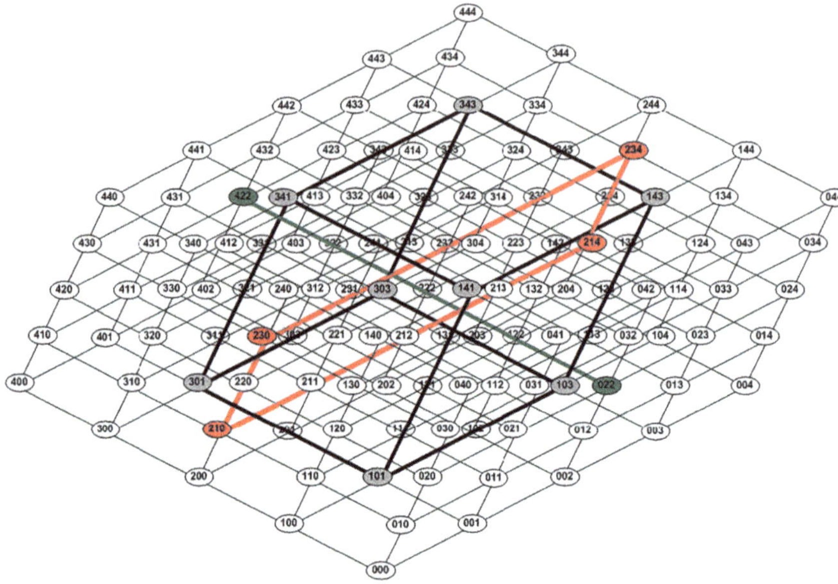

Fig. 8 Highlighted vertices are elements of $V(3,4,3)$, $V(2,3,4)$ and $V(4,2,2)$, *respectively*. $V(3,4,3)$ composes a 3-dimensional, $V(2,3,4)$—a 2-dimensional, and $V(4,2,2)$—a 1-dimensional binary cube

Retaining monotonicity

Consider V, some vertical equivalence class in Ξ^n_{m+1}, and let $E(V)$ denote the corresponding sub-cube in B^n. Splitting of $E(V)$ into Hansel chains will induce the splitting of V: each chain $\alpha_1 < \alpha_2 < \cdots < \alpha_t$ in $E(V)$ will induce a sequence $a_1 < a_2 < \cdots < a_t$ of corresponding vertices in Ξ^n_{m+1}, where two neighbor vertices in the sequence can belong to non neighbor levels. Henceforth these sequences will be referred to as *associated-chains*, a_i as *origin* vertices, and α_i–as *induced* vertices. The above description is summarized by the following Theorem.

Theorem 20 ([7]) *Let Ξ^n_{m+1} be partitioned into vertical equivalence classes, and V is one of them. If $\alpha_1 < \alpha_2 < \cdots < \alpha_t$ is a chain in $E(V)$, then $a_1, a_2, \ldots, a_t$, is the associated chain in Ξ^n_{m+1}, where $a_1, a_2, \ldots, a_t$ are origins of $\alpha_1, \alpha_2, \ldots, \alpha_t$. Conversely, if $a_1 < a_2 < \cdots < a_t$ belong to Ξ^n_{m+1}, and induced vertices $\alpha_1, \alpha_2, \ldots, \alpha_t$ belong to the same binary cube, then $\alpha_1, \alpha_2, \ldots, \alpha_t$ is a chain there.*

A corollary is obtained in the form of:

Theorem 21 ([7]) *Let Ξ^n_{m+1} be partitioned into vertical equivalence classes, and V is one of them. If $F : \Xi^n_{m+1} \to \{0, 1\}$ is a monotone function then $f : E(V) \to \{0, 1\}$ is monotone, where f is defined as follows: for every $\alpha \in E(V)$, $f(\alpha) = 1$ if and only if $F(a) = 1$, where a is the origin of α in Ξ^n_{m+1}.*

5.3 Identification with Cube-Partition

In this section, we present two cube-type partition algorithms [7] for identification of monotone functions defined on Ξ^n_{m+1}. Let $f : \Xi^n_{m+1} \to \{0, 1\}$ be a monotone function defined with the help of an oracle Ω_f.

Algorithm ACS.

- Partition Ξ^n_{m+1} into binary cubes $B_1, \ldots, B_{|\hat{H}|}$ as it is described in the previous section.
- In each B_i, apply Hansel's algorithm for identification of monotone functions $f_i : E_i \to \{0, 1\}$, defined as follows: for every $\alpha \in B_i$, $f_i(\alpha) = 1$ if and only if $f(a) = 1$, where a is the origin of α in Ξ^n_{m+1}.
- Integrate the results of $|\hat{H}|$ binary recognitions into f.

It is worth mentioning, that in each particular step we consider all induced cubes and treat them independently, although comparison of vertices could improve the result of recognition. The complexity of recognition (by an oracle) for one separate cube is estimated as the sizes of its two middle layers. This is an overestimate in global terms, but we pay this cost for parallelization.

Theorem 22 ([7]) *Let $f : \Xi^n_{m+1} \to \{0, 1\}$ be a monotone function defined with the help of an oracle Ω_f. Let Δ_{ACS} denote the Shannon complexity of ACS. Then,*

$$\Delta_{ACS} = \sum_{k=0}^{n} \left(\binom{n}{k} \cdot (m/2)^k \cdot \left(\binom{k}{\lfloor k/2 \rfloor} + \binom{k}{\lfloor k/2 \rfloor + 1} \right) \right) \tag{1}$$

for even m and

$$\Delta_{ACS} = ((m+1)/2)^n \cdot \left(\binom{n}{\lfloor n/2 \rfloor} + \binom{n}{\lfloor n/2 \rfloor + 1} \right) \tag{2}$$

for odd m.

The proof is obtained by Theorem 21 and results of [1].

Below we estimate the real numeric value of these complexities. Consider the case of odd m. First we estimate the term $\binom{n}{\lfloor n/2 \rfloor}$ asymptotically. For this we apply the well known formula from [79]

$$\binom{n}{k} \sim \frac{2^{n+1}}{\sqrt{2\pi n}} \cdot \exp\left(-\frac{(2k-n)^2}{2n}\right), \text{ when } n, k \to \infty, \text{ and } k - \frac{n}{2} = o(n^{3/4}).$$

Inserting this into (2) we obtain that

$$\Delta_{ACS} \to ((m+1)/2)^n \cdot 2 \cdot \frac{2^{n+1}}{\sqrt{2\pi n}} = 4\frac{(m+1)^n}{\sqrt{2\pi n}} \text{ for odd } m \text{ when } n \to \infty.$$

For even m consider the summand of (1) with $k = m$. It is easy to see that it is asymptotic to $\frac{4m^n}{\sqrt{2\pi n}}$ when $n \to \infty$. The sub-sum for small k, say up to $n/2$ is small. The maximal summand is located close to n. To check this, we form the fraction of terms $\binom{n}{k} \cdot (m/2)^k \cdot \frac{2^{k+1}}{\sqrt{2\pi k}}$ for $k+1$ and k. Equaling this to 1 we get $k = n - \frac{(k+1\sqrt{1+1/k}}{m})$ which proves the claim. What is important is the order of the complexity which is $\frac{4m^n}{\sqrt{2\pi n}}$.

Simple structures, and easily calculable quantities used in *Algorithm ACS* can serve as important elements in practical identification algorithms. On the other hand, parallel computation of this recognition is possible, as we deal with disjoint binary cubes.

Algorithm ACS′.

- Partition Ξ^n_{m+1} into binary cubes $B_1, \ldots, B_{|\hat{H}|}$ as it is described in the previous section and let G be the associated set of sub-grids of this partition.
- In each $g \in G$, apply Alekseev's algorithm for identification of monotone function $f_g : g \to \{0, 1\}$, defined as follows: for every $\alpha \in g$, $f_g(\alpha) = 1$ if and only if $f(a) = 1$, where a is the origin of α in Ξ^n_{m+1}.
- Integrate the results of $|\hat{H}|$ binary recognitions.

Theorem 23 ([7]) *Let $f : \Xi^n_{m+1} \to \{0, 1\}$ be a monotone function defined with the help of an oracle. Consider m to be odd. f can be identified by Algorithm ACS′ using $(|M_{\hat{H}}| + \lfloor log_2 m \rfloor \cdot |N_{\hat{H}}|) \cdot (\binom{n}{\lfloor n/2 \rfloor} + \binom{n}{\lfloor n/2 \rfloor + 1})$ accesses to the oracle, where $M_{\hat{H}}$ and $N_{\hat{H}}$ denote the sets of vertices of middle levels of $\hat{H}$ (levels which contain vertices of rank $\lfloor mn/4 \rfloor$, $\lfloor mn/4 \rfloor + 1$ respectively). The formula for even m can be derived in a similar way.*

The claim follows from Theorem 21 and [1].

5.4 Estimates of the Middle Layer Related Quantities

Complexity estimates of algorithms of monotone binary function recognition are based on the sizes of the middle layers. In this section, we present simplified and usable formulas to these quantities. Here are some known results that we integrate.

On Ordered and Restricted Partitions of Integers

Ratsaby [66] considers $w_{m,n}(N)$,–the notion of the number of ordered integer partitions of the given integer N into fixed n $(1 \leq n \leq N)$ parts such that each summand is at least 0 but not larger than m. The following very useful formula is derived:

$$w_{n,m}(N) = \sum_{i=0}^{n} (-1)^i \binom{n}{i} \binom{N+n-1-i(m+1)}{n-1}. \tag{3}$$

It is easy to check that the following three concepts are identical: formula of $w_{n,m}(N)$ given in (3), ordered integer partitions of N into n parts with summands between 0 and m, and the size of N-th layer of Ξ_{m+1}^{n}. In this way, the given equivalency provides us with exact formula to the numbers of our interest–size of N-th layer of Ξ_{m+1}^{n}.

One technical note. The formula in (3) implicitly controls the indices and the part of positive summands. When $N+n-1-i(m+1)$ is negative or 0, or when it is less than $n-1$ then the corresponding summands become 0. Anyhow, it is correct to add these restrictions in an explicit way.

Ratsaby [66], after the proof of (1) and (2) derives another, asymptotic estimate for $w_{n,m}(N)$ by the use of saddle point evaluation of integrals. Here is the final postulation:

Let $N \geq 1,\ m \geq 1$, and $n = \frac{2N}{m}\left(1 + c_1 N^{-1/2}\right)$ for some absolute constant $c_1 > 0$. Let $\sigma = \frac{m}{6}\left(1 + \frac{m}{2}\right)$ and $N \to \infty$, then,

$$w_{n,m}(N) = \frac{1}{\sqrt{2\pi n\sigma}} (m+1)^n \exp\left(\frac{-(N - \frac{mn}{2})^2}{2n\sigma}\right) \cdot \left(1 - 2\phi\left(-n^{-\frac{1}{10}}\sqrt{\sigma}\right) + O\left(exp(-n^{1/5})\right)\right) \tag{4}$$

where $\phi(x)$ is the Gaussian cumulative distribution. Formula (4) is useful of course, and surely we may use it to know the asymptotic value of sizes of N-th layers of Ξ_{m+1}^{n}. But there are conditions, and to understand the range of layers where it works, let us perform the following.

Rewrite the pre-condition $n = \frac{2N}{m}\left(1 + c_1 N^{-1/2}\right)$ of the formula (4) in an equivalent form: $\frac{mn}{2} = N + c_1\sqrt{N}$, mentioning that $c_1 > 0$. This shows that the formula does not work around the value $N = \frac{mn}{2}$, which is the middle layer of Ξ_{m+1}^{n}. For $N \to \infty$ consider any 2 functions, with $\varphi(N) \to 0$ and $\psi(N) \to \infty$. Areas $N \geq \frac{mn}{2} - \varphi(N)\sqrt{\frac{mn}{2}}$ and $N \leq \frac{mn}{2} - \psi(N)\sqrt{\frac{mn}{2}}$ are out of the desired domain $\frac{mn}{2} = N + c_1\sqrt{N}$.

So the applied value of formula (4) is limited and from [66] we use mainly the entire formula (3) but not the asymptotic part.

The basic complexity component of Alekseev's algorithm, the algorithm CS as well as the entire sizes of sets $M_{\widehat{H}}$ and $N_{\widehat{H}}$ are out of the mentioned domain so that to overcome this obstacle and to estimate the real size of $M_{\widehat{H}}$ and $N_{\widehat{H}}$ we apply to one more well-known result obtained by V. Korobkov.

POSET Probabilistic Model Over the Lattice and Its Extensions

In [50] the basic research objective is a generalized model of monotone Boolean function and its recognition. A sub-case of this investigation, as we will see, completes estimates of [66], and help to understand clearly the complexity of monotone recognition on Ξ^n_{m+1} and thus,–the cost that we pay for parallelization in algorithm CS. To justify this, we need to enter into some elements of the proof of (1) from [39].

Consider a partially ordered set (poset) $T = \{r_1, r_2, \ldots, r_s\}$. Map the elements of T onto several layers over the plane so that comparable elements of T are placed only in nearby layers, see [13]. This can be explained by a step by step process over the set $\{r_1, r_2, \ldots, r_s\}$. At first, on the 0 layer we place all those vertices that have no preceding elements. The first layer is composed of vertices that are following ***only*** the vertices of the layer 0. And iteratively, the $k+1$ layer will include those elements that follow ***only*** the elements from layers $0, 1, \ldots, k$. Let $r, r \le s$ be the number of the used layers. Let $N(T, 1, k) = s_k$ denote the number of elements of k-th layer, $0 \le k \le r$. Generalizing, consider the n-th Cartesian degree T^n of T and denote by $N(T, n, k)$ the number of elements of layers k, $k = 0, 1, \ldots, nr$. Layer k of T^n is formed as the sum of layers of individual layers of n composing direction of T^n.

In accord to [50] consider n random variables ξ_i on T^n, distributed (corresponding to coordinates) independently and identically so that they attain integer values j, $0 \le j \le r$ by probabilities $\frac{s_j}{s}$, respectively. We transform this event system into a simpler and equivalent model. Consider probability distribution over the elements of coordinates, i (partially ordered set T), evaluating probabilities of elements to $1/s$ in accord to homogeneous probability distribution. The composite event about appearance of the given elements on each coordinate, that is the vector $t_1, t_2, \ldots, t_n$ will have the probability s^{-n}. ξ_i, in this interpretation is the number of the random element layer with probability as share of elements belonging to that layer. We see that ξ_i corresponds to a composite event. Another composite event is "the sum of numbers of coordinate layers is equal to k". The quantity of such events is just the number of partitions of k into n summands restricted by $0 \div s$ in coordinates, that equals to $N(T, n, k)$ and to $w_{s,n}(k)$ simultaneously. Due to independent and homogeneous probabilities s^{-n} of composing events it can be easily concluded that $P\left\{\sum_{i=1}^{n} \xi_i = k\right\} \cdot s^n = N(T, n, k)$.

Asymptotic of Lattice Layer Sizes

At this point we apply the known V. Glagolev's methodology [38] in obtaining the asymptotic of lattice layer sizes. Probabilistic model of Sect. 5.4 will be considered and in this concern the B. Gnedenko's theorem [39] will be applied:

$$P\left\{\sum_{i=1}^{n} \xi_i = k\right\} \to \frac{1}{\sqrt{2\pi n}\,\sqrt{D\xi}} \cdot \exp^{-\frac{(k-nM\xi)^2}{2nD\xi}}. \tag{5}$$

That held uniformly by k and with $n \to \infty$. Mξ and Dξ denote the expected value and dispersion of the random number ξ [79]. (1) is a consequence of this formula.

Consider, in terms of Sect. 5.4 the case of Ξ^n_{m+1} in detail.

In Ξ^n_{m+1} T is just the simple linearly ordered set $\Xi_{m+1} = \{0, 1, \dots, m\}$. It is easy to compute the main value of random variable ξ in the probabilistic distribution over the Ξ^n_{m+1}:

$M\xi = \sum_{k=0}^{m} \frac{k}{m+1} = \frac{m}{2}$.

Here each layer consists of one element and all probabilities are equal to $\frac{1}{m+1}$. In a similar way it can be computed the dispersion of ξ:

$D\xi = M\xi^2 - (M\xi)^2 = \sum_{k=0}^{m} \frac{k^2}{m+1} - \frac{m}{4}^2 = \frac{m(2m+1)}{6} - \frac{m}{4}^2 = \frac{m(m+2)}{12}$.

Substituting Mξ and Dξ into (5), we obtain exactly the first three multipliers of expression (4):

$$\frac{1}{\sqrt{2\pi n\sigma}}(m+1)^n \exp\left(\frac{-(k-\frac{\mathrm{mn}}{2})^2}{2n\sigma}\right).$$

Remind that $\sigma = \frac{m}{6}\left(1 + \frac{m}{2}\right)$.

In this way we eliminate the constraint $c_1 > 0$ from expression (4), and obtain a simple form asymptotic estimate of layer sizes of lattices instead of inequality type estimates at [1]. Obtained formulas work for middle layers, explaining complexity of V. Alekseev's algorithm of multi-dimensional multi-valued monotone recognition, and the complexity of chain split type algorithm CS on monotone recognition of binary functions of Ξ^n_{m+1}. In middle layers the expression $(k - \frac{\mathrm{mn}}{2})^2$ is zero or it is small, moreover in $\exp\left(\frac{-(k-\frac{\mathrm{mn}}{2})^2}{2n\sigma}\right)$ this part of formula is divided on growing n and m and the obtained formula for middle layers appears in the foolowing simple form:

$$\frac{1}{\sqrt{2\pi n\sigma}}(m+1)^n.$$

In the simplest binary case $m = 1$ this converts to the formula for $C_n^{\lfloor \frac{n}{2} \rfloor}$ that we brought above. Additional conditions appear when we require that $\frac{-(k-\frac{\mathrm{mn}}{2})^2}{2n\sigma} \to 0$.

6 Concluding Remarks

Monotone Boolean functions are not only a means of computing functional dependencies of databases, frequent sets of items in data mining, or some other application, but also represent fundamental mathematical apparatus that models the entire systems of data science. The fundamental limitations of these models and the structure of their data and relations are reducible to the apparatus of Boolean functions that have the attractive property of monotonicity. However, decoding and analysing monotone Boolean functions is a multifaceted problem, and many unsolved or inefficiently solved problems remain in this framework. Combinatorial constructions related to this topic have been considered in some detail, but they are complex and often reduced to enumeration (brute-force). One fundamentally new approach is to involve a new resource in the process, namely, artificial intelligence with its machine learning (ML) tools [69, 77, 81, 82]. In this formulation, the emphasis is placed on solving a large number of problems of the classes under consideration, accumulating the results of their solutions in the form of a database, training on them by ML, and not solving but recognizing the solution of the problem under consideration by analyzing the parameters of the problem and the database information. The problem in this formulation is already becoming popular, and our first goal is collect and survey all pros and cons related to monotone Boolean function to understand the right models of their ML applications [69].

Another possible approach continues the first one and seeks ways of refining, and reconstructing the problem constraints, with considering subtypes of monotone Boolean functions, the decoding of which requires refined approaches, and the associated algorithms, combinatorial or ML-based, that can be practically implemented. There is a list of practical problems in big data analytics that reduce to the diverse classes of monotone Boolean functions. Our effort in this direction activated the study of new classes–shadow minimized Boolean functions, k-reason monotone functions and k-distance functions. This survey summarises the basic knowledge related to monotone Boolean functions in a concise form, and presents complementary, new and less known knowledge in a more detailed form. We hope that this will help both in organizing a systematic study of individual classes of monotonic data structures and in preparing the necessary conditions for the transition of research to an artificial intelligence platform.

References

1. Alekseev, V.B.: Deciphering of some classes of monotonic many-valued functions. USSR Comput. Math. Math. Phys. **16**(1), 180–189 (1976)
2. Aslanyan, L., Aslanyan, H., Khosravi, H.: Optimal node scheduling for integrated connected-coverage in wireless sensor networks. In: Ninth International Conference on Computer Science and Information Technologies Revised Selected Papers, pp. 1–13. IEEE (2013)
3. Aslanyan, L., Castellanos, J.: Logic based pattern recognition-ontology content (1). Inf. Theor. Appl. **14**(3), 206–210 (2007)
4. Aslanyan, L., Gishyan, K., Sahakyan, H.: Target class classification recursion preliminaries. Balt. J. Mod. Comput. **11**(3), 398–410 (2023)
5. Aslanyan, L., Katona, G., Sahakyan, H.: Shadow minimization Boolean function reconstruction. Informatica **35**(1), 1–20 (2024)
6. Aslanyan, L., Sahakyan, H.: Chain split and computations in practical rule mining. In: International Book Series, Information Science and Computing, Classification, forecasting, Data Mining, Book, vol. 8, pp. 132–135 (2009)
7. Aslanyan, L., Sahakyan, H.: The splitting technique in monotone recognition. Discret. Appl. Math. **216**, 502–512 (2017). January

8. Aslanyan, L., Sahakyan, H., Romanov, V., Da Costa, G., Kacimi, R.: Large network target coverage protocols. In: 2019 Computer Science and Information Technologies (CSIT), pp. 57–64. IEEE (2019)
9. Aslanyan, L.H.: The discrete isoperimetry problem and related extremal problems for discrete spaces. Problemy Kibernetiki **36**, 85–128 (1979)
10. Aslanyan, L.H., Danoyan, H.E.: Complexity of hash-coding type search algorithms with perfect codes. J. Next Gener. Inf. Technol. **5**(4), 26–35 (2014)
11. Balogh, J., Katona, G.O.H., Linz, W., Tuza, Z.: The domination number of the graph defined by two levels of the n-cube, II. Eur. J. Comb. **91**, 103201 (2021)
12. Bezrukov, S.L., Kuzmanovski, N., Lim, J.: Pull-push method: a new approach to edge-isoperimetric problems. Discret. Math. **346**(12), 113632 (2023)
13. Birkhoff, G.: Lattice Theory, vol. 25. American Mathematical Society (1940)
14. Black, H., Chakrabarty, D., Seshadhri, C.: Directed isoperimetric theorems for boolean functions on the hypergrid and an $o(n\sqrt(d))$ monotonicity tester. In: Proceedings of the 55th Annual ACM Symposium on Theory of Computing, pp. 233–241 (2023)
15. Blum, A.: Learning a function of r relevant variables. In: Learning Theory and Kernel Machines: 16th Annual Conference on Learning Theory and 7th Kernel Workshop, COLT, Kernel 2003, Washington, DC, USA, August 24–27, 2003, Proceedings, pp. 731–733. Springer (2003)
16. Boros, E., Hammer, P.L.: Pseudo-Boolean optimization. Discret. Appl. Math. **123**(1–3), 155–225 (2002)
17. Boros, E., Hammer, P.L., Ibaraki, T., Kawakami, K.: Identifying 2-monotonic positive Boolean functions in polynomial time. In: ISA'91 Algorithms: 2nd International Symposium on Algorithms Taipei, Republic of China, December 16–18, 1991. Proceedings 2, pp. 104–115. Springer (1991)
18. Boros, E., Hammer, P.L., Ibaraki, T., Kawakami, K.: Polynomial-time recognition of 2-monotonic positive Boolean functions given by an oracle. SIAM J. Comput. **26**(1), 93–109 (1997)
19. Braverman, M., Khot, S., Kindler, G., Minzer, D.: Improved monotonicity testers via hypercube embeddings (2022). arXiv:2211.09229
20. Carlet, C., Joyner, D., Stănică, P., Tang, D.: Cryptographic properties of monotone Boolean functions. J. Math. Cryptol. **10**(1), 1–14 (2016)
21. Chakrabarty, D., Seshadhri, C.: A o (n) monotonicity tester for boolean functions over the hypercube. In: Proceedings of the Forty-Fifth Annual ACM Symposium on Theory of Computing, pp. 411–418 (2013)
22. Chao, T.W., Yu, H.H.H.: Kruskal–Katona-type problems via entropy method (2023). arXiv:2307.15379
23. Clements, G.F.: A minimization problem concerning subsets of a finite set. Discret. Math. **4**(2), 123–128 (1973)
24. Crawford-Kahrl, P., Cummins, B., Gedeon, T.: Joint realizability of monotone Boolean functions. Theor. Comput. Sci. **922**, 447–474 (2022)
25. Damásdi, G., Gerbner, D., Katona, G.O.H., Keszegh, B., Lenger, D., Methuku, A., Nagy, D.T, Pálvölgyi, D., Patkós, B., Vizer, M., et al.: Adaptive majority problems for restricted query graphs and for weighted sets. Discret. Appl. Math. **288**, 235–245 (2021)
26. Daykin, D.E., Godfrey, J., Hilton, A.J.W.: Existence theorems for Sperner families. J. Comb. Theory Ser. A **17**(2), 245–251 (1974)
27. de Bruijn, N.G., van Tengbergen, C.E., Kruyswijk, D.: On the set of divisors of a number. Nieuw Arch. Wiskunde **23**(2), 191–193 (1951)
28. Dedekind, R.: Uber die begründung der idealtheorie. Gött. Nachr 106–11 (1895)
29. Dedekind, R.: Ueber Zerlegungen von Zahlen durch ihre grössten gemeinsamen Theiler. F. Vieweg & Sohn (1897)
30. Dodis, Y., Goldreich, O., Lehman, E., Raskhodnikova, S., Ron, D., Samorodnitsky, A.: Improved testing algorithms for monotonicity. In: International Workshop on Randomization and Approximation Techniques in Computer Science, pp. 97–108. Springer (1999)

31. Elgot C.C.: Truth functions realizable by single threshold organs. In: 2nd Annual Symposium on Switching Circuit Theory and Logical Design (SWCT 1961), pp. 225–245. IEEE (1961)
32. Engel, K.: Sperner Theory. Number 65 in Encyclopedia of Mathematics and its Application. Cambridge University Press (1997)
33. Fischer, E., Lehman, E., Newman, I., Raskhodnikova, S., Rubinfeld, R., Samorodnitsky, A.: Monotonicity testing over general poset domains. In: Proceedings of the Thirty-Fourth Annual ACM Symposium on Theory of Computing, pp. 474–483 (2002)
34. Frankl, P., Katona, G.O.H.: On strengthenings of the intersecting shadow theorem. J. Comb. Theory Ser. A **184**, 105510 (2021)
35. Freixas, J.: On the enumeration of some inequivalent monotone Boolean functions. Optimization 1–14 (2022)
36. Damir Nasibullovich Gainanov: On one criterion of the optimality of an algorithm for evaluating monotonic Boolean functions. Zhurnal Vychislitel'noi Matematiki i Matematicheskoi Fiziki **24**(8), 1250–1257 (1984)
37. Gerbner, D., Imolay, A., Katona, G.O.H., Nagy, D.T., Nagy, K., Patkós, B., Stadler, D., Zólomy, K.: Identification of a monotone Boolean function with *k* "reasons" as a combinatorial search problem (2024). arXiv:2411.19833
38. Glagolev, V.: Some estimates of disjunctive normal forms in the algebra of logic. Probl. Cybern. **19**, 75–94 (1967)
39. Gnedenko, B.V.: On the local limit theorem of probability theory. Russian Math. Surveys **3**(3), 187–194 (1948)
40. Goldreich, O., Goldwassert, S., Lehman, E., Ron, D.: Testing monotonicity. In: Proceedings 39th Annual Symposium on Foundations of Computer Science (Cat. No. 98CB36280), pp. 426–435. IEEE (1998)
41. Griggs, J.R., Kalinowski, T., Leck, U., Roberts, I.T., Schmitz, M.: The saturation spectrum for antichains of subsets. Order 1–38 (2023)
42. Hansel, G.: Sur le nombre des fonctions booléennes monotones de n variables. CR Acad. Sci. Paris **262**(20), 1088–1090 (1966)
43. Jackson, J.C., Lee, H.K., Servedio, R.A., Wan, A.: Learning random monotone DNF. Discret. Appl. Math. **159**(5), 259–271 (2011)
44. Jung, A.: Shadow ratio of hypergraphs with bounded degree. Graphs Comb. **39**(3), 40 (2023)
45. Kabulov, A., Berdimurodov, M.: Parametric algorithm for searching the minimum lower unity of monotone boolean functions in the process synthesis of control automates. In: 2021 International Conference on Information Science and Communications Technologies (ICISCT), pp. 1–4. IEEE (2021)
46. Natal'ya Nikolaevna Katerinochkina: Sets containing a maximal number of pairwise incomparable n-dimensional k-ary sets. Math. Notes Acad. Sci. USSR **24**(3), 696–700 (1978)
47. Katona, G.: A theorem of finite sets. In: Theory of Graphs: Proceedings of the Colloquium Held at Tihany, Hungary, September 1966, (1968)
48. Katona, G.: A theorem of finite sets. Classic Papers in Combinatorics, pp. 381–401 (1987)
49. Kleitman, D.J.: On an extremal property of antichains in partial orders. The LYM property and some of its implications and applications. In: Combinatorics: Proceedings of the NATO Advanced Study Institute held at Nijenrode Castle, Breukelen, The Netherlands 8–20 July 1974, pp. 277–290. Springer (1975)
50. Korobkov, V.K.: On monotone functions of the algebra of logic. Problemy kibernetiki **13**, 5–28 (1965)
51. Korshunov, A.D.: Monotone boolean functions. Russ. Math. Surv. **58**(5), 929 (2003)
52. Korshunov, A.D.: On the number of monotone Boolean menge. Problemy kibernetiki **38**, 5–109 (1981)
53. Kovalerchuk, B., Delizy, F.: Visual data mining using monotone Boolean functions. In: Visual and Spatial Analysis: Advances in Data Mining, Reasoning, and Problem Solving, pp. 387–406. Springer (2005)
54. Kruskal, J.B.: The number of simplices in a complex. Math. Optim. Tech. **10**, 251–278 (1963)

55. Kulhandjian, M., Aslanyan, L., Sahakyan, H., Kulhandjian, H., D'Amours, C.: Multidisciplinary discussion on 5g from the viewpoint of algebraic combinatorics. In: 2019 Computer Science and Information Technologies (CSIT), pp. 69–76. IEEE (2019)
56. Lange, J., Vasilyan, A.: Agnostic proper learning of monotone functions: beyond the black-box correction barrier. In: 2023 IEEE 64th Annual Symposium on Foundations of Computer Science (FOCS), pp. 1149–1170. IEEE (2023)
57. Lovász, L.: Combinatorial Problems and Exercises, vol. 361. American Mathematical Society (2007)
58. Madden, G.N.: Shadows of colored complexes and cycle decompositions of equipartite hypergraphs. Ph.D. thesis, Illinois State University (2023)
59. Makino, K., Ibaraki, T.: A fast and simple algorithm for identifying 2-monotonic positive Boolean functions. J. Algorithms **26**(2), 291–305 (1998)
60. de Bruijn, N.G.: Asymptotic Methods in Analysis. Courier Corporation (2014)
61. Nosov, M.V.: Estimates of the degrees of separating polynomials for monotone and self-dual functions. Intell. Syst. Theory Appl. **27**(2), 79–82 (2023)
62. Alfréd Rényi Institute of Mathematics, editor. Summit280 Book of Abstracts. János Bolyai Mathematical Society (2023)
63. O'Donnell, R., Servedio, R.: Learning Monotone Functions from Random Examples in Polynomial Time. CMU School of Computer Science (2005)
64. Pawelski, B.: On the number of inequivalent monotone Boolean functions of 9 variables. IEEE Trans. Inf. Theory (2024)
65. Post, E.L.: The Two-Valued Iterative Systems of Mathematical Logic. (AM-5). Princeton University Press (1941)
66. Ratsaby, J.: Estimate of the number of restricted integer-partitions. Appl. Anal. Discret. Math. 222–233 (2008)
67. Sahakyan, H.: On nonconvexity of the set of hypergraphic sequences. In: 2023 Computer Science and Information Technologies (CSIT), pp. 47–52. NAS RA (2023)
68. Sahakyan, H., Aslanyan, L.: Discrete tomography with distinct rows: relaxation. In: 2017 Computer Science and Information Technologies (CSIT), pp. 117–120. IEEE (2017)
69. Sahakyan, H., Katona, G., Aslanyan, L.: Study on using reinforcement learning for the Monotone Boolean Reconstruction. Informatica **48**(4), (2025)
70. Sales, M., Schülke, B.: A local version of Katona's intersection theorem (2022). arXiv:2206.04278
71. Schröder, B.: Ordered Sets: An Introduction with Connections from Combinatorics to Topology. Birkhäuser (2016)
72. Serjantov, A.: Optimal algorithm for deciphering of some classes of monotone functions. J. Comput. Math. Math. Phys. **23**(1), 206–212 (1983)
73. Nikolay Aleksandrovich Sokolov: Optimal reconstruction of monotone Boolean functions. USSR Comput. Math. Math. Phys. **27**(6), 181–187 (1987)
74. Sperner, E.: Ein satz über untermengen einer endlichen menge. Mathematische Zeitschrift **27**(1), 544–548 (1928)
75. Tennakoon, R., Suter, D., Zhang, E., Chin, T.J., Bab-Hadiashar, A.: Consensus maximisation using influences of monotone boolean functions. In: Proceedings of the IEEE/CVF Conference on Computer Vision and Pattern Recognition, pp. 2866–2875 (2021)
76. Tonoyan, G.P.: Chain partitioning of n-cube vertices and deciphering of monotone Boolean functions. J. Comput. Math. Math. Phys. **19**(6), (1979)
77. Valiant, L.G.: A theory of the learnable. Commun. ACM **27**(11):1134–1142 (1984)
78. Van Hirtum, L., De Causmaecker, P., Goemaere, J., Kenter, T., Riebler, H., Lass, M., Plessl, C.: A computation of D (9) using FPGA supercomputing (2023). arXiv:2304.03039
79. Feller, W.: Introduction to the Probability Theory and its Applications, 3rd edn. John Wiley & Sons, Inc. (1968)
80. Zhang, E., Suter, D., Tennakoon, R., Chin, T.J., Bab-Hadiashar, A., Truong, G., Gilani, S.Z.: Maximum consensus by weighted influences of monotone Boolean functions. In: Proceedings of the IEEE/CVF Conference on Computer Vision and Pattern Recognition, pp. 8964–8972 (2022)

81. Zhuravlev, Y.I., Ryazanov, V.V., Ryazanov, V.V., Aslanyan, L.H., Sahakyan, H.A.: Comparison of different dichotomous classification algorithms. Pattern Recognit. Image Anal. **30**, 303–314 (2020)
82. Zhuravlev, Y.I., Ryazanov, V.V., Aslanyan, L.H., Sahakyan, H.A.: On a classification method for a large number of classes. Pattern Recognit Image Anal. **29**, 366–376 (2019)
83. Zolotykh, N.Y., Shevchenko, V.N.: Estimating the complexity of deciphering a threshold functions in a k-valued logic. Comput. Math. Math. Phys. **39**(2), 328–334 (1999)

A Survey on Ordered Ramsey Numbers

Martin Balko

Abstract The ordered Ramsey number of a graph $G^<$ with a linearly ordered vertex set is the smallest positive integer N such that any two-coloring of the edges of the ordered complete graph on N vertices contains a monochromatic copy of $G^<$ in the given ordering. The study of the quantitative behavior of ordered Ramsey numbers is a relatively new theme in Ramsey theory full of interesting and difficult problems. In this survey paper, we summarize recent developments in the theory of ordered Ramsey numbers. We point out connections to other areas of combinatorics and some well-known conjectures. We also list several new and challenging open problems and highlight the often strikingly different behavior from the unordered case.

1 Introduction

In a broad sense, *Ramsey theory* refers to any result whose underlying philosophy can be captured by the statement "Every structure of a given kind contains a large well-organized substructure". This part of discrete mathematics has developed enormously in the last few decades, emerging into a field with important statements in many areas, including combinatorics, geometry, logic, and number theory.

Our focus will be on graph theory where one of the cornerstones of Ramsey theory, from which Ramsey theory derives its name, is *Ramsey's theorem* [71]. This classical result states that for any two graphs G and H, there is a positive integer N such that every red-blue coloring of the complete graph K_N on N vertices contains a red copy of G or a blue copy of H as a subgraph of K_N. The smallest such N is called the *Ramsey number* of G and H and is denoted by $R(G, H)$. When G and H are isomorphic, we write $R(G)$ instead of $R(G, G)$.

Estimating the growth rate of Ramsey numbers is a notoriously difficult problem in general with a remarkable influence in combinatorics. Despite several attempts

M. Balko (✉)
Department of Applied Mathematics, Faculty of Mathematics and Physics, Charles University, Prague, Czech Republic
e-mail: balko@kam.mff.cuni.cz

G. O. H. Katona et al. (eds.), *Sum(m)it280*, Bolyai Society Mathematical Studies 32,
https://doi.org/10.1007/978-3-032-18810-6_4

over the last 70 years, see for example [20, 73, 75], Ramsey numbers are not fully understood even for complete graphs. For a long time, the best-known bounds on $R(K_n)$ essentially were

$$2^{n/2} \leq R(K_n) \leq 2^{2n}. \tag{1}$$

The lower bound is due to Erdős [33] while the upper bound was proved by Erdős and Szekeres [37] in their seminal paper from 1935 where they independently rediscovered Ramsey's theorem. Very recently, Campos et al. [15] made a breakthrough and improved the exponential upper bound to $R(K_n) \leq (4 - \varepsilon)^n$ for some small constant $\varepsilon > 0$.

For sparser graphs, Ramsey numbers grow at a significantly slower rate. For example, a classical result by Chvátal et al. [18] says that Ramsey numbers of bounded-degree graphs are only linear in the number of vertices. More precisely, for every positive integer Δ, there is a positive integer $C = C(\Delta)$ such that every n-vertex graph G with maximum degree Δ satisfies $R(G) \leq C \cdot n$.

In this survey, we focus on a more recent variant of Ramsey numbers defined for *ordered graphs*, that is, graphs with linearly ordered vertex sets. Studying these so-called *ordered Ramsey numbers* became quite an active area in graph Ramsey theory [27] which offers numerous applications in several parts of combinatorics such as discrete geometry and extremal theory of matrices. We try to summarize the recent developments in the theory of ordered Ramsey numbers, pointing out connections to other areas of combinatorics and some well-known conjectures. We also list several new and challenging open problems and highlight the often strikingly different behavior from the unordered case.

In Sect. 2, we survey the recent developments about ordered Ramsey numbers of ordered graphs. We start by mentioning general bounds on ordered Ramsey numbers. Then, we will focus on specific classes of ordered graphs, where we can prove much more precise estimates and, in some cases, even the exact formulas for the ordered Ramsey numbers. We also discuss extensions of ordered Ramsey numbers for more than two colors. In Sect. 3, we consider *hypergraph ordered Ramsey numbers*. Besides general bounds, we focus on so-called monotone paths, a case with close connections to the classical Erdős–Szekeres theorem. Finally, in Sect. 4 we discuss *edge-ordered Ramsey numbers* where we work with the ordering of the edges, not vertices.

We note that this paper is not intended to serve as an exhaustive survey of the subject. For example, we treat problems of an asymptotic nature rather than being concerned with the computation of exact ordered Ramsey numbers. We apologize in advance for any particularly glaring omissions.

For the sake of clarity of presentation, we will maintain some conventions throughout the paper. For a positive integer n, we use $[n]$ to denote the set $\{1, \ldots, n\}$. We omit floor and ceiling signs whenever they are not crucial and we use log to denote base 2 logarithm.

2 Graph Ordered Ramsey Numbers

An *ordered graph* $G^<$ is a pair $(G, <)$ consisting of a graph G and a linear ordering $<$ of the vertex set of G. We call $G^<$ an *ordering* of G. Two ordered graphs $G^<$ and $H^<$ are *isomorphic* if their underlying graphs G and H are isomorphic via a one-to-one correspondence $V(G) \to V(H)$ that preserves the vertex orderings. Note that, for every positive integer n, there is only one ordered complete graph on n vertices up to isomorphism. We use $K_n^<$ to denote such an ordered complete graph.

An ordered graph $G^<$ is an *ordered subgraph* of an ordered graph $H^<$ if G is a subgraph of H and the vertex ordering of G is a suborder of the vertex ordering of H. If an ordered graph $H^<$ contains an ordered subgraph isomorphic to an ordered graph $G^<$, then we say that $H^<$ contains a *copy* of $G^<$.

We can now define an analog of Ramsey numbers for ordered graphs. The *ordered Ramsey number* $R_<(G^<, H^<)$ of two ordered graphs $G^<$ and $H^<$ is the minimum positive integer N such that every red-blue coloring of the edges of $K_N^<$ contains a red copy of $G^<$ or a blue copy of $H^<$. If $G^<$ and $H^<$ are isomorphic, then we simply write $R_<(G^<)$ instead of $R_<(G^<, H^<)$ and we call it the *diagonal* case. We refer to $R_<(G^<, H^<)$ with non-isomorphic $G^<$ and $H^<$ as the *off-diagonal* case.

Note that we have $R(G) \leq R_<(G^<)$ for every graph G and each ordering $G^<$ of G. Moreover, $R(K_n) = R_<(K_n^<)$ for every positive integer n. Since every ordered graph $G_<$ on n vertices is an ordered subgraph of $K_n^<$, we thus obtain the bounds

$$R(G) \leq R_<(G^<) \leq R(K_n).$$

It follows that the number $R_<(G^<)$ is finite for every ordered graph $G^<$. Similarly, $R_<(G^<, H^<)$ is also finite for any ordered graphs $G^<$ and $H^<$ and thus ordered Ramsey numbers are well-defined.

From some point of view, the study of ordered Ramsey numbers is as old as Ramsey theory itself. For example, one of the oldest Ramsey-type results, the famous *Erdős–Szekeres lemma* on monotone subsequences proved by Erdős and Szekeres [37] in 1935, can be derived from estimates on ordered Ramsey numbers. This classical result states that for every positive integer n, every sequence of at least $(n-1)^2 + 1$ distinct real numbers contains an increasing or a decreasing subsequence of length n. Moreover, this bound is sharp. To see that it follows from results in ordered Ramsey theory, consider the following ordering $MP_n^<$, called the *monotone path*, of the path P_n on n vertices. If $v_1 < \cdots < v_n$ are the vertices of $MP_n^<$, then the edges of $MP_n^<$ are the pairs $\{v_i, v_{i+1}\}$ for every $i = 1, \ldots, n-1$; see part (a) of Fig. 1. Given a sequence $S = (s_1, \ldots, s_N)$ of distinct real numbers, we construct an ordered graph $K_n^<$ with vertex set S and the ordering of the vertices given by their positions in S. Then, we color an edge $\{s_i, s_j\}$ with $i < j$ red if $s_i < s_j$ and blue otherwise. Afterward, red monotone paths on n vertices correspond to increasing subsequences of S of length n and blue monotone paths on n vertices to decreasing subsequences of S of length n. The Erdős–Szekeres lemma now follows from the fact

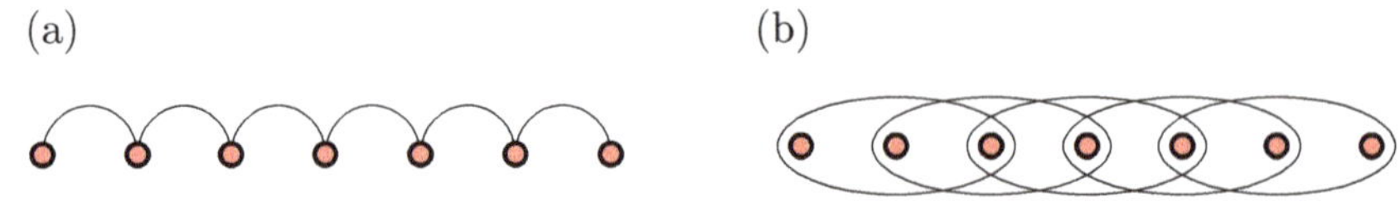

Fig. 1 **a** The monotone path $MP_n^<$ for $n = 7$. **b** The monotone path $MP_n^{<(3)}$ for $n = 7$. In all figures of ordered graphs, the vertices are ordered from left to right

$$R_<(MP_n^<) = (n-1)^2 + 1 \tag{2}$$

proved, for example, by Choudum and Ponnusamy [16] or by Milans et al. [59].

Similarly, the Erdős–Szekeres theorem on convex polygons can be derived from estimates on ordered Ramsey numbers of 3-uniform monotone hyperpaths; see Sect. 3.1. The ordered Ramsey numbers are also closely connected to the extremal theory of $\{0, 1\}$-matrices [6, 53] and variants of Ramsey numbers from discrete geometry, for example, to so-called *geometric Ramsey numbers* [54, 55]. Given these results and connections, there is a strong motivation to study ordered Ramsey numbers and their variants.

While there had been a lot of results about ordered Ramsey numbers of monotone hyperpaths [16, 31, 45, 59, 62], there was surprisingly little work on ordered Ramsey numbers of more general ordered graphs and hypergraphs. The first systematic study of ordered Ramsey numbers was conducted by Balko et al. [6] and independently by Conlon et al. [22]. Since then there has been much progress in understanding the ordered Ramsey numbers, which we describe in this section.

2.1 Bounded Degrees and Interval Chromatic Number

Since $R_<(G^<) \le R(K_n)$ for every ordered graph $G^<$ on n vertices, we see from (1) that $R_<(G^<)$ grows at most exponentially in n. For dense ordered graphs, a standard probabilistic argument shows that this is asymptotically tight. However, the ordered Ramsey numbers may differ substantially from the usual Ramsey numbers for sparser graphs. This result was proved independently by Balko et al. [6] and by Conlon et al. [22] who showed that there are ordered matchings with superpolynomial ordered Ramsey numbers. A *matching* is a graph consisting of pairwise disjoint edges.

Theorem 1 ([6, 22]) *There is a constant $C > 0$ such that for every even $n \ge 2$, there is an ordered matching $M_n^<$ on n vertices satisfying*

$$R_<(M_n^<) \ge n^{C \log n / \log \log n}.$$

Note that this result is in sharp contrast with the classical result by Chvátal et al. [18], according to which Ramsey numbers of bounded-degree graphs are only linear. For ordered matching, this is a rather typical behavior, as Conlon et al. [22]

proved that this superpolynomial lower bound holds for almost all ordered matchings. Very recently, Fox and He [43] applied Theorem 1 to obtain superpolynomial lower bounds on Ramsey numbers of sparse digraphs.

Given the lower bound from Theorem 1, it is clear that to obtain polynomial upper bounds on ordered Ramsey numbers, bounding the maximum degree is not enough and we need to use some other graph parameter. Balko et al. [6] and independently Conlon et al. [22] showed that if we additionally bound the following analog of the chromatic number, then we obtain polynomial upper bounds on ordered Ramsey numbers. We will state the bound obtained in [22], as it is stronger than the one in [6], but first, we need to introduce some definitions.

A subset I of vertices of an ordered graph $G^<$ is an *interval* in $G^<$ if for every pair u, v of vertices of I with $u < v$, every vertex w of $G^<$ satisfying $u < w < v$ is contained in I. The *interval chromatic number* of $G^<$ is the minimum number of intervals the vertex set of $G^<$ can be partitioned into so that there is no edge between vertices of the same interval. We note that there is a variant of the *Erdős–Stone–Simonovits theorem* for ordered graphs proved by Pach and Tardos [70], which is expressed in terms of the interval chromatic number. For a positive integer d, a graph G is *d-degenerate* if every subgraph of G contains a vertex of degree at most d. For positive integers $t, n_1, \ldots, n_t$, let $K^<_{n_1,\ldots,n_t}$ be an ordering of the complete t-partite graph $K_{n_1,\ldots,n_t}$ in which the vertices of the color class of size n_i form the ith interval. If $n_1 = \cdots = n_t = n$, we simply write $K^<_t(n)$ instead of $K^<_{n_1,\ldots,n_t}$.

Theorem 2 ([22]) *Let $G^<$ be an ordered d-degenerate graph on n vertices with maximum degree Δ. For positive integers n' and χ, let $s = \lceil \log \chi \rceil$ and $D = 8\chi^2 n'$. Then,*

$$R_<(G^<, K^<_\chi(n')) \leq 2^{s^2 d+s} \Delta^s n^s D^{ds+1}.$$

In particular, if $G^<$ is an ordered d-degenerate graph with n vertices and with interval chromatic number χ, then Theorem 2 implies

$$R_<(G^<) \leq n^{32d \log \chi}. \tag{3}$$

Note that for fixed d this bound almost matches the lower bound from Theorem 1. Although the bounds are quite close, there is still a small gap in the exponent, even for $d = 1$. Conlon et al. [22] thus posed the following problem.

Problem 1 ([22]) Close the gap between the upper and lower bounds for ordered Ramsey numbers of matchings.

It follows from (3) that bounded-degree ordered graphs with bounded interval chromatic numbers have polynomial ordered Ramsey numbers. However, we have no non-trivial lower bounds for this case and thus Balko et al. [6] stated the following problem.

Problem 2 ([6]) Is there a constant $c > 0$ such that for every fixed Δ there is a sequence $\{G^<_n\}_{n\in\mathbb{N}}$ of ordered Δ-regular graphs $G^<_n$ with n vertices and interval chromatic number 2 such that $R_<(G^<_n) \geq n^{c\Delta}$?

Using a result of Erdős and Szemerédi [39], Conlon et al. [22] derived the following result from the proof of Theorem 2, which shows that the ordered Ramsey numbers behave more like the usual Ramsey numbers for dense ordered graphs.

Theorem 3 ([22]) *There is a constant $C > 0$ such that every ordered d-degenerate graph $G^<$ on n vertices satisfies*

$$R_<(G^<) \leq 2^{Cd \log^2 (2n/d)}.$$

This result is close to sharp for very small d by Theorem 1 and for very large d by (1).

A simple argument shows that if an n-vertex ordered matching $M^<$ has the interval chromatic number 2, then the bound $R_<(M^<) \leq n^{\lceil \log n \rceil}$ proved by Conlon et al. [22] for any n-vertex ordered matching $M^<$ can be significantly improved to $O(n^2)$ [22]. This is quite close to the truth, as Conlon et al. [22] constructed an ordered matching $M^<$ on n vertices with interval chromatic number 2 such that $R_<(M^<) \geq \frac{cn^2}{\log^2 (n) \log \log n}$ for some constant $c > 0$. This was improved by Balko et al. [8], who proved a stronger bound, which additionally holds for almost all ordered matchings with interval chromatic number 2. To state this result, we need to introduce some definitions first, which will also be useful later.

For a positive integer n, the *random n-permutation* is a permutation of the set $[n]$ chosen independently uniformly at random from the set of all $n!$ permutations of the set $[n]$. For a positive integer n and the random n-permutation π, the *random ordered matching $M^<(\pi)$* on n vertices is the ordered matching with the vertex set $[2n]$ and with edges $\{i, n + \pi(i)\}$ for every $i \in [n]$. Note that the interval chromatic number of every random ordered matching is 2. The random ordered matching satisfies an event A *asymptotically almost surely* if the probability that A holds tends to 1 as n goes to infinity.

Theorem 4 ([8]) *There is a constant $C > 0$ such that the random ordered matching $M^<(\pi)$ on n vertices asymptotically almost surely satisfies*

$$R_<(M^<(\pi)) \geq C \cdot \left(\frac{n}{\log n}\right)^2.$$

Similarly, as in the case of general ordered matchings, there is a gap between the lower and upper bound. Thus, Conlon et al. [22] posed the following problem for ordered matchings with interval chromatic number 2.

Problem 3 ([22]) Close the gap between the upper and lower bounds for ordered Ramsey numbers of ordered matchings with interval chromatic number 2.

Geneson et al. [46] posed a similar problem for ordered paths with interval chromatic number 2. Here, it follows from Theorem 2 that the ordered Ramsey number $R_<(P^<)$ of any ordered path $P^<$ on n vertices with interval chromatic number 2 is at most $O(n^3)$ while the best known lower bound $\Omega((n/\log n)^2)$ follows from Theorem 4.

Problem 4 ([46]) Is it true that $R_<(P^<) \leq O(n^2)$ for every ordering $P^<$ of the path on n vertices with interval chromatic number 2?

2.2 Bounded Bandwidth

Besides the interval chromatic number, another natural parameter for ordered graphs is their bandwidth. For an ordered graph $G^<$, the *bandwidth* of $G^<$ is the length of the longest edge in $G^<$. That is, the maximum of $|i - j|$ taken over all edges $\{u, v\}$ of $G^<$, where i is the position of u and j is the position of v in $<$. We call the number $|i - j|$ the *length* of the edge $\{u, v\}$.

Conlon et al. [22] proved that, for every positive integer k, every ordered matching $M^<$ on n vertices with bandwidth at most k satisfies $R_<(M^<) \leq n^{\lceil \log k \rceil + 2}$. They also asked whether this result can be extended by proving a polynomial upper bound on ordered Ramsey numbers of all ordered graphs with bounded bandwidth. This problem was solved by Balko et al. [6] who proved that, for every positive integer k, there is a constant $C = C(k)$ such that every n-vertex ordered graph $G^<$ with bandwidth k satisfies $R_<(G^<) \leq C \cdot n^{128k}$.

Observe that every n-vertex ordered graph $G^<$ with bandwidth at most k is an ordered subgraph of the n-vertex ordered graph $P^<_{k,n}$ that contains all edges of length at most k. In particular, $R_<(G^<) \leq R_<(P^<_{k,n})$. Note that $P^<_{1,n} = MP^<_n$ and thus $R_<(P^<_{1,n}) = (n-1)^2 + 1$ by (2). Mubayi [63] improved the upper bound on $R_<(P^<_{k,n})$ in the case $k = 2$ by showing $R_<(P^<_{2,n}) \leq O(n^{19.487})$ for every $n \geq 2$. He also used this result to determine the correct tower growth rate of the k-uniform hypergraph Ramsey number of a $(k+1)$-clique versus a monotone k-uniform path; see Sect. 3 for the definitions. After that, Gishboliner et al. [49] proved that $R_<(P^<_{k,n}) \leq cn^{4k-2}$ for some constant $c = c(k) > 0$ and asked whether in fact $R_<(P^<_{k,n}) \leq c'n^C$ for a constant $c' = c'(k) > 0$ and an absolute constant $C > 0$. Very recently, Girão et al. [48] gave a positive answer to this problem and derived the currently strongest bound.

Theorem 5 ([48]) *For every $k \in \mathbb{N}$ and $\varepsilon > 0$, there is $C = C(k, \varepsilon) > 0$ such that for every positive integer n we have*

$$R_<(P^<_{k,n}) \leq C \cdot n^{4+\varepsilon}.$$

Motivated by Mubayi's work [63] on hypergraph Ramsey numbers, Mubayi and Suk [66] considered the off-diagonal setting and proved that $R_<(P^<_{k,n}, K^<_n) \leq (2n)^{k(k+1)\log n}$ for all positive integers k and n. Gishboliner et al. [49] proved that $R_<(P^<_{k,n}, K^<_n) \leq cn^{k(2k-1)}$ for some constant $c = c(t) > 0$ and conjectured that the exponent can even be chosen to be linear in k. This conjecture was verified by Girão et al. [48] who proved that there exists $C > 0$ such that, for all positive integers $n > k$,

$$R_<(P^<_{k,n}, K^<_n) \leq n^{Ck}.$$

In fact, Girão et al. [48] derived an essentially tight bound

$$R_<(P^<_{k,s}, K^<_n) \leq R(K_s, K_k)^C \cdot n$$

which is valid for all s, t and n, improving earlier estimate by Mubayi and Suk [66].

Despite this remarkable progress, there are still some open problems left. For example, there is still a gap between the lower and upper bounds on $R_<(P^<_{k,n})$. While the best upper bound is of order $n^{4+o(1)}$, we only have a trivial quadratic lower bound. Girão et al. [48] believed that the truth is closer to the lower bound and posed the following problem.

Problem 5 ([48]) Is $R_<(P^<_{k,n}) = n^{2+o(1)}$ as $n \to \infty$?

2.3 *Off-Diagonal Ordered Ramsey Numbers*

For the off-diagonal ordered Ramsey numbers, Conlon et al. [22] investigated the ordered Ramsey numbers $R_<(M^<, K^<_3)$, where $M^<$ is an ordered matching on n vertices. It follows from the well-known bound $R(K_n, K_3) \leq O(n^2/\log n)$ by Ajtai et al. [1] that

$$R_<(M^<, K^<_3) \leq O\left(\frac{n^2}{\log n}\right).$$

For $R(K_n, K_3)$, this bound is tight as shown by Kim [56], but Conlon et al. [22] expect that this upper bound is far from optimal for $R_<(M^<, K^<_3)$. They constructed an ordered matching $M^<$ on n vertices satisfying

$$R_<(M^<, K^<_3) \geq O\left(\left(\frac{n}{\log n}\right)^{4/3}\right)$$

and posed the following problem.

Problem 6 ([22]) Does there exist an $\varepsilon > 0$ such that every ordered matching $M^<$ on n vertices satisfies $R_<(M^<, K^<_3) \leq O(n^{2-\varepsilon})$?

In case there is an affirmative answer to this problem, one has to employ the sparsity of ordered matching to distinguish it from K_n. However, ordered matchings on n vertices typically contain an edge between any two intervals of length $3\sqrt{n}\log n$ [22] and thus can be used to "simulate" complete graphs. This makes Problem 6 particularly interesting. However, it is still open and seems to be difficult.

Nevertheless, there has been some partial progress. Rohatgi [72] proved that asymptotically almost surely the random ordered matching $M^<(\pi)$ on n vertices with interval chromatic number 2 satisfies $R_<(M^<(\pi), K^<_3) \leq O(n^{24/13})$. He also showed that if $M^<$ is a non-crossing ordered matching, then, for every $\varepsilon > 0$, we have

$R_<(M^<, K_3^<) \leq cn^{1+\varepsilon}$ for some constant $c = c(\varepsilon) > 0$. Here, an ordered graph $G^<$ is *non-crossing* if it does not contain two edges $\{u, v\}$ and $\{x, y\}$ with $u < x < v < y$. Balko and Poljak [10] later showed that $R_<(M^<(\pi), K_3^<) \geq \Omega((n/\log n)^{5/4})$ and $R_<(M^<(\pi), K_3^<) \leq O(n^{7/4})$ asymptotically almost surely, improving the result by Rohatgi [72].

Since Problem 6 seems difficult, some weaker variants of this problem were introduced in the literature. For example, Rohatgi [72] posed the following interesting conjecture.

Conjecture 1 ([72]) *For a positive integer χ, there is a constant $\varepsilon(\chi) > 0$ such that*

$$R_<(M^<, K_3^<) \leq O(n^{2-\varepsilon(\chi)})$$

for almost every ordered matching $M^<$ on n vertices with interval chromatic number χ.

Even in the case of $M^<(\pi)$ the bounds are roughly by the factor $\sqrt{n}$ apart, so Balko and Poljak [10] asked the following question.

Problem 7 ([10]) What is the growth rate of $R_<(M^<(\pi), K_3^<)$ for the random ordered matchings $M^<(\pi)$ on n vertices with interval chromatic number 2?

2.4 *Minimum Ordered Ramsey Numbers*

Conlon et al. [22] characterized graphs G for which the ordered Ramsey number $R_<(G^<)$ is linear in the number of vertices of G for every ordering $G^<$ of G. These are precisely graphs G whose edges can be covered by a constant number of vertices. A similar problem is to determine graphs that admit an ordering for which the corresponding ordered Ramsey number is linear. This motivated the following definition.

For an unordered graph G, the *minimum ordered Ramsey number of* G is defined as

$$\min R_<(G) = \min\{R_<(G^<) : G^< \text{ is an ordering of } G\}.$$

Since Ramsey numbers of bounded-degree graphs are linear in the number of vertices, it is natural to ask whether the minimum ordered Ramsey numbers of bounded-degree graphs are always at most linear. Conlon et al. [22] considered this unlikely and asked whether random 3-regular graphs have superlinear ordered Ramsey numbers for all orderings.

Balko et al. [8] gave an affirmative answer to this problem. For a positive integer d, we let $G(d, n)$ denote the random d-regular graph on n vertices drawn uniformly and independently from the set of all d-regular graphs on the vertex set $[n]$.

Theorem 6 ([8]) *For every fixed integer $d \geq 3$, asymptotically almost surely*

$$\min R_<(G(d,n)) \geq \frac{n^{3/2-1/d}}{4 \log n \log \log n}.$$

Theorem 6 shows that random d-regular graphs have superlinear minimum ordered numbers for any fixed integer number $d \geq 3$. In particular, almost every 3-regular graph G on n vertices satisfies

$$\min R_<(G) \geq n^{7/6}/(4 \log n \log \log n).$$

It can be easily seen that the minimum ordered Ramsey numbers of ordered matchings are linear. Balko et al. [8] showed that this is the case for 2-regular graphs as well by proving that there is a constant C such that for every graph G on n vertices with maximum degree 2, we have $\min R_<(G) \leq Cn$. In fact, Balko et al. [8] proved a stronger Turán-type statement for bipartite graphs of maximum degree 2.

For the upper bounds in the case of a larger maximum degree, a simple corollary of Theorem 2 states that every graph G on n vertices with constant maximum degree Δ admits an ordering $G^<$ with $R_<(G^<)$ polynomial in n. More precisely, every graph G with n vertices and with maximum degree Δ satisfies

$$\min R_<(G) \leq O(n^{(\Delta+1)\lceil \log(\Delta+1) \rceil + 1}). \tag{4}$$

Thus, the currently best-known bounds on minimum ordered Ramsey numbers of 3-regular graphs are of order $\Omega(n^{7/6}/(\log n \log \log n))$ and $O(n^{4\lceil \log(4) \rceil + 1}) = O(n^9)$ by Theorem 6 and by (4), respectively. Since the gap is rather large, Balko et al. [8] posed the following problem.

Problem 8 ([8]) Close the gap between the upper and lower bounds for minimum ordered Ramsey numbers of 3-regular graphs.

2.5 Specific Classes of Ordered Graphs

Obtaining precise formulas for ordered Ramsey numbers is difficult, essentially the only non-trivial class of graphs with fully characterized ordered Ramsey numbers are stars [6, 16]. However, there are some specific and natural orderings for which it is possible to derive such formulas or at least very close bounds that often lead to interesting applications. In this section, we focus on three classes of ordered graphs: nested matchings, alternating paths, and monotone cycles.

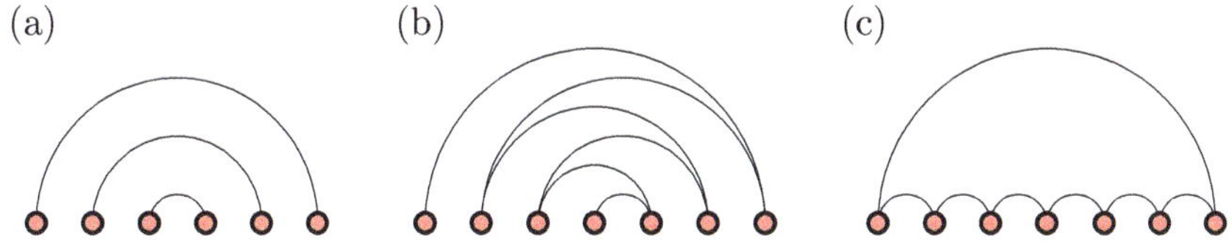

Fig. 2 **a** The nested matching NM_k for $k = 3$. **b** The alternating path $AP_n^<$ for $n = 7$. **c** The monotone cycle $MC_n^<$ for $n = 7$

Nested Matchings

A basic building block in the proof of the result by Rohatgi [72] about the ordered Ramsey numbers $R_<(M^<, K_3^<)$ for a non-crossing ordered matching $M^<$ is based on so-called nested matchings. A *nested ordered matching* $NM_k^<$ on vertices $v_1 < \cdots < v_{2k}$ has edges $\{v_i, v_{n-i+1}\}$ for every $i = 1, \ldots, k$; see part (a) of Fig. 2.

Rohatgi [72] proved $4k - 1 \le R_<(NM_k^<, K_3^<) \le 6k$ for every positive integer k. He believed that the lower bound was tight and conjectured that $R_<(NM_k^<, K_3^<) = 4k - 1$. This conjecture is true for $k \le 3$, but Balko and Poljak [9] disproved it for any $k \ge 4$ by showing the following bounds.

Theorem 7 ([9]) *For every positive integer k, we have*

$$R_<(NM_k^<, K_3^<) \le (3 + \sqrt{5})k < 5.3k.$$

If $k \ge 6$, *we have*

$$R_<(NM_k^<, K_3^<) \ge 4k + 1.$$

Using the lower bounds from Theorem 7, Balko and Poljak improved the best-known bounds on the maximum chromatic number of so-called *k-queue graphs*, which addresses a problem posed by Dujmovič and Wood [30]. Despite this progress, no precise formula for $R_<(NM_k, K_3^<)$ is known.

Alternating Paths

The *alternating path* $AP_n^<$ is the ordering of P_n on vertices $v_1 < \cdots < v_n$ with edges $\{v_i, v_j\}$ for $i + j \in \{n + 1, n + 2\}$; see part (b) of Fig. 2. Note that the alternating path $AP_n^<$ is an ordered subgraph of $K^<_{\lceil n/2 \rceil, \lfloor n/2 \rfloor}$, and so it has interval chromatic number 2.

Using a result from the extremal theory of $\{0, 1\}$-matrices, Balko et al. [6] proved the following linear bounds on $R_<(AP_n^<)$.

Proposition 1 ([6]) *For every integer* $n > 2$, *we have*

$$5\lfloor n/2 \rfloor - 4 \le R_<(AP_n^<) \le 2n - 3 + \sqrt{2n^2 - 8n + 11}.$$

The proof of the lower bound was later extended by Neidinger and West [68] to obtain various linear lower bounds for various classes of ordered graphs with interval chromatic number 2.

The precise multiplicative factor in $R_<(P_n^<)$ is unknown. Computer experiments by Balko et al. [6] indicate that ordered Ramsey numbers of alternating paths might be minimal among all path orderings, making it a particularly interesting class of ordered paths.

Problem 9 ([6]) For some positive integer n, is there an ordering $P_n^<$ of the path P_n on n vertices such that $R_<(P_n^<) < R_<(AP_n^<)$?

From the cases with $n \leq 8$, it seemed that $R_<(P_n^<)$ could be equal to $\lfloor (n-2)\frac{1+\sqrt{5}}{2} \rfloor + n$, but this was later disproved by Balko and Poljak [10]. We also note that the alternating path $AP_n^<$ is contained in every ordered graph on $N \geq n/\varepsilon$ vertices with at least εN^2 edges [8].

Monotone Cycles

We do not know any precise formulas for ordered Ramsey numbers of ordered cycles except one particularly natural ordering of the cycle C_n, called the *monotone cycle* $MC_n^<$, which is obtained from the monotone path $MP_n^<$ by adding an edge between the first and the last vertex in the vertex ordering; see part (c) of Fig. 2.

The precise formula for ordered cycles covers even the off-diagonal case and is given by the following result by Balko et al. [6].

Theorem 8 ([6]) *For all integers $r \geq 2$ and $s \geq 2$, we have*

$$R_<(MC_r^<, MC_s^<) = 2rs - 3r - 3s + 6.$$

Theorem 8 has an interesting application in discrete geometry. The *geometric Ramsey numbers* [19, 54, 55] are natural analogs of ordered Ramsey numbers. For a finite set of points $P \subset \mathbb{R}^2$ in general position, let K_P be the *complete geometric graph on P*, which is a complete graph drawn in the plane so that its vertices are represented by the points in P and the edges are drawn as straight-line segments between the pairs of points in P. The graph K_P is *convex* if P is in convex position. The *geometric Ramsey number* of a graph G, denoted by $Rg(G)$, is the smallest N such that every complete geometric graph K_P on N vertices with edges colored by two colors contains a non-crossing monochromatic drawing of G. If we consider only convex complete geometric graphs K_P in the definition, then we get so-called *convex geometric Ramsey number* $Rc(G)$. Note that these numbers are finite only if G is outerplanar and that $Rc(G) \leq Rg(G)$ for every outerplanar graph G.

Balko et al. [6] observed that the geometric and convex geometric Ramsey numbers of cycles are equal to the ordered Ramsey numbers of monotone cycles, which together with Theorem 8 gives the following formula for $Rc(C_n)$ and $Rg(C_n)$.

Corollary 1 ([6]) *For every integer $n \geq 3$, we have $Rc(C_n) = Rg(C_n) = 2n^2 - 6n + 6$.*

So in this case, vertex orderings are enough to determine these numbers, no deeper geometric structure is needed, which sometimes seems to be the case; see Sect. 3.1 for other examples.

2.6 Multicolor Ordered Ramsey Numbers

The definition of Ramsey numbers extends to a multicolor case. For any integer $q \geq 2$ and a graph G, we can define the *q-color Ramsey number* $R(G; q)$ as the smallest positive integer N such that every q-coloring of K_N contains a monochromatic copy of G as a subgraph of K_N. A straightforward modification of the Erdős and Szekeres neighborhood-chasing argument [37] yields $R(K_n; q) \leq q^{qn}$. On the other hand, a product argument by Lefmann [58] shows that $R(K_n; q) \geq 2^{qn/4}$. Although there has recently been remarkable progress in obtaining better lower and upper bounds on $R(K_n; q)$ [4, 21, 74, 79], the q-color Ramsey numbers are still far from being completely understood.

The multicolor setting can be introduced for ordered Ramsey numbers as well. For any integer $q \geq 2$ and ordered graphs $G_1^<, \ldots, G_q^<$, the *q-color ordered Ramsey number* $R_<(G_1^<, \ldots, G_q^<; q)$ is the smallest positive integer N such that for every q-coloring of $K_N^<$ by colors from $[q]$ there is $i \in [q]$ and a copy of G_i in color i forming a ordered subgraph of $K_N^<$. In the case $G^< = G_1^< = \cdots = G_q^<$, we simply write $R_<(G^<; q)$ instead of $R_<(G_1^<, \ldots, G_q^<; q)$.

Similarly as before, it can be shown that the numbers $R_<(G_1^<, \ldots, G_q^<; q)$ are always bounded and that $R_<(G^<; q) \leq R(K_n; q)$ for any n-vertex ordered graph $G^<$. However, we only have a poor understanding of ordered Ramsey numbers for more than two colors, even for ordered matchings. Conlon et al. [22] proved the upper bound $R_<(M^<; q) \leq n^{(2 \log n)^{(q-1)}}$ for any ordered matching $M^<$ on n vertices. They believe that a much stronger upper bound should hold and pose the following problem.

Problem 10 ([22]) For any integer $q \geq 3$, show that there exists a constant c_q such that $R_<(M^<; q) \leq n^{c_q \log n}$ for any ordered matching $M^<$ on n vertices.

For ordered matchings on n vertices with interval chromatic number two and $q \geq 2$ colors, Balko and Grinerová [7] showed that $R_<(M^<; q) = n^{\Theta(q)}$; see also [50]. It would be interesting to determine the leading constant in the exponent.

Problem 11 ([7, 50]) For any integer $q \geq 3$, determine the constant c such that $R_<(M^<; q) = n^{cq+o(1)}$ for any ordered matching $M^<$ on n vertices with interval chromatic number two.

In the off-diagonal setting, Balko and Grinerová [7, 50] also introduced a multicolor variant of Problem 6 about numbers $R_<(K_3^<, \ldots, K_3^<, M^<; q)$, where $M^<$ is an ordered matching on n vertices and $q \geq 3$ is an integer. Using a result by Alon and Rödl [2], which gave a solution to a longstanding conjecture of Erdős and Sós [32], the known bounds are between $\Omega((n/\log n)^{4/3})$ and $O(n^q \text{ poly} \log n)$. These bounds are far apart and it is not clear where the truth should lie.

Problem 12 ([7, 50]) Improve the upper and lower bounds on off-diagonal multicolor ordered Ramsey numbers $R_<(K_3^<, \ldots, K_3^<, M^<; q)$.

In the similar spirit of the conjecture by Erdős and Sós [32], Balko and Grinerová [7, 50] asked the following question.

Problem 13 ([7, 50]) Are there ordered matchings $M^<$ on n vertices such that

$$\lim_{n\to\infty} \frac{R_<(K_3^<, K_3^<, M^<)}{R_<(K_3^<, M^<)} = \infty?$$

For ordered graphs with bounded bandwidth, Girão et al. [48] proved that $R_<(P_{k,n}^<; q) \leq Dn^{Cq \log q}$ for some constants $D = D(k, q) > 0$ and $C > 0$. They also mention some interesting consequences for the problem of finding monochromatic directed paths in multicolored tournaments and conjecture that their upper bound can be improved.

Conjecture 2 ([48]) *For all positive integers k, n, q, there are constants $C > 0$ and $D = D(k, q) > 0$ such that $R_<(P_{k,n}^<; q) \leq Dn^{Cq}$.*

The multicolor setting was also considered by Neidinger and West [68], who provided various lower bounds on ordered Ramsey numbers of so-called *2-ichromatic ordered graphs*, and by Balko et al. [6], who showed the following dichotomy: $R_<(G^<; q)$ is either polynomial in q or exponential in q for every ordered graph $G^<$. Barát et al. [14] also considered multicolor ordered Ramsey numbers of various classes of ordered mathchings and their connections to graph drawings. However, overall, the q-color ordered Ramsey numbers remain quite unexplored.

3 Hypergraph Ordered Ramsey Numbers

In full generality, Ramsey's theorem applies not only for graphs but also for *k-uniform hypergraphs*, that is, pairs (V, E) where V is a finite set of *vertices* and E is a set of k-tuples from V called *edges*. We use $K_n^{(k)}$ to dentote the complete k-uniform hypergraph $(V, \binom{V}{k})$. Ramsey's theorem then states that, for all integers $k, q \geq 2$ and every k-uniform hypergraph H, there is a positive integer N such that in every q-coloring of $K_N^{(k)}$ there a monochromatic copy of H forming a subhypergraph of $K_N^{(k)}$. We denote the smallest such integer N by $R(H; q)$ and call it the *Ramsey number of* H. If $q = 2$, we write $R(H)$ instead of $R(H; 2)$.

For $k \geq 3$, Ramsey numbers $R(K_n^{(k)})$ are even less understood than their graph counterparts. For example, it is only known that

$$2^{\Omega(n^2)} \leq R(K_n^{(3)}) \leq 2^{2^{O(n)}}, \tag{5}$$

as shown by Erdős et al. [35]. A famous conjecture of Erdős, for whose proof Erdős offered \$500 reward, states that there is a constant $c > 0$ such that $R(K_n^{(3)}) \geq 2^{2^{cn}}$. The case $k = 3$ is of particular importance as if one determines the growth rate of $R(K_n^{(3)})$ precisely, then the so-called *Stepping-up lemma* by Erdős and Rado [36] would determine the growth rate of $R(K_n^{(k)})$ for every $k \geq 4$.

The *tower function* $t_h(x)$ of *height* $h - 1$ is defined by the recursive formula $t_1(x) = x$ and $t_h(x) = 2^{t_{h-1}(x)}$ for every $h \geq 2$. Using (5) together with the Stepping-up lemma, one can derive lower bounds on $R(K_n^{(k)})$ for $k \geq 4$. However, there is a difference of one exponential between known upper and lower bounds. More precisely, we have

$$t_{k-1}(\Omega(n^2)) \leq R(K_n^{(k)}) \leq t_k(O(n)) \tag{6}$$

for every fixed $k \geq 3$.

Perhaps surprisingly, the Ramsey number $R(H)$ of every k-uniform hypergraph H with bounded k and with bounded maximum degree is at most linear in the number of vertices of H [18, 23, 28, 29, 52, 67].

The *ordered k-uniform hypergraph* $H^<$ is a pair $(H, <)$ consisting of a k-uniform hypergraph H and a linear ordering $<$ of its vertex set. The notion of an *ordered subhypergraph* is analogous to its graph counterpart. Again, there is a unique ordered complete k-uniform hypergraph on n vertices up to isomorphism and we denote it by $K_n^{<(k)}$.

The *ordered Ramsey number* $R_<(H^<, G^<)$ of two ordered k-uniform hypergraphs $H^<$ and $G^<$ is the smallest $N \in \mathbb{N}$ such that every red-blue coloring of the hyperedges of $K_N^{<(k)}$ contains a blue copy of $H^<$ or a red copy of $G^<$ forming an ordered subhypergraph. In the *diagonal case* $H^< = G^<$, we just write $R_<(H^<)$ instead of $R_<(H^<, H^<)$. We can also naturally extend the definition to $q \geq 2$ colors, in which case we use $R_<(H^<; q)$ to denote the q-color variant of $R_<(H^<)$.

Similarly as for ordered graphs, the ordered Ramsey number of every n-vertex k-uniform hypergraph is bounded from above by $R(K_n^{(k)})$. In particular, these numbers are always finite by (6). It is also easy to see that the ordered Ramsey numbers of k-uniform hypergraphs grow at least as fast as the standard Ramsey numbers.

3.1 Monotone Paths and the Erdős–Szekeres Theorem

We start with one example, the original motivation for a systematic study of ordered Ramsey numbers. We showed in Sect. 2 that the Erdős–Szekeres lemma is a consequence of a stronger Ramsey statement about monotone paths. There is a similar connection between the celebrated *Erdős–Szekeres theorem* and ordered Ramsey numbers of 3-uniform hypergraphs.

To state the Erdős–Szekeres theorem, a foundational result in discrete geometry, we first need to introduce some definitions. A finite set P of points in the plane is in *general position* if no three points from P lie on a common line. A finite set of points is in *convex position* if its points form vertices of a convex polygon.

Theorem 9 (The Erdős–Szekeres theorem [37]) *For every positive integer n, there is a positive integer N such that every set of at least N points in general position in the plane contains n points in convex position.*

We use $ES(n)$ to denote the smallest such integer N. The statement of the Erdős–Szekeres theorem is a generalization of Esther Klein's problem, which was named the *Happy Ending Problem* by Paul Erdős, as it eventually led to the marriage of George Szekeres and Esther Klein. Erdős and Szekeres provided two proofs of this famous result. One is an application of Ramsey's theorem and yields a rather poor upper bound on the function $ES(n)$. The other proof uses more geometry and gives the estimate

$$ES(n) \leq \binom{2n-4}{n-2} + 1 \tag{7}$$

for every $n \geq 2$. Already in 1935, Erdős and Szekeres believed that this bound could be significantly improved. Based on their results for $n = 2, 3, 4$, they posed the famous and still open *Erdős–Szekeres conjecture*, for whose proof Erdős offered \$500 reward.

Conjecture 3 (The Erdős–Szekeres conjecture [37]) *For every integer $n \geq 2$, we have*

$$ES(n) = 2^{n-2} + 1.$$

In the 1960s, Erdős and Szekeres [38] supported Conjecture 3 with the lower bound

$$ES(n) \geq 2^{n-2} + 1. \tag{8}$$

Despite several attempts over the years [17, 57, 61, 69, 77, 78], it is still open to decide whether the upper bound $ES(n) \leq 2^{n-2} + 1$ holds. It is only known that the conjecture is true for $n \leq 6$ [77]. However, there was a breakthrough by Suk [76] in 2017, who proved a very close estimate $ES(n) \leq 2^{n+o(n)}$.

As mentioned, the Erdős–Szekeres theorem, more precisely the bound (7), follows from known estimates on ordered Ramsey numbers of monotone 3-uniform paths. For an integer $k \geq 2$, the *monotone k-uniform path* on n vertices, denoted by $MP_n^{<(k)}$, is an ordered k-uniform n-vertex hypergraph with edges formed by k-tuples of consecutive vertices in $<$; see part (b) of Fig. 1. Note that the monotone path $MP_n^{<}$ corresponds to $MP_n^{<(2)}$. The monotone paths are also called *tight paths* in the literature.

We can again rather easily show that $ES(n) \leq R_<(MP_n^{<(3)})$ by coloring triples of points according to their orientations with the ordering given by x-coordinates; see [6, 22]. The bound (7) then follows from the fact

$$R_<(MP_n^{<(3)}) = \binom{2n-4}{n-2} + 1 \tag{9}$$

for every $n \geq 2$, which was proved by Moshkovitz and Shapira [62]. Moreover, several other interesting geometric applications of estimates on $R_<(MP_n^{<(k)})$ for $k \geq 3$ appeared, for example, variants of the Erdős–Szekeres Theorem for convex bodies [45], higher-order Erdős–Szekeres theorems [31], and Ramsey-type results [5] for *k-intersecting pseudoconfigurations of points* [60], *C_d-arrangements of n pseudohyperplanes* in $\mathbb{R}^d$ [41], and extensions of the cyclic arrangement of hyperplanes with a pseudohyperplane [80]. Formulations of the Erdős–Szekeres conjecture in terms of hypergraph ordered Ramsey numbers were also used in computer attacks on this conjecture and its generalizations [3, 11, 77].

Given this motivation, the ordered Ramsey numbers $R_<(MP_n^{<(k)})$ have been quite intensively studied [5, 16, 31, 45, 59, 62] even for higher uniformities and their growth rate is nowadays well understood. Moshkovitz and Shapira [62] showed the following strong result.

Theorem 10 ([62]) *For all positive integers n and $k \geq 3$,*

$$R_<(MP_{n+k-1}^{<(k)}) = t_{k-1}((2-o(1))n),$$

where $t_{k-1}(\cdot)$ is the tower function of height $k-2$.

Note that the tower is of height one smaller than the tower-type upper bound on $R(K_n^{(k)})$ from (6) obtained with the Stepping-up lemma. In fact, Moshkovitz and Shapira [62] proved

$$R_<(MP_{n+k-1}^{<(k)}) = \rho_k(n) + 1,$$

where $\rho_k(n)$ is the number of *line partitions of n of order k* (see [62] for definitions). For $k = 3$, this gives the exact formula $R_<(MP_n^{<(3)}) = \binom{2n-4}{n-2} + 1$.

Moshkovitz and Shapira [62] note that $R_<(MP_n^{<(3)}; q) - 1$ is the number of antichains in $[n-2]^q$. In particular, for $n = 4$, we get that $R_<(MP_4^{<(3)}; q) - 1$ equals the *Dedekind numbers*, the numbers of antichains in the Boolean poset $[2]^q$. The problem of estimating the number of antichains in $[n]^q$ is a natural extension of the classical problem of estimating the Dedekind numbers. This problem from enumerative combinatorics was recently explored by Falgas-Ravry et al. [40] who resolved a conjecture of Moshkovitz and Shapira [62] about the number of antichains in $[n]^q$ and, consequently, about the growth rate of $R_<(MP_n^{<(3)}; q)$. Their bound is very strong and even determines the leading constant in the exponent.

Theorem 11 ([40]) *For every $\varepsilon > 0$ if q is a sufficiently large integer, then for every integer $n \geq 2$*

$$(1-\varepsilon)\sqrt{\frac{6}{\pi(n^2-1)q}} \cdot n^q \leq \log R_<(MP_{n+2}^{<(3)}; q) \leq (1+\varepsilon)\sqrt{\frac{6}{\pi(n^2-1)q}} \cdot n^q.$$

We conclude this subsection with a problem about $R_<(MP_n^{<(4)})$ mentioned by Moshkovitz and Shapira [62]. While we know the exact formula for $R_<(MP_n^{<(3)})$ by (9), Moshkovitz and Shapira [62] argue that it might be hard to determine the asymptotic of the logarithm of $R_<(MP_n^{<(4)})$. This problem is connected to the extremal theory of partially ordered sets as $R_<(MP_n^{<(4)}) - 1$ is exactly the number of antichains in the restricted *Young's lattice* [62].

3.2 Connections to Hypergraph Ramsey Numbers

Here, we mention two interesting connections between ordered Ramsey numbers and hypergraph Ramsey numbers. A surprising relation between off-diagonal ordered Ramsey numbers of monotone k-uniform paths and the classical Ramsey numbers $R(K_n^{<(k-1)})$ was discovered by Mubayi and Suk [64]. They showed the following estimate, which in the case $q = 2$, $k = O(1)$, and n tending to infinity shows that estimating $R_<(K_n^{<(k)}, MP_{k+1}^{<(k)})$ is equivalent to the famous problem of determining the tower growth rate of the Ramsey numbers $R(K_n^{(k-1)})$ of complete $(k-1)$-uniform hypergraphs (6).

Theorem 12 ([64]) *For all integers $n > k \geq 2$,*

$$R(K_{\lfloor n/q \rfloor}^{(k-1)}; q) \leq R_<(K_n^{<(k)}, MP_{k-q+1}^{<(k)}) \leq R(K_n^{(k-1)}; q).$$

Mubayi and Suk [64] also introduced the following strengthening of a conjecture of Erdős and Hajnal [34] about off-diagonal Ramsey numbers of complete k-uniform hypergraphs.

Conjecture 4 ([64]) *For any fixed integer $k \geq 4$,*

$$R_<(K^{<(k)}, MP_{k+1}^{<(k)}) \geq t_{k-1}(\Omega(n)).$$

For more problems and results about off-diagonal ordered Ramsey numbers of monotone k-uniform paths, we recommend a recent survey by Mubayi and Suk [65]; see also [49, 66] for some newer results.

Another connection was discovered by Conlon et al. [22] who showed that for any 3-uniform hypergraph H, there is a family of ordered graphs $\mathcal{S}_H$ such that the Ramsey number of H is bounded in terms of the ordered Ramsey number of the family $\mathcal{S}_H$. Here, the *ordered Ramsey number $R_<(\mathcal{F})$ of a family $\mathcal{F}$* of ordered graphs is the smallest positive integer N such that every 2-coloring of the edges of $K_N^<$ contains a monochromatic ordered copy of some ordered graph from $\mathcal{F}$.

For an ordered graph $G^<$ with the vertex set $[n]$, let $T(G^<)$ be a 3-uniform hypergraph on vertex set $[n+1]$ obtained by taking all triples whose first pair is an edge of $G^<$. For a 3-uniform hypergraph H on $n+1$ vertices, we let $\mathcal{S}_H$ be the collection of ordered graphs $G^<$ on $[n]$ such that H is a subhypergraph of $T(G^<)$.

Conlon et al. [22] then related upper bounds on Ramsey numbers of 3-uniform hypergraphs to ordered Ramsey numbers by proving the following result.

Theorem 13 ([22]) *Every 3-uniform hypergraph H satisfies*

$$R(H) \leq 2^{\binom{R_<(\mathcal{S}_H)}{2}} + 1.$$

Conlon et al. [22] expect that the bound from Theorem 13 is close to sharp in many cases. For example, the choice $H = K_{n+1}^{(3)}$ satisfies $\mathcal{S}_H = \{K_n^<\}$, for which Theorem 13 produces the double-exponential bound from (5), which is believed to be tight. However, Conlon et al. [22] constructed some cases where the bound from Theorem 13 is far from the truth.

3.3 Bounded Degrees and Interval Chromatic Number

Very little is known about ordered Ramsey numbers of general ordered k-uniform hypergraphs with $k \geq 3$ as ordered Ramsey numbers have been studied mostly for ordered graphs only. We primarily focus on the case $k = 3$ here.

A simple probabilistic argument provides the lower bound $R_<(H^<) \geq 2^{\Omega(n^2)}$ for every ordered 3-uniform hypergraph with n vertices and $\Omega(n^3)$ hyperedges, which is of the same asymptotic growth rate as we have in the lower bound on $R_<(K_n^{<(3)})$ by (5). Thus, we consider mostly sparse 3-uniform hypergraphs.

The *degree* of a vertex v in a hypergraph H is the number of hyperedges of H that contain v. It follows from a result by Moshkovitz and Shapira [62] that there are ordered 3-uniform hypergraphs $H^<$ on n vertices with maximum degree 3 such that $R_<(H^<) \geq 2^{\Omega(n)}$.

Therefore, to obtain smaller upper bounds on the ordered Ramsey numbers, it is necessary to bound other parameters besides the maximum degree. A natural choice is again the interval chromatic number, which is defined analogously as for ordered graphs. Recall that for ordered graphs, bounding both parameters indeed helps, as the ordered Ramsey number $R_<(G^<)$ of every n-vertex ordered graph $G^<$ with bounded maximum degree d and bounded interval chromatic number χ is at most polynomial in the number of vertices by (3). Theorem 2 actually gives a stronger estimate

$$R_<(G^<, K_\chi^<(n)) \leq n^{32d \log \chi}.$$

A natural question is whether we can get similar bounds for ordered k-uniform hypergraphs with $k \geq 3$. For integers $k \geq 2$ and $\chi \geq k$, we use $K_\chi^{(k)}(n)$ to denote the *complete k-uniform χ-partite hypergraph*, that is, the vertex set of $K_\chi^{(k)}(n)$ is partitioned into χ sets of size n and every k-tuple with at most one vertex in each of these parts forms a hyperedge. Let $K_\chi^{<(k)}(n)$ be the ordering of $K_\chi^{(k)}(n)$ in which the color classes form consecutive intervals. Conlon et al. [25] showed that, for all positive integers $\chi \geq 3$ and n,

$$R(K_{\chi}^{(3)}(n)) \leq 2^{2^{2R}n^2},$$

where $R = R(K_{\chi-1})$. Since every ordering of $K_{\chi}^{(3)}(\chi n)$ contains an ordered subhypergraph isomorphic to $K_{\chi}^{<(3)}(n)$ and every ordered 3-uniform hypergraph on n vertices with interval chromatic number χ is an ordered subhypergraph of $K_{\chi}^{<(3)}(n)$, we obtain $R_<(H^<) \leq 2^{2^{2R}\chi^2 n^2}$, for every ordered 3-uniform hypergraph $H^<$ on n vertices with interval chromatic number χ where $R = R(K_{\chi-1})$. In particular, if the interval chromatic number χ of $H^<$ is fixed, we have $R_<(H^<) \leq 2^{O(n^2)}$.

Note that the last bound is asymptotically tight for dense ordered hypergraphs with bounded interval chromatic number. Thus, Balko and Vizer [13] considered the sparse case in which we additionally bound the maximum degree, similarly as in the setting from Sect. 2.1. More specifically, the first nontrivial case for ordered 3-uniform hypergraphs with interval chromatic number 3, for which they obtained an estimate with a subquadratic exponent.

Theorem 14 ([13]) *Let $H^<$ be an ordered 3-uniform hypergraph on n vertices with maximum degree d. Then there are constants $C = C(d)$ and $c > 0$ such that*

$$R_<(H^<, K_3^{<(3)}(n)) \leq 2^{O(n^{2-1/(1+cd^2)})}. \tag{10}$$

The main idea of the proof of Theorem 14 is based on an embedding lemma from [26], where the authors study Erdős–Hajnal-type theorems for 3-uniform tripartite hypergraphs. Theorem 14 immediately gives $R_<(H^<) \leq 2^{O(n^{2-\varepsilon})}$ for any ordered 3-uniform hypergraph $H^<$ on n vertices with maximum degree d and with interval chromatic number 3.

The upper bound (10) is quite close to the truth, as even when $H^<$ is fixed we get a superexponential lower bound on $R_<(H^<, K_3^{<(3)}(n))$, as shown by Fox and He [42] and independently by Balko and Vizer [13].

Theorem 15 ([13, 42]) *For every $t \geq 3$ and every positive integer n, we have*

$$R_<(K_{t+1}^{<(3)}, K_3^{<(3)}(n)) \geq 2^{\Omega(n \log n)}.$$

For ordered hypergraphs of uniformity $k > 3$, we recall that it follows from (6) that their ordered Ramsey numbers can grow as a tower of height $k-1$. By modifying a result of Conlon et al. [24], Balko and Vizer [13] showed that we do not have a tower-type growth rate for $R_<(H^<)$ once the uniformity and the interval chromatic number of $H^<$ are bounded.

Proposition 2 ([13]) *Let χ, k be fixed integers with $\chi \geq k \geq 2$ and let $H^<$ be an ordered k-uniform hypergraph on n vertices with interval chromatic number χ. Then,*

$$R_<(H^<) \leq 2^{O(n^{\chi-1})}.$$

The ordered Ramsey numbers for ordered k-uniform hypergraphs with $k \geq 3$ and the edge-ordered Ramsey numbers are quite unexplored, so there is a plenty of open problems.

Theorems 14 and 15 give estimates on the ordered $R_<(G^<, K_3^{<(3)}(n))$ and although the exponents in the bounds are reasonably close, there is still a gap between them and it would be interesting to close it.

Problem 14 ([13]) Let d be a fixed positive integer and let $H^<$ be an ordered 3-uniform hypergraph on n vertices with maximum degree d. Close the gap between the lower and upper bounds on $R_<(H^<, K_3^{<(3)}(n))$.

Another interesting problem is to extend the upper bound with subquadratic exponent from Theorem 14 to ordered 3-uniform hypergraphs with bounded maximum degree and fixed interval chromatic number that is larger than 3.

Problem 15 ([13]) Let d and χ be fixed positive integers. Is there an $\varepsilon = \varepsilon(d, \chi) > 0$ such that, for every ordered 3-uniform hypergraph $H^<$ on n vertices with maximum degree d and with interval chromatic number χ, we have

$$R_<(H^<) \leq 2^{O(n^{2-\varepsilon})}?$$

In general, we are not aware of any nontrivial upper bounds on ordered Ramsey numbers of ordered 3-uniform hypergraphs with bounded maximum degree.

Problem 16 ([13]) What is the upper bound on ordered Ramsey numbers of ordered 3-uniform hypergraphs with bounded maximum degree?

4 Edge-Ordered Ramsey Numbers

Besides vertices, we can also order the edges of a given graph and then study its extremal properties. An *edge-ordered graph* $G^\prec$ is a pair $(G, \prec)$ consisting of a graph $G = (V, E)$ and a linear ordering $\prec$ of the set of edges E. We use the term *edge-ordering of* G for the ordering $\prec$ and also for $G^\prec$. An edge-ordered graph $G^\prec$ is an *edge-ordered subgraph* of an edge-ordered graph $H^\prec$ if G is a subgraph of H and the edge-ordering of G is a suborder of the edge-ordering of H. We say that $G^\prec$ and $H^\prec$ are *isomorphic* if there is a graph isomorphism between G and H that also preserves the edge-orderings.

Motivated by extremal results for edge-graphs obtained by Gerbner et al. [47], Balko and Vizer [12] introduced Ramsey numbers for edge-ordered graphs. The *edge-ordered Ramsey number* $R_\prec(G^\prec)$ of an edge-ordered graph $G^\prec$ is the minimum positive integer N such that there exists an edge-ordering $K_N^\prec$ of K_N such that every 2-coloring of the edges of $K_N^\prec$ contains a monochromatic copy of $G^\prec$ as an edge-ordered subgraph of $K_N^\prec$.

Note that the definition of edge-ordered Ramsey numbers is defined quite differently than ordered Ramsey numbers as the edge-ordering of the complete graph whose edges are being colored depends on the given edge-ordered graphs. This is necessary, as otherwise there might be an edge-ordered graph $G^{\prec}$ and an edge-order of $K_N^{\prec}$ such that $G^{\prec}$ is not an edge-ordered subgraph of $K_N^{\prec}$.

The finiteness of ordered Ramsey numbers was quite easy to show, as it followed from the finiteness of standard Ramsey numbers. This is not the case for edge-ordered Ramsey numbers, where it takes some effort to prove that these numbers are finite for any pair of edge-ordered graphs. This was proved by Balko and Vizer [12].

Theorem 16 ([12]) *For every edge-ordered graph $G^{\prec}$, the edge-ordered Ramsey number $R_{\prec}(G^{\prec})$ is finite.*

Theorem 16 also follows from a recent deep result of Hubička and Nešetřil [51, Theorem 4.33] about Ramsey numbers of general relational structures. In comparison, the proof of Theorem 16 is less general, but it is much simpler and produces better and more explicit bound on $R_{\prec}(G^{\prec})$. The proof of Theorem 16 yields a stronger induced-type statement where additionally the ordering of the vertex set is fixed and the colorings can use an arbitrary number of colors.

However, the bound on the edge-ordered Ramsey numbers obtained in the proof of Theorem 16 is enormous, it grows faster than, for example, a tower function of any fixed height. Fox and Li [44] improved the bound on edge-ordered Ramsey numbers to a single exponential type.

Theorem 17 ([44]) *For each positive integer n, there is an edge-ordered graph $G^{\prec}$ on $N = 2^{100n^2 \log^2 n}$ vertices such that, for every 2-coloring of the edges of $G^{\prec}$, there exists a monochromatic subgraph containing a copy of every n-vertex edge-ordered graph.*

In particular, if $G^{\prec}$ is an edge-ordered graph on n vertices, then

$$R_{\prec}(G^{\prec}) \leq 2^{100n^2 \log^2 n}.$$

Fox and Li [44] also extended their result to more colors and proved the following polynomial upper bound on edge-ordered Ramsey numbers of edge-ordered graphs of bounded degeneracy, improving earlier estimates by Balko and Vizer [12].

Theorem 18 ([44]) *If $G^{\prec}$ is an edge-ordered d-degenerate graph on n vertices, then*

$$R_{\prec}(G^{\prec}) \leq n^{600d^3 \log (d+1)}.$$

For edge-ordered graphs, a major open question is to find a better estimate for edge-ordered Ramsey numbers of edge-ordered complete graphs on n vertices. In particular, Fox and Li [44] asked whether the exponential lower bound is tight for general edge-orderings of K_n. For sparser edge-ordered graphs, Fox and Li [44] conjectured that the upper bound from Theorem 18 can be improved.

Conjecture 5 ([44]) *If $H^{\prec}$ is an edge-ordered d-degenerate graph on n vertices, then $R_{\prec}(H^{\prec}) \leq n^{O(d)}$.*

Currently, there are no known edge-ordered d-degenerate graphs with superlinear edge-ordered Ramsey numbers. However, Fox and Li [44] believe that there are such examples and conjectured that the upper bound from Conjecture 5 is tight up to the constant in the exponent.

The edge-ordered Ramsey numbers can be naturally extended to edge-ordered k-uniform hypergraphs with any $k \geq 2$. The existence of such numbers follows from a result of Hubička and Nešetřil [51], but the resulting bounds are enormous. Fox and Li [44] posed a natural problem to give a better estimate on such hypergraph edge-ordered Ramsey numbers.

Competing Interests M. Balko was supported by grant no. 23-04949X of the Czech Science Foundation (GAČR) and by the Center for Foundations of Modern Computer Science (Charles Univ. project UNCE 24/SCI/008).

References

1. Ajtai, M., Komlós, J., Szemerédi, E.: A note on Ramsey numbers. J. Combin. Theory Ser. A **29**(3), 354–360 (1980)
2. Alon, N., Rödl, V.: Sharp bounds for some multicolor Ramsey numbers. Combinatorica **25**(2), 125–141 (2005)
3. Baek, J., Balko, M.: The Erdős—Szekeres conjecture revisitied (2024)
4. Balister, P., Bollobás, B., Campos, M., Griffiths, S., Hurley, E., Morris, R., Sahasrabudhe, J., Tiba, M.: Upper bounds for multicolour Ramsey numbers (2024). arXiv:2410.17197
5. Balko, M.: Ramsey numbers and monotone colorings. J. Combin. Theory Ser. A **163**, 34–58 (2019)
6. Balko, M., Cibulka, J., Král, K., Kynčl, J.: Ramsey numbers of ordered graphs. Electron. J. Combin. **27**, P1.16 (2020)
7. Balko, M., Grinerová, K.: Estimating multicolor ordered Ramsey numbers (2024)
8. Balko, M., Jelínek, V., Valtr, P.: On ordered Ramsey numbers of bounded-degree graphs. J. Combin. Theory Ser. B **134**, 179–202 (2019)
9. Balko, M., Poljak, M.: On off-diagonal ordered Ramsey numbers of nested matchings. Discret. Math. **346**(2), 13 (2023). Paper No. 113223
10. Balko, M., Poljak, M.: On ordered Ramsey numbers of matchings versus triangles. Electron. J. Combin. **31**(2), 16 (2024). Paper No. 2.23
11. Balko, M., Valtr, P.: A SAT attack on the Erdős-Szekeres conjecture. European J. Combin. **66**, 13–23 (2017)
12. Balko, M., Vizer, M.: Edge-ordered Ramsey numbers. European J. Combin. **87**, 103100 (2020)
13. Balko, M., Vizer, M.: On ordered Ramsey numbers of tripartite 3-uniform hypergraphs. SIAM J. Discrete Math. **36**(1), 214–228 (2022)
14. Barát, J., Gyárfás, A., Tóth, G.: Monochromatic spanning trees and matchings in ordered complete graphs. J. Graph Theory **105**(4), 523–541 (2024)
15. Campos, M., Griffiths, S., Morris, R., Sahasrabudhe, J.: An exponential improvement for diagonal Ramsey (2023). arXiv:2303.09521
16. Choudum, S.A., Ponnusamy, B.: Ordered Ramsey numbers. Discret. Math. **247**(1–3), 79–92 (2002)

17. Chung, F.R.K., Graham, R.L.: Forced convex n-gons in the plane. Discrete Comput. Geom. **19**(3), 367–371 (1998). Dedicated to the memory of Paul Erdős
18. Chvátal, V., Rödl, V., Szemerédi, E., Trotter, W.T., Jr.: The Ramsey number of a graph with bounded maximum degree. J. Combin. Theory Ser. B **34**(3), 239–243 (1983)
19. Cibulka, J., Gao, P., Krčál, M., Valla, T., Valtr, P.: On the geometric Ramsey number of outerplanar graphs. Discrete Comput. Geom. **53**(1), 64–79 (2015)
20. Conlon, D.: A new upper bound for diagonal Ramsey numbers. Ann. Math. (2) **170**(2), 941–960 (2009)
21. Conlon, D., Ferber, A.: Lower bounds for multicolor Ramsey numbers. Adv. Math. **378**, 5 (2021). Paper No. 107528
22. Conlon, D., Fox, J., Lee, C., Sudakov, B.: Ordered Ramsey numbers. J. Combin. Theory Ser. B **122**, 353–383 (2017)
23. Conlon, D., Fox, J., Sudakov, B.: Ramsey numbers of sparse hypergraphs. Random Struct. Algorithms **35**(1), 1–14 (2009)
24. Conlon, D., Fox, J., Sudakov, B.: Hypergraph Ramsey numbers. J. Am. Math. Soc. **23**(1), 247–266 (2010)
25. Conlon, D., Fox, J., Sudakov, B.: Large almost monochromatic subsets in hypergraphs. Israel J. Math. **181**, 423–432 (2011)
26. Conlon, D., Fox, J., Sudakov, B.: Erdős-Hajnal-type theorems in hypergraphs. J. Combin. Theory Ser. B **102**(5), 1142–1154 (2012)
27. Conlon, D., Fox, J., Sudakov, B.: Recent developments in graph Ramsey theory. In: Surveys in Combinatorics 2015. London Mathematical Society Lecture Note Series, vol. 424, pp 49–118. Cambridge University Press, Cambridge (2015)
28. Cooley, O., Fountoulakis, N., Kühn, D., Osthus, D.: Embeddings and Ramsey numbers of sparse k-uniform hypergraphs. Combinatorica **29**(3), 263–297 (2009)
29. Cooley, O., Fountoulakis, N., Kühn, D., Osthus, D.: 3-uniform hypergraphs of bounded degree have linear Ramsey numbers. J. Combin. Theory Ser. B **98**(3), 484–505 (2008)
30. Dujmović, V., Wood, D.R.: On linear layouts of graphs. Discrete Math. Theor. Comput. Sci. **6**(2), 339–357 (2004)
31. Eliáš, M., Matoušek, J.: Higher-order Erdős-Szekeres theorems. Adv. Math. **244**, 1–15 (2013)
32. Erdős, P., Sós, Vera T.: Problems and results on Ramsey-Turán type theorems (preliminary report). In: Proceedings of the West Coast Conference on Combinatorics, Graph Theory and Computing (Humboldt State University, Arcata, California, 1979). Congressus Numerantium, vol. 26, pp. 17–23. Utilitas Mathematica, Winnipeg, MB (1980)
33. Erdős, P.: Some remarks on the theory of graphs. Bull. Am. Math. Soc. **53**, 292–294 (1947)
34. Erdős, P., Hajnal, A.: On Ramsey like theorems. Problems and results. In: Combinatorics (Proceedings of the Conference on Combinatorial Mathematics held at the Mathematical Institute, Oxford, in 1972), pp. 123–140. Institute of Mathematics and its Applications, Southend-on-Sea (1972)
35. Erdős, P., Hajnal, A., Rado, R.: Partition relations for cardinal numbers. Acta Math. Acad. Sci. Hungar. **16**, 93–196 (1965)
36. Erdős, P., Rado, R.: Combinatorial theorems on classifications of subsets of a given set. Proc. London Math. Soc. **3**(2), 417–439 (1952)
37. Erdős, P., Szekeres, G.: A combinatorial problem in geometry. Compositio Math. **2**, 463–470 (1935)
38. Erdős, P., Szekeres, G.: On some extremum problems in elementary geometry. Ann. Univ. Sci. Budapest. Eötvös Sect. Math. **3**(4), 53–62 (1960)
39. Erdős, P., Szemerédi, A.: On a Ramsey type theorem. Period. Math. Hungar. **2**, 295–299 (1972)
40. Falgas-Ravry, V., Räty, E., Tomon, I.: Dedekind's problem in the hypergrid (2023). arXiv:2310.12946
41. Felsner, S., Weil, H.: Sweeps, arrangements and signotopes. Discrete Appl. Math. **109**(1–2), 7–94 (2001). 14th European Workshop on Computational Geometry CG'98 (Barcelona)
42. Fox, J., He, X.: Independent sets in hypergraphs with a forbidden link. Proc. Lond. Math. Soc. (3) **123**(4), 384–409 (2021)

43. Fox, J., He, X.: Ramsey numbers of sparse digraphs. Israel J. Math. 1–48 (2024)
44. Fox, J., Li, R.: On edge-ordered Ramsey numbers. Random Struct. Algorithms **57**(4), 1174–1204 (2020)
45. Fox, J., Pach, J., Sudakov, B., Suk, A.: Erdős–Szekeres-type theorems for monotone paths and convex bodies. Proc. Lond. Math. Soc. (3) **105**(5), 953–982 (2012)
46. Geneson, J., Holmes, A., Liu, X., Neidinger, D., Pehova, Y., Wass, I.: Ramsey numbers of ordered graphs under graph operations (2019). arXiv:1902.00259
47. Gerbner, D., Methuku, A., Nagy, D.T., Pálvölgyi, D., Tardos, G., Vizer, M.: Edge ordered Turán problems. Acta Math. Univ. Comenian. (N.S.) **88**(3), 717–722 (2019)
48. Girão, A., Janzer, B., Janzer, O.: Ordered Ramsey numbers of powers of paths (2024). arXiv:2401.02360
49. Gishboliner, L., Jin, Z., Sudakov, B.: Ramsey problems for monotone paths in graphs and hypergraphs. Combinatorica **44**(3), 509–529 (2024)
50. Grinerová, K.: Multi-colored ordered Ramsey numbers, Bachelor's thesis (2024)
51. Hubička, J., Nešetřil, J.: All those Ramsey classes (Ramsey classes with closures and forbidden homomorphisms). Adv. Math. **356**, 106791 (2019)
52. Ishigami, Y.: Linear Ramsey numbers for bounded-degree hypergrahps. Electron. Notes Discret. Math. **29**, 47–51 (2007)
53. Janzer, B., Janzer, O., Magnan, V., Methuku, A.: Tight general bounds for the extremal numbers of 0–1 matrices. Int. Math. Res. Not. IMRN **2024**(15), 11455–11463 (2024)
54. Károlyi, Gy., Pach, J., Tóth, G.: Ramsey-type results for geometric graphs. I. Discrete Comput. Geom. **18**(3), 247–255 (1997). ACM Symposium on Computational Geometry (Philadelphia, PA, 1996)
55. Károlyi, Gy., Pach, J., Tóth, G., Valtr, P.: Ramsey-type results for geometric graphs. II. Discrete Comput. Geom. **20**(3), 375–388 (1998). ACM Symposium on Computational Geometry (Nice, 1997)
56. Kim, J.H.: The Ramsey number $R(3, t)$ has order of magnitude $t^2/\log t$. Random Struct. Algorithms **7**(3), 173–207 (1995)
57. Kleitman, D., Pachter, L.: Finding convex sets among points in the plane. Discrete Comput. Geom. **19**(3), 405–410 (1998). Dedicated to the memory of Paul Erdős
58. Lefmann, H.: A note on Ramsey numbers. Studia Sci. Math. Hungar. **22**(1–4), 445–446 (1987)
59. Milans, K.G., Stolee, D., West, D.B.: Ordered Ramsey theory and track representations of graphs. J. Comb. **6**(4), 445–456 (2015)
60. Miyata, H.: On combinatorial properties of points and polynomial curves (2017). arxiv:1703.04963
61. Nassajian Mojarrad, H., Vlachos, G.: An improved upper bound for the Erdős–Szekeres conjecture. Discrete Comput. Geom. **56**(1), 165–180 (2016)
62. Moshkovitz, G., Shapira, A.: Ramsey theory, integer partitions and a new proof of the Erdős-Szekeres theorem. Adv. Math. **262**, 1107–1129 (2014)
63. Mubayi, D.: Variants of the Erdős-Szekeres and Erdős-Hajnal Ramsey problems. European J. Combin. **62**, 197–205 (2017)
64. Mubayi, D., Suk, A.: Off-diagonal hypergraph Ramsey numbers. J. Combin. Theory Ser. B **125**, 168–177 (2017)
65. Mubayi, D., Suk, A.: A survey of hypergraph Ramsey problems. In: Discrete Mathematics and Applications. Springer Optimization and Its Applications, vol. 165, pp. 405–428. Springer, Cham (2020)
66. Mubayi, D., Suk, A.: Ramsey numbers of cliques versus monotone paths. Eur. J. Comb. **118**, 7 (2024). Paper No. 103922
67. Nagle, B., Olsen, S., Rödl, V., Schacht, M.: On the Ramsey number of sparse 3-graphs. Graphs Combin. **24**(3), 205–228 (2008)
68. Neidinger, D., West, D.B.: Ramsey numbers of interval 2-chromatic ordered graphs. Graphs Combin. **35**(5), 1065–1076 (2019)
69. Norin, S., Yuditsky, Y.: Erdős–Szekeres without induction. Discrete Comput. Geom. **55**(4), (2016)

70. Pach, J., Tardos, G.: Forbidden paths and cycles in ordered graphs and matrices. Israel J. Math. **155**, 359–380 (2006)
71. Ramsey, F.P.: On a Problem of Formal Logic. Proc. London Math. Soc. (2) **30**(4), 264–286 (1929)
72. Rohatgi, D.: Off-diagonal ordered Ramsey numbers of matchings. Electron. J. Combin. **26**(2), 18 (2019). Paper No. 2.21
73. Sah, A.: Diagonal Ramsey via effective quasirandomness. Duke Math. J. **172**(3), 545–567 (2023)
74. Sawin, W.: An improved lower bound for multicolor Ramsey numbers and a problem of Erdős. J. Combin. Theory Ser. A **188**, 11 (2022). Paper No. 105579
75. Spencer, J.: Ramsey's theorem–a new lower bound. J. Comb. Theory Ser. A **18**, 108–115 (1975)
76. Suk, A.: On the Erdős-Szekeres convex polygon problem. J. Am. Math. Soc. **30**(4), 1047–1053 (2017)
77. Szekeres, G., Peters, L.: Computer solution to the 17-point Erdős-Szekeres problem. ANZIAM J. **48**(2), 151–164 (2006)
78. Tóth, G., Valtr, P.: Note on the Erdős–Szekeres theorem. Discrete Comput. Geom. **19**(3), 457–459 (1998). Dedicated to the memory of Paul Erdős
79. Wigderson, Y.: An improved lower bound on multicolor Ramsey numbers. Proc. Am. Math. Soc. **149**(6), 2371–2374 (2021)
80. Ziegler, G.: Higher Bruhat orders and cyclic hyperplane arrangements. Topology **32**(2), 259–279 (1993)

Structure and Noise in Dense and Sparse Random Graphs: Percolated Stochastic Block Model via the EM Algorithm and Belief Propagation with Non-backtracking Spectra

Marianna Bolla, Hannu Reittu, and Runtian Zhou

Abstract In this survey paper it is illustrated how spectral clustering methods for unweighted graphs are adapted to the dense and sparse regimes. Whereas Laplacian and modularity based spectral clustering is apt to dense graphs, recent results show that for sparse ones, the non-backtracking spectrum is the best candidate to find assortative clusters of nodes. Here belief propagation in the sparse stochastic block model is derived with arbitrarily given model parameters that results in a non-linear system of equations; with linear approximation, the spectrum of the non-backtracking matrix is able to specify the number k of clusters. Then the model parameters themselves can be estimated by the EM algorithm. Bond percolation in the assortative model is considered in the following two senses: the within- and between-cluster edge probabilities decrease with the number of nodes and edges coming into existence in this way are retained with probability β. As a consequence, the optimal k is the number of the structural real eigenvalues (greater than $\sqrt{c}$, where c is the average degree) of the non-backtracking matrix of the graph. Assuming, these eigenvalues $\mu_1 > \cdots > \mu_k$ are distinct, the multiple phase transitions obtained for β are $\beta_i = \frac{c}{\mu_i^2}$; further, at β_i the number of detectable clusters is i, for $i = 1, \ldots, k$. Inflation–deflation techniques are also discussed to classify the nodes themselves, which can be the base of the sparse spectral clustering. Simulation results, as well as real life examples are presented.

M. Bolla (✉)
Department of Stochastics, Budapest University of Technology and Economics, Budapest, Hungary
e-mail: marib@math.bme.hu

H. Reittu
NVTT Technical Research Center of Finland, Espoo, Finland
e-mail: hannu.reittu@vtt.fi

R. Zhou
Department of Mathematics, Duke University, Durham, NC, USA
e-mail: rz169@duke.edu

G. O. H. Katona et al. (eds.), *Sum(m)it280*, Bolyai Society Mathematical Studies 32,
https://doi.org/10.1007/978-3-032-18810-6_5

1 Introduction

In [10] we considered generalized quasirandom properties of expanding quasirandom graph sequences, which are deterministic counterparts of generalized random graphs, where the probabilities of edges coming into existence only depend on the cluster memberships of the endpoints, and are fixed during the expansion. If the number of the underlying clusters (number of nodes of the so-called model graph of [25]) is k, then a dense graph sequence is obtained in this way (the average degree of nodes is proportional to n, where n is the number of nodes at a stage of the expansion). We proved that the adjacency matrix of such a random or quasirandom graph G_n coming from the k-cluster model has k structural eigenvalues (of order n), while the others are $o(n)$ in absolute value. Also, the normalized modularity matrix defined in [7] (its spectrum is in $[-1, 1]$) has k eigenvalues separated from 0, whereas the others are $o(1)$. These graph-based matrices are symmetric, and hence, they have real eigenvalues. Further, the clusters can be recovered by the weighted k-means algorithm applied to the $(k-1)$-dimensional representatives of the nodes obtained via the eigenvectors corresponding to the $k-1$ non-trivial structural eigenvalues (more precisely, the objective function of the k-means clustering is $o(1)$).

The paper [4] examines random d-regular graphs with d fixed and the number of nodes tending to infinity. Hence, it is a sparse graph sequence, and it is proved that the coordinates of any (suitably normalized) adjacency eigenvector (except the trivial one) asymptotically follow Gaussian distribution, keeping the orthogonality. More generally, in case of sparse graphs, the eigenvectors corresponding to the largest modulus eigenvalues of the usual graph-based matrices are concentrated on the high degree nodes, see, e.g., [5], and so, they are not able to cluster the nodes.

Consequently, the adjacency, Laplacian, or modularity matrix based methodology is not applicable in the sparse case, see [36]. To consider this situation, the edge probabilities are rescaled, with n or $\log n$. These possibilities are discussed in [1]. In [29, 30], the authors consider scaling with n, and refine the block model threshold conjecture; they specify when information theoretically is possible or impossible to distinguish between the clusters, depending on the model parameters. This so-called Kesten–Stigum threshold will be discussed in Sect. 2. Bickel and Chen [6] give a nonparametric statistical view of network models, and consider consistency in the strong or weak sense, when the estimated cluster assignment approaches the true one as $n \to \infty$. They prove that the likelihood modularity is always consistent, while the Newman–Girvan modularity is consistent only under stronger assumptions. They also consider different scalings. Other papers, instead of consistency, define different measures for the agreement of the true and the recovered clusterings (correlation, proportions of misclassifications, and information theoretical measures, e.g., the Kullbach–Leibler divergence). In [5], the transition happens at $o(\log n)$ average degrees. At $\Omega(\log n)$ (e.g., poly-log n) average degrees the paper [16] establishes results in this intermediate regime.

In the Laplacian or modularity based spectral clustering [8], k is specified via spectral gaps, and the distance between the eigen-subspace spanned by the eigenvectors corresponding to the structural eigenvalues and that of the step-vectors is estimated by the Davis–Kahan theorem (see, e.g., [37]). As the square of this distance is the objective function of the k-means algorithm, this is at our disposal to find the clusters approximately in the dense case. This concept can be extended to non-symmetric matrices, like to the non-backtracking one, and due to theorems of [36] that use Bauer–Fike type perturbation results, similar subspace perturbations are applicable to sparse graphs too.

In the sparse case we scale with n, so the average degree of the graph is of constant order. The so obtained sparse stochastic block model (called SBM_k) is assortative in the sense that the within-cluster probability of two nodes being connected is higher than the between-cluster one. Here bond percolation has two meanings: the first is that the within- and between-cluster edge probabilities decrease with n, and the second is that edges coming into existence in this way are retained with probability β. A real-life social network is an instance of such an expanding sequence, where two persons are connected randomly, with probability depending on their cluster memberships and this probability is the "smaller" the "larger" the network is; further, along these connections an information (e.g., an infection) is transmitted only with probability β (which is related to the seriousness of the epidemic, akin to the recombination ratio).

The question is that for which values of the model parameters and β the k clusters can be distinguished with high probability (w.h.p.). If we knew k, the model parameters could be estimated with the EM algorithm, see [8]. The method of belief propagation (BP) is able to relate the model parameters to the non-backtracking spectrum of the graph, see [17, 20, 23, 39]. However, the detectability thresholds have been given only in special cases (e.g., in the symmetric case of [17, 28], to be described in Sect. 2). In the present paper, BP is derived in the generic case; it provides a non-linear system of equations for the conditional probabilities of the nodes belonging to the clusters (given their neighbors), where the number of equations is $2mk$ (m is the number of edges in the underlying graph). Though, there are numerical algorithms to solve this system of equations, BP is mainly important from theoretical point of view: via linearization and stability issues, our Theorem 1 establishes a close relation between the model parameters and the structural eigenvalues of the non-backtracking matrix $\boldsymbol{B}$ of the graph under mild assumptions. In this way, the number k of clusters is approximately equal to the number of the structural eigenvalues of $\boldsymbol{B}$; but for given k, it is the EM algorithm that has the massive theoretical background to estimate the model parameters themselves.

We also experienced that, surprisingly, the leading k (structural) eigenvalues of the expected adjacency matrix of a random graph coming from the sparse SBM_k model are (w.h.p. as $n \to \infty$) closer to the structural eigenvalues of $\boldsymbol{B}$ than to those of the adjacency matrix $\boldsymbol{A}$ of a randomly generated graph from this model, under some balancing conditions for the cluster sizes and average degrees of the clusters. This seems to contradict to the laws of large numbers, but in the sparse case (possessing nearly constant average degrees) it is supported by theoretical considerations as for deformed Wigner matrices (see, e.g., [15, 36]). Roughly speaking, here the amount of the noise in the data suppresses the structure, and it is impossible to find the clusters with the traditional spectral clustering methods. Luckily, in this case, the non-backtracking matrix is at our disposal.

Since the structural eigenvalues of $\boldsymbol{B}$ and those of the expected adjacency matrix are closely aligned, the so-called detectability threshold and the phase transitions for β also depend on these eigenvalues. However, in practical situations, only the non-backtracking spectrum is at our disposal and we do not know the model parameters. Based on this, we conclude the value of k and derive multiple phase transitions in the general SBM_k model, where, in addition, the edges of the graph coming from this model are retained with probability β, and we have the further percolated sparse stochastic block model SBM_k^β too.

The rest of the paper is organized as follows. In Sect. 2, properties of the non-backtracking matrix are discussed together with basic notions and notation; further, the sparse stochastic block model and bond-percolation models are introduced. In Sect. 3, we perform BP theoretically which shows that k is about the number of the structural eigenvalues of the non-backtracking matrix $\boldsymbol{B}$ that are larger than $\sqrt{c}$, where c is the average degree of the random graph $G_n \in SBM_k$; it is Theorem 1 that supports this choice of k. In Sect. 5, we run the EM algorithm to estimate the parameters of the SBM_k model that fits to our observed graph G_n. In the possession of the new estimates for the clusters and their average degrees, we adjust the number k so that possible eigenvalues of $\boldsymbol{B}$ around $\sqrt{c}$ are added or deleted, and k_0 denotes the new number of clusters. In Sect. 4, we use inflation–deflation techniques for this purpose and apply a theorem of [36] to classify the majority of nodes by the k-means algorithm; this issue is summarized in Theorem 2, further, by a corollary and remark after it. Eventually, we find number k of phase transitions for β in the SBM_k^β model, when $1, 2, \ldots, k$ clusters can be detected and supported by simulations in Sect. 6. In Sect. 7, real life application to quantum chemistry graphs is presented. Section 8 is devoted to conclusions and further perspectives.

2 Preliminaries

2.1 Non-backtracking Matrix of Unweighted Graphs

The *Hashimoto edge-incidence matrix*, with other words, *non-backtracking matrix* $\boldsymbol{B}$ of a simple graph G on n nodes and m edges is defined as a $2m \times 2m$ non-symmetric matrix of 0–1 entries, see [20, 23, 26] in context of non-backtracking random walks, community detection, and centrality. The general entry of $\boldsymbol{B}$ is

$$b_{ef} = \delta_{e\to f}\delta_{f\neq e^{-1}},$$

for $e, f \in E^{\to}$, where $E^{\to}$ is the set of bidirected edges of G (each existing edge is considered in both possible directions), and for $e = (e_1, e_2)$ the reversely directed edge is denoted e^{-1}, so $e^{-1} = (e_2, e_1)$; further, the $e \to f$ relation means that $e_2 = f_1$

and δ is the (1-0) indicator of the event in its lower index. Therefore, $b_{ef} = 1$ exactly when $e_2 = f_1$ and $f_2 \neq e_1$.

With other notation (used in Sect. 3) referring to the nodes, we have

$$b_{l \to s,\, j \to i} = \delta_{sj}(1 - \delta_{il}),$$

where δ is now the Kronecker-delta. So $b_{l \to s,\, j \to i} = 1$ if and only if for the quadruple in the lower indices, $l \to s = j \to i$ holds with $l \neq i$. Since $\boldsymbol{B}$ is a real matrix of non-negative entries, it has a largest absolute value eigenvalue which is positive real, so it is the spectral radius $\rho(\boldsymbol{B})$ of $\boldsymbol{B}$; and $\boldsymbol{B}$ can also have some other so-called structural real eigenvalucs (those are positive in assortative networks). Since the characteristic polynomial of $\boldsymbol{B}$ has real coefficients, its complex eigenvalues occur in conjugate pairs in the bulk of its spectrum. Note that the underlying simple graph is not directed, just we consider its edges as bidirected for the purpose of the message passing equations of Sect. 3.

Here we enumerate some results about the spectral properties of $\boldsymbol{B}$, see [19] for more details and proofs. Note that the spectrum of $\boldsymbol{B}$ is not sensitive to high-degree nodes, because a walk starting at a node cannot return to it immediately. Also trees, disconnected from the graph or dangling off it, contribute zero eigenvalues to the non-backtracking spectrum, as a non-backtracking random walk when forced to a leaf, must stop there.

Some important properties of $\boldsymbol{B}$ are summarized here.

- The frequently used Ihara's formula (see [19, 23, 26]) implies that, whenever G is not a tree ($m \geq n$), $\boldsymbol{B}$ has $m - n$ eigenvalues equal to 1 and $m - n$ eigenvalues equal to -1, whereas its further eigenvalues are those of the $2n \times 2n$ matrix

$$\boldsymbol{K} = \begin{pmatrix} \boldsymbol{O} & \boldsymbol{D} - \boldsymbol{I}_n \\ -\boldsymbol{I}_n & \boldsymbol{A} \end{pmatrix}, \tag{1}$$

  where $\boldsymbol{A}$ is the adjacency- and $\boldsymbol{D} = \mathrm{diag}(d_1, \ldots, d_n)$ is the diagonal degree-matrix of the graph (it contains the row-sums of $\boldsymbol{A}$, i.e., the node-degrees d_i's of G in its main diagonal).

  Note that $\boldsymbol{K}$ always has at least one additional eigenvalue 1, the geometric multiplicity of which is equal to the number of the connected components of the underlying graph (see Proposition 5.2 of [19]).
- For the spectral radii of $\boldsymbol{A}$ and $\boldsymbol{B}$, the relation

$$\rho(\boldsymbol{B}) \leq \rho(\boldsymbol{A}) \tag{2}$$

  holds (see [26]).
- If G is connected with $d_{min} \geq 2$, then for all eigenvalues of $\boldsymbol{B}$, $|\mu| \geq 1$ holds. In particular, if G has no nodes of degree 2, then the eigenvalues of $\boldsymbol{B}$ with $|\mu| = 1$ are ± 1's.

- Proposition 5.5 of [19] states that if G is a connected graph which is not a tree or cycle and $d_{\min} \geq 2$, then $\rho(\boldsymbol{K}) > 1$. In general (except for trees and cycles), the spectral radius $\rho(\boldsymbol{B}) > 1$. Also, by the Gershgorin theorem, for a connected graph, $\rho(\boldsymbol{B}) \leq d_{\max} - 1$ (with equality if and only if G is regular).
- If G is a connected graph that is not a cycle and $d_{\min} \geq 2$, then $\boldsymbol{B}$ is irreducible. Therefore, the Frobenius theorem is applicable to $\boldsymbol{B}$, and under the above conditions, it has a single positive real eigenvalue among its maximum absolute value ones with corresponding eigenvector of all positive real coordinates.
- Usually, only the eigenvalue 0 of $\boldsymbol{B}$ is defected (its geometric and algebraic multiplicity is not the same), like trees, otherwise the eigenvectors corresponding to non-zero eigenvalues (even to multiple ones) are linearly independent. In case of random graphs, e.g., in the stochastic block model to be discussed, there is a bulk of the spectrum (containing ± 1's and complex conjugate pairs), the other so-called structural eigenvalues (greater than $\sqrt{c}$) are real (positive in the assortative case) and the corresponding eigenvectors are nearly orthogonal (see [13]) and the one with the largest absolute value, giving $\rho(\boldsymbol{B})$, is single.
- Though $\boldsymbol{B}$ is not a normal matrix, even not always diagonalizable (the algebraic and geometric multiplicity of some of its eigenvalues may not be the same), it exhibits some symmetry. Observe that

$$b^*_{ef} = b_{fe} = b_{e^{-1} f^{-1}}$$

for each directed pair of edges and for every entry b^* of the transpose $\boldsymbol{B}^*$ of $\boldsymbol{B}$. This phenomenon can be described by involution and swapping.
In the sequel, the vectors are column vectors and * denotes the adjoint of a matrix or a vector (in case of real entries, it is the usual transposition).
- Introduce the notation

$$\breve{x}_e := x_{e^{-1}}, \quad e \in E^{\rightarrow}$$

for relating the coordinates of the $2m$-dimensional vectors $\mathbf{x}$ and $\breve{\mathbf{x}}$ of $\mathbb{R}^{E^{\rightarrow}}$. Now $\mathbf{x}$ is partitioned into two m-dimensional vectors $\mathbf{x}_1$ and $\mathbf{x}_2$, where the coordinates of $\mathbf{x}_1$ correspond to $j \to i$ edges with $j < i$ and those of $\mathbf{x}_2$ correspond to their inverses. If $\mathbf{x} = (\mathbf{x}_1^*, \mathbf{x}_2^*)^*$, then $\breve{\mathbf{x}} = (\mathbf{x}_2^*, \mathbf{x}_1^*)^*$, i.e., $\mathbf{x}$ and $\breve{\mathbf{x}}$ can be obtained from each other by swapping the set of their first m and second m coordinates.
Let $\boldsymbol{V}$ denote the following involution on $\mathbb{R}^{2m}$ ($\boldsymbol{V} = \boldsymbol{V}^{-1}$, $\boldsymbol{V}^2 = \boldsymbol{I}$, $\boldsymbol{V}$ is an orthogonal and symmetric matrix at the same time):

$$\boldsymbol{V} = \begin{pmatrix} \boldsymbol{O} & \boldsymbol{I}_m \\ \boldsymbol{I}_m & \boldsymbol{O} \end{pmatrix},$$

where the blocks are of size $m \times m$. With it, $\boldsymbol{V}\mathbf{x} = \breve{\mathbf{x}}$ and $\boldsymbol{V}\breve{\mathbf{x}} = \mathbf{x}$.
- Relation to the line-graph:

Proposition 1 *Let us partition the non-backtracking matrix* $\boldsymbol{B}$ *of the connected simple graph* G *on* n *nodes and* m *edges into the following four* $m \times m$ *quadrants:*

$$B = \begin{pmatrix} B_{11} & B_{12} \\ B_{21} & B_{22} \end{pmatrix},$$

where the first m rows (columns) correspond to the $j \to i$ edges with $j < i$, and the next m rows (columns) to their inverses, in the same order. Then

$$B_{11}^* = B_{22}, \quad B_{22}^* = B_{11}, \quad B_{12}^* = B_{12}, \quad and \quad B_{21}^* = B_{21}, \tag{3}$$

where B_{11} is upper- and B_{22} is lower-triangular. Further, $B_{11} + B_{12} + B_{21} + B_{22}$ is equal to the $m \times m$ adjacency matrix of the line-graph of G.

Note that the nodes of the line-graph of G are its edges, and two edges are adjacent if and only if they have a node in common. Let N denote the $n \times m$ node–edge (0-1) incidence matrix of the simple graph G (except for trees, $n \le m$). Then NN^* and N^*N are Gramians (covariance matrices), so they are positive semidefinite, have the same rank and the same set of positive eigenvalues. $NN^* = D + A$ is called signless Laplacian (whereas, the Laplacian matrix is $D - A$, see [8]). On the other hand, $N^*N - 2I_m$ is the adjacency matrix of the line-graph of G (because the valence of each edge is 2). Consequently, its eigenvalues are those of the positive eigenvalues of $D + A$ decreased by 2, and the others are equal to -2, see [14] for more explanation. Furthermore (see, e.g., [24]) it is known that if the line-graphs of two simple graphs, provided they both have node-degrees at least 4, are isomorphic, then they are isomorphic too. However, if this condition does not hold, it can happen that two not isomorphic graphs have isomorphic line-graphs. For example, if we have a triangle and a star on 4 vertices, then the adjacency matrix of their line-graphs is the same 3×3 matrix (adjacency matrix of a complete graph, as any two edges of them have a node in common). The same happens if we have graphs with such dangling triangles or stars. Therefore, in case of sparse graphs, B carries more information for the graph than its line-graph; further, considering its edges in both directions makes it possible to perform non-backtracking random walks and message passing along them, see Sect. 3.

In [31] it is proved that two simple graphs are isomorphic if and only if their corresponding non-backtracking graphs are isomorphic (their non-backtracking matrices are the same if we consider the bidirected edges in the same order). Under non-backtracking graph we understand the directed graph on $2m$ vertices with adjacency relation corresponding to the definition of the non-backtracking matrix.

Proof (Proposition 1) Let us label the rows (and columns) of B with the edges, ordered lexicographically: the first m rows correspond to the $j \to i$ edges ($1 \le j < i \le n$), whereas the second m ones to the inverse edges $i \to j$, in the same order. Obviously, B_{12} is a symmetric matrix as its corresponding upper- and lower-diagonal entries, producing an entry 1, contain the matching edges in both directions; B_{21} is a symmetric matrix for the same reason. Obviously, B_{11} is upper-, B_{22} is lower-triangular, and they are transposes of each other, because there is a one-to one correspondence between the upper-diagonal entries equal to 1 of B_{11} and the lower-diagonal entries equal to 1 of $\boldsymbol{B}_{22}$ such that one contains an $e \to f$ and the other an $f \to e$ relation. The other entries are equal to 0.

The above argument shows that in the same position, at most only one entry can be equal to 1 in the four quadrants of $\boldsymbol{B}$, which means that the corresponding adjacency entry of the line-graph of G is 1; the others are zeros. It is also obvious that the undirected $\{l, j\}$ and $\{j, i\}$ edges can be joined at j in four possible ways: $i \to j \to l$, $l \to j \leftarrow i$, $i \leftarrow j \to l$, and $l \to j \to i$. Only one of them results in the true relation $\{l \to j\} \to \{j \to i\}$. This proves the second statement. The sum of the four quadrants of $\boldsymbol{B}$ is of course symmetric, because of the first statement. □

As a consequence of Proposition 1,

$$(\boldsymbol{B}^{\ell})^* \boldsymbol{V} = \boldsymbol{V}\boldsymbol{B}^{\ell}$$

holds true for any natural number ℓ. Applying this for $\ell = 1$, the relation $\boldsymbol{B}^* = \boldsymbol{V}\boldsymbol{B}\boldsymbol{V}$ and

$$\boldsymbol{B}^*\check{\mathbf{x}} = \overbrace{\boldsymbol{B}\mathbf{x}}^{\smile}$$

is valid for any vector $\mathbf{x} \in \mathbb{R}^{2m}$. This phenomenon is called PT (parity-time) invariance in physics.

This implies the following: if $\mathbf{x}$ is a right eigenvector of $\boldsymbol{B}$ with a real eigenvalue μ, then $\check{\mathbf{x}}$ is a right eigenvector of $\boldsymbol{B}^*$ with the same eigenvalue. Consequently, if $\mathbf{x}$ is a right eigenvector of $\boldsymbol{B}$, then $\check{\mathbf{x}}$ is a left eigenvector of it (and vice versa), with the same real eigenvalue.

Another easy implication is that $(\boldsymbol{B}\boldsymbol{V})^* = \boldsymbol{V}\boldsymbol{B}^* = \boldsymbol{V}\boldsymbol{V}\boldsymbol{B}\boldsymbol{V} = \boldsymbol{B}\boldsymbol{V}$, so $\boldsymbol{B}\boldsymbol{V}$ and $\boldsymbol{V}\boldsymbol{B}$ are symmetric matrices that also follows by (3). Indeed,

$$\boldsymbol{B}\boldsymbol{V} = \begin{pmatrix} \boldsymbol{B}_{12} & \boldsymbol{B}_{11} \\ \boldsymbol{B}_{22} & \boldsymbol{B}_{21} \end{pmatrix}, \quad \boldsymbol{V}\boldsymbol{B} = \begin{pmatrix} \boldsymbol{B}_{21} & \boldsymbol{B}_{22} \\ \boldsymbol{B}_{11} & \boldsymbol{B}_{12} \end{pmatrix}.$$

Consequently, $\boldsymbol{B}\boldsymbol{V}$ is diagonalizable in an orthogonal basis, and so, the singular values of $\boldsymbol{B}$ are the absolute values of its real eigenvalues, whereas, the singular vector pairs are the eigenvectors and their swappings. Namely, the leading singular values are $d_i - 1$ for $i = 1, \dots, n$, and so, they are completely determined by the degree sequence of G. However, the eigenvalues of $\boldsymbol{B}$ are quite different.

- But what is the relation between the eigenvectors corresponding to the leading (real) eigenvalues of $\boldsymbol{B}$ and $\boldsymbol{K}$? To ease the discussion, two auxiliary matrices, defined in [36] will be used: the $2m \times n$ *end matrix* $\boldsymbol{End}$ has entries $end_{ei} = 1$ if i is the end-node of the (directed) edge e and 0, otherwise; the $2m \times n$ *start matrix* $\boldsymbol{Start}$ has entries $start_{ei} = 1$ if i is the start-node of the (directed) edge e and 0, otherwise. Then for any vector $\mathbf{u} \in \mathbb{R}^n$ and for any edge $e = \{i \to j\}$, the following holds:
$$(\boldsymbol{End}\,\mathbf{u})_e = u_j \quad \text{and} \quad (\boldsymbol{Start}\,\mathbf{u})_e = u_i.$$

Consequently, $\boldsymbol{End}\,\mathbf{u}$ is the $2m$-dimensional inflated version of the n-dimensional vector $\mathbf{u}$, where the coordinate u_j of $\mathbf{u}$ is repeated as many times, as many edges have end-node j; likewise, in the $2m$-dimensional inflated vector $\boldsymbol{Start}\,\mathbf{u}$, the coordinate u_i of $\mathbf{u}$ is repeated as many times, as many edges have start-node i. As each edge is considered in both possible directions, these numbers are the node-degrees d_j and d_i, respectively. Note that

$$\boldsymbol{End}^*\,\boldsymbol{End} = \boldsymbol{Start}^*\,\boldsymbol{Start} = \mathrm{diag}(d_1, \ldots, d_n) = \boldsymbol{D}. \tag{4}$$

Akin to [23], for any vector $\mathbf{x} \in \mathbb{R}^{2m}$ define

$$x_i^{out} := \sum_{j:\, j\sim i} x_{i\to j} \quad \text{and} \quad x_i^{in} := \sum_{j:\, j\sim i} x_{j\to i}, \quad i = 1, \ldots, n.$$

We put these coordinates into the n-dimensional (column) vectors $\mathbf{x}^{in}$ and $\mathbf{x}^{out}$. Trivially,

$$\mathbf{x}^{out} = \boldsymbol{Start}^*\mathbf{x}, \quad \text{and} \quad \mathbf{x}^{in} = \boldsymbol{End}^*\mathbf{x}, \quad i = 1, \ldots, n. \tag{5}$$

For them, we have the following relations (now more comfortably, we use $\boldsymbol{B}^*$ instead of $\boldsymbol{B}$):

$$\begin{aligned}
(\boldsymbol{B}^*\mathbf{x})_i^{out} &= \sum_{e:\, e_1 = i} (\boldsymbol{B}^*\mathbf{x})_e = \sum_{e:\, e_1=i} \sum_{f\to e,\, f\neq e^{-1}} x_f \\
&= \sum_{e:\, e_1=i} [\sum_{f\to e} x_f - x_{e^{-1}}] \\
&= \sum_{f:\, f_2=i} x_f \sum_{e:\, e_1=i} 1 - \sum_{e:\, e_1=i} x_{e^{-1}} \\
&= x_i^{in} d_i - \sum_{e:\, e_2^{-1}=i} x_{e^{-1}} = d_i x_i^{in} - x_i^{in} = (d_i - 1) x_i^{in}.
\end{aligned} \tag{6}$$

Also,

$$\begin{aligned}
(\boldsymbol{B}^*\mathbf{x})_i^{in} &= \sum_{e:\, e_2 = i} (\boldsymbol{B}^*\mathbf{x})_e = \sum_{e:\, e_2=i} \sum_{f\to e,\, f\neq e^{-1}} x_f \\
&= \sum_{j=1}^{n} a_{ji} \sum_{f:\, f_2=j,\, f_1\neq i} x_f \\
&= \sum_{j=1}^{n} a_{ji} \sum_{f:\, f_2=j} x_f - \sum_{j=1}^{n} a_{ji} x_{i\to j} \\
&= \sum_{j=1}^{n} a_{ij} x_j^{in} - \sum_{j:\, j\sim i} x_{i\to j} = (\mathbf{A}\mathbf{x}^{in})_i - x_i^{out},
\end{aligned} \tag{7}$$

where we used that the $(0-1)$ adjacency matrix $\boldsymbol{A}$ of the graph is symmetric with entries $a_{ij} = a_{ji} = \delta_{i\sim j}$.
Summarizing, if $\mathbf{x}$ is a right eigenvector of $\boldsymbol{B}^*$ with (real) eigenvalue μ, then

$$\mu \begin{pmatrix} \mathbf{x}^{out} \\ \mathbf{x}^{in} \end{pmatrix} = \begin{pmatrix} (\boldsymbol{B}^*\mathbf{x})^{out} \\ (\boldsymbol{B}^*\mathbf{x})^{in} \end{pmatrix} = \begin{pmatrix} \boldsymbol{O} & \boldsymbol{D}-\boldsymbol{I}_n \\ -\boldsymbol{I}_n & \boldsymbol{A} \end{pmatrix} \begin{pmatrix} \mathbf{x}^{out} \\ \mathbf{x}^{in} \end{pmatrix} \tag{8}$$

and

$$\begin{pmatrix} (\boldsymbol{B}^*\mathbf{x})^{out} \\ (\boldsymbol{B}^*\mathbf{x})^{in} \end{pmatrix} = \begin{pmatrix} (\boldsymbol{D}-\boldsymbol{I}_n)\mathbf{x}^{in} \\ \boldsymbol{A}\mathbf{x}^{in} - \mathbf{x}^{out} \end{pmatrix} = \boldsymbol{K} \begin{pmatrix} \mathbf{x}^{out} \\ \mathbf{x}^{in} \end{pmatrix}.$$

In particular, if $\mathbf{x}$ is a right eigenvector of $\boldsymbol{B}^*$ with a real eigenvalue $\mu \neq 0$, then the $2n$-dimensional vector comprised of parts $\mathbf{x}^{out}$ and $\mathbf{x}^{in}$ is a right eigenvector of $\boldsymbol{K}$ with the same eigenvalue. Indeed,

$$\mu \begin{pmatrix} \mathbf{x}^{out} \\ \mathbf{x}^{in} \end{pmatrix} = \begin{pmatrix} (\mu\mathbf{x})^{out} \\ (\mu\mathbf{x})^{in} \end{pmatrix} = \begin{pmatrix} (\boldsymbol{B}^*\mathbf{x})^{out} \\ (\boldsymbol{B}^*\mathbf{x})^{in} \end{pmatrix} = \boldsymbol{K} \begin{pmatrix} \mathbf{x}^{out} \\ \mathbf{x}^{in} \end{pmatrix}.$$

According to the previous remarks, the vector $\mathbf{x}$ is a left eigenvector, and $\check{\mathbf{x}}$ is a right eigenvector of $\boldsymbol{B}$ with the same eigenvalue. To both of them the two segments, $\mathbf{x}^{out}$ and $\mathbf{x}^{in}$ of the right eigenvector of $\boldsymbol{K}$ are responsible. In view of the relation $\mathbf{x}^{out} = \frac{1}{\mu}(\boldsymbol{D}-\boldsymbol{I}_n)\mathbf{x}^{in}$, obtained by (8), it suffices to consider only $\mathbf{x}^{in} \in \mathbb{R}^n$ for further clustering purposes, see Sect. 4.

2.2 Random Graphs Obtained by Bond-Percolation

In the simple graph scenario, in [33], it is proved that the so-called message passing (in other words, belief propagation) system of equations (briefly, BP) has a non-trivial solution if and only if, for the edge retention probability β, $\beta > \frac{1}{\rho(\boldsymbol{B})}$ holds, where $\rho(\boldsymbol{B})$ is the spectral radius of the non-backtracking matrix $\boldsymbol{B}$ of a given sparse social network. So $\frac{1}{\rho(\boldsymbol{B})}$ is the *bond-percolation threshold* for the giant component to appear in this simple case; which in view of (2) is greater than $\frac{1}{\rho(\boldsymbol{A})}$ of [12] in the dense case.

Now our purpose is to characterize the non-backtracking spectrum of a random graph G_n on n nodes coming from the SBM_k model, and via perturbations, try to use these properties to find out the number of clusters and the clusters themselves. The parameters of the sparse SBM_k model are as follows.

The $k \times k$ *probability matrix* $\boldsymbol{P}$ of the percolated sparse random graph $G_n \in SBM_k$ has entries

$$p_{ab} = \frac{c_{ab}}{n},$$

where the $k \times k$ symmetric *affinity matrix* $\boldsymbol{C} = (c_{ab})$ stays constant as $n \to \infty$. An edge between $i < j$ comes into existence, independently of the others, with probability p_{ab} if $i \in V_a$ and $j \in V_b$, where $(V_1, \dots, V_k)$ is a partition of the node-set V into k disjoint clusters. This will produce the upper-diagonal part of the $n \times n$ random adjacency matrix $\boldsymbol{A}$, and $a_{ji} := a_{ij}$. It can be extended to the $i = j$ case when self-loops are allowed, or else, the diagonal entries of the adjacency matrix are zeros.

Let $\bar{\boldsymbol{A}}$ denote the $n \times n$ inflated matrix of the $k \times k$ matrix $\boldsymbol{P}$: $\bar{a}_{ij} = p_{ab}$ if $i \in V_a$ and $b \in V_b$. When loops are allowed, then $\mathbb{E}(a_{ij}) = \bar{a}_{ij}$ for all $1 \le i, j \le n$. In the loopless case, the expected adjacency matrix $\mathbb{E}\boldsymbol{A}$ differs from $\bar{\boldsymbol{A}}$ with respect to the the main diagonal, but the diagonal entries are negligible as will be shown in Sect. 4.

Specifically, sometimes $c_{ab} = c_{in}$ is the within-cluster $(a = b)$ and $c_{ab} = c_{out}$ is the between-cluster $(a \ne b)$ affinity. In [32], the network is called *assortative* if $c_{in} > c_{out}$, and *disassortative* if $c_{in} < c_{out}$. Of course, remarkable difference is needed between the two, to recognize the clusters. The cluster sizes are $n_1, \dots, n_k$ $(\sum_{i=1}^{k} n_i = n)$, so the $k \times k$ diagonal matrix $\boldsymbol{R} := \mathrm{diag}(r_1, \dots, r_k)$, where $r_a = \frac{n_a}{n}$ is the relative size of cluster a $(a = 1, \dots, k)$, is also a model parameter $(\sum_{a=1}^{k} r_a = 1)$. The model SBM_k is called *symmetric* if $r_1 = \dots = r_k = \frac{1}{k}$ and all diagonal entries of the affinity matrix are equal to c_{in}, whereas the off-diagonal ones to c_{out}.

The average degree of a real world graph on m edges and n nodes is $\frac{2m}{n}$. The expected average degree of the random graph G_n generated from the SBM_k model is

$$c = \frac{1}{n} \sum_{i=1}^{n} \sum_{j=1}^{n} \bar{a}_{ij} = \frac{1}{n} \sum_{a=1}^{k} \sum_{b=1}^{k} n_a n_b p_{ab} = \frac{1}{n^2} \sum_{a=1}^{k} \sum_{b=1}^{k} n_a n_b c_{ab} = \sum_{a=1}^{k} r_a c_a, \tag{9}$$

where $c_a = \sum_{b=1}^{k} r_b c_{ab}$ is the average degree of cluster a. It is valid only if self-loops are allowed. Otherwise, c_a and c should be decreased with a term of order $\frac{1}{n}$, but it will not make too much difference in the subsequent calculations.

In [13], the case when $c_a = c$ for all a is considered. In this case $\frac{1}{c}\bar{\boldsymbol{A}}$ is a stochastic matrix, and so, the spectral radius of $\bar{\boldsymbol{A}}$ is c.

The symmetric case is a further special case of this, when

$$c = \frac{c_{in} + (k-1)c_{out}}{k}.$$

In this case, the separation of the clusters only depends on the c_{in}, c_{out} relation. If c_{in} is "close" to c_{out}, then the groups cannot be distinguished. The detectability threshold (Kesten–Stigum threshold) in the symmetric case is

$$|c_{in} - c_{out}| > k\sqrt{c}, \tag{10}$$

see [13, 17, 28].

In the symmetric $k = 2$ case, the authors of [29] prove that under condition (10), their algorithm gives a clustering that has a correlation with the true one separated from zero; furthermore, this correlation tends to 1 as the ratio of the left and right hand sides of (10) increases. They also write that the $k \geq 2$ and $c_1 = \cdots = c_k$ case is the hardest one, as otherwise the clusters could be distinguished by sorting the node-degrees.

Note that Theorem 4 of [13] allows a small fluctuation for n_a's, and so, for r_a's and c_a's too. We make use of this when compare the structural eigenvalues of the non-backtracking matrix to the eigenvalues of the so-called transition matrix assessed during the BP process. Observe that these assumptions are natural in social networks: the relative sizes of communities are balanced (tend to a limit as $n \to \infty$); further, the average degrees of nodes of the clusters do not differ much as $n \to \infty$, which means that members of different communities have on average closely the same number of bonds to members of their own or other communities.

Stephan and Massoulié [36] investigate more general random graph models, also in the edge-weighted case. In the special unweighted situation we shall use the following (informal) statement of them for the SBM_k model, where the matrix $\bar{\boldsymbol{A}}$ is not only the diagonal-corrected expected adjacency matrix, but it is approximately the variance matrix of the random adjacency matrix $\boldsymbol{A}$.

Proposition 2 (Based on Theorem 1 of [36]) *Assume that* $k = \text{rank}\, \bar{\boldsymbol{A}} = n^{o(1)}$, *the graph is sparse enough, and the eigenvectors corresponding to the non-zero eigenvalues of the matrix* $\bar{\boldsymbol{A}}$ *are sufficiently delocalized. Let* k_0 *denote the number of eigenvalues of* $\bar{\boldsymbol{A}}$ *whose absolute value is larger than* $\sqrt{\rho}$, *where* ρ *is the spectral radius of* $\bar{\boldsymbol{A}}$: *these are* $\nu_1 \geq \cdots \geq \nu_{k_0}$ *with corresponding eigenvectors* $\mathbf{u}_1, \dots, \mathbf{u}_{k_0}$ *(they form an orthonormal system as* $\bar{\boldsymbol{A}}$ *is a real symmetric matrix). Then for* $i \leq k_0 \leq k$, *the* i*th largest eigenvalue* μ_i *of* $\boldsymbol{B}$ *is asymptotically (as* $n \to \infty$*) equal to* ν_i *and all the other eigenvalues of* $\boldsymbol{B}$ *are constrained to the circle (in the complex plane) of center* 0 *and radius* $\sqrt{\rho}$. *Further, if* $i \leq k_0$ *is such that* ν_i *is a sufficiently isolated eigenvalue of* $\bar{\boldsymbol{A}}$, *then the standardized eigenvector of* $\boldsymbol{B}$ *corresponding to* μ_i *has inner product close to 1 with the standardized inflated version of* $\mathbf{u}_i$, *namely, with* $\frac{\boldsymbol{End}\,\mathbf{u}_i}{\|\boldsymbol{End}\,\mathbf{u}_i\|}$.

Note that if our random graph comes from the SBM_k model with $k \times k$ parameter matrices $\boldsymbol{R}$ and $\boldsymbol{C}$, then $\bar{\boldsymbol{A}}$ has rank at most k, and its nonzero eigenvalues are identical to the real eigenvalues of the matrix $\boldsymbol{RC}$ (same as those of $\boldsymbol{R}^{\frac{1}{2}}\boldsymbol{C}\boldsymbol{R}^{\frac{1}{2}}$), and the eigenvectors of it are inflated versions of those of $\boldsymbol{R}^{\frac{1}{2}}\boldsymbol{C}\boldsymbol{R}^{\frac{1}{2}}$; consequently, they are step-vectors on k different steps and so, the k-means algorithm is applicable to the k_0-dimensional representatives of the nodes constructed with the first k_0 shrunken, normalized eigenvectors of $\boldsymbol{B}$ corresponding to its leading eigenvalues $\mu_1, \dots, \mu_{k_0}$. The point is that we do not need the model parameters and the spectral decomposition of $\bar{\boldsymbol{A}}$ itself for the k-means clustering. This issue together with inflation–deflation techniques will be discussed in details in Sect. 4. Also note that in the SBM_k model, when $c_a = c$ $(a = 1, \dots, k)$, then $\rho = c$ and $k_0 = k$.

3 Belief Propagation (BP) in the General SBM

To derive the percolation threshold analytically, we use the BP or cavity method of [28] in the generic case. In [28], it was developed for the symmetric case. Akin to the BP, called message passing in [33], let us introduce the following notation: for $a = 1, \dots, k$,

$$\psi_i^a \propto \mathbb{P}(i \text{ is in the cluster } a),$$

given the observed graph on n nodes, coming from the SBM_k model, where ψ_i^a for $a = 1, \dots, k$ defines the marginal membership (state) distribution of node i.

We assume that our neighbors are independent of each other, when conditioned on our own state. Equivalently, we assume that our neighbors are correlated only through us. This can be modeled by having each node j send a message $\psi_{j\to i}$ to i, which is an estimate of j's marginal (in other words, state or membership) if i were not there (more precisely, if we did not know whether or not there is an edge between i and j). Therefore, the conditional probability

$$\psi_{j\to i}^a := \mathbb{P}(j \text{ is in cluster } a \text{ when } i \text{ is not present})$$

(or when we do not know the membership of i) can be computed through the neighbors of j that are different from i (in the realization of the random graph coming from the SBM_k model) as follows:

$$\psi_{j\to i}^a = C_a^{ij} r_a \prod_{l\sim j,\, l\neq i} \sum_{b=1}^{k} \psi_{l\to j}^b \, p_{ab}, \quad a = 1, \dots, k, \tag{11}$$

where C_a^{ij} is a normalizing factor. Here $\psi_{l\to j}^b$ is the conditional probability that node l (which is not i and to which j is connected) belongs to cluster b even in the absence of j, and the summation is an application of the rule of completely disjoint events with probabilities p_{ab} for $b = 1, \dots, k$.

This so-called BP or *message-passing equation* is a system of $2mk$ non-linear equations with the same number of unknowns. It can be solved by initializing messages randomly, then repeatedly updating them via (11). This procedure usually converges quickly and the resulting fixed point gives a good estimate of the marginals:

$$\psi_i^a \propto r_a \prod_{j\sim i} \sum_{b=1}^{k} \psi_{j\to i}^b \, p_{ab},$$

where the constant of proportionality is chosen according to $\sum_{a=1}^{k} \psi_i^a = 1$. However, the system of equations contains the model parameters, so it can be solved only if the model parameters are known.

In [28], the symmetric case is treated, when BP has a trivial fixed point $\psi^a_{j\to i} = \frac{1}{k}$, for $a = 1, \dots, k$. If it gets stuck there, then BP does no better than chance. It happens when this trivial fixed point of this discrete dynamical system is asymptotically stable.

In the generic case, we want to have an unstable fixed point via linearization:

$$\psi^a_{j\to i} := r_a + \varepsilon^a_{j\to i}.$$

If we substitute this in (11) and expand to first order in $\boldsymbol{\varepsilon}$ (vector of the $2mk$ coordinates $\varepsilon^a_{j\to i}$'s), the updated system of equations is

$$\begin{aligned}
\varepsilon^a_{j\to i} = \psi^a_{j\to i} - r_a &= r_a \left\{ C^{ij}_a \prod_{l\sim j,\, l\neq i} \left[\sum_{b=1}^{k} \psi^b_{l\to j}\, p_{ab} \right] - 1 \right\} \\
&= r_a \left\{ C^{ij}_a \prod_{l\sim j,\, l\neq i} \left[\sum_{b=1}^{k} (r_b + \varepsilon^b_{l\to j})\, p_{ab} \right] - 1 \right\} \\
&= r_a \left\{ C^{ij}_a \prod_{l\sim j,\, l\neq i} \left[\sum_{b=1}^{k} r_b p_{ab} + \sum_{b=1}^{k} \varepsilon^b_{l\to j}\, p_{ab} \right] - 1 \right\} \\
&= r_a \left\{ C^{ij}_a \prod_{l\sim j,\, l\neq i} \left[\frac{c_a}{n} + \sum_{b=1}^{k} \varepsilon^b_{l\to j} \frac{c_{ab}}{n} \right] - 1 \right\} \\
&= r_a \left\{ C^{ij}_a (\frac{1}{n})^{s_j - 1} \left[\sum_{b=1}^{k} \sum_{l\sim j,\, l\neq i} \varepsilon^b_{l\to j}\, c_{ab}\, c_a^{s_j-2} + c_a^{s_j-1} \right] - 1 \right\} + O(\boldsymbol{\varepsilon}^2),
\end{aligned} \tag{12}$$

where s_j denotes the number of neighbors of j and $s_j - 1$ is the number of its neighbors that are different from i (this number is frequently 0 or 1, as we have a sparse graph). If $s_j < 2$ happens, then the corresponding entry of the non-backtracking matrix is 0. To specify the normalizing factor C^{ij}_a, we substitute zeros for ε's that provide the trivial solution. This approximately yields

$$C^{ij}_a \left(\frac{1}{n}\right)^{s_j-1} c_a^{s_j-1} - 1 = 0,$$

so

$$C^{ij}_a = \left(\frac{n}{c_a}\right)^{s_j-1}.$$

Substituting this into the last equation of (12), we get

$$\begin{aligned}\varepsilon^a_{j\to i} &= r_a \left\{ \left(\frac{n}{c_a}\right)^{s_j-1} \left(\frac{1}{n}\right)^{s_j-1} c_a^{s_j-2} \left[\sum_{b=1}^{k} \sum_{l\sim j, l\neq i} \varepsilon^b_{l\to j}\, c_{ab} + c_a \right] - 1 \right\} + O(\boldsymbol{\varepsilon}^2) \\ &= r_a \left\{ \frac{1}{c_a} \left[\sum_{b=1}^{k} \sum_{l\sim j, l\neq i} \varepsilon^b_{l\to j}\, c_{ab} + c_a \right] - 1 \right\} + O(\boldsymbol{\varepsilon}^2).\end{aligned}$$

The linear approximation of the above system of equations is

$$\boldsymbol{\varepsilon} = \boldsymbol{M}\boldsymbol{\varepsilon}, \tag{13}$$

where $\boldsymbol{\varepsilon}$ is a $2mk$-dimensional vector and the $(j \to i,\ a)$, $(l \to s,\ b)$ entry of $\boldsymbol{M}$ is $\partial\varepsilon^a_{j\to i}/\partial\varepsilon^b_{l\to s}$. It is not zero only for $l \to s$, $j \to i$ pairs with $s = j$ and $l \neq i$, i.e., when the corresponding entry of $\boldsymbol{B}$ is 1. If it is not zero, then the partial derivative of $\varepsilon^a_{j\to i}$ with respect to $\varepsilon^b_{l\to j}$ is $\frac{r_a}{c_a} c_{ab}$ (the second order terms are disregarded). Therefore, it can easily be seen that $\boldsymbol{M}$ is a Kronecker-product:

$$\boldsymbol{M} = \boldsymbol{B} \otimes \boldsymbol{T},$$

i.e., $\boldsymbol{T}$ is substituted for each entry 1 of the non-backtracking matrix $\boldsymbol{B}$, where $\boldsymbol{T} = \boldsymbol{GRC}$ is the *transmission matrix* with $\boldsymbol{G} = \mathrm{diag}(\frac{1}{c_1}, \dots, \frac{1}{c_k})$. Note that $\boldsymbol{T}$ will be a stochastic matrix (with largest eigenvalue 1) only if $r_1 = \dots = r_k$.

For the solution of (13), the fixed point of the linear dynamical system

$$\boldsymbol{\varepsilon}^{(t+1)} = \boldsymbol{M}\boldsymbol{\varepsilon}^{(t)}, \quad t = 1, 2, \dots$$

is looked for. We can find a fixed point other than the trivial $\mathbf{0}$ if $\mathbf{0}$ is an unstable solution, for which a sufficient condition is that the spectral radius of the matrix $\boldsymbol{M}$ is greater than 1. In this way, we have proved the following.

Theorem 1 *Assume that the graph G comes from the SBM_k model with positive integer k and $k \times k$ parameter matrices $\boldsymbol{R} = \mathrm{diag}(r_1, \dots, r_k)$ (cluster proportions) and $\boldsymbol{C}$ (symmetric affinity matrix). The approximating linear dynamical system behind the BP system (12) is $\boldsymbol{\varepsilon}^{(t+1)} = \boldsymbol{M}\boldsymbol{\varepsilon}^{(t)}$, where $\boldsymbol{\varepsilon}$ is a $2mk$-dimensional vector and the $2mk \times 2mk$ matrix of the system is $\boldsymbol{M} = \boldsymbol{B} \otimes \boldsymbol{T}$. Here $\boldsymbol{B}$ is the non-backtracking matrix of the graph G and $\boldsymbol{T} = \boldsymbol{GRC}$ is the transmission matrix with $\boldsymbol{G} = \mathrm{diag}(\frac{1}{c_1}, \dots, \frac{1}{c_k})$, where $c_a = \sum_{b=1}^{k} r_b c_{ab}$ is the average degree of cluster a, for $a = 1, \dots, k$. If the trivial fixed point $\mathbf{0}$ is an unstable fixed point of the above system, then there is at least one eigenvalue of the matrix $\boldsymbol{B} \otimes \boldsymbol{T}$ (product of eigenvalues of $\boldsymbol{B}$ and $\boldsymbol{T}$) that is greater than 1 (in absolute value).*

Note that in [28] only the symmetric case is treated. In [13] the special case $c_1 = \dots = c_k = c$ is considered when the matrix $\boldsymbol{T}$ becomes $\boldsymbol{T} = \frac{1}{c}\boldsymbol{RC}$; then the leading eigenvalues of $\boldsymbol{B}$ and $\boldsymbol{RC}$ are w.h.p. "close" to each other. Also, the largest eigenvalue of $\boldsymbol{RC}$ is c, which is trivially the case if $k = 1$ and we have the Erdős–Rényi random graph. More precisely, the authors of [13] allow "small" fluctuations of the cluster membership proportions that causes the same order of fluctuations in the average degrees of the clusters. For the membership proportion of cluster a, denoted by $r_a^{(n)}$, the relation $\lim_{n\to\infty} r_a^{(n)} = r_a$ is assumed with $\sum_{a=1}^k r_a = 1$. Specifically, in [13], the assumption

$$\max_{a\in\{1,\dots,k\}} |r_a^{(n)} - r_a| = O(n^{-\gamma}) \tag{14}$$

is made with some $\gamma \in (0, 1]$, where n denotes the number of nodes. This assumption implies that in the $c_1 = \cdots = c_k = c$ case, $\max_{a\in\{1,\dots,k\}} |c_a^{(n)} - c| = O(n^{-\gamma})$.

Theorem 4 of [13] states that if

$$\max_a c_a^{(n)} = c + O(n^{-\gamma}), \tag{15}$$

with some $\gamma \in (0, 1]$, and the relative proportions of the clusters converge, then w.h.p.

$$\mu_i = \nu_i + o(1) \quad (i = 1, \dots k_0) \quad \text{and} \quad \mu_i < \sqrt{c} + o(1) \quad (i > k_0),$$

where μ_i's and ν_i's ($i = 1, \dots, k_0$) are the structural eigenvalues of $\boldsymbol{B}$ and $\boldsymbol{RC}$, respectively, whereas $k_0 \leq k$ is the positive integer for which $\nu_i^2 \geq \nu_1$ ($i = 1, \dots k_0$) and $\nu_{k_0+1}^2 < \nu_1$ holds. Therefore, in the SBM_1 (Erdős–Rényi) model, $\mu_1 = c + o(1)$ and $\mu_2 \leq \sqrt{c} + o(1)$.

However, in the generic case too, simulations show that sometimes the leading eigenvalues of $\boldsymbol{B}$ and $\boldsymbol{RC}$ are closer to each other than those of $\boldsymbol{B}$ and $c\boldsymbol{T}$, but always closer than those of $\boldsymbol{A}$ and any of them. The following considerations can be made. If (15) holds, then

$$\min_a \frac{c}{c_a^{(n)}} = \frac{c}{\max_a c_a^{(n)}} = 1 + O(n^{-\gamma}). \tag{16}$$

By the rules of the Kronecker-products,

$$\boldsymbol{B} \otimes c\boldsymbol{T} = (\boldsymbol{I}_{2m}\boldsymbol{B}) \otimes (c\boldsymbol{GRC}) = (\boldsymbol{I}_{2m} \otimes c\boldsymbol{G})(\boldsymbol{B} \otimes \boldsymbol{RC}).$$

But

$$\boldsymbol{I}_{2m} \otimes c\boldsymbol{G} = \mathrm{diag}(\frac{c}{c_1}, \dots . \frac{c}{c_k}, \dots, \frac{c}{c_1}, \dots . \frac{c}{c_k})$$

repeated $2m$ times. By minimax theorems, the eigenvalues of $(\boldsymbol{I}_{2m} \otimes c\boldsymbol{G})(\boldsymbol{B} \otimes \boldsymbol{RC})$ are within factors u and v of those of $\boldsymbol{B} \otimes \boldsymbol{RC}$, where

$$u = \min_a \frac{c}{c_a^{(n)}} \quad \text{and} \quad v = \max_a \frac{c}{c_a^{(n)}}.$$

Here u and v depend on n, but we do not denote this dependence. Also, for fixed "large" n, the eigenvalues of $c\boldsymbol{GRC}$ are within a factor u and v of those of $\boldsymbol{RC}$.

Then the condition $\lambda(\boldsymbol{B})\,\lambda(\boldsymbol{T}) > 1$, i.e.,

$$\lambda(\boldsymbol{B})\,\lambda(c\boldsymbol{T}) = \lambda(\boldsymbol{B})\,\lambda(c\boldsymbol{GRC}) > c$$

holds if

$$\lambda(\boldsymbol{B})\,(u\lambda(\boldsymbol{B})) > c.$$

This implies that

$$\lambda(\boldsymbol{B}) \geq \frac{\sqrt{c}}{\sqrt{u}} \quad \text{and} \quad \lambda(c\boldsymbol{GRC}) \geq u\lambda(\boldsymbol{B}) \geq \sqrt{u}\sqrt{c}$$

should hold, where $u \leq 1$.

Consequently, real eigenvalues of $\boldsymbol{B}$ that are larger than $\sqrt{c}$ should be taken into account. However, $\boldsymbol{R}$ and $\boldsymbol{C}$ contain the unknown model parameters, $\boldsymbol{G}$ depends on them, and there is only $\boldsymbol{B}$ at our disposal from a sample. But the model itself is adapted to the data, i.e., the underlying graph. In [33], the author makes the following crucial observation: "The calculation, we have described is based on a fit of the stochastic block model to a network that was itself generated from the same model."

So we should consider the real eigenvalues of $\boldsymbol{B}$ that are greater than $\sqrt{c}$. Their number will be denoted by k_0. Also, the non-zero eigenvalues of the expected adjacency matrix $\bar{\boldsymbol{A}}$ of the next section are the same as those of $\boldsymbol{RC}$, so they are in the neighborhood of the leading eigenvalues of $\boldsymbol{B}$ within a factor between u and v. On the contrary, their closeness to $\boldsymbol{A}$ is up to an additive constant (less than $\sqrt{c}$) as will be shown in Sect. 4.

Note that when the average degrees c_a's of the clusters differ significantly, then the leading eigenvalues of $\boldsymbol{B}$ are not closely aligned with those of $\boldsymbol{RC}$. However, this case is relatively simple, as then the clusters could be formed by sorting the node degrees.

4 Inflation–Deflation

Here we use the inflation–deflation technique of [8].

With the notation of the SBM_k model: $\boldsymbol{C} = (c_{ab})$ is the $k \times k$ symmetric affinity matrix (assume that its rank is k) and $\boldsymbol{P} = \frac{1}{n}\boldsymbol{C}$ is the corresponding probability matrix of the random graph $G_n \in SBM_k$. If self-loops are allowed, then $\bar{\boldsymbol{A}}$ is the expected adjacency matrix of G_n and it contains p_{ab}'s in the (a, b)-th block. Therefore, it is the inflated matrix of $\boldsymbol{P}$ with blow-up sizes $n_1, \dots, n_k$ ($\sum_{a=1}^{k} n_a = n$). The blow-up ratios are $r_a = \frac{n_a}{n}$ for $a = 1, \dots, k$. The expected average degree of G_n is

$$c = \frac{1}{n}\sum_{i=1}^{n}\sum_{i=1}^{n}\bar{a}_{ij} = \frac{1}{n}\sum_{a=1}^{k}\sum_{b=1}^{k} n_a n_b c_{ab} = \sum_{a=1}^{k} r_a c_a,$$

as in (9). Note that c depends only on the model parameters and it is of constant order. If the diagonal entries of $\boldsymbol{A}$ are zeros, then $\bar{\boldsymbol{A}}$ differs from the expected adjacency matrix by $\mathrm{diag}(p_{11}, \ldots, p_{11}, \ldots, p_{kk}, \ldots, p_{kk})$ with n_a entries equal to each other in the a-th block. However, by the Weyl's perturbation theorem, the k leading eigenvalues of $\bar{\boldsymbol{A}}$ and $\mathbb{E}(\boldsymbol{A})$ differ only with a term of order $\frac{1}{n}$.

Proposition 3 *The matrix* $\bar{\mathbf{A}}$ *has rank k and its non-zero eigenvalues (*ν*'s) with unit norm eigenvectors (*$\mathbf{u}$*'s) satisfy*

$$\bar{A}\mathbf{u} = \nu\mathbf{u}, \tag{17}$$

where $\mathbf{u}$ *is the inflated vector (step-vector) of* $\tilde{\mathbf{u}} = (u(1), \ldots, u(k))^*$ *with block-sizes* $n_1, \ldots, n_k$*. With the notation* $\boldsymbol{R} = \frac{1}{n}\mathrm{diag}(n_1, \ldots, n_k) = \mathrm{diag}(r_1, \ldots, r_k)$*, the deflated equation of* (17) *is equivalent to*

$$\boldsymbol{R}^{\frac{1}{2}}\boldsymbol{C}\boldsymbol{R}^{\frac{1}{2}}\mathbf{v} = \nu\mathbf{v}, \tag{18}$$

where $\mathbf{v} = \sqrt{n}\boldsymbol{R}^{\frac{1}{2}}\tilde{\mathbf{u}}$*. Further, if* $\mathbf{u}_1, \ldots \mathbf{u}_k$ *is the set of orthonormal eigenvectors, corresponding to the eigenvalues* $\nu_1, \ldots, \nu_k$ *of* $\bar{\boldsymbol{A}}$*, then* $\mathbf{v}_i = \sqrt{n}\boldsymbol{R}^{\frac{1}{2}}\tilde{\mathbf{u}}_i$ $(i = 1, \ldots, k)$ *is the set of orthonormal eigenvectors of* $\boldsymbol{R}^{\frac{1}{2}}\boldsymbol{C}\boldsymbol{R}^{\frac{1}{2}}$ *with the same eigenvalues. Also,* $\boldsymbol{R}^{\frac{1}{2}}\mathbf{v}_i = \sqrt{n}\boldsymbol{R}\tilde{\mathbf{u}}_i$ *is a right eigenvector of* $\boldsymbol{RC}$ *and* $\boldsymbol{R}^{-\frac{1}{2}}\mathbf{v}_i = \sqrt{n}\tilde{\mathbf{u}}_i$ *is a left eigenvector of* $\boldsymbol{RC}$ *with eigenvalue* ν_i*, for* $i = 1, \ldots, k$*.*

Proof The matrix $\bar{\boldsymbol{A}}$ has k different rows (columns), so it is of rank k and its non-zero eigenvalues (ν's) have corresponding unit norm eigenvectors ($\mathbf{u}$'s) in (17), which are trivially step-vectors, i.e., $\mathbf{u}$ is the inflated vector of, say, $\tilde{\mathbf{u}} = (u(1), \ldots, u(k))^*$ with block-sizes $n_1, \ldots, n_k$. The deflated equation of (17) is equivalent to

$$\boldsymbol{R}^{\frac{1}{2}}\boldsymbol{P}\boldsymbol{R}^{\frac{1}{2}}\mathbf{v} = \frac{\nu}{n}\mathbf{v},$$

that in turn is equivalent to (18), where the eigenvector $\mathbf{v} = \sqrt{n}\boldsymbol{R}^{\frac{1}{2}}\tilde{\mathbf{u}}$ also has unit norm.

If $\mathbf{u}_1, \ldots \mathbf{u}_k$ is the set of orthonormal eigenvectors of $\bar{A}$, then $\mathbf{v}_i = \sqrt{n}\boldsymbol{R}^{\frac{1}{2}}\tilde{\mathbf{u}}_i$ $(i = 1, \ldots, k)$ is the set of orthonormal eigenvectors of $\boldsymbol{R}^{\frac{1}{2}}\boldsymbol{C}\boldsymbol{R}^{\frac{1}{2}}$ with the same eigenvalues. Also,

$$\boldsymbol{RC}(\boldsymbol{R}^{\frac{1}{2}}\mathbf{v}_i) = \boldsymbol{R}^{\frac{1}{2}}(\boldsymbol{R}^{\frac{1}{2}}\boldsymbol{C}\boldsymbol{R}^{\frac{1}{2}})\mathbf{v}_i = \boldsymbol{R}^{\frac{1}{2}}\nu_i\mathbf{v}_i = \nu_i\boldsymbol{R}^{\frac{1}{2}}\mathbf{v}_i,$$

so $\boldsymbol{R}^{\frac{1}{2}}\mathbf{v}_i = \sqrt{n}\boldsymbol{R}\tilde{\mathbf{u}}_i$ is a right eigenvector of $\boldsymbol{RC}$ for $i = 1, \ldots, k$. Further,

$$(\boldsymbol{R}^{-\frac{1}{2}}\mathbf{v}_i)^*\boldsymbol{RC} = \mathbf{v}_i^*(\boldsymbol{R}^{\frac{1}{2}}\boldsymbol{C}\boldsymbol{R}^{\frac{1}{2}})\boldsymbol{R}^{-\frac{1}{2}} = \nu_i\mathbf{v}_i^*\boldsymbol{R}^{-\frac{1}{2}} = \nu_i(\boldsymbol{R}^{-\frac{1}{2}}\mathbf{v}_i)^*,$$

so $\boldsymbol{R}^{-\frac{1}{2}}\mathbf{v}_i = \sqrt{n}\tilde{\mathbf{u}}_i$ is a left eigenvector of $\boldsymbol{RC}$ with the same eigenvalue ν_i, for $i = 1, \dots, k$.

Observe that $(\boldsymbol{RC})^* = \boldsymbol{CR}$, so the non-symmetric matrices $\boldsymbol{RC}$ and $\boldsymbol{CR}$ have the same set of eigenvalues, which are also real, as they are identical to the eigenvalues of the symmetric matrix $\boldsymbol{R}^{\frac{1}{2}}\boldsymbol{C}\boldsymbol{R}^{\frac{1}{2}}$. Further, right eigenvectors of $\boldsymbol{RC}$ are left eigenvectors of $\boldsymbol{CR}$, and vice versa. This finishes the proof. □

Remark 1 We remark the following.

- Let the real eigenvalues of $\bar{\boldsymbol{A}}$ be ordered in non-increasing order. Since $\bar{\boldsymbol{A}}$ is an irreducible matrix of positive entries, by the Frobenius theorem, ν_1 is a single positive eigenvalue and $|\nu_i| < \nu_1$ for $i = 2, \dots, k$. Further, the eigenvector $\mathbf{u}_1$ or $\bar{\boldsymbol{A}}$, corresponding to ν_1, has all positive coordinates.
- In particular, if $c_1 = \cdots = c_k = c$, then $\frac{1}{c}\boldsymbol{CR}$ is a stochastic matrix with largest eigenvalue 1 and corresponding right eigenvector $\mathbf{1} \in \mathbb{R}^k$. Consequently, the largest eigenvalue of $\frac{1}{c}\boldsymbol{RC}$ is also 1 with left eigenvector $\mathbf{1} \in \mathbb{R}^k$. Therefore, $\nu_1 = c$ and $\tilde{\mathbf{u}}_1 = \frac{1}{\sqrt{n}}\mathbf{1}$ also has equal coordinates. So $\mathbf{u}_1$ is such too, and $\mathbf{u}_1 = \frac{1}{\sqrt{n}}\mathbf{1} \in \mathbb{R}^n$ is the unit-norm eigenvector of $\bar{\boldsymbol{A}}$, corresponding to $\nu_1 = c$ in this special case.
- If we further assume the c_{in} versus c_{out} scenario, then in the symmetric case, $\nu_2 = \cdots = \nu_k = c\lambda$ with $0 < |\lambda| < 1$ (see Sect. 6).
- Note that under the condition (14) of Sect. 3, the eigenvalues of $\boldsymbol{RC}$ and $\bar{\boldsymbol{A}}$ may differ within a factor approaching 1 as $n \to \infty$.
- In the unpercolated situation, the edge-probabilities are kept constant, $\boldsymbol{P} = \boldsymbol{C}$ and the $n \times n$ $\bar{\boldsymbol{A}}$ is the blown-up matrix of the $k \times k$ probability matrix $\boldsymbol{P}$. So, its eigenvalues are n times the eigenvalues of $\boldsymbol{R}^{-1/2}\boldsymbol{C}\boldsymbol{R}^{-1/2}$. The Wigner-noise added has spectral norm $O(\sqrt{n})$(see [8]), therefore it cannot suppress the structure, unlike in the sparse SBM_k model, discussed in the following.

The (random) adjacency matrix $\boldsymbol{A}$ of (the random graph) G_n coming from the sparse SBM_k model is $\boldsymbol{A} = \bar{\boldsymbol{A}} + \boldsymbol{W}$, where $\boldsymbol{W}$ is an appropriate (random) error matrix and all the matrices are $n \times n$ symmetric (though we do not use an index n for this fact). We can achieve that the matrix $\boldsymbol{A}$ contains 1's in the (a, b)-th block with probability p_{ab}, and 0's otherwise. Indeed, for indices $1 \le a < b \le k$ and $i \in V_a$, $j \in V_b$ let

$$w_{ji} = w_{ij} := \begin{cases} 1 - p_{ab} & \text{with probability} \quad p_{ab} \\ -p_{ab} & \text{with probability} \quad 1 - p_{ab} \end{cases} \tag{19}$$

be independent random variables; further, for $a = 1, \dots, k$ and $i, j \in V_a$ $(i \le j)$ let

$$w_{ji} = w_{ij} := \begin{cases} 1 - p_{aa} & \text{with probability} \quad p_{aa} \\ -p_{aa} & \text{with probability} \quad 1 - p_{aa} \end{cases} \tag{20}$$

be also independent, otherwise $\boldsymbol{W}$ is symmetric. If self-loops are not allowed, then $w_{ii} = -p_{aa}$ with probability 1 could be defined for $i \in V_a$. It won't change significantly the subsequent calculations.

This $\boldsymbol{W}$ is not a Wigner noise (see [8]) as it does not have a nested structure. However, if $i \in V_a$ and $j \in V_b$, then

$$\mathrm{Var}(\sqrt{n}w_{ij}) = np_{ab}(1 - p_{ab}) = n\frac{c_{ab}(n - c_{ab})}{n^2} = c_{ab}(1 - \frac{c_{ab}}{n}) \leq c_{ab},$$

where, without hurting the generality, we may assume $0 < c_{ab} < 1$. (Even if for some a, b, $c_{ab} \geq 1$, with large enough n, $\frac{c_{ab}}{n} < 1$ holds.) Therefore, $\sqrt{n}\boldsymbol{W}$ is a general Wigner-type matrix with standard deviation of its entries bounded by $\sigma = \max_{a,b}\sqrt{c_{ab}}$. With constant σ for each entries of $\sqrt{n}\boldsymbol{W}$, in [15] it is proved that the spectrum of $\boldsymbol{W}$ obeys the semicircle law with radius 2σ. More generally, the authors of [22] establish similar result for a Wigner-type matrix with independent (in and above the main diagonal), but usually not identically distributed entries; also, the authors of [2] have analogous findings for the spectral norms of rectangular matrices. Latter results only require the fourth moments of the entries to exist. (Note that sometimes all the moments are required to exist, or the Poincaré inequality is required to be satisfied.) In [35], universality for the leading eigenvalues of a Wigner-type matrix is established.

Now, the finite rank matrix $\bar{\boldsymbol{A}}$ is considered as a perturbation on $\boldsymbol{W}$. By the theory of deformed Wigner matrices [15], the relation between σ and the non-zero eigenvalues of $\bar{\boldsymbol{A}}$ decides the situation. Since the largest eigenvalue of the matrix $\bar{\boldsymbol{A}}$ of rank k is between $\min_a c_a$ and $\max_a c_a$, it is typically smaller than $\sigma = \max_{a,b}\sqrt{c_{ab}}$, therefore the largest eigenvalue of $\bar{\boldsymbol{A}} + \boldsymbol{W}$ is within the semicircle, and the structure in $\bar{\boldsymbol{A}}$ is suppressed by the noise $\boldsymbol{W}$. By [15], a non-zero eigenvalue ν_j of $\bar{\boldsymbol{A}}$ is perturbed as $\nu_j + \frac{\sigma^2}{\nu_j}$, if $\nu_j > \sigma$. The deviation from ν_j is of constant order, it does not decrease with n. So it is either swallowed by the noise or it is not changed with n. Therefore, the relation between the non-zero eigenvalues of $\bar{\boldsymbol{A}}$ and $\boldsymbol{W}$ depends on the individual c_{ab}s, but in general, it is impossible to distinguish between them.

Luckily, this is not the case with the eigenvalues of $\boldsymbol{B}$. As we saw, $\boldsymbol{B}$'s leading eigenvalues are within a factor $\max_a \frac{c}{c_a}$ from those of $\boldsymbol{RC}$. In this way, the additive errors can be larger than the multiplicative ones.

As for the eigenvectors, in the SBM_k model, the eigenvectors of the expected adjacency matrix $\bar{\boldsymbol{A}}$ are n-dimensional step-vectors having k different coordinates. If $\mathbf{u}$ denotes such an eigenvector, then the corresponding $2m$-dimensional inflated vector $\boldsymbol{End}\,\mathbf{u}$ has the same coordinates with multiplicities equal to the node-degrees ($k \leq n \leq m$). Those are also step-vectors on k steps, and by Proposition 2, are close to the corresponding eigenvector of $\boldsymbol{B}$ (with the close eigenvalue). Therefore, the deflated vector is

$$\boldsymbol{End}^*\boldsymbol{End}\,\mathbf{u} = (\boldsymbol{End}\,\mathbf{u})^{in}$$

and, by Equation (4), $\boldsymbol{D}^{-1}\,(\boldsymbol{End}\,\mathbf{u})^{in}$ approximates $\mathbf{u}$; i.e., the first n coordinates of the corresponding $\boldsymbol{K}$-eigenvector, divided with the corresponding node-degrees, will approximate $\mathbf{u}$.

Recall Proposition 2, according to which the leading positive real eigenvalues of $\boldsymbol{B}$ (those of $\boldsymbol{K}$) are close to those of the expected adjacency matrix. In case of the conditions of Proposition 2, the corresponding normalized eigenvectors of $\boldsymbol{B}$ have an inner product with the standardized $\boldsymbol{End}$-transforms of the analogous eigenvectors of the expected adjacency matrix very close to 1. But the eigenvectors of the latter one are step-vectors, and the eigenvectors of the former have a large inner product with them, so they are within a "small" distance from the subspace spanned by the two sets of the k leading eigenvectors. This implies that without knowing the model parameters, the sum of the inner variances of the k-dimensional representatives of the directed edges is "small". In this way, we obtain k clusters of the bidirected edges, but via the majority of their end-nodes we can classify the majority of the nodes too.

To decrease the computational complexity, we can as well use the right eigenvectors of $\boldsymbol{K}$ (only one half of them), more precisely, the vectors $\boldsymbol{D}^{-1}\,\mathbf{x}_i^{in}$ for $i = 1, \ldots, k$. For the representatives, based on them, the k-means algorithm is applicable.

More precisely, by [36], let $\mathbf{x}$ be a unit-norm eigenvector of $\boldsymbol{B}$, corresponding to the eigenvalue μ that is close to the eigenvalue ν of the expected adjacency matrix, with corresponding eigenvector $\mathbf{u} \in \mathbb{R}^n$. If our graph is from the SBM_k model, then (without knowing its parameters) we know that $\mathbf{u}$ is a step-vector with at most k different coordinates. In [36] it is proved that

$$\left\langle \mathbf{x}, \frac{\boldsymbol{End}\,\mathbf{u}}{\|\boldsymbol{End}\,\mathbf{u}\|} \right\rangle \geq \sqrt{1-\varepsilon} \geq 1 - \frac{1}{2}\varepsilon,$$

where ε can be arbitrarily "small" with increasing n. Consequently,

$$\left\| \mathbf{x} - \frac{\boldsymbol{End}\,\mathbf{u}}{\|\boldsymbol{End}\,\mathbf{u}\|} \right\|^2 \leq 2 - 2(1 - \frac{1}{2}\varepsilon) = \varepsilon$$

and, as $\mathbf{x}^{in} = \boldsymbol{End}^*\mathbf{x}$ holds by (5) and $\boldsymbol{End}^*\boldsymbol{End} = \boldsymbol{D}$ holds by (4),

$$\left\| \boldsymbol{End}^*\mathbf{x} - \boldsymbol{End}^* \frac{\boldsymbol{End}\,\mathbf{u}}{\|\boldsymbol{End}\,\mathbf{u}\|} \right\|^2 = \left\| \mathbf{x}^{in} - \boldsymbol{D} \frac{\mathbf{u}}{\|\boldsymbol{End}\,\mathbf{u}\|} \right\|^2$$

also holds. Hence,

$$\left\| \boldsymbol{D}^{-1}\mathbf{x}^{in} - \frac{\mathbf{u}}{\|\boldsymbol{End}\,\mathbf{u}\|} \right\|^2 \leq \|\boldsymbol{D}^{-1}\boldsymbol{End}^*\|^2\varepsilon \leq \varepsilon.$$

Indeed, the largest eigenvalue of $(\boldsymbol{D}^{-1}\boldsymbol{End}^*)(\boldsymbol{End}\,\boldsymbol{D}^{-1}) = \boldsymbol{D}^{-1}\boldsymbol{D}\boldsymbol{D}^{-1} = \boldsymbol{D}^{-1}$ is $\max_i \frac{1}{d_i}$, so the largest singular value (spectral norm) of $\boldsymbol{D}^{-1}\boldsymbol{End}^*$ is estimated from above with $\left(\max_i \frac{1}{d_i}\right)^{\frac{1}{2}}$. Therefore,

$$\|\boldsymbol{D}^{-1}\boldsymbol{End}^*\|^2 \leq \max_i \frac{1}{d_i} = \frac{1}{\min_i d_i} \leq 1.$$

This results can be summarized in the following theorem.

Theorem 2 *Assume that the expected adjacency matrix of the underlying random graph on n nodes and m edges has rank k with k single eigenvalues and corresponding unit-norm eigenvectors* $\mathbf{u}_1, \ldots, \mathbf{u}_k \in \mathbb{R}^n$. *Assume that the non-backtracking matrix* $\boldsymbol{B}$ *of the random graph has k structural (real) eigenvalues (aligned with those of the expected adjacency matrix) with eigenvectors* $\mathbf{x}_1, \ldots, \mathbf{x}_k \in \mathbb{R}^{2m}$ *such that*

$$\left\langle \mathbf{x}_j, \frac{\boldsymbol{End}\,\mathbf{u}_j}{\|\boldsymbol{End}\,\mathbf{u}_j\|} \right\rangle \geq \sqrt{1-\varepsilon}, \quad j = 1, \ldots, k.$$

Then for the transformed vectors $\boldsymbol{D}^{-1}\mathbf{x}_j^{in} \in \mathbb{R}^n$, *the relation*

$$\sum_{j=1}^{k} \left\| \boldsymbol{D}^{-1}\mathbf{x}_j^{in} - \frac{\mathbf{u}_j}{\|\boldsymbol{End}\,\mathbf{u}_j\|} \right\|^2 \leq k\varepsilon \tag{21}$$

holds, where $\boldsymbol{D}$ *is the diagonal degree-matrix of the underlying graph.*

Corollary 1 *If* $\mathbf{u}_j$*'s are step-vectors on k steps (e.g., if our graph comes from the* SBM_k *model), then the sum of the inner variances of the node representatives*

$$(\frac{1}{d_i}x_{1i}^{in}, \ldots, \frac{1}{d_i}x_{ki}^{in}), \quad i = 1, \ldots, n$$

is estimated from above with $k\varepsilon$ *too. Indeed, the sum of the inner variances is the minimum squared distance between the subspace spanned by the vectors* $\boldsymbol{D}^{-1}\mathbf{x}_j^{in}$ $(j = 1, \ldots, k)$ *and that of the k-step vectors, see [7] for the proof. Since in the* SBM_k *model,* $\mathbf{u}_j$*'s are also step-vectors on the same k steps (as was proved in Proposition 3), this sum of the inner variances is less than the left-hand side of inequality* (21)*, so it is at most* $k\varepsilon$.

Also, if the underlying graph is sparse and the conditions of Theorem 1 of [36] hold, then $\varepsilon > 0$ *can be arbitrarily small with increasing n. To minimize the objective function of the k-means algorithm, there are polynomial time approximating schemes, e.g., [34], so without knowing the model parameters and the vectors* $\mathbf{u}_j$*s, we can find the clusters with the help of the leading* $\boldsymbol{B}$*-eigenvectors only.*

It is important that here k is the number of the structural eigenvalues of $\boldsymbol{B}$*, and the expected adjacency matrix must have a "good" k-rank approximation.*

This gives rise to a future research on sparse spectral clustering.

Remark 2 In case of an unweighted graph, the n-dimensional vector $\mathbf{x}_j^{in}$, obtained from a right eigenvector of $\boldsymbol{B}^*$, is the second segment of the right eigenvector of the matrix $\boldsymbol{K}$, corresponding to the same eigenvalue. Consequently, by the swapping issues of Sect. 2, the n-dimensional vector $\mathbf{x}_j^{in}$, obtained from a right eigenvector of $\boldsymbol{B}$, is the first segment of the right eigenvector of $\boldsymbol{K}$, for $j = 1, \ldots, k$. So, we have to perform the spectral decomposition of a $2n \times 2n$ matrix only instead of a $2m \times 2m$

one, which fact has further computational benefit (except for trees, $n \le m$ holds, but usually n is much smaller than m).

Also note that, by Remark 1, in the $c_1 = \cdots = c_k = c$ special case, $\mathbf{u}_1 = \frac{1}{\sqrt{n}}\mathbf{1} \in \mathbb{R}^n$, and therefore, the normalized $\boldsymbol{End}\,\mathbf{u}_1 \in \mathbb{R}^k$ also has equal coordinates. Consequently, the vectors $\mathbf{x}_1$ and $\boldsymbol{D}^{-1}\mathbf{x}_1^{in}$ also nearly have the same coordinates, so the first coordinates of the k-dimensional node representatives can be disregarded, and instead we use $(k-1)$-dimensional representatives.

5 EM Algorithm for Estimating the Parameters of the SBM_k Model

Given a simple graph $G = (V, \boldsymbol{A})$ $(|V| = n)$ with adjacency matrix $\boldsymbol{A}$ and an integer k $(1 < k < n)$, we are looking for the hidden k-partition $(V_1, \dots, V_k)$ of the nodes such that

- nodes are independently assigned to cluster V_a with probability r_a, $a = 1, \dots, k$; $\sum_{a=1}^k r_a = 1$;
- given the cluster memberships, nodes of V_a and V_b are connected independently, with probability

$$\mathbb{P}(i \sim j \mid i \in V_a, j \in V_b) = p_{ab}, \quad 1 \le a, b \le k.$$

The parameters are collected in the probability vector $\mathbf{r} = (r_1, \dots, r_k)$ and the $k \times k$ symmetric probability matrix $\boldsymbol{P} = (p_{ab})$. We could as well consider the affinity matrix, but it does not make much difference as n is fixed now.

Here the $n \times n$ symmetric, 0-1 adjacency matrix $\boldsymbol{A} = (a_{ij})$ of G represents a statistical sample. There are no loops, so the diagonal entries are zeros, whereas the independent off-diagonal entries are considered as an i.i.d. sample. Based on $\boldsymbol{A}$, we want to estimate the parameters $\mathbf{r}$ and $\boldsymbol{P}$ of the above SBM_k model. In fact, $\boldsymbol{A}$ is the incomplete data specification as the cluster memberships are missing. Therefore, it is straightforward to use the EM algorithm for parameter estimation from incomplete data.

First we complete our data matrix $\boldsymbol{A}$ with latent membership vectors $\boldsymbol{\Delta}_1, \dots, \boldsymbol{\Delta}_n$ of the nodes that are k-dimensional i.i.d. $Poly(1, \mathbf{r})$ (multinomially distributed) random vectors. More precisely, $\boldsymbol{\Delta}_i = (\Delta_{1i}, \dots, \Delta_{ki})$, where $\Delta_{ai} = 1$ if $i \in V_a$ and zero otherwise. Thus, the sum of the coordinates of any $\boldsymbol{\Delta}_i$ is 1, and $\mathbb{P}(\Delta_{ai} = 1) = r_a$.

The likelihood is the square-root of

$$\prod_{a=1}^{k}\prod_{i=1}^{n}\prod_{b=1}^{k}[p_{ab}^{\sum_{j:\,j\ne i}\Delta_{bj}a_{ij}} \cdot (1-p_{ab})^{\sum_{j:\,j\ne i}\Delta_{bj}(1-a_{ij})}]^{\Delta_{ai}}, \tag{22}$$

which is maximized by the following iteration.

Starting with initial parameter values $\mathbf{r}^{(0)}$, $\boldsymbol{P}^{(0)}$ and membership vectors $\boldsymbol{\Delta}_1^{(0)}, \ldots, \boldsymbol{\Delta}_n^{(0)}$, the t-th step of the iteration is the following ($t = 1, 2, \ldots$).

- **E**-step: we calculate the conditional expectation of each Δ_i conditioned on the model parameters and on the other cluster assignments obtained in step $t-1$ and collectively denoted by $M^{(t-1)}$. By the Bayes rule, the responsibility of node i for cluster a is

$$r_{ai}^{(t)} = \mathbb{E}(\Delta_{ai} \mid M^{(t-1)}) = \frac{\mathbb{P}(M^{(t-1)}|\Delta_{ai} = 1) \cdot r_a^{(t-1)}}{\sum_{b=1}^k \mathbb{P}(M^{(t-1)}|\Delta_{bi} = 1) \cdot r_b^{(t-1)}}$$

($a = 1, \ldots, k$; $i = 1, \ldots, n$). For each i, $r_{ai}^{(t)}$ is proportional to the numerator, where

$$\mathbb{P}(M^{(t-1)}|\Delta_{ai} = 1) = \prod_{b=1}^k (p_{ab}^{(t-1)})^{\sum_{j\neq i} \Delta_{bj}^{(t-1)} a_{ij}} \cdot (1 - p_{ab}^{(t-1)})^{\sum_{j\neq i} \Delta_{bj}^{(t-1)} (1-a_{ij})}$$

is the part of the likelihood (22) that affects node i under the condition $\Delta_{ai} = 1$.
- **M**-step: we maximize the truncated binomial likelihood

$$p_{ab}^{\sum_{i\neq j} r_{ai}^{(t)} r_{bj}^{(t)} a_{ij}} \cdot (1 - p_{ab})^{\sum_{i\neq j} r_{ai}^{(t)} r_{bj}^{(t)} (1-a_{ij})}$$

with respect to the parameter p_{ab}, for all a, b pairs separately. Obviously, the maximum is attained when the following estimators of p_{ab}'s comprise the symmetric matrix $\boldsymbol{P}^{(t)}$: $p_{ab}^{(t)} = \frac{\sum_{i,j:i\neq j} r_{ai}^{(t)} r_{bj}^{(t)} a_{ij}}{\sum_{i,j:i\neq j} r_{ai}^{(t)} r_{bj}^{(t)}}$ ($1 \le a \le b \le k$), where edges connecting nodes of clusters a and b are counted fractionally, multiplied by the membership probabilities of their endpoints.

The maximum likelihood estimator of $\mathbf{r}$ in the t-th step is $\mathbf{r}^{(t)}$ of coordinates $r_a^{(t)} = \frac{1}{n} \sum_{i=1}^n r_{ai}^{(t)}$ ($a = 1, \ldots, k$), while that of the membership vector $\boldsymbol{\Delta}_i$ is obtained by discrete maximization: $\Delta_{ai}^{(t)} = 1$ if $r_{ai}^{(t)} = \max_{b\in\{1,\ldots,k\}} r_{bi}^{(t)}$ and 0, otherwise. (In case of ties, nodes are classified arbitrarily.) This choice of $\mathbf{r}$ will increase (better to say, not decrease) the likelihood function. Note that it is not necessary to assign nodes uniquely to the clusters, the responsibility r_{ai} of a node i can as well be regarded as the intensity of node i belonging to cluster a, akin to the fuzzy clustering.

According to the general theory of the EM algorithm [18], in exponential families (as in the present case), convergence to a local maximum can be guaranteed (depending on the starting values), but it runs in polynomial time in the number of nodes n. However, the speed and limit of the convergence depend on the starting clustering, which can be chosen by means of preliminary application of some nonparametric multiway cut algorithm.

6 Special Cases and β-Percolation

In the most special "symmetric" case, the transition matrix is $\boldsymbol{T} = \boldsymbol{GRC} = \frac{1}{ck}\boldsymbol{C}$, where $\boldsymbol{C}$ contains c_{in} in its main diagonal and c_{out}, otherwise; see the BP method of Sect. 3. Now $\boldsymbol{T}$ is a stochastic matrix, so its largest eigenvalue is 1 with corresponding eigenvector $\mathbf{1}$ (the all 1's vector). The other eigenvalue is

$$\lambda = \frac{c_{in} - c_{out}}{kc} \tag{23}$$

with multiplicity $k-1$. In the assortative case, $\lambda > 0$; further, $\lambda < 1$, as in the symmetric case

$$c = \frac{c_{in} + (k-1)c_{out}}{k}$$

holds. Consequently, the eigenvalues of $\boldsymbol{RC} = c\boldsymbol{T}$ are c and $c\lambda$, latter one has multiplicity $k-1$.

With the BP method of [28], that treats this special case, the eigenvalues of $\boldsymbol{B} \otimes \boldsymbol{T}$ greater than 1 are considered that boils down to the condition $c\lambda^2 > 1$, which gives the Kesten–Stigum threshold (10).

Now assume that $G_n \in SBM_k^{\beta}$. Then the $k \times k$ probability matrix is $(p_{ab}) = \beta\frac{1}{n}\boldsymbol{C} = \frac{\beta\boldsymbol{C}}{n}$. Consequently, $\boldsymbol{C}$ and c is multiplied by β, but $\boldsymbol{T} = \boldsymbol{GRC}$ remains unchanged. So we consider $\beta c\boldsymbol{T}$ as for the model side, but the underlying graph and its $\boldsymbol{B}$ is the same as before. Therefore, the eigenvalues of $\boldsymbol{B} \otimes \beta c\boldsymbol{T} = \beta c(\boldsymbol{B} \otimes \boldsymbol{T})$ should be greater than c if a non-trivial solution is required:

$$\beta\,\lambda(\boldsymbol{B} \otimes c\boldsymbol{T}) > c.$$

If the eigenvalues of $\boldsymbol{B}$ and $c\boldsymbol{T}$ are aligned, then this gives that $\lambda(\boldsymbol{B}) > \frac{\sqrt{c}}{\sqrt{\beta}}$ is needed for detectability; equivalently,

$$\beta > \frac{c}{\lambda^2(B)} = \left(\frac{\sqrt{c}}{\lambda(\boldsymbol{B})}\right)^2. \tag{24}$$

This is in accord with the fact, that in the $k = 1$ case, in the Erdős–Rényi model, when the largest eigenvalue μ_1 of $\boldsymbol{B}$ is around c, then $\beta = \frac{1}{\mu_1}$ is the percolation threshold, see [33]. In the multiclass scenario, $\frac{c}{\mu_i^2}$ are further phase transitions, leading to i clusters, for $i = 1, \dots, k_0$ until $\mu_{k_0} \geq \sqrt{c}$, but $\mu_{k_0+1} < \sqrt{c}$. Also note that inequality (24) has relevance only if $\lambda(\boldsymbol{B}) > \sqrt{c}$, so eigenvalues of $\boldsymbol{B}$ greater than $\sqrt{c}$ give the phase transitions.

It is also in accord with the findings in the special $k = 2$ case. Let us consider the symmetric SBM_2 model: $r_1 = r_2 = \frac{1}{2}$, $c_1 = c_2 = \frac{c_{in}+c_{out}}{2} = c$. Then condition (10) in the assortative case boils down to

$$c_{in} - c_{out} = |c_{in} - c_{out}| > 2\sqrt{c} = \sqrt{2}\sqrt{c_{in} + c_{out}}. \tag{25}$$

If $\boldsymbol{C}$ and c are multiplied with β, we get $\beta(c_{in} - c_{out}) > \sqrt{2}\sqrt{\beta(c_{in} + c_{out})}$. This means that

$$c_{in} - c_{out} > \sqrt{2}\frac{\sqrt{c_{in} + c_{out}}}{\sqrt{\beta}}.$$

Since $\beta < 1$, the right hand side gives a higher lower bound than $\beta = 1$ in Equation (25). This is also equivalent to

$$\beta > \frac{2(c_{in} + c_{out})}{(c_{in} - c_{out})^2}, \tag{26}$$

which makes sense if $\frac{2(c_{in}+c_{out})}{(c_{in}-c_{out})^2} < 1$, i.e., if

$$c_{in} - c_{out} > 2\sqrt{c} = \sqrt{2}\sqrt{c_{in} + c_{out}},$$

in accord with (25). So, until equality is attained in (26), additional β-percolation can give detectable clusters too.

In the special case when $c_1 = \cdots = c_k = c$, the above is also equivalent to

$$c\lambda > \sqrt{c},$$

where $c\lambda$ is the second largest eigenvalue (of multiplicity $k-1$) of $c\boldsymbol{T} = \boldsymbol{RC}$, and it is closely related to the eigenvalues $\mu_2, \dots, \mu_k$ of $\boldsymbol{B}$.

The above is also in accord with (24), since the threshold for β is

$$\beta > \frac{c}{\mu_2^2} = \frac{c}{(\frac{c_{in}-c_{out}}{2})^2} = \frac{\frac{c_{in}+c_{out}}{2}}{\frac{(c_{in}-c_{out})^2}{4}} = \frac{2(c_{in} + c_{out})}{(c_{in} - c_{out})^2},$$

which is the same as inequality (26).

Observe that inequality (10) implies

$$\beta|c_{in} - c_{out}| > k\sqrt{\beta c},$$

which yields

$$\beta > \frac{k^2 c}{(c_{in} - c_{out})^2} = \frac{k[c_{in} + (k-1)c_{out}]}{(c_{in} - c_{out})^2} = \frac{c}{(c\lambda)^2} = \left(\frac{\sqrt{c}}{\lambda}\right)^2$$

in the symmetric case, where λ is the second largest eigenvalue of $\boldsymbol{RC}$ with multiplicity $k-1$ and it is aligned with $\mu_2, \dots, \mu_k$. So in the symmetric case, as the second largest eigenvalue of $\boldsymbol{RC}$ has multiplicity $k-1$, it gives rise to $k-1$ very close phase transitions. In other cases, those can be farther apart.

Another implication is that

$$\sqrt{\beta}|c_{in} - c_{out}| > k\sqrt{c},$$

where the left-hand side decreases with $\beta < 1$. It can be decreased until equality is attained, when $\beta = \frac{c}{\lambda^2}$.

In the generic scenario, it means that at $\frac{c}{\mu_i^2}$, for $i = 2, \dots, k$, there are phase transitions for β producing $2, \dots, k$ detectable clusters:

$$|c_{in} - c_{out}| > k\frac{\sqrt{c}}{\sqrt{\beta}} > k\sqrt{c},$$

so there is a leverage of the percolation threshold with $1/\sqrt{\beta}$.

In the c_{in} versus c_{out} scenario, $c_1 = \dots = c_k = c$ is equivalent to $r_1 = \dots = r_k$, and so,

$$\beta|c_{in} - c_{out}| > k\sqrt{\beta c} \quad \text{and} \quad |c_{in} - c_{out}| > k\frac{\sqrt{c}}{\sqrt{\beta}}$$

which allows more detectable clusters if β is increased, see the subsequent illustrations.

Eventually, this phenomenon is illustrated by simulations. First a random graph was generated on $n = 900$ nodes, with parameter matrices $\boldsymbol{R} = \text{diag}(\frac{3}{10}, \frac{1}{3}, \frac{11}{30})$ and

$$\boldsymbol{C} = \begin{pmatrix} 30 & 12 & 10 \\ 12 & 32 & 9 \\ 10 & 9 & 27 \end{pmatrix}.$$

Then we percolated (retained) the edges with probability β. For small β's the graph was not connected, which fact caused the eigenvalue 1 of the matrix $\boldsymbol{K}$ to have several geometric, and even more algebraic multiplicities. With increasing β, Figs. 1, 2, 3, and 4 show the adjacency matrices and the $\boldsymbol{K}$-spectrum of the β-percolated random graph, when 1,2, and 3 structural (positive real) eigenvalues emerge out of the bulk (within the encircled region). This happens about when β exceeds β_1, β_2, and β_3 ($\beta_i = \frac{c}{\mu_i^2}$), but the phase transitions are not strict, as the alignment of the structural eigenvalues of the non-backtracking matrix $\boldsymbol{B}$ (those of $\boldsymbol{K}$) and those of the model matrix $\boldsymbol{RC}$ is just asymptotic (for "large" n) and guaranteed only when the expected degrees of the clusters are approximately the same.

Next a random graph was generated on $n = 900$ nodes, with parameter matrices $\boldsymbol{R} = \text{diag}(\frac{35}{107}, \frac{42}{107}, \frac{30}{107})$ and

$$\boldsymbol{C} = \begin{pmatrix} 30 & 11.28 & 7.728 \\ 11.28 & 25 & 10.36 \\ 7.728 & 10.36 & 35 \end{pmatrix}$$

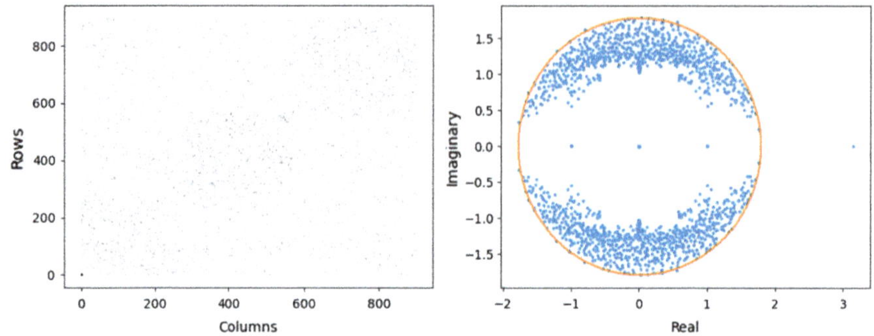

(a) Adjacency matrix (black: 1, white: 0). (b) Spectrum of matrix $\boldsymbol{K}$.

Fig. 1 $\beta_1 < \beta = 0.191 < \beta_2$

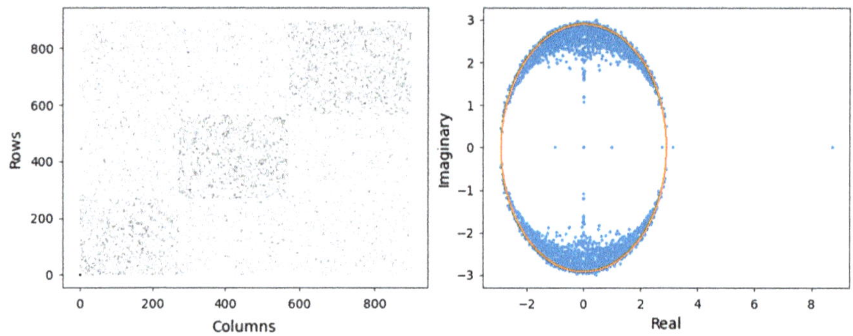

(a) Adjacency matrix (black: 1, white: 0). (b) Spectrum of matrix

Fig. 2 $\beta = \beta_3 = 0.505$

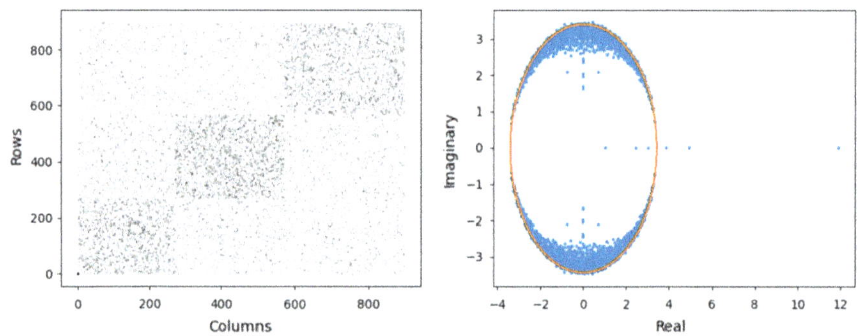

(a) Adjacency matrix (black: 1, white: 0). (b) Spectrum of matrix $\boldsymbol{K}$.

Fig. 3 $\beta_3 < \beta = 0.7$

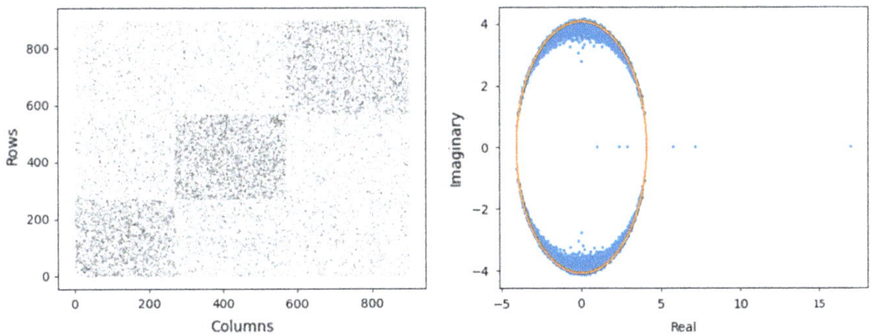

(a) Adjacency matrix (black: 1, white: 0). (b) Spectrum of matrix $\boldsymbol{K}$.

Fig. 4 $\beta = 1$

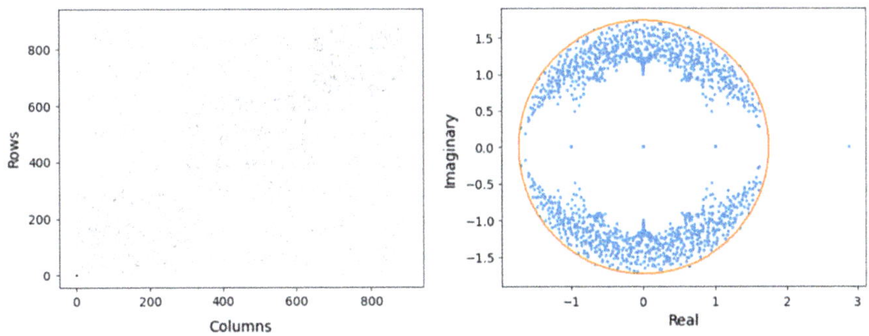

(a) Adjacency matrix (black: 1, white: 0). (b) Spectrum of matrix $\boldsymbol{K}$.

Fig. 5 $\beta_1 < \beta = 0.183 < \beta_2$

constructed so that the average degrees of the clusters be the same, i.e., $c_a = \sum_{b=1}^{3} r_b c_{ab}$ is the same for $a = 1, 2, 3$.

Figures 5, 6, 7, and 8 show the adjacency matrices and the $\boldsymbol{K}$-spectrum of the β-percolated random graph, when 1,2, and 3 structural (positive real) eigenvalues appear. Here the phase transitions are closer to β_i's than in the previous generic case.

Summarizing, at β_1, a big giant component is formed. In [21], the effect of different seeding patterns is investigated in context of the COVID19 pandemic in Hungary. The authors use the SIR model and assume the existence of a denser "central region" and a sparser "periphery". They compare the total number of nodes that get infected if a given number of seeds (initial infections) are distributed uniformly and randomly in the central region and in the whole graph, respectively. Under switchover phenomenon they mean that for a low infection rate, an epidemics starting in the central region is worse (results in a larger epidemics), but this switches over so that the epidemics starting uniformly over the whole country is worse. They do not consider spectral connections, other parameters are used.

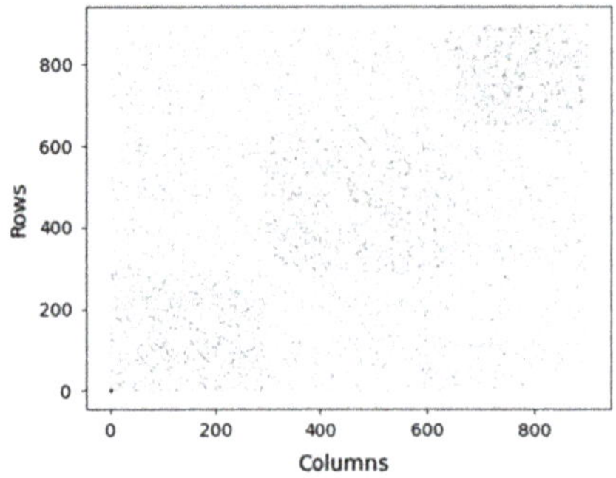

(a) Adjacency matrix (black: 1, white: 0).

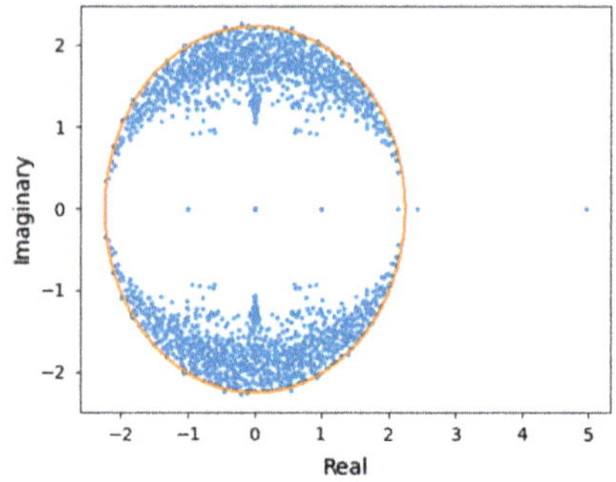

(b) Spectrum of matrix $\boldsymbol{K}$.

Fig. 6 $\beta = \beta_2 = 0.305$

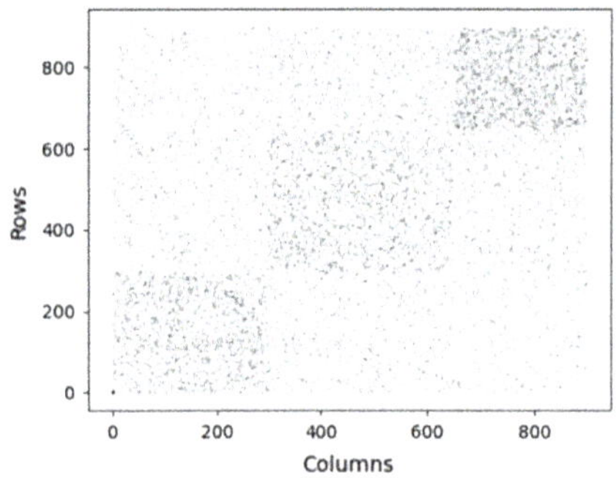

(a) Adjacency matrix (black: 1, white: 0).

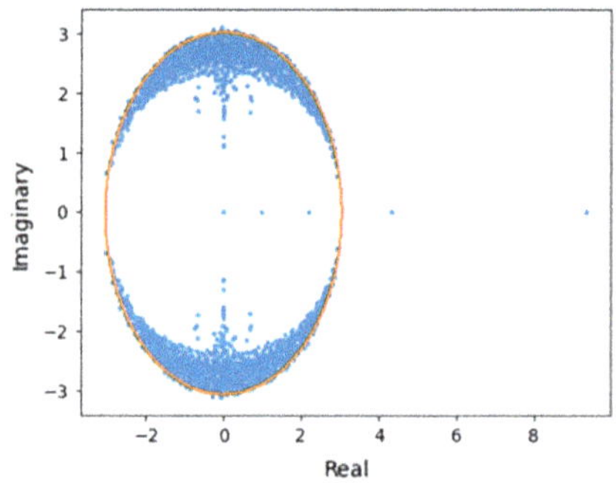

(b) Spectrum of matrix $\boldsymbol{K}$.

Fig. 7 $\beta_2 < \beta = 0.563 < \beta_3$

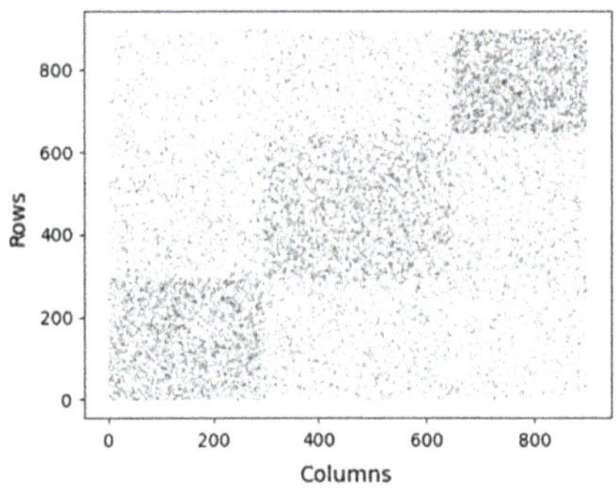

(a) Adjacency matrix (black: 1, white: 0).

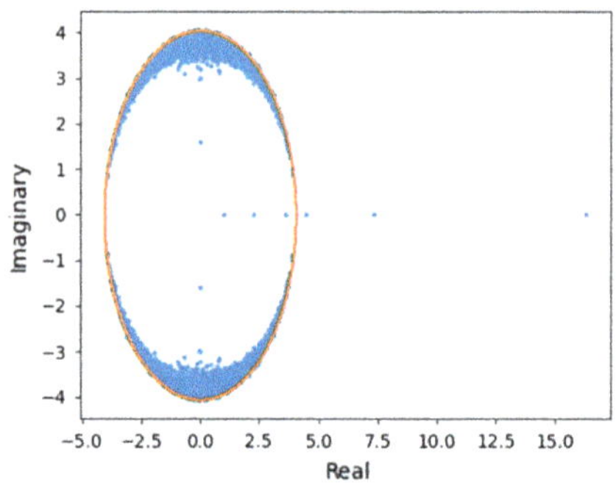

(b) Spectrum of matrix $\boldsymbol{K}$.

Fig. 8 $\beta = 1$

According to this, below a certain β, the central seeding is more dangerous than the uniform one. Indeed, below β_1 there are many small components, among which there is a central one. Above that, the uniform seeding is more dangerous than the central one, as it is able to infect the giant component and other components too. However, if the seeds arrive in time, then actions can be done to separate the most infected clusters as follows.

At β_2, there are 2 well-distinguishable clusters, ..., at β_k, there are k well-distinguishable ones, since the within-cluster bonds are strengthened. It does not mean that there are no between-cluster bonds, but those are negligible compared to the within-cluster ones. Also, by deleting the between-cluster bonds, the infection can be confined to the clusters. If the seed is in one cluster, the epidemic can be confined to that cluster. So with increasing β, there is a chance to localize the epidemic to one large city and to save the others.

For example, if β passes over β_2, then 2 clusters can be separated, and if the seed is in one of them, then the infection can be localized to that cluster and we can save the other by deleting the connecting bonds (even if we do not delete all, the epidemic will spread to the other cluster with small probability only). If β passes over β_3, then 3 clusters can be separated, and if the seed is in one of them, then the infection can be localized to that cluster and we can save the others by deleting the connecting bonds, etc. Above β_k the situation is similar to that of $\beta = 1$, the uniform seeding is dangerous, but seeds can be localized to the first infected cluster. In the symmetric case, $\beta_2 = \cdots = \beta_k$, so all these phase transitions occur at the same time from the giant cluster to the k-cluster scenario. Multiple transitions are spectacular if the eigenvalues of $\boldsymbol{B}$ greater than $\sqrt{c}$ are separated from each other and from $\sqrt{c}$. Therefore, β can as well be considered as a tuning parameter. Also note that in the SBM_k model when $c_a = c$ $(a = 1, \ldots, k)$, then μ_1 is close to c and so, β_1 is close to $\frac{1}{\mu_1}$ (w.h.p.). This is always the case if $k = 1$, and so, $\frac{c}{\mu_1^2} = \frac{1}{\mu_1}$ as in [33].

Here we assume that our starting graph G_n is from the SBM_k model, but k is not specified at the beginning, we conclude for it from the spectrum of $\boldsymbol{B}$ of G_n. For "large" n the graph is sparse enough, but it can have k "giant" components (of sizes proportional with n) in the sense that the within-cluster edge densities are much higher than the between-cluster ones.

Here we rather use the results of [33], where only one giant component is assumed (corresponding to the $k = 1$ case). If $k > 1$, further giant components and phase-transitions are possible.

7 Application to Large Quantum Chemistry Networks

Computational quantum chemistry aims at computing properties of molecules and materials based on quantum mechanics of many-electron systems, see, e.g. [3, 38]. In quantum mechanics, states of a system are states in a Hilbert space and relevant physical quantities correspond to some linear operators in such spaces. In a molecule, energy levels of a many-electron system are eigenvalues of the Hamiltonian operator, which is quantum version of a classical energy function also known as Hamilton function in classical mechanics. In case of many-electron molecules, the corresponding Hilbert space is an infinite dimensional functional space of so called many-electron wave-functions. For numerical computations, various finite dimensional subspaces may be used (see [38]). In such subspaces, Hamiltonian operator is represented by the Hamiltonian matrix. The approximate energy levels of a molecule are the real eigenvalues of this matrix, which is self-adjoint.

When a molecule has more than just few electrons, the corresponding subspace needs to be very large to achieve a good approximation. To a Hamiltonian matrix one can associate a simple graph, in which the nodes correspond to basis functions in the subspace, and two nodes are connected if the squared modulus of the corresponding Hamiltonian matrix-entry exceeds a certain threshold. Finding structures in such a network can be useful. In [27] it was suggested that assortative network communities allow reduction of the matrix size needed for eigenvalue computations. The Hamiltonian matrix is typically sparse, each row with n elements has approximately poly-log n non-zero elements. As a result, in chemistry we naturally encounter very large and sparse networks.

As a sample we consider the Hamiltonian matrix of the LiH (Lithium Hydride) molecule with 7638 rows and columns. The corresponding skeleton graph has average degree about 459.559, which means that on average a row of the adjacency matrix has 94% zero elements. The largest real eigenvalue of the matrix $\boldsymbol{K}$ is about 464.473, which is similar in magnitude to the average degree, an indicator of sparsity of the skeleton graph. The results are shown in Fig. 9. In panel (a), the 18 largest eigenvalues

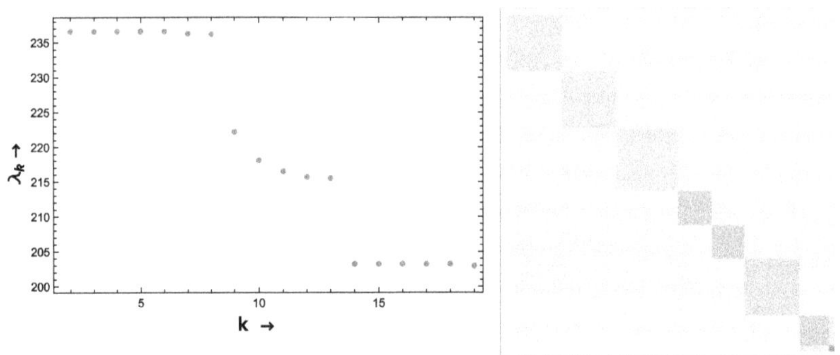

(a) 18 largest real eigenvalues (λ_k) of the matrix $\boldsymbol{K}$, excluding the top eigenvalue ($\lambda_1 \approx 464.473$). A large gap between the 8th and 9th largest eigenvalues reveal number of clusters, which is 8.

(b) Clustering the vertices into 8 clusters by the in-vectors, illustrated by the corresponding subdivision of the adjacency matrix. Rows and columns of each cluster are stacked together. Intensity of gray-level corresponds to density of 1s, darker areas having larger density than lighter ones.

Fig. 9 Spectral clustering of the skeleton graph of the LiH molecule chemical network with 7638 nodes

of the matrix $\boldsymbol{K}$ are presented, excluding the trivial top eigenvalue ($\lambda_1 \approx 464.473$) that mirrors the average degree of the skeleton graph. A large gap between the 8th and 9th largest eigenvalues reveals the number of clusters, which is 8. Panel (b) shows the clustering of the vertices in 8 clusters, by the in-vectors, see Theorem 2. Rows and columns of each cluster are stacked together. Intensity of gray-level corresponds to density of 1s, darker areas having larger density than lighter areas.

8 Conclusions and Further Perspectives

The sparse stochastic block model behind an expanding graph sequence was considered. Here the spectral properties of the adjacency matrix give less information for the clusters than in the dense case of [10, 25]; instead, the non-backtracking spectra are used.

Bond-percolation is considered in two senses: the first is that the within- and between-cluster edge probabilities decrease with n, and the second is that edges coming into existence in this way are retained with probability β. Similarities and differences between the BP and EM algorithms are discussed. Both are based on an iteration algorithm, but former one solves a system of equations with the transmission probabilities between the node pairs ($2mk$ non-linear equations and unknowns, where k is the number of clusters), while keeping the model parameters fixed; latter one, for fixed k, estimates the model parameters and the missing memberships of the nodes by using the classical EM iteration for mixtures. Furthermore, via inflation–deflation techniques and findings of [36], we are able to classify the nodes of the sparse graph with representation based k-means clustering, which gives rise to a sparse spectral clustering procedure. Simulation results for the so-called β-percolation are also considered, and strategies are suggested for possible localization of a pandemic.

If the c_{in} versus c_{out} scenario does not hold, then in generic assortative networks there are k different within-cluster and $\binom{k}{2}$ different between-cluster affinities; former ones, denoted by c_w's, are significantly "larger" than the latter ones, denoted by c_b's. In this case, $c_1 = \cdots = c_k = c$ is not necessarily equivalent to $r_1 = \cdots = r_k$, and the Kesten–Stigum threshold should be developed by giving lower bounds for every $c_w - c_b$ difference, which are positive in assortative networks. Finding such dense diagonal blocks is an important problem of quantum chemistry, for which purpose, a large network is analyzed. In the estimation, discrepancy techniques (see, e.g., [9]) may help.

More generally, we plan to investigate non-backtracking spectra of edge-weighted graphs, that can be considered as a generalized β-model: the pandemic is submitted with probability w_{ij} between individual pairs, instead of the unique β, that gives rise to investigate an $SBM_k^{\boldsymbol{W}}$ model with edge-weight matrix $\boldsymbol{W} = (w_{ij})$. Since then, this problem was partially solved in [11].

Acknowledgements This research was supported by the NKFIH project Dynamical systems and fractals, no.142169 and by the Business Finland project HaQuPrA (Harnessing Quantum for Practical Applications).
The research was partly done under the auspices of the Budapest Semesters in Mathematics program, in the framework of a research course in the spring semester 2023, where the third author made many simulation experiments. The first author, who supervised this course, thanks Josephine Walk for helping to organize the early simulation results into digestible form. She is indebted to László Lovász for valuable conversations on this topic, and also thanks Tamás Móri, Katalin Friedl, and Daniel Keliger for their useful remarks.

References

1. Abbe, E., Sandon, C.: Community detection in general stochastic block models: fundamental limits and efficient algorithms for recovery. In: 2015 IEEE 56th Annual Symposium on Foundations of Computer Science, pp. 670–688. IEEE (2015)
2. Achlioptas, D., McSherry, F.: Fast computation of low-rank matrix approximations. J. ACM (JACM) **54**(2), 9–es (2007)
3. Árendás, P., Furtenbacher, T., Császár, A.G.: On spectra of spectra. J. Math. Chem. **54**, 806–822 (2016)
4. Backhausz, Á., Szegedy, B.: On the almost eigenvectors of random regular graphs. Ann. Probab. **47**(3), 1677–1725 (2019)
5. Benaych-Georges, F., Bordenave, C., Knowles, A.: Largest eigenvalues of sparse inhomogeneous erdős-rényi graphs. Ann. Probab. **47**(3), 1653–1676 (2019)
6. Bickel, P.J., Chen, A.: A nonparametric view of network models and newman-girvan and other modularities. Proc. Natl. Acad. Sci. **106**(50), 21068–21073 (2009)
7. Bolla, M.: Penalized versions of the newman-girvan modularity and their relation to normalized cuts and k-means clustering. Phys. Rev. E-Stat. Nonlinear Soft Matter Phys. **84**(1), 016108 (2011)
8. Bolla, M.: Spectral Clustering and Biclustering: Learning Large Graphs and Contingency Tables. Wiley (2013)
9. Bolla, M.: Relating multiway discrepancy and singular values of nonnegative rectangular matrices. Discret. Appl. Math. **203**, 26–34 (2016)
10. Bolla, M.: Generalized quasirandom properties of expanding graph sequences. Nonlinearity **33**(4), 1404 (2020)
11. Bolla, M., Reittu, H., Abdelkhalek, F.: Clustering the nodes of sparse edge-weighted graphs via non-backtracking spectra. In: Future of Information and Communication Conference, pp. 81–99. Springer (2025)
12. Bollobás, B., Borgs, C., Chayes, J., Riordan, O.: Percolation on dense graph sequences. Ann. Probab. **38**, 150–183 (2010)
13. Bordenave, C., Lelarge, M., Massoulié, L.: Non-backtracking spectrum of random graphs: community detection and non-regular ramanujan graphs. Ann. Probab. **46**(1), 1–71 (2018)
14. Brouwer, A.E., Haemers, W.H.: Spectra of Graphs. Springer Science and Business Media (2011)
15. Capitaine, M., Donati-Martin, C., Féral, D.: The largest eigenvalues of finite rank deformation of large wigner matrices: convergence and nonuniversality of the fluctuations. Ann. Probab. **37**(1), 1–47 (2009)
16. Coste, S., Zhu, Y.: Eigenvalues of the non-backtracking operator detached from the bulk. Random Matr.: Theory Appl. **10**(03), 2150028 (2021)
17. Decelle, A., Krzakala, F., Moore, C., Zdeborová, L.: Inference and phase transitions in the detection of modules in sparse networks. Phys. Rev. Lett. **107**(6), 065701 (2011)

18. Dempster, A.P., Laird, N.M., Rubin, D.B.: Maximum likelihood from incomplete data via the em algorithm. J. Roy. Stat. Soc.: Ser. B (Methodol.) **39**(1), 1–22 (1977)
19. Glover, C., Kempton, M.: Some spectral properties of the non-backtracking matrix of a graph. Linear Algebra Appl. **618**, 37–57 (2021)
20. Karrer, B., Newman, M.E., Zdeborová, L.: Percolation on sparse networks. Phys. Rev. Lett. **113**(20), 208702 (2014)
21. Keliger, D., Lovász, L., Móri, T.F., Ódor, G.: Switchover phenomenon for general graphs. J. Graph Theory **108**(3), 560–581 (2025)
22. Khorunzhy, A.M., Khoruzhenko, B.A., Pastur, L.A.: Asymptotic properties of large random matrices with independent entries. J. Math. Phys. **37**(10), 5033–5060 (1996)
23. Krzakala, F., Moore, C., Mossel, E., Neeman, J., Sly, A., Zdeborová, L., Zhang, P.: Spectral redemption in clustering sparse networks. Proc. Natl. Acad. Sci. **110**(52), 20935–20940 (2013)
24. Lovász, L.: Combinatorial Problems and Exercises, vol. 361. American Mathematical Soc. (2007)
25. Lovász, L., Sós, V.T.: Generalized quasirandom graphs. J. Combin. Theory Ser. B **98**(1), 146–163 (2008)
26. Martin, T., Zhang, X., Newman, M.E.: Localization and centrality in networks. Phys. Rev. E **90**(5), 052808 (2014)
27. Mniszewski, S.M., Dub, P.A., Tretiak, S., Anisimov, P.M., Zhang, Y., Negre, C.F.: Reduction of the molecular hamiltonian matrix using quantum community detection. Sci. Rep. **11**(1), 4099 (2021)
28. Moore, C.: The computer science and physics of community detection: Landscapes, phase transitions, and hardness. Bulletin of the European Association for Theoretical Computer Science **121** (2017)
29. Mossel, E., Neeman, J., Sly, A.: Reconstruction and estimation in the planted partition model. Probab. Theory Relat. Fields **162**, 431–461 (2015)
30. Mossel, E., Neeman, J., Sly, A.: A proof of the block model threshold conjecture. Combinatorica **38**(3), 665–708 (2018)
31. Mulas, R., Zhang, D., Zucal, G.: There is no going back: properties of the non-backtracking laplacian. Linear Algebra Appl. **680**, 341–370 (2024)
32. Newman, M.E.J.: Networks. Oxford University Press, An Introduction (2010)
33. Newman, M.E.J.: Message passing methods on complex networks. Proc. R. Soc. A **479**(2270), 20220774 (2023)
34. Ostrovsky, R., Rabani, Y., Schulman, L.J., Swamy, C.: The effectiveness of lloyd-type methods for the k-means problem. J. ACM (JACM) **59**(6), 1–22 (2013)
35. Soshnikov, A.: Universality at the edge of the spectrum in wigner random matrices. Commun. Math. Phys. **207**, 697–733 (1999)
36. Stephan, L., Massoulié, L.: Non-backtracking spectra of weighted inhomogeneous random graphs. Math. Stat. Learn. **5**(3), 201–271 (2022)
37. Stewart, G.W., guang Sun, J.: Matrix Perturbation Theory. Academic Press (1990)
38. Szabo, A., Ostlund, N.S.: Modern quantum chemistry: introduction to advanced electronic structure theory. Courier Corporation (1996)
39. Torres, L., Chan, K.S., Tong, H., Eliassi-Rad, T.: Nonbacktracking eigenvalues under node removal: X-centrality and targeted immunization. SIAM J. Math. Data Sci. **3**(2), 656–675 (2021)

Disjoint Zero-Sum Subsets in Abelian Groups and Their Application: A Survey

Sylwia Cichacz

Abstract We provide a summary of research on disjoint zero-sum subsets in finite Abelian groups, which is a branch of additive group theory and combinatorial number theory. An orthomorphism of a group Γ is defined as a bijection φ of Γ such that the mapping $g \mapsto g^{-1}\varphi(g)$ is also bijective. In 1981, Friedlander, Gordon, and Tannenbaum conjectured that when Γ is Abelian, for any $k \geq 2$ dividing $|\Gamma| - 1$, there exists an orthomorphism of Γ fixing the identity and permuting the remaining elements as products of disjoint k-cycles. Using the idea of disjoint zero-sum subsets, we provide a solution of this conjecture for $k = 3$ and $|\Gamma| \equiv 4 \pmod{24}$. We also present some applications of zero-sum sets in graph labeling.

1 Introduction

Assume Γ is a finite Abelian group with the operation denoted by $+$. For convenience we will write ka to denote $a + a + \cdots + a$ where the element a appears k times, $-a$ to denote the inverse of a, and we will use $a - b$ instead of $a + (-b)$. Moreover, the notation $\sum_{a \in S} a$ will be used as a short form for $a_1 + a_2 + a_3 + \ldots$, where $a_1, a_2, a_3, \ldots$ are all elements of the set S. The identity element of Γ will be denoted by 0. A subset S of Γ is referred to as a *zero-sum subset* if $\sum_{g \in S} g = 0$. Zero-sum problems typically examine the conditions under which given sequences contain non-empty zero-sum subsequences with particular properties [26]. The concept was introduced by Erdős, Ginzburg, and Ziv, who demonstrated that $2n - 1$ is the smallest integer such that every sequence over a cyclic group $\mathbb{Z}_n$ has a zero-sum subsequence of length n [20]. This finding sparked extensive research in the area.

For example, a well-studied aspect of zero-sum problems for graphs is zero-sum Ramsey theory, which can be framed as follows: What is the smallest number n such that any complete graph K_m ($m \geq n$) with edges labeled by elements of a finite group Γ contains a subgraph of a specified type where the total weight of the edges is zero in Γ?

S. Cichacz (✉)
AGH University of Krakow, Kraków, Poland
e-mail: cichacz@agh.edu.pl

G. O. H. Katona et al. (eds.), *Sum(m)it280*, Bolyai Society Mathematical Studies 32,
https://doi.org/10.1007/978-3-032-18810-6_6

Another problem for zero-sum sets is the issue of disjoint subsets in Γ. This approach was inspired by research on Steiner triple systems and initiated by Skolem in 1957, who posed the following question [39]:

Is there a partition of the set of nonzero elements of the cyclic group $\mathbb{Z}_n$ into three-element subsets such that the sum of the elements in each subset is $0 \pmod{n}$, for $n \equiv 1 \pmod{6}$?

This question received an affirmative answer [29, 39, 40] and served as the starting point for research of Steiner triple systems in Abelian groups [41] and a more general problem posed by Tannenbaum:

Problem 1 ([42]) Let Γ be an Abelian group of order m. Let t and m_i for $i \in \{1, \ldots, t\}$ be positive integers such that $\sum_{i=1}^{t} m_i = m - 1$. Let $w_i \in \Gamma$ for $i \in \{1, \ldots t\}$. Determine when there exists a partition of the set $\Gamma^* = \Gamma \backslash \{0\}$ into subsets $\{S_i\}_{i=1}^{t}$ such that $|S_i| = m_i$ and $\sum_{s \in S_i} s = w_i$ for each $i \in \{1, \ldots, t\}$.

If a respective subset partition of Γ^* exists, then we say that $\{m_i\}_{i=1}^{t}$ is *realizable* in Γ^* with $\{w_i\}_{i=1}^{t}$.

The most extensively studied case in the literature is the *Zero-Sum Partition* (ZSP) problem, which occurs when $w_i = 0$ for every $i \in \{1, \ldots, t\}$. In this paper, we will present a survey on this concept as well as an application of it to graph labeling problems. Note that many results on disjoint zero-sum sets have been obtained independently due to their application to various graph problems.

2 Preliminaries

For positive integers a and b such that $a < b$ let $[a, b] = \{a, a+1, \ldots, b\}$. Recall that any group element $\iota \in \Gamma$ of order 2 (i.e., $\iota \neq 0$ and $2\iota = 0$) is called an *involution*. Let us denote the number of involutions in Γ by $|I(\Gamma)|$. The sum of all elements of a group Γ is equal to the sum of its involutions. Therefore $\sum_{g \in \Gamma} g = \iota$ if $|I(\Gamma)| = 1$ (where ι is the involution) and $\sum_{g \in \Gamma} g = 0$, otherwise (see [18], Lemma 8).

3 Zero-Sum Partition of Abelian Group

The case most studied in the literature is the *Zero-Sum Partition (ZSP)* problem, i.e., when $w_i = 0$ for every $i \in [1, t]$ in Problem 1. Note that, for an Abelian group Γ with $|I(\Gamma)| = 1$, the set Γ^* does not admit zero-sum partitions. Moreover, given a group Γ, if $\{m_i\}_{i=1}^{t}$ is realizable in Γ^* (with $w_i = 0$ for every $i \in [1, t]$), then necessarily $m_i \geq 2$ for every $i \in [1, t]$. It was proved that this condition is sufficient when the group Γ is cyclic of odd order [24, 31], for groups $(\mathbb{Z}_p)^n$ where $p > 2$ is a prime [25], for any Abelian group of odd order [42, 47], and for Γ having exactly three involutions [47].

Let us generalize this situation with the following definition.

Definition 1 Let Γ be a finite Abelian group of order m and x be a positive integer. We say that Γ has *x-Zero-Sum Partition Property (x-ZSPP)* if, for every positive integer t and every integer partition $\{m_i\}_{i=1}^{t}$ of $m-1$ (i.e., $\sum_{i=1}^{t} m_i = m-1$) with $m_i \geq x$ for every $i \in [1, t]$, there exists a subset partition $\{S_i\}_{i=1}^{t}$ of Γ^* (i.e., subsets S_i are pairwise disjoint and their union is Γ^*) with $|S_i| = m_i$ and $\sum_{s\in S_i} s = 0$ for every $i \in [1, t]$.

The following theorem was first conjectured by Kaplan et al. [31], who also demonstrated its necessity and was later proved by Zeng in [47].

Theorem 1 ([47]) *A finite Abelian group Γ has 2-ZSPP if and only if $|I(\Gamma)| \in \{0, 3\}$.*

It was shown independently by a few authors that $\Gamma \cong (\mathbb{Z}_2)^n$, $n > 1$ has 3-ZSPP.

Theorem 2 ([7, 19, 43]) *If $\Gamma \cong (\mathbb{Z}_2)^n$ for some integer n with $n > 1$, then Γ has 3-ZSPP.*

The above theorems confirm Tannenbaum's conjecture in the cases where $|I(\Gamma)| \in \{3, |\Gamma^*|\}$:

Conjecture 1 ([42]) *Let Γ be a finite Abelian group of order m with $|I(\Gamma)| > 1$. Let $R = \Gamma\backslash(I(\Gamma) \cup \{0\})$. Then, for every positive integer t and every integer partition $\{m_i\}_{i=1}^{t}$ of $m-1$, with $m_i \geq 2$ for every $i \in [1, |R|/2]$, and $m_i \geq 3$ for every $i \in [|R|/2+1, t]$, there is a subset partition $\{S_i\}_{i=1}^{t}$ of Γ^* such that $|S_i| = m_i$ and $\sum_{s\in S_i} s = 0$ for every $i \in [1, t]$.*

Later, Tannenbaum showed that (2, 2, 2, 3, 3, 3) is not realizable in $(\mathbb{Z}_4 \oplus (\mathbb{Z}_2)^2)^*$ (see [43]). More recently, Cichacz and Suchan proved that Conjecture 1 is false in general, and it is only true in the cases covered by Theorems 1 and 2.

Theorem 3 ([16]) *Let Γ be a finite Abelian group with $|I(\Gamma)| > 1$. Let $R = \Gamma^*\backslash I(\Gamma)$. For any positive integer t and any integer partition $\{m_i\}_{i=1}^{t}$ of $|\Gamma^*|$, with $m_i \geq 2$ for all i, $1 \leq i \leq |R|/2$, and $m_i \geq 3$ for all i, $|R|/2+1 \leq i \leq t$, there is a subset partition $\{S_i\}_{i=1}^{t}$ of Γ^* such that $|S_i| = m_i$ and $\sum_{s\in S_i} s = 0$ for all $1 \leq i \leq t$ if and only if $|I(\Gamma)| \in \{3, |\Gamma^*|\}$.*

The necessity of the condition $|I(\Gamma)| \in \{3, |\Gamma^*|\}$ in the above theorem is justified by the following constructions, which show that the required zero-sum partitions do not exist in the remaining cases:

Let $|I(\Gamma)| = 2^m - 1 < |\Gamma^*|$ with $m \geq 3$ and $R = \Gamma \ (I(\Gamma) \cup \{0\})$.

- If m is even, then $(\underbrace{2, \ldots, 2}_{|R|/2-1}, \underbrace{3, \ldots, 3}_{|I(\Gamma)|/3-2}, 4, 4)$ is not realizable in Γ^*.
- If m is odd, then $(\underbrace{2, \ldots, 2}_{|R|/2-1}, \underbrace{3, \ldots, 3}_{(|I(\Gamma)|+2)/3})$ is not realizable in Γ^*.

For every Abelian group Γ with more than one involution and large enough order, first Cichacz and Tuza [17] showed that it has 4-ZSPP. This result was improved by Cichacz and Suchan [15], for every Γ of order 2^n for some integer $n > 1$ such that $|I(\Gamma)| \neq 1$, with the following theorem.

Theorem 4 ([15])*Let* Γ *be such that* $|I(\Gamma)| > 1$ *and* $|\Gamma| = 2^n$ *for some integer* $n > 1$. *Then* Γ *has* 3-*ZSPP.*

Later, Cichacz and Suchan proved the following:

Theorem 5 ([16])*Let* Γ *be such that* $|I(\Gamma)| > 1$. *Then* Γ *has* 4-*ZSPP.*

Recall that any group Γ can be factorized as $\Gamma \cong L \oplus H$, where L is the Sylow 2-group of Γ and the order of H is odd. Cichacz and Suchan proved that if ($|H|$ mod 6) $\in \{1, 3\}$, then Γ has 3-ZSPP [16]. Thus, the following conjecture remains open for groups $\Gamma \cong L \oplus H$, where L is the Sylow 2-group of Γ and $|H| \equiv 5(\mathrm{mod} 6)$.

Conjecture 2 ([11])*Let* Γ *be a finite Abelian group with* $|I(\Gamma)| > 1$. *Then* Γ *has* 3-*ZSPP.*

Note that the motivation for Conjecture 2 was the following result:

Theorem 6 ([11]) *Let* Γ *be a finite Abelian group with* $|I(\Gamma)| > 1$. *If* $m > 2$ *divides* $|\Gamma|$, *then there is a partition of* Γ *into pairwise disjoint subsets* $S_1, S_2, \ldots, S_t$, *such that* $|S_i| = m$ *and* $\sum_{s \in S_i} s = 0$ *for every* $i \in [1, t]$.

The most recent result is due to Müyesser and Pokrovskiy, who showed that if an Abelian group Γ with more than one involution has sufficiently large order, then Conjecture 2 holds [33].

3.1 Skolem Partition of Abelian Groups

A Skolem partition is a concept related to combinatorial design, particularly involving sequences and partitions of sets with certain properties. It is named after the Norwegian mathematician Thoralf Skolem [39].

A *Skolem sequence* of order n is a sequence of $2n$ elements that consists of the numbers $1, 1, 2, 2, \ldots, n, n$ arranged in such a way that for each k (where $1 \leq k \leq n$), the two occurrences of k are exactly k positions apart. For example, the sequence 42324311 is a Skolem sequence.

Skolem provided the method to derive Steiner triple system from a Skolem sequence [40] and acted as a foundation for a concept introduced by Tannenbaum. Namely, we call a 6-subset C of an Abelian group Γ *good* if $C = \{c, d, -c-d, -c, -d, c+d\}$ for some c and d in Γ. Notice that the sum of elements of a good 6-subset is 0. Moreover, it can be partitioned into three zero-sum 2-subsets or two zero-sum 3-subsets.

The following definition was given by Tannenbaum [42].

Definition 2 Let Γ be a finite Abelian group of order $m = 6k + s$ for a non-negative integer k and $s \in \{1, 3, 5\}$. A partition of Γ^* into k good 6-subsets and $(s-1)/2$ zero-sum 2-subsets is called a *Skolem partition of* Γ^*.

Note that, if Γ is an Abelian group of order m such that every integer partition $\{m_i\}_{i=1}^t$ of $m-1$ with $m_i \in \{2,3\}$ for every $1 \le i \le t$ is realizable in Γ^*, then also every integer partition $\{m'_i\}_{i=1}^{t'}$ of $m-1$ with $m'_i \ge 2$ for every $1 \le i \le t'$ is realizable. Similarly, it is easy to see that if a Skolem partition of Γ^* exists, then every integer partition $\{m_i\}_{i=1}^t$ of $m-1$ with $m_i \in \{2,3\}$ for every $1 \le i \le t$ is realizable. So the following theorem by Tannenbaum indeed offers a stronger version of 2-ZSPP.

Theorem 7 ([42])*If* Γ *is a finite Abelian group such that* $|I(\Gamma)| = 0$, *then* Γ^* *has a Skolem partition.*

It is worth mentioning that 6-good subsets have significant applications in proving the x-ZSPP [15, 16, 47] of groups, thus in [16] the authors generalized Definition 2.

Let S be a subset of cardinality $m = 6k + s$, for a positive integer k and $s \in \{0, 2, 4\}$, of a finite Abelian group Γ. A partition of S into k good 6-subsets and $s/2$ zero-sum 2-subsets is called a *Skolem partition of* S. It is known that $\mathbb{Z}_m \setminus I(\mathbb{Z}_m)$, for $m = 6k + s$ with any positive integer k, has a Skolem partition when $s \in \{0, 4\}$, but the situation is slightly different for $s = 2$. For $m \equiv 2$ or $8 \pmod{24}$, there exists a Skolem partition, whereas for $m \equiv 14$ or $20 \pmod{24}$, such a partition does not exist (see Lemma 4 in [42]).

Note that there exist groups Γ with $|I(\Gamma)| > 1$ such that the set $R = \Gamma^* \setminus I(\Gamma)$ has a Skolem partition. For instance, let $\Gamma \cong (\mathbb{Z}_2)^\eta \oplus H$ for some integer $\eta > 1$ and $|H| \equiv 1 \pmod{6}$ [16].

We finish this section with the open problem.

Problem 2 ([16])Characterize finite Abelian groups Γ for which $R = \Gamma^* \setminus I(\Gamma)$ has a Skolem partition.

4 Disjoint Zero-Sum Subsets

Recall that for an Abelian group Γ with $|I(\Gamma)| = 1$, the set Γ^* does not have zero-sum partitions. Therefore, for such groups, one can consider the set $R = \Gamma^* \setminus I(\Gamma)$.

For cyclic groups, it was proved:

Corollary 1 ([42, 47]) *Let* n *be an even positive integer and* m, l *be non-negative integers such that* $3m + 2l = n - 2$. *Then the set* $R = \mathbb{Z}_n \setminus \{0, \frac{n}{2}\}$ *can be partitioned into* m *zero-sum* 3*-sets* $A_1, A_2, \ldots, A_m$ *and* l *zero-sum* 2*-sets* $B_1, B_2, \ldots, B_l$.

For general groups with one involution, a weaker result was obtained recently:

Theorem 8 ([13]) *Let* Γ *be an Abelian group of order* n *such that it has a unique involution* ι_Γ *and consider any integers* $r_1, r_2, \ldots, r_t$ *with* $n - 2 = r_1 + r_2 + \cdots + r_t$ *and* $r_1 \ge 3$, $r_i \ge 4$ *for all* $2 \le i \le t$. *Then there exist pairwise disjoint zero-sum subsets* $S_1, S_2, \ldots, S_t$ *in* $\Gamma \setminus \{0, \iota_\Gamma\}$ *such that* $|S_i| = r_i$ *for all* $1 \le i \le t$.

Motivated by the above theorem and Corollary 1, we propose the following conjecture for groups Γ having a unique involution.

Conjecture 3 ([13]) *Let* Γ *be a finite Abelian group of order* m *with* $|I(\Gamma)| = 1$. *Let* $R = \Gamma \setminus (\{0\} \cup I(\Gamma))$. *For every positive integer* t *and integer partition* $\{m_i\}_{i=1}^t$ *of* $m-2$ *with* $m_i \geq 2$ *for every* $1 \leq i \leq t$, *there is a subset partition* $\{S_i\}_{i=1}^t$ *of* R *such that* $|S_i| = m_i$ *and* $\sum_{s \in S_i} s = 0$ *for every* $1 \leq i \leq t$.

5 Group Orthomorphism

A *complete mapping* of a group Γ is defined as a bijection φ of Γ such that the mapping $\theta\colon g \mapsto g\varphi(g)$ is also bijective [27]. A related concept is an *orthomorphism* of Γ, which is a bijection η of Γ such that the mapping $\tilde{\theta}\colon g \mapsto g^{-1}\eta(g)$ is also bijective. See [21] for a more detailed introduction to the theory of complete mappings and orthomorphisms. Originally, complete mappings were introduced by Mann in 1942 as a tool for constructing mutually orthogonal Latin squares [32].

For finite Abelian groups, the following was proved:

Theorem 9 ([27, 28]) *A finite Abelian* Γ *group has a complete mapping if and only if* $|I(\Gamma)| \neq 1$.

The most famous conjecture in the area is the Hall-Paige conjecture [28], which states that a group Γ admits a complete mapping if and only if Γ is a finite group and the Sylow 2-subgroups of Γ are either trivial or non-cyclic. Hall and Paige showed that a finite group with a nontrivial cyclic Sylow 2-subgroup does not admit complete mappings. The converse was established in 2009 by Wilcox [46], Evans [22], and Bray et al. [6].

A group Γ of order n is said to be *R-sequenceable* if the nonidentity elements of the group can be listed in a sequence $g_1, g_2, \ldots, g_{n-1}$ such that $g_1^{-1}g_2, g_2^{-1}g_3, \ldots, g_{n-1}^{-1}g_1$ are all distinct. This idea was introduced in 1974 by Ringel [37], who used this concept in his solution of Heawood's map coloring problem. Friedlander, Gordon and Tannenbaum generalized the notion of R-sequenceability by stating the following conjecture:

Conjecture 4 ([24]) *Let* Γ *be an Abelian group of order* n *such that* $|I(\Gamma)| \neq 1$. *Suppose for some integer* $k \geq 2$ *that* k *divides* $n-1$. *Then, there exists an orthomorphism of* Γ *that fixes the identity element and permutes the remaining elements as products of disjoint cycles of length* k.

The above conjecture is still open. There are several partial results. Friedlander, Gordon, and Tannenbaum confirmed their conjecture for groups of order at most 15, and Abelian p-groups where $p \geq 3$ [24]. Recently, the conjecture was confirmed for sufficiently large groups [33].

We will show the following:

Theorem 10 *Let* Γ *be an Abelian group such that* $|\Gamma| \cong 1 \pmod 3$, $|I(\Gamma)| \neq 1$. *If* $\Gamma \cong L \oplus H$, *with* $|L| = 2^\eta \equiv 1 \pmod 3$ *for some positive integer* η, *and* $|H| \equiv 1 \pmod 6$, *then there exists an orthomorphism of* Γ *that fixes the identity element and permutes the remaining elements as products of disjoint cycles of length* 3.

Proof The group Γ has 3-ZSPP (see Theorem 2.2, [16]), thus it can be written as:

$$\Gamma = \{0\} \cup \bigcup_{i=1}^{(|\Gamma|-1)/3} \{x_0^i, x_1^i, x_2^i\},$$

where $x_0^i + x_1^i + x_2^i = 0$ for any $i = 1, 2, \ldots, (|\Gamma| - 1)/3$. Set now $\varphi(x_j^i) = -x_{j+2}^i$ for $j = 0, 1, 2$, $i = 1, 2, \ldots, (|\Gamma| - 1)/3$, where the subscripts are taken modulo 3. Note that $\varphi(x_j^i) - x_j^i = -x_{j+2}^i - x_j^i = x_{j+1}^i$. □

6 Some Applications in Graph Labeling

In this section, we explore applications of zero-sum sets in Abelian groups to problems involving magic-type labelings of graphs. Generally, such a labeling for a graph $G = (V, E)$ is defined as a mapping from either V, E, or their union $V \cup E$ to a set of labels, which is usually a set of integers or group elements. The weight of a graph element is typically calculated as the sum of labels of adjacent or incident elements, either of one type or both. When the weight of all elements is required to be identical, we refer to it as magic-type labeling, whereas if all weights must be distinct, it is called antimagic-type labeling. Perhaps the best-known problem in this field is the anti-magic conjecture (see [38]), which posits that for every graph except K_2, there exists a bijective labeling of edges with integers $1, 2, \ldots, |E|$ such that each vertex has a unique weight. This conjecture remains unsolved.

6.1 Antimagic Labeling of Trees

In [31], Kaplan, Lev and Roditty considered the following generalization of the concept of an antimagic graph.

Let Γ be an Abelian group (not necessarily finite) and A be a finite subset of $\Gamma^* = \Gamma \backslash \{0\}$ with $|A| = |E(G)|$. An A-labeling of $G = (V, E)$ is a one-to-one mapping $f \colon E(G) \to A$. The weight of every vertex is calculated as the sum (taken in Γ) of the labels of incident edges. We say that G is *A-antimagic* if all the weights differ.

As in [31], we define *k-tree T* to be a rooted tree, where every vertex that is not a leaf has at least k children. We have the following corollary:

Corollary 2 ([31]) *Let Γ be a finite Abelian group with the 2-ZSPP. Then every 2-tree with $|\Gamma|$ vertices is Γ^*-antimagic.*

Let $v_1, v_2, \ldots, v_t$ be the vertices of 2-tree which are not leaves, and denote their corresponding numbers of children by $r_1, r_2, \ldots, r_t$. Since $r_i \geq 2$ there exists a zero-sum partition of Γ^* into sets $A_1, A_2, \ldots, A_t$ of cardinalities $r_1, r_2, \ldots, r_t$, respectively. Label now the edges of the set $N_i = \{v_i w \colon w \text{ is a child of } v_i\}$ by the elements of A_i. Since each vertex of T, except the root, has a unique parent, it easily follows that all the weights are distinct.

Kaplan, Lev, and Roditty proved that every 2-tree has a $(\mathbb{Z}_n)^*$-antimagic labeling if n is odd. Moreover, from this result it follows that every 2-tree of odd order is antimagic. Zeng proved that an Abelian finite group Γ has 2-ZSPP if and only if $|I(\Gamma)| \in \{0, 3\}$ [47]. As a corollary, he obtained that every 2-tree has a Γ^*-antimagic labeling if $|I(\Gamma)| \in \{0, 3\}$. Theorem 5 implies that every 4-tree of order Γ is Γ^*-antimagic.

Kaplan, Lev, and Roditty conjectured that a tree with $|\Gamma|$ vertices is Γ^*-antimagic if and only if $|I(\Gamma)| \neq 1$. However, they showed that $|I(\Gamma)| \neq 1$ is the necessary condition; the conjecture is not true. It is enough to consider a path P_{2^m}, $m > 1$ and a group $A \cong (\mathbb{Z}_2)^m$. Observe that there do not exist two non-zero elements in A that sum up to 0. Note that paths are not the only examples of such trees, others are trees of order n with maximum degree $n-2$ [12]. The following open problem was posted in [12].

Problem 3 ([12]) Characterize trees such that they have Γ^*-antimagic labeling for any Γ with $|I(\Gamma)| \neq 1$.

6.2 *Group Irregular Labeling of Graphs*

It is well known that in any simple graph G, at least two vertices with the same degree exist. However, the situation changes if we consider an edge labeling $f : E(G) \to [1, k]$ (where labels need not be distinct) and calculate the *weighted degree* of each vertex v as the sum of labels of all edges incident to v. The labeling f is called *irregular* if all vertex weighted degrees are unique, making it an antimagic-type labeling. The smallest value of k that permits an irregular labeling is known as the *irregularity strength of* G, denoted by $s(G)$. This problem was introduced in [8] and has attracted considerable interest. Generally, it is known that $s(G) \leq n-1$ for any graph G of order n with no isolated edges and at most one isolated vertex, except for K_3, as shown by Aigner and Triesch [1] and Nierhoff [35]. Although the upper bound is tight for the family of star graphs, this result can be refined for graphs with sufficiently large minimum degree δ. The best general upper bound was proved by Przybyło and Wei, who proved that for any $\varepsilon \in (0, 0.25)$ there exist absolute constants c_1, c_2 such that for all graphs G on n vertices with minimum degree $\delta \geq 1$ and without isolated edges, $s(G) \leq n/\delta(1 + c_1/\delta^{\varepsilon}) + c_2$ [36].

Tuza began considering irregular labeling of a graph $G = (V, E)$ by the group $\Gamma = (\mathbb{Z}_2)^m$. In this labeling, edges receive weights from elements of Γ and the weighted degrees of vertices are computed using operations in Γ. Tuza denoted $m_M(G)$ as the smallest m for which a $(\mathbb{Z}_2)^m$-irregular labeling of a graph G exists, and showed that $m_M(G) \leq 3\log_2|V|$ [44]. This result was later improved by Aigner and Triesch in [2] to $m_M(G) \leq \lceil \log_2|V| \rceil + 4$. These authors utilized the observation that this problem is related to subsets with zero-sum in the group $(\mathbb{Z}_2)^m$.

It is noteworthy that if there exists a $(\mathbb{Z}_2)^m$-irregular labeling of a graph G with connected components $\{C_i\}_{i=1}^t$, then the sum of weighted degrees in each component equals zero (this follows because the label of each edge is added twice to the weighted degrees). Aigner and Triesch conjectured $m_M(G) \leq \lceil \log_2 |V| \rceil + 1$ for any graph G. For large $|V| = n$, this was proved by Tuza using probabilistic methods in [45]. It was subsequently proved for any graph by Caccetta and Jia [7], as well as independently by Egawa [19], who focused on partitioning the group Γ into subsets with zero-sum (we will put it into more details in the next section).

In [4] other groups were considered. The authors introduced *group irregularity strength* (denoted $s_g(G)$) of G as the smallest k such that for every Abelian group Γ of order k there exists a Γ-irregular labeling of G. Note that $s(G) \leq s_g(G)$ for every graph G.

Anholcer, Cichacz and Milanič have shown the following theorem that describes the value of $s_g(G)$ for all connected graphs G of order $n \geq 3$.

Theorem 11 ([4]) *Let G be an arbitrary connected graph of order $n \geq 3$. Then*

$$s_g(G) = \begin{cases} n+2 & \text{if } G \cong K_{1,3^{2q+1}-2} \text{ for some integer } q \geq 1 \\ n+1 & \text{if } n \equiv 2 (\text{mod } 4) \wedge G \ncong K_{1,3^{2q+1}-2} \text{ for any integer } q \geq 1 \\ n & \text{otherwise.} \end{cases}$$

It is easy to see that to distinguish all n vertices in an arbitrary graph of that order, we need at least n different elements of Γ. However, the following lemma shows that an Abelian group of order n is not always enough to have a Γ-irregular labeling of G.

Lemma 1 *([4]) Let G be a graph of order n. If $n \equiv 2 \pmod 4$, then there is no Γ-irregular labeling of G for any Abelian group Γ of order n.*

So far, the best upper bound for any graph was given by Anholcer, Cichacz, and Przybyło in [5]

Theorem 12 ([5]) *Let G be an arbitrary graph of order n having no component of order less than 3. Then $s_g(G) \leq 2n$.*

In [3], the 2-ZSPP of groups of odd order is used to bound the group irregularity strength of disconnected graphs without a star as a connected component. Roughly speaking, the authors divide every connected component into 2-subsets and 3-subsets of vertices and partition the set of non-zero elements of the corresponding group into the same number of zero-sum 2-subsets and 3-subsets, and later use the method of augmented paths to find the labeling. Namely, for any two given vertices v_1 and v_2 from the same connected component of a graph G, there exist walks from v_1 to v_2. Begin by labeling all edges of G by 0. Next, select v_1 and v_2 and modify all of the edges of a chosen walk from v_1 to v_2 by adding some element of an Abelian group Γ. Now, add some element $a \in \Gamma$ to all the labels of the edges in an odd position on the walk, starting from v_1, and add $-a$ to the rest of the labels of edges. This modification of labels increases the weighted degree of v_1 by a and increases the weighted degree of v_2 by a or $-a$. We should also note that in both cases, the operation does not modify the weighted degree of any other vertices of the walk.

In [14], the authors used the above method and Theorem 7 to improve Theorem 12 for graphs without small stars.

Theorem 13 ([14]) *Let G be a graph of order n having neither a component of order less than 3 nor a $K_{1,1+2u}$ component for $u \in \{1, 2\}$. Then $s_g(G) \leq n + 7$.*

Moreover, it was shown that for a graph G of order n having neither a component of order less than 3 nor a $K_{1,1+2u}$ component for $u \in \{1, 2\}$, there exists a Γ-irregular labeling for any group Γ of odd order $t \geq n + 7$.

6.3 Group Irregular Labeling of Directed Graphs

Let $\overrightarrow{G}$ be a directed graph of order n. Suppose that there exists a mapping ψ from the arc set $E(\overrightarrow{G})$ to an Abelian group Γ such that a mapping φ_ψ from the vertex set $V(\overrightarrow{G})$ defined as

$$\varphi_\psi(x) = \sum_{y \in N^+(x)} \psi(xy) - \sum_{y \in N^-(x)} \psi(yx), \quad (x \in V(\overrightarrow{G})),$$

is injective. Then we say that $\overrightarrow{G}$ is *realizable* in Γ, and that the mapping ψ is *Γ-irregular.*

Since for $\Gamma = (\mathbb{Z}_2)^m$, it does not matter whether we consider Γ-irregular labeling of a simple graph G or a directed graph $\overrightarrow{G}$, this problem for $(\mathbb{Z}_p)^m$ (where p is an odd prime number) was defined by Fukuchi in [25] as a generalization of the above-mentioned problem of Tuza [44]. Analogously to the results in [7, 19], Fukuchi showed that there exists a $(\mathbb{Z}_p)^m$-irregular labeling of the directed graph $\overrightarrow{G}$ with weakly connected components $\{\overrightarrow{G}_i\}_{i=1}^t$ if and only if there are pairwise disjoint subsets $\{S_i\}_{i=1}^t$ in $(\mathbb{Z}_p)^m$ such that $|S_i| = |V(\overrightarrow{G}_i)|$ and $\sum_{s \in S_i} s = 0$ for $i \in [1, t]$. This result was then extended to any Abelian group Γ by Cichacz and Tuza [17][1].

Fukuchi demonstrated that $(\mathbb{Z}_p)^m$ can be partitioned into zero-sum subsets $\{S_i\}_{i=1}^t$, thereby achieving the result stating that any directed graph $\overrightarrow{G}$ of order n is realizable in any group $(\mathbb{Z}_p)^m$ provided $n \leq p^m$.

Based on the results of Tannenbaum [42] and Zeng [47], we immediately conclude that if $|I(\Gamma)| \in \{0, 3\}$ and $|\Gamma| \geq |V(\overrightarrow{G})|$, then any $\overrightarrow{G}$ without components of order less than 3 is realizable in Γ.

Moreover, using the result from [13, 15, 16] we obtain the following.

Theorem 14 ([13, 15, 16]) *Any digraph $\overrightarrow{G}$ of order n with no weakly connected components of order less than 4 has a Γ-irregular labeling for every Γ such that $|\Gamma| \geq n + 5$.*

[1] This observation also follows from a more general result on the group connectivity of graphs obtained by Jaeger, Linial, Payan, and Tarsi in [30].

Cichacz and Tuza proved that if n is large enough with respect to an arbitrarily fixed $\varepsilon > 0$ then $\overrightarrow{G}$ has a Γ-irregular labeling for any Γ such that $|\Gamma| > (1+\varepsilon)n$ [17]. Therefore, Cichacz and Suchan stated the conjecture.

Conjecture 5 ([16])*There exists a constant K such that any digraph $\overrightarrow{G}$ of order n with no weakly connected components of order less than 3 has a Γ-irregular labeling for every Γ such that $|\Gamma| \geq n + K$.*

6.4 Group Distance Magic Labeling

A Γ-*distance magic labeling* of a graph $G = (V, E)$ with $|V| = n$ is a bijection ℓ between V and an Abelian group Γ of order n such that the weight $w(x) = \sum_{y\in N(x)} \ell(y)$ of every vertex $x \in V$ is equal to the same element $\mu \in \Gamma$, called the *magic constant*.

Notice that the constant sum partitions of a group Γ lead to complete multipartite Γ-distance magic labeled graphs. For instance, the partition $\{0\}, \{1, 2, 4\}, \{3, 5, 6\}$ of the group $\mathbb{Z}_7$ with constant sum 0 leads to a $\mathbb{Z}_7$-distance magic labeling of the complete tripartite graph $K_{1,3,3}$ (see [11]). In general, suppose we have a constant sum partition of Γ: $\{S_i\}_{i=1}^t$ of Γ with $|S_i| = m_i$ and $\sum_{s\in S_i} s = \nu$ for every $i \in [1, t]$ and some $\nu \in \Gamma$. Let G be a complete t-partite graph with the color classes $\{A_i\}_{i=1}^t$, where $|A_i| = m_i$ for every $i \in [1, t]$. Let us label the vertices of A_i with distinct elements of S_i for every $i \in [1, t]$, and compute the weight of every vertex as the sum (in Γ) of the labels of its neighbors. It is easy to see that all the weights are equal, and thus we have a Γ-distance magic labeling. On the other hand, suppose $G = K_{m_1,\ldots,m_t}$ is a complete t-partite graph of order m with the color classes $\{A_i\}_{i=1}^t$ that is Γ-distance magic with a labeling ℓ. So $\sum_{i=1,i\neq j}^t \sum_{x\in A_i} \ell(x) = \mu$ for every $j \in [1, t]$, which implies that $\sum_{x\in A_j} \ell(x) = \nu$ for every $j \in [1, t]$, and some $\nu \in \Gamma$.

Let $G = K_{m_1,\ldots,m_t}$ be a complete t-partite graph of order m. Let now $1 \leq m_1 \leq m_2 \leq \cdots \leq m_t$. Using some constant-sum partition properties of every finite Abelian group Γ of order m, the following results are proved:

1. For $t = 2$, if $m_1 + m_2 \not\equiv 2 \pmod 4$ then G is Γ-distance magic [9].
2. For $t = 3$, if $m_2 > 1$ and $m_1 + m_2 + m_3 \neq 2^p$ for any positive integer p, or $m_1 \neq 2$ and $m_2 > 2$, then G admits a Γ-distance magic labeling [10].
3. If $m_1 = m_2 = \cdots = m_t > 2$ and $|I(\Gamma)| \neq 1$, then G admits a Γ-distance magic labeling [11].
4. If $m_1 \geq 3$ and $m_t \geq \frac{1}{2}(m + \sqrt{2m+1}) - 1$, and $|I(\Gamma)| \neq 1$, then G admits a Γ-distance magic labeling [17].
5. If $m_1 \geq 4$, m is large enough, and $|I(\Gamma)| \neq 1$, then G admits a Γ-distance magic labeling [17],
6. If $m_2 \geq 2$, t is odd, and $\Gamma \cong \mathbb{Z}_n$, then G admits a Γ-distance magic labeling [23].

7 Statements and Declarations

This work was partially supported by program "Excellence initiative - research university" for the AGH University of Krakow as well as by AGH University of Krakow grant no. 16.16.420.054, funded by the Polish Ministry of Science and Higher Education. This manuscript does not rely on any datasets. The author has no relevant financial or non-financial interests to disclose.

References

1. Aigner, M., E., T.: Irregular assignments of trees and forests. SIAM J. Disc. Math. **3**(4), 439–449 (1990)
2. Aigner, M., Triesch, E.: Codings of graphs with binary edge labels. Graphs Comb. **10**, 1–10 (1994)
3. Anholcer, M., Cichacz, S.: Group irregular labelings of disconnected graphs. Contrib. Discret. Math. **12**, 2 (2017)
4. Anholcer, M., Cichacz, S., Milanič, M.: Group irregularity strength of connected graphs. J. Comb. Optim. **30**(1), 1–17 (2015)
5. Anholcer, M., Cichacz, S., Przybyło, J.: Linear bound on nowhere-zero group irregularity strength and nowhere-zero group sum chromatic number of graphs. Appl. Math. Comput. **343**, 149–155 (2019)
6. Bray, J.N., Cai, Q., Cameron, P.J., Spiga, P., Zhang, H.: The Hall-Paige conjecture, and synchronization for affine and diagonal groups. J. Algebra **545**, 27–42 (2020)
7. Caccetta, L., Jia, R.-Z.: Binary labeling of graphs. Graphs Comb. **13**(2), 119–137 (1997)
8. Chartrand, G., Jacobson, M., Lehel, J., Oellermann, O., Ruiz, S., Saba, F.: Irregular networks. Congr. Numer. **64**, 187–192 (1988)
9. Cichacz, S.: Note on group distance magic complete bipartite graphs. Open Math. **12**(3), 529–533 (2014)
10. Cichacz, S.: On zero sum-partition of abelian groups into three sets and group distance magic labeling. Ars Mathematica Contemporanea **13**(2), 417–425 (2017)
11. Cichacz, S.: Zero sum partition of abelian groups into sets of the same order and its applications. Electron. J. Comb. **25**(1), 1–20 (2018)
12. Cichacz, S.: E_A -cordial labeling of graphs and its implications for A-antimagic labeling of trees. Discret. Math. **348**, 114493 (2025)
13. Cichacz, S.: Irregular graph labelings in abelian groups. Ars Mathematica Contemporanea 26 (2026) #P2.09. https://doi.org/10.26493/1855-3974.3380.cb8
14. Cichacz, S., Krupińska, B.: Group irregularity strength of disconnected graphs. Discret. Math. **348**, 114548 (2025)
15. Cichacz, S., Suchan, K.: Zero-sum partitions of abelian groups of order 2^n. Disc. Math. Theor. Comput. Sci. **25**, 1, #6 (2023)
16. Cichacz, S., Suchan, K.: Zero-sum partitions of abelian groups and their applications to magic- and antimagic-type labelings. Discret. Math. Theor. Comput. Sci. **26**, 3, #14 (2024)
17. Cichacz, S., Tuza, Z.: Realization of digraphs in abelian groups and its consequences. J. Graph Theory 1–15 (2021)
18. Combe, D., Nelson, A.M., Palmer, W.D.: Magic labellings of graphs over finite abelian groups. Australasian J. Comb. **29**, 259–272 (2004)
19. Egawa, Y.: Graph labelings in elementary abelian 2-groups. Tokyo J. Math. **20**(2), 365–379 (1997)
20. Erdős, P., Ginzburg, A., Ziv, A.: Theorem in the additive number theory. Bull. Res. Council Israel **10**, 41–43 (1961)

21. Evans, A.: Orthomorphism Graphs of Groups. Springer, Berlin (1992)
22. Evans, A.: The admissibility of sporadic simple groups. J. Algebra **321**(1), 105–116 (2009)
23. Freyberg, B.: On constant sum partitions and applications to distance magic-type graphs. AKCE Int. J. Graphs Comb. **17**(1), 373–379 (2020)
24. Friedlander, R., Gordon, B., Tannenbaum, P.: Partitions of groups and complete mappings. Pac. J. Math. **92**(2), 283–293 (1981)
25. Fukuchi, Y.: Graph labelings in elementary abelian groups. Discret. Math. **189**(1–3), 117–122 (1998)
26. Gao, W., Geroldinger, A.: Zero-sum problems in finite abelian groups: A survey. Expo. Math. **24**(4), 337–369 (2006)
27. Hall, M.: A combinatorial problem on abelian groups. Proc. Am. Math. Soc. **3**(4), 584–587 (1952)
28. Hall, M., Paige, L.: Complete mappings of finite groups. Pac. J. Math. **5**, 541–549 (1955)
29. Hanani, H.: Note on steiner triple systems. Math. Scand. **8**, 154–156 (1960)
30. Jaeger, F., Linial, N., Payan, C., Tarsi, M.: Group connectivity of graphs-a nonhomogeneous analogue of nowhere-zero flow properties. J. Comb. Theory Ser. B **56**(2), 165–182 (1992)
31. Kaplan, G., Lev, A., Roditty, Y.: On zero-sum partitions and anti-magic trees. Discret. Math. **309**(8), 2010–2014 (2009)
32. Mann, H.B.: On orthogonal latin squares. Bull. Am. Math. Soc. **50**, 249–257 (1944)
33. Müyesser, A. Pokrovskiy, A.: Random hall-paige conjecture. Invent. Math. **240**, 779–867 (2025)
34. Müyesser, A.: Cycle type in Hall-Paige: a proof of the Friedlander-Gordon-Tannenbaum. Forum Math. Sigma **14**, e50 (2026). https://doi.org/10.1017/fms.2026.10197
35. Nierhoff, T.: A tight bound on the irregularity strength of graphs. SIAM J. Discret. Math. **13**(3), 313–323 (2000)
36. Przybyło, J., Wei, F.: On the asymptotic confirmation of the Faudree-Lehel conjecture for general graphs. Combinatorica **43**, 791–826 (2023)
37. Ringel, G.: Cyclic arrangements of the elements of a group. Notices Am. Math. Soc. **21**, A95–A96 (1974)
38. Ringel, G., Hartsfield, N.: Pearls in Graph Theory. Academic Press Limited, UK (1994)
39. Skolem, T.: On certain distributions of integers in pairs with given differences. Mathematica Scandinavica 57–68 (1957)
40. Skolem, T.: Some remarks on the triple systems of steiner. Mathematica Scandinavica, 273–280 (1958)
41. Tannenbaum, P.: Abelian steiner triple systems. Canad. J. Math. **28**, 1251–1268 (1976)
42. Tannenbaum, P.: Partitions of abelian groups into sets with zero sums. Congr. Numer. **33**, 341–348 (1981)
43. Tannenbaum, P.: Partitions of $\mathbb{Z}_2^n$. SIAM J. Algebr. Discret. Methods **4**(1), 22–29 (1983)
44. Tuza, Z.: Encoding the vertices of a graph with binary edge labels. Sequences 287–299 (1990)
45. Tuza, Z.: Zero-sum block designs and graph labelings. J. Comb. Des. **3**, 89–99 (1995)
46. Wilcox, S.: Reduction of the hall-paige conjecture to sporadic simple groups. J. Algebra **321**(5), 1407–1428 (2009)
47. Zeng, X.: On zero-sum partitions of abelian groups. Integers **15**, Paper No. A44: 16 pp (2015)

A Stopping Game on Zero-Sum Sequences

Adrian Dumitrescu and Arsenii Sagdeev

Abstract We introduce and analyze a natural game formulated as follows. In this one-person game, the player is given a random permutation $A = (a_1, \dots, a_n)$ of a multiset M of n reals that sum up to 0, where each of the $n!$ permutation sequences is equally likely. The player only knows the value of n beforehand. The elements of the sequence are revealed one by one and the player can stop the game at any time. Once the process stops, say, after the ith element is revealed, the player collects the amount $\sum_{j=i+1}^{n} a_j$ as his/her payoff and the game is over (the payoff corresponds to the unrevealed part of the sequence). Three online algorithms are given for maximizing the expected payoff in the binary case when M contains only 1's and -1's. ALGORITHM 1 is slightly suboptimal, but is easier to analyze. Moreover, it can also be used when n is only known with some approximation. ALGORITHM 2 is exactly optimal but not so easy to analyze on its own. ALGORITHM 3 is the simplest of all three. It turns out that the expected payoffs of the player are $\Theta(\sqrt{n})$ for all three algorithms. In the end, we address the general problem and deal with an arbitrary zero-sum multiset, for which we show that our ALGORITHM 3 returns a payoff proportional to $\sqrt{n}$, which is worst case-optimal.

A. Dumitrescu (✉)
Algoresearch L.L.C., Milwaukee, WI, USA
e-mail: ad.dumitrescu@algoresearch.org

Research Institute of the University of Bucharest, Bucharest, Romania

A. Dumitrescu · A. Sagdeev
Alfréd Rényi Institute of Mathematics, Budapest, Hungary
e-mail: sagdeevarsenii@gmail.com

A. Sagdeev
Karlsruhe Institute of Technology, Karlsruhe, Germany

G. O. H. Katona et al. (eds.), *Sum(m)it280*, Bolyai Society Mathematical Studies 32,
https://doi.org/10.1007/978-3-032-18810-6_7

1 Introduction

Stopping rules are of key importance in the study of random processes and/or online algorithms. Here is an illustrative example concerning the behavior of the largest payoff expectation in a game proposed by Leo Moser almost 70 years ago [10].

A real number is drawn at random from [0, 1]. We may either keep the number selected, or reject it and draw again. We can then either keep the second number chosen or reject it, and draw again, and so on. Suppose we have at most n choices, and the payoff is the number we keep. What stopping rule gives the largest payoff expectation E_n and how can we estimate E_n?

Moser showed that, for any distribution, given at most n choices, one should stop after the first choice if and only if the outcome is at least E_{n-1}. In particular, for the uniform distribution, $f_1(x) = 1$ for $0 \leq x \leq 1$, E_n satisfies the recurrence $E_{n+1} = \frac{1}{2}(1 + E_n^2)$, with $E_0 = 0$, $E_1 = 1/2$. For illustration, we list some values in Table 1. Observe that $\lim_{n\to\infty} E_n = 1$. The case of the normal distribution was studied by Guttman [5]. Other examples of stopping rules, mostly from the area of selection, are mentioned in the "Related work" paragraph later in this section. See also [8, Chap. 13].

Table 1 E_n in Moser's procedure for the uniform distribution (the first three digits)

n	E_n	n	E_n	n	E_n
1	0.500	5	0.775	100	0.981
2	0.625	10	0.861	200	0.990
3	0.695	20	0.919	300	0.993

Here we consider the following one-person game. Let M be a (multi)set of n reals that sum up to 0. The value of n is known to the player beforehand. The player is given a random permutation $A = (a_1, \ldots, a_n)$ of M, where each of the $n!$ permutation sequences is equally likely. The elements of the sequence are revealed one by one and the player can stop the game at any step, i.e., after the a_i is revealed, for any $i = 1, 2, \ldots, n$, or the stop is automatic at the end of the process. Once the process stops, say, after the ith element is revealed, the player collects the amount $\sum_{j=i+1}^{n} a_j$ as his/her payoff and the game is over. Note that the payoff corresponds to the unrevealed part of the sequence.

Observe that any rational strategy for the game achieves a nonnegative outcome, since the player can always reduce its losses by continuing to play until the end, where he/she could force a zero payoff.

In the *binary case*, i.e., if $n = 2m$ and M consists of m copies of -1's and 1's, respectively, we define the *state* of the process as the pair (i, j), where i and j denote the numbers of -1 and $+1$, revealed so far, respectively.

Example.

Let $A = (-1, 1, 1, 1, -1, 1, -1, -1)$, with $n = 8$. If the player stops the game before the first element (or after the last one) is revealed, his/her payoff is zero. If the player stops the game before the second element is revealed, his/her payoff is 1 (the total amount of the unrevealed subsequence). Stopping the game at any other step would be irrational—and is therefore omitted from discussion.

Comment.

This game played on the multiset $-M$ is equivalent to the dual version of the game in which the payoff is the sum of the revealed values up to the stopping point. Indeed, for each zero-sum sequence, we have $\sum_{j=1}^{i} a_j = 0 - \sum_{j=i+1}^{n} a_j = \sum_{j=i+1}^{n} -a_j$.

Our results.

We present three online algorithms for maximizing the expected payoff in the binary case. ALGORITHM 1 (in Sect. 3) is slightly suboptimal, but is easier to analyze. Moreover, it can also be used when n is only known with some approximation (in Sect. 7).

ALGORITHM 2 (in Sect. 4) is optimal but not so easy to analyze on its own. This algorithm requires quadratic preprocessing time but is otherwise very simple to use.

ALGORITHM 3 (in Sects. 5 and 6) is the simplest of all three. Moreover, it is applicable to any zero-sum multiset.

We show that the three algorithms share the same asymptotic performance, $\Theta(\sqrt{n})$, which is worst-case optimal. Our results are summarized in the following.

Theorem 1 *Let $M = \{a_1, a_2, \ldots, a_n\}$ be an arbitrary zero-sum multiset of reals. There is an online algorithm that reads a random permutation of M and stops with an expected payoff of $\Omega(\mu\sqrt{n})$, where $\mu = \sum_{i=1}^{n} |a_i|/n$ is the average absolute value for the input sequence. This bound is worst-case optimal, i.e., equality is attained (apart from the hidden constant factor) for a uniform sequence, with $n/2$ values equal to μ and $-\mu$, respectively.*

Related work.

The *best choice* problem for partially ordered sets considers a selection process in an online setting [3, 12]. Let X be a finite set equipped with a strict partial order $\prec$, i.e., an asymmetric and transitive relation. The objects (elements) of X are shuffled—each of the $|X|!$ permutations is equally likely—and appear one by one before a selector **S**, who must make a single online choice. The tth element x_t appears at time $t = 1, 2 \ldots, n$. The selector knows the cardinality $n = |X|$ but is *ignorant* of $\prec$. **S** has access to the comparisons between elements that appeared so far; thus after time t, **S** only knows the relation $\{(i, j) \colon i, j \leq t, x_i \prec x_j\}$.

The selector's goal is to pick an element that is *maximal* with respect to $\prec$. That is, his/her goal is to stop the process at some time τ (that depends on the permutation) so that the currently examined element x_t is *maximal* with respect to $\prec$. Such a random variable τ is called a *stopping time*. The question that arises is: (Q) Can the selector obtain a maximal element with probability at least δ, for some constant δ?—irrespective of $(X, \prec)$. Preater [12] was the first to obtain such an algorithm: the selector can pick a maximal element with probability at least $1/8$. Georgiou, Kuchta, Morayne, and Niemiec [3] improved the analysis of Preater's algorithm and showed that the success probability is at least $1/4$, which is tight for that algorithm.

The case when $\prec$ is a total (i.e., linear) order and **S** knows that beforehand, is the classic *secretary problem* [2, 6]. In the language of online algorithms, one can refer to the problem as the *online maximum* problem: the elements in a random sequence A of n (distinct) numbers from an unknown distribution are revealed one by one, and the goal is to select the maximum by an irrevocable decision made immediately after examining the ith element for some $i = 1, 2, \dots, n$. Each of the $n!$ permutations of the ranks is equally likely. Note that inaction leads to the automatic, i.e., forced, selection of the last (the nth) element in the sequence.

The optimal algorithm achieves this task with a probability of success bounded from below by a constant, see [11, pp. 73–77] and [6, pp. 90–95]. For large n, the optimal strategy passes approximately a fraction of $1/e$ [1] of the sequence and chooses the first element that is larger than every element of the initial prefix. Then the probability of success is approximately $1/e = 0.367\dots$. Similar algorithms, whose probabilities of success are bounded from below by a constant, can be designed for finding the *online minimum*, *online minimum and maximum*, *online smallest two and largest two*, and for other variants.

Under the additional assumption that the random numbers are uniformly distributed between 0 and 1, namely in Moser's problem described in the introduction, the probability of success of the optimal strategy is about 0.580 (for large n); see [11, pp. 77–79], [2].

The success of an algorithm for optimal selection can be also quantified by the rank of the selected element, where, say, the best rank is 1 and the worst rank is n. Chow, Moriguti, Robbins, and Samuels [1] obtained the elegant result that for the stopping rule which minimizes the expectation of the absolute rank of the selected element, as $n \to \infty$, the expected rank tends to the value of the infinite product $\prod_{j=1}^{\infty} \left(\frac{j+2}{j}\right)^{\frac{1}{j+1}} \approx 3.8695$.

Finally, a card guessing game that exhibits some similarity to the one we study here can be found in the articles of Levasseur [7] and Zagier [13].

2 Preliminaries

A *lattice path* in $\mathbb{Z}^2$ is a path that moves right or up by one unit at each step. Consider the integer lattice points in the square $[0, m] \times [0, m]$ and the set L_m of lattice paths

[1] Euler's number $e = 2.718\dots$ is the base of the natural logarithm.

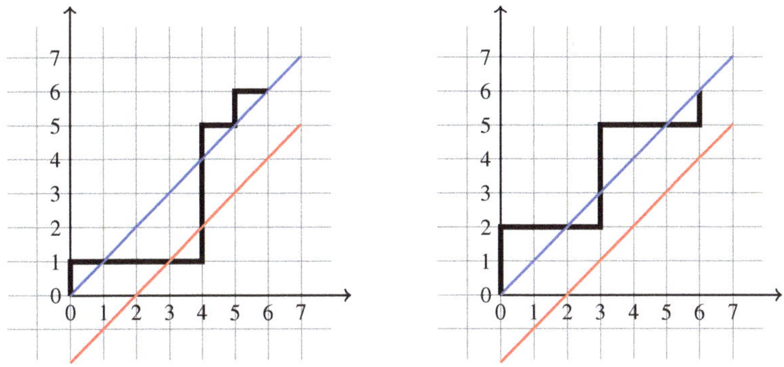

Fig. 1 Left: a lattice path from $(x, y) = (0, 0)$ to $(6, 6)$ that reaches the red line $y = x - 2$ (left), and one that does not (right). Here $n = 6$. The line $y = x$ is in blue

connecting the lower left corner of the square $(0, 0)$ with the upper right corner of the square (m, m).

In the binary case, the evolution of the game corresponds to a prefix of a path in L_m. The path moves a step to the right if a revealed element is -1, and moves a step up if it is $+1$. If the player stops the game after the ith element is revealed ($i \geq 1$), the path length is i. See Fig. 1.

Recall the following result regarding lattice paths—obtained by the so-called *reflection principle*:

Lemma 1 *[9, Ch. 1.3] Let $t \geq 1$ be a positive integer. The number of lattice paths from $(0, 0)$ to (m, m) that reach line $\ell : y = x - t$ is $\binom{2m}{m-t}$.*

Let E_t be the event that a random lattice path in L_m reaches line $\ell : y = x - t$. Recall that the total number of paths in L_m is $|L_m| = \binom{2m}{m}$. Then by Lemma 1 we obtain:

$$\mathrm{Prob}\,(E_t) = \frac{\binom{2m}{m-t}}{\binom{2m}{m}}. \tag{1}$$

3 A Simple Threshold Algorithm for the Binary Case

Algorithm outline.

An integer threshold value $t = \Theta(\sqrt{n})$ is set; where $n = 2m$. The algorithm waits until the lattice path reaches the line $\ell : y = x - t$, if at all, when it stops the game. If the lattice path describing the evolution of the game doesn't reach the line ℓ, the algorithm continues until the end, i.e., the entire input sequence is revealed. Next we proceed with the details.

ALGORITHM 1

Input: the elements of a zero-sum sequence A of length $2m$ over $\Sigma = \{-1, +1\}$ are revealed one by one

1 Set $t = \left\lfloor \sqrt{\frac{m+1}{2}} \right\rfloor$

2 Follow the lattice path from $(0, 0)$ to (m, m) ($\rightarrow$ or $\uparrow$) as dictated by the entries of A that have been revealed (step by step) by repeating Step 3

3 If the path has reached the line $\ell : y = x - t$ stop; else continue

Algorithm analysis.

We first deduce a lower bound on the expected payoff. Set $t = \left\lfloor \sqrt{\frac{m+1}{2}} \right\rfloor$. From (1) we further obtain by applying the standard inequality $1 - x \geq e^{-2x}$ for $0 \leq x \leq 1/2$ in the last step:

$$\begin{aligned}
\text{Prob}\,(e_t) &= \frac{\binom{2m}{m-t}}{\binom{2m}{m}} = \frac{m(m-1)\cdots(m-t+1)}{(m+t)(m+t-1)\cdots(m+1)} \\
&= \prod_{i=1}^{t}\left(1 - \frac{t}{m+i}\right) && (2) \\
&\geq \left(1 - \frac{t}{m+1}\right)^t \geq \exp\left(\frac{-2t^2}{m+1}\right) && (3) \\
&\geq \exp(-1).
\end{aligned}$$

Let W_1 denote the expected payoff taken by the player. For the above value of t we have:

$$W_1 = t \cdot \text{Prob}\,(E_t) \geq \frac{1}{e}\left\lfloor \sqrt{\frac{m+1}{2}} \right\rfloor = \Omega(\sqrt{m}). \tag{4}$$

As for the upper bound, observe that the payoff is at least t with probability only at most Prob (E_t) even if the player has complete information about a given random sequence, and chooses the best moment to stop the game. Hence, the expected payoff of *any* algorithm is at most the sum of these probabilities (see [8, Lemma 2.9]):

$$\sum_{t=1}^{m} \text{Prob}\,(e_t) = \sum_{t=1}^{m} \frac{\binom{2m}{m-t}}{\binom{2m}{m}} = \frac{2^{2m-1}}{\binom{2m}{m}} - \frac{1}{2} = \frac{\sqrt{\pi m}}{2} - \frac{1}{2} + O(m^{-1/2}),$$

where the last equality is a routine series expansion. Hence, $W_1 = \Theta(\sqrt{m})$, as claimed.

Executing one step of ALGORITHM 1 takes constant time.

Remarks.

Our findings apply with some adjustments to the following variants:

- The player is required to stop the game before the last element in the sequence is revealed. Since the state of the process is known at any step, the player can always reduce its losses, if any are perceived at the current step, by continuing to play until exactly one element, the last one, remains, where he/she can stop the game and so the loss can be limited to -1. The analysis of this variant is similar.
- The length n of the input sequence is only known approximately, say, $N \le n \le 2N$, for a given N. For this variant, the player can use ALGORITHM 1 with a similar expected performance, where n is replaced by N.

4 An Optimal Algorithm for the Binary Case

The algorithm constructs two square matrices, whose content is described below.

Let $[T[i, j]]_{0 \le i, j \le m}$ be a $(m+1) \times (m+1)$ matrix with nonnegative entries, where $T[i, j]$ is the expected value of the player's payoff if he/she starts playing optimally in state (i, j), i.e., after i elements of -1 and j elements of $+1$ have been revealed. As such, $T[0, 0]$ is the expected *value* of the game in an optimal play.

The optimal strategy is encoded in the *stopping* square matrix S as described below. Let $[S[i, j]]_{0 \le i, j \le m}$ be a $(m+1) \times (m+1)$ matrix with 0-1 entries, where

$$S[i, j] = \begin{cases} 1 & \text{if stop at this step—in state } (i, j), \\ 0 & \text{if continue at this step—in state } (i, j). \end{cases}$$

The matrix T is constructed in a *backwards* manner by using dynamic programming. Indeed, the last row and column of T are known beforehand: $T[i, m] = 0$, for $i = 0, 1, \ldots, m$, and $T[m, j] = m - j$ for $j = 0, 1, \ldots, m$. If the entries $T[i+1, j]$ and $T[i, j+1]$ are known, the entry $T[i, j]$ can be computed as follows:

First, the transition probabilities p_0 and p_1 to reach states $(i+1, j)$ and $(i, j+1)$, respectively, by revealing a new element in state (i, j) are computed:

$$p_0 = \frac{m-i}{2m-i-j}, \quad p_1 = \frac{m-j}{2m-i-j}.$$

Then

$$T[i, j] = \max\left(i - j,\, p_0 \cdot T[i+1, j] + p_1 \cdot T[i, j+1]\right). \tag{5}$$

As such, the entries of T can be computed row by row, from top to bottom, and each row from right to left.

Consider an entry $T[i, j]$; If the maximum in (5) is given by the first term, $i - j$, the corresponding entry in S is set to 1: $S[i, j] = 1$. Indeed, this means that the

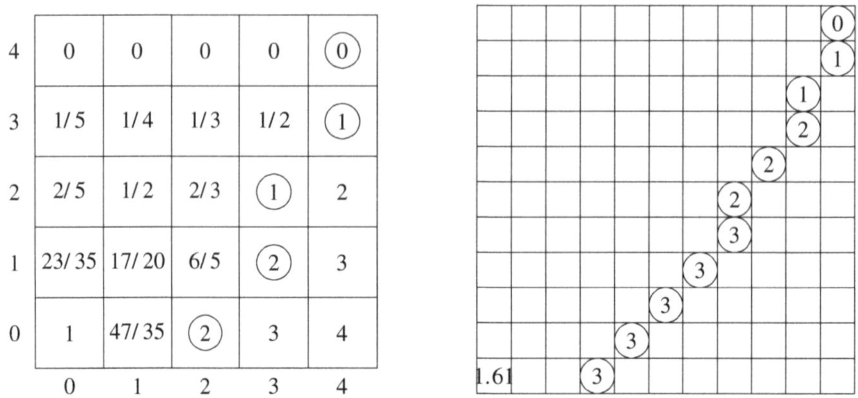

Fig. 2 Left: an optimal strategy for $n = 8$: $T[0, 0] = 1$, $T[1, 0] = 47/35$, $T[2, 0] = 2$, …, $T[4, 4] = 0$. Right: an optimal strategy for $n = 20$; to avoid the clutter, only the reachable stopping states are shown. Note that all stopping states are below the main diagonal (with one exception, the zero entry for the automatic stop in the end)

expected payoff if the algorithm continues from state (i, j) does not exceed the payoff available in state (i, j) if the algorithm stops. Otherwise, if the maximum in (5) is given by the second term, set $S[i, j] = 0$. Some of the stopping entries are unreachable (for example, in the last column), but this is not an issue.

In summary, the algorithm works as follows:

ALGORITHM 2

Input: the elements of a zero-sum sequence A of length $2m$ over $\Sigma = \{-1, +1\}$ are revealed one by one

1 Compute the matrices T and S as specified above

2 Follow the path in S starting from the entry $(0, 0)$ by repeating Step 3

3 If $S[i, j] = 1$ (i.e., if (i, j) is a stopping state), stop; else continue to the next entry in S ($S[i + 1, j]$ or $S[i, j + 1]$) as dictated by the most recently revealed element of A

Two examples with the optimal strategy as computed by ALGORITHM 2 appear in Fig. 2.

Algorithm analysis.

For any given state (i, j) the algorithm has two options: stop or continue. We may assume w.l.o.g. that $i \geq j$. Note that payoff of value $i - j$ is immediately available for collection if the algorithm stops in state (i, j). Recall that each zero-sum sequence is equally likely and thus, for a given prefix, every suffix of it is equally likely. Note that the second term in (5) is equal to the expected payoff if the algorithm continues from state (i, j).

If the maximum in (5) is given by the first term, $i - j$, this means that the expected payoff if the algorithm continues from state (i, j) does not exceed the payoff available in state (i, j) if the algorithm stops. Recall that $S[i, j] = 1$ in this case and thus the algorithm correctly stops in this state. If the maximum in (5) is given by the second term, this means that the expected payoff if the algorithm continues from state (i, j) exceeds those available in state (i, j) if the algorithm stops. Recall that $S[i, j] = 0$ in this case and thus the algorithm correctly continues from this state.

Note that both T and S can be computed in quadratic time. Executing one step of ALGORITHM 2 takes constant time.

Let W_2 be the expected payoffs of the player under ALGORITHM 2. Since ALGORITHM 2 is optimal, it is stronger than ALGORITHM 1 from Sect. 3, and thus $W_2 \geq W_1 = \Omega(\sqrt{n})$. In addition, the general upper bound analysis in Sect. 3 yields that every algorithm, including ALGORITHM 2, has a payoff at most $O(\sqrt{n})$. Consequently, $W_2 = \Theta(\sqrt{n})$, as claimed.

Example.

With a shuffled standard deck of 52 cards ($n = 52$), the player using ALGORITHM 2 is guaranteed \$2.62 in expectation (which is optimal), if the payoff is, say, the number of black cards minus the number of red cards in what is left unturned from the deck, with \$1 per card. For comparison, the player using ALGORITHM 1 is guaranteed \$1.54 in expectation, by (4) (however, no optimization was attempted in its analysis).

5 A 'Stop in the Middle' Algorithm for the Binary Case

In this section we offer yet another algorithm for the binary case, ALGORITHM 3. Its expected payoff is $\Omega(\sqrt{n})$, asymptotically the same as for the previous two algorithms. This algorithm is extremely simple as it does not perform any bookkeeping. However, the main reason for presenting it is its applicability to arbitrary zero-sum sequences as shown in Sect. 6.

For convenience, here we formulate the algorithm for the dual version of the game, in which, if the player stops the game after the ith element of A is revealed, his/her payoff is $\sum_{j=1}^{i} a_j$.

Consider the function $g : \mathbb{R} \to \mathbb{R}$ defined as

$$g(x) = \begin{cases} x & \text{if } x > 0 \\ 0 & \text{otherwise.} \end{cases} \tag{6}$$

The algorithm's payoff is $g(\sum_{j=1}^{n/2} a_j)$. Note that this only requires computing a running sum up to index $n/2$ and making a single sign determination.

Table 2 A few values of $W_3(n)$

n	2	4	6	8	10	12	14	16	32	64
$W_3(n)$	$\frac{1}{2}$	$\frac{1}{3}$	$\frac{3}{5}$	$\frac{18}{35}$	$\frac{5}{7}$	$\frac{50}{77}$	$\frac{350}{429}$	$\frac{980}{1287}$	≈ 1.10	≈ 1.57

ALGORITHM 3

Input: the elements of a zero-sum sequence A of length n are revealed one by one

1 Compute the running sum $S = \sum_{j=1}^{n/2} a_j$ step by step

2 If $S > 0$ stop; else continue to the end

Assume for convenience in the analysis that $n = 4m$. Then it is easy to see that in the binary case ($\Sigma = \{-1, +1\}$), the total number of zero-sum sequences is $\binom{4m}{2m}$, whereas there are exactly $\binom{2m}{m-k}\binom{2m}{m+k}$ of them for which the payoff is $2k$, $0 \le k \le m$. Hence, the expected payoff $W_3(n)$ is

$$W_3(n) = \binom{4m}{2m}^{-1} \sum_{k=1}^{m} 2k \binom{2m}{m-k}\binom{2m}{m+k}. \tag{7}$$

A few values of $W_3(n)$ for some small even n are listed in Table 2.

Algorithm analysis.

The following standard identity, which we prove in Section 7, leads to a closed form expression of the expected payoff.

Lemma 1

$$\sum_{k=1}^{m} 2k \binom{2m}{m-k}\binom{2m}{m+k} = 2m \binom{2m-1}{m-1}^2.$$

Now a routine series expansion yields

$$W_3(n) = 2m \binom{2m-1}{m-1}^2 \binom{4m}{2m}^{-1} = \sqrt{\frac{m}{2\pi}} + O(m^{-1/2}) = \sqrt{\frac{n}{8\pi}} + O(n^{-1/2}).$$

Therefore, $W_3(n) = \Theta(\sqrt{n})$, asymptotically the same as for the previous two algorithms.

6 The 'Stop in the Middle' Algorithm for the General Case

In this section we generalize our results to an *arbitrary* zero-sum multiset $M = \{a_1, \ldots, a_n\}$ and thereby prove Theorem 1. Assume for convenience that the input size n is even (the treatment for the case of odd n is similar, but somewhat more involved, and is therefore omitted). As in Sect. 5, here we consider the dual version of the game, in which the payoff is the sum of the revealed elements.

We show that the expected payoff achieved by ALGORITHM 3 on a random permutation of M is $f(M) = \Omega(\mu\sqrt{n})$, where $\mu = \sum_{i=1}^{n} |a_i|/n$ is the *average absolute value* for the input multiset. Let n_+ and n_- denote the number of positive and respectively negative entries in M; where $n = n_+ + n_-$. Let $m = n/2$. Note that $f(\cdot)$ is a linear function, i.e., $f(\lambda M) = \lambda f(M)$, where $\lambda M = \lambda\{a_1, \ldots, a_n\} = \{\lambda a_1, \ldots, \lambda a_n\}$.

The key lemma below shows that the expected payoff can only decrease when replacing two distinct elements by two copies of their average. If the two elements are both positive or both negative, this type of replacement maintains n_+ and n_-. If the two elements have different signs, the replacement can be used to adjust the counts n_+ and n_-, as illustrated after the proof of the lemma.

Lemma 2 *Let $a, b \in M$, $a < b$, be two distinct elements of M and $M' = M \setminus \{a, b\}$. Then*

$$f\left(M' \cup \{a, b\}\right) \geq f\left(M' \cup \left\{\frac{a+b}{2}, \frac{a+b}{2}\right\}\right).$$

Proof Let $F \subset M$ denote the first half of the random input permutation of M. If both $a, b \in F$, or both $a, b \notin F$, then the corresponding payoffs satisfy the required inequality (as an equality, since $a + b = \frac{a+b}{2} + \frac{a+b}{2}$). Suppose now that exactly one element of $\{a, b\}$ appears in F, and let x denote the sum of the other $m - 1$ elements in F. Note that a precedes b with probability $1/2$ and vice-versa, so it remains to check that

$$\frac{1}{2}g(x + a) + \frac{1}{2}g(x + b) \geq g\left(x + \frac{a+b}{2}\right). \tag{8}$$

Note that $-b < -\frac{a+b}{2} < -a$ by the assumption. Consider the four possible intervals where x can lie in: $x \leq -b$, $-b \leq x \leq -\frac{a+b}{2}$, $-\frac{a+b}{2} \leq x \leq -a$, $-a \leq x$. In each of these cases, the verification of (8) is straightforward:

1. $x \leq -b$: $\frac{1}{2} \cdot 0 + \frac{1}{2} \cdot 0 = 0$.
2. $-b \leq x \leq -\frac{a+b}{2}$: $0 + \frac{1}{2}(x + b) \geq 0$.
3. $-\frac{a+b}{2} \leq x \leq -a$: $0 + \frac{1}{2}(x + b) \geq x + \frac{a+b}{2}$.
4. $-a \leq x$: $\frac{1}{2}(x + a) + \frac{1}{2}(x + b) = x + \frac{a+b}{2}$,

and the verification is complete. □

If M satisfies $n_+ = n_- = n/2$, we invoke Lemma 2 with a pair of distinct elements of M that are either both positive or both negative, while maintaining the zero-sum property and the average absolute value μ. It is not hard to see that $f(\cdot)$ is continuous as a function of a given n-element multiset. Therefore, repeated invocation of Lemma 2 allows us to average all the elements of M of same sign by compactness. In other words, we have $f(M) \geq f(B)$, where B is a 'binary' multiset consisting of two elements μ and $-\mu$, each with multiplicity $n/2$. Besides, recall from Sect. 5 that $f(B) = \Theta(\mu\sqrt{n})$.

If M fails to satisfy $n_+ = n_- = n/2$, we assume without loss of generality that $n_- < n_+$ and proceed as follows. On each step, we pick a and b as the most negative and the least positive elements of M, respectively. We have that $|a| > b$ by the zero-sum property. One application of Lemma 2 increases n_- by one, and reduces n_+ by one while maintaining the zero-sum property. Note, however, that this operation decreases the average absolute value μ. After $(n_+ - n_-)/2$ such steps, we get a multiset M' that satisfies $n_+ = n_- = n/2$. Since

$$n_+ - \frac{n_+ - n_-}{2} > \frac{n_+}{2}$$

of the largest positive elements of the initial multiset M remain in M', its average absolute value μ' satisfies $\mu' > \mu/2$ by the zero-sum property. Moreover, the argument from the previous paragraph yields that $f(M') = \Omega(\mu'\sqrt{n})$. Therefore, we conclude that $f(M) \geq f(M') = \Omega(\mu\sqrt{n})$, as desired.

Example.

Let $M = \{-5, -3, -3, +1, +1, +1, +2, +6\}$, with $n = 8$ and $\mu = 11/4$. In the first step, we average -5 with $+1$ to balance the number of positive and negative elements. Then we average the elements of the same sign in arbitrary order. The elements of the relevant action pairs are underlined.

$$\begin{aligned} f(\{\underline{-5}, -3, -3, \underline{1}, 1, 1, 2, 6\}) &\geq f(\{-3, -3, -2, -2, 1, 1, \underline{2}, \underline{6}\}) \\ &\geq f(\{-3, -3, -2, -2, 1, 1, 4, 4\}) \\ &\vdots \\ &\geq f(\{-2.5, -2.5, -2.5, -2.5, 2.5, 2.5, 2.5, 2.5\}) \\ &= 2.5 \cdot f(\{-1, -1, -1, -1, 1, 1, 1, 1\}) \\ &= \frac{5 \cdot 18}{2 \cdot 35} = \frac{9}{7}. \end{aligned}$$

7 Concluding Remarks

Zero-sum sequences are relevant in the study of dynamic or random processes and/or optimization under uncertainty. We formulate two questions for further investigation. The first one is more technical but its solution may offer additional insight on the problem. The second one appears to be interesting and motivating for real applications, such as simulation or trading.

1. Can the entries of T describing the payoffs of an optimal strategy for zero-sum sequences over $\Sigma = \{-1, +1\}$ be derived in closed form?
2. Can a better expected payoff be guaranteed in the general case, by taking advantage of the specificity of the input?

Acknowledgements Work on this paper was partially supported by ERC Advanced Grant no. 882971 "GeoScape". The second author thanks Fedor Petrov for a helpful discussion of Lemma 1.

Appendix: Proof of Lemma 1

Our argument relies on a standard Vandermonde identity, see [4, Section 5.2],

$$\sum_{i=0}^{r} \binom{s}{i}\binom{t}{r-i} = \binom{s+t}{r},$$

and a handful of simple algebraic manipulations:

$$\begin{aligned}
\sum_{k=1}^{m} 2k\binom{2m}{m-k}\binom{2m}{m+k} &= m\sum_{k=0}^{m} 2\binom{2m}{m-k}\binom{2m}{m+k} - 2\sum_{k=0}^{m-1}(m-k)\binom{2m}{m-k}\binom{2m}{m+k} \\
&= m\binom{2m}{m}^2 + m\sum_{k=-m}^{m}\binom{2m}{m-k}\binom{2m}{m+k} - 2m\sum_{k=0}^{m-1} 2\binom{2m-1}{m-k-1}\binom{2m}{m+k} \\
&= m\binom{2m}{m}^2 + m\sum_{i=0}^{2m}\binom{2m}{i}\binom{2m}{2m-i} - 2m\sum_{k=0}^{m-1} 2\binom{2m-1}{m-k-1}\binom{2m-1}{m+k-1} \\
&\qquad - 2m\sum_{k=0}^{m-1} 2\binom{2m-1}{m-k-1}\binom{2m-1}{m+k} \\
&= m\binom{2m}{m}^2 + m\binom{4m}{2m} - 2m\sum_{k=-m}^{m-1}\binom{2m-1}{m-k-1}\binom{2m-1}{m+k-1} \\
&\qquad - 2m\binom{2m-1}{m-1}^2 - 2m\sum_{k=-m+1}^{m-1}\binom{2m-1}{m-k-1}\binom{2m-1}{m+k} \\
&= m\binom{2m}{m}^2 + m\binom{4m}{2m} - 2m\sum_{i=0}^{2m-1}\binom{2m-1}{i}\binom{2m-1}{2m-1-i} \\
&\qquad - 2m\binom{2m-1}{m-1}^2 - 2m\sum_{i=0}^{2m-2}\binom{2m-1}{i}\binom{2m-1}{2m-2-i}
\end{aligned}$$

$$= m\binom{2m}{m}^2 - 2m\binom{2m-1}{m-1}^2 + m\binom{4m}{2m} - 2m\binom{4m-2}{2m-1} - 2m\binom{4m-2}{2m-2}$$

$$= m\binom{2m}{m}^2 - 2m\binom{2m-1}{m-1}^2 = 2m\binom{2m-1}{m-1}^2.$$

References

1. Yuan Shih Chow, Sigaiti Moriguti, Herbert Robbins, and Stephen M. Samuels, Optimal selection based on relative rank (the "secretary problem"), *Israel Journal of Mathematics* **2(2)** (1964), 81–90
2. Thomas, S.: Ferguson, Who solved the secretary problem? Stat. Sci. **4**(3), 282–296 (1989)
3. Georgiou, N., Kuchta, M., Morayne, M., Niemiec, J.: On a universal best choice algorithm for partially ordered sets. Random Structures and Algorithms **32**(3), 263–273 (2008)
4. Graham, R.L., Knuth, D.E., Patashnik, O., Mathematics, C.: A Foundation for Computer Science. Addison-Wesley, Reading, MA (1990)
5. Guttman, I.: On a problem of L. Moser, Canadian Mathematical Bulletin **3**(1), 35–39 (1960)
6. Richard J. Larsen and Morris L. Marx, *An Introduction to Probability and its Applications*, Prentice–Hall, Englewood Cliffs, New Jersey, 1985
7. Kenneth, M.: Levasseur, How to beat your kids at their own game. Math. Mag. **61**(5), 301–305 (1988)
8. Michael Mitzenmacher and Eli Upfal, *Probability and Computing: Randomized Algorithms and Probabilistic Analysis*, 2nd edition, Cambridge University Press, 2017
9. Sri Gopal Mohanty: Lattice Path Counting and Applications. Academic Press, New York (1979)
10. Moser, L.: On a problem of Cayley, Scripta. Mathematica **22**, 289–292 (1956)
11. Mosteller, F.: Fifty Challenging Problems in Probability with Solutions. Addison-Wesley, Reading, MA (1965)
12. Preater, J.: The best-choice problem for partially ordered objects. Oper. Res. Lett. **25**(4), 187–190 (1999)
13. Zagier, D.: How often should you beat your kids? Math. Mag. **63**(2), 89–92 (1990)

Effective MC-Finiteness

Yuval Filmus, Eldar Fischer, and Johann A. Makowsky

Abstract An integer sequence $(a_n)_{n\in\mathbb{N}}$ is *MC-finite* if for all m, the sequence $a_n \bmod m$ is eventually periodic. There are MC-finite sequences $(a_n)_{n\in\mathbb{N}}$ such that the function $F : (m, n) \mapsto a_n \bmod m$ is not computable. In Filmus et al. (2023) [3] we presented concrete examples of MC-finite sequences taken from the Online Encyclopedia of Integer Sequences (OEIS) without discussing the computability of F. In this paper we discuss cases when this F is effectively computable.

1 Polynomial Recurrence Sequences

One type of MC-finite sequences is those obtained from polynomial recurrence sequences (PRS). These are sequences of vectors $\mathbf{a}_n \in \mathbb{Z}^k$ (for fixed k) where $\mathbf{a}_1, \dots, \mathbf{a}_c$ are given, and for $n > c$, each entry of $\mathbf{a}_n$ is a fixed polynomial with integer coefficients which depends on the c preceding vectors $\mathbf{a}_{n-1}, \dots, \mathbf{a}_{n-c}$. We call k the *dimension* and c the *depth*.

We can extract an integer sequence by looking at the first entry of all vectors. We call a sequence obtained in this way a polynomial recurrence sequence (PRS). These are MC-finite.

Theorem 1 *If $(a_n)_{n\in\mathbb{N}}$ is given by a PRS, then it is MC-finite and $a_n \bmod m$ is FPT. Moreover, we can take $T(m) = m^{O(1)}$, and so $a_n \bmod m$ can be computed in polynomial time if m is encoded in unary.*

Proof Suppose that the PRS is obtained from the recurrence $(\mathbf{a}_n)_{n\in\mathbb{N}}$ of dimension k and depth c.

Y. Filmus · E. Fischer · J. A. Makowsky (✉)
Faculty of Computer Science, Technion-IIT, Haifa, Israel
e-mail: janos@cs.technion.ac.il

Y. Filmus
e-mail: yuvalfi@technion.ac.il

E. Fischer
e-mail: eldar@cs.technion.ac.il

G. O. H. Katona et al. (eds.), *Sum(m)it280*, Bolyai Society Mathematical Studies 32,
https://doi.org/10.1007/978-3-032-18810-6_8

There is an integer polynomial mapping $F\colon (\mathbb{Z}^k)^c \to (\mathbb{Z}^k)^c$ such that for all $n > c$,

$$F(\mathbf{a}_{n-c}, \ldots, \mathbf{a}_{n-1}) = (\mathbf{a}_{n-c+1}, \ldots, \mathbf{a}_n).$$

Let $\mathbf{a}_n^{(m)}$ be obtained by taking every entry modulo m. We can project F to a mapping $F^{(m)}\colon (\mathbb{Z}_m^k)^c \to (\mathbb{Z}_m^k)^c$ which can be computed explicitly. The sequence

$$\begin{aligned}&(\mathbf{a}_1^{(m)}, \ldots, \mathbf{a}_n^{(m)}),\\ &F^{(m)}(\mathbf{a}_1^{(m)}, \ldots, \mathbf{a}_n^{(m)}),\\ &F^{(m)}(F^{(m)}(\mathbf{a}_1^{(m)}, \ldots, \mathbf{a}_n^{(m)})),\\ \ldots\end{aligned}$$

is eventually periodic since it results from applying the mapping $F^{(m)}$ repeatedly, and $F^{(m)}$ has a finite domain. In fact, it must repeat within the first $m^{kc} + 1$ steps by the pigeonhole principle.

We can now explicitly extract the representation of $(\mathbf{a}_n^{(m)})_{n \in \mathbf{N}}$ which is eventually periodic and which allows us to compute $a_n \bmod m$ in time polynomial in n. □

2 Specker's Method

2.1 *Density Function*

Let τ be a finite relational vocabulary $\bar{R} = (R_1^{\rho(1)}, R_2^{\rho(2)}, \ldots, R_s^{\rho(s)})$ where $\rho(i)$ is the arity of $R_i^{\rho(i)}$ and $\rho = \max_i(\rho(i))$. Let C be a class of τ-structures and $C(n)$ be the set of structures of the form $([n], \bar{R})$. We define the *density function* $d_C(n)$ *of* C to be the number of ways the set $[n]$ can be made into a structure $([n], \bar{R}) \in C(n)$.

We want to study cases of C where $d_C(n)$ is (effectively) MC-finite.

2.2 *Structures of Bounded Degree*

Let $\mathfrak{A} = \langle A, \bar{R}\rangle$ be a τ-structure. We define a symmetric relation E_A on A, and call $\langle A, E_A\rangle$ the *Gaifman-graph of* $\mathfrak{A}$.

- Let $a, b \in A$. $(a, b) \in E_A$ iff there exists a relation $R \in \bar{R}$ and some $\bar{a} \in R$ such that both a and b appear in $\bar{a}$ (possibly with other elements of A as well).
- For any element $a \in A$, the *G-degree* of a is the number of elements $b \neq a$ for which $(a, b) \in E_A$.
- $\mathfrak{A}$ is of *bounded G-degree* d if every $a \in A$ has G-degree at most d.
- We say that $\mathfrak{A}$ is *G-connected* if its Gaifman-graph is connected.

- For a class of structures $\mathcal{P}$ we say it is of bounded degree d (resp. connected) iff all its structures are of bounded degree d (resp. connected).

We note that the Gaifman graph may have loops.

2.3 DU-Matrices

For two τ-structures $\mathfrak{A}$, $\mathfrak{B}$ we denote by $\mathfrak{A} \sqcup \mathfrak{B}$ their disjoint union. Let $\mathfrak{A}_i, i \in \mathbb{N}$ be an enumeration of all finite τ-structures with universe $[n]$ for some $n \in \mathbb{N}$ and let C be a class of τ-structures. The DU-matrix of C, $DU(C)$ is the infinite (0, 1)-matrix where columns and rows are labeled $\mathfrak{A}_i$. We write $DU(C)_{i,j}$ for the entry $(\mathfrak{A}_i, \mathfrak{A}_j)$.

$$DU(C)_{i,j} = \begin{cases} 1 & \mathfrak{A}_i \sqcup \mathfrak{A}_j \in C \\ 0 & \mathfrak{A}_i \sqcup \mathfrak{A}_j \notin C \end{cases}.$$

Let $rk(DU(C))$ be the rank of $DU(C)$ over the field $\mathbb{Z}/\mathbb{Z}_2$.

Below we show that there are uncountably many graph properties of fixed DU-rank Hence graph properties of fixed finite DU-rank are not necessarily definable even in Second Order Logic, or any other logical formalism in which there are only countably many formulas.

Proposition 1 *Let C be a class of finite τ-structures.*

(i) *Assume all τ-structures in C are G-connected. If we allow the empty set as a τ-structure, then $rk(DU(C)) = 2$, but otherwise $rk(DU(C)) = 0$.*
(ii) *Let $A \subseteq \mathbb{N}^+$ a set of non-negative integers and* Circ *the set of circles C_n : $n \in \mathbb{N}+$. Define $\mathrm{Circ}_A = \{C_i : i \in A\}$. Then $rk(DU(\mathrm{Circ}) = rk(DU(\mathrm{Circ}_A)$ irrespective of the choice of A. We conclude that there are uncountably many graph classes of the same DU-rank as for* Circ.
(iii) *Now let Circ^m_A be the class of graphs which consist of k disjoint unions of graphs in Circ_A. If we allow the empty graph the rank $rk(DU(\mathrm{Circ}^m_A)$ is $m + 1$ irrespective of A. We conclude that there are uncountably many graph classes of the same DU-rank as for* Circ^m.

From now on we allow structures to have an empty universe. We next look at variations of graph classes C closed under disjoint union. A graph property C is a *Gessel class of graphs* iff C is closed under disjoint unions and closed under taking G-connected components.

Facts 2 *Let C be a graph property.*

(i) *If C is a Gessel class of graphs then $rk(DU(C)) \leq 2$.*
(ii) *Let G_1, G_2, H be graphs and H be connected. Then $G_1 \sqcup G_2$ does not contain H as a subgraph (induced subgraph, minor) iff neither G_1 nor G_2 contains H as a subgraph (induced subgraph, minor).*

(iii) *Let C be a class of connected graphs and* $\text{Forb}_{sub}(C)$ *the class of graphs which do not contain a graph $H \in C$ as a subgraph. Then* $\text{Forb}_{sub}(C)$ *is a Gessel class of graphs.*
(iv) *The same holds if we replace* $\text{Forb}_{sub}(C)$ *by* $\text{Forb}_{ind}(C)$*, respectively* $\text{Forb}_{min}(C)$ *where forbidden subgraphs are replaced by forbidden induced subgraphs, respectively forbidden minors.*
(v) *Let C a class of τ-structures. Assume that for all finite τ-structures $\mathfrak{A}, \mathfrak{B}$ we have $\mathfrak{A} \sqcup \mathfrak{B} \in C$ iff both $\mathfrak{A}$ and $\mathfrak{B}$ are in C. Then $rk(DU(C))$ is finite.*
(vi) *Let $\mathfrak{D}$ be a class of finite τ-structures. we denote by $cl_{DU}(\mathfrak{D})$ the closure of $\mathfrak{D}$ under disjoint unions. If $C = cl_{DU}(\mathfrak{D})$, then $rk(DU(C))$ is finite.*
(vii) *If C consists of graphs which are the disjoint union of two equal-sized cliques, $rk(DU(C))$ is infinite.*

Theorem 3 ([6, 7]) *Let C be a class of τ-structures such that*

(a) *every $\mathfrak{A} \in C$ is of G-degree at most d, and*
(b) *$rk(DU(C)) = r$ is finite.*

Then

(i) *the density function $d_C(n)$ of C is MC-finite.*
(ii) *If additionally all structures in C are G-connected, then $d_C(n)$ modulo m is ultimately* 0.

In [6] and its expanded version [7], the authors also exhibit a modular recurrence relation which depends only on the G-degree d of C, the modulus m, and a fixed finite submatrix of $DU(C)$ containing r rows. In the sequel we derive this recurrence relation in detail.

2.4 Describing $DU(C)$

In the sequel we assume that C has finite DU-rank. Two structures $\mathfrak{A}_i$ and $\mathfrak{A}_j$ are *DU-equivalent for C, equivalent for short*, if the columns i and j are identical in $DU(C)$. In this case we write $\mathfrak{A}_i \sim \mathfrak{A}_j$.

Observation 4 *(i)* *If $\mathfrak{A}_i \sim \mathfrak{A}_j$, then for every $\mathfrak{B}$ we have $\mathfrak{B} \sqcup \mathfrak{A}_i \sim \mathfrak{B} \sqcup \mathfrak{A}_j$.*
(ii) *As the rank of $DU(C)$ is finite, there are only finitely many equivalence classes.*

Let C_d be the class of structures of C whose maximal degree is $\leq d$. Let $\mathfrak{D}_0, \mathfrak{D}_1, \ldots, \mathfrak{D}_s$ be an enumeration of all DU-equivalence classes of C_d with $\mathfrak{D}_0$ the only equivalence class which contains a structure with maximal degree $\geq d+1$. Let $\mathcal{E} = (\mathfrak{D}_1, \ldots, \mathfrak{D}_s)$ and let $d_{\mathcal{E}}(n)$ be the vector-valued function $d_{\mathcal{E}}(n) = (d_{\mathfrak{D}_1}(n), \ldots, d_{\mathfrak{D}_s}(n))$.

Observation 5 *Let $DU(C)$ of finite DU-rank. Modulo m there are only finitely many possible values for $d_{\mathcal{E}}(n)$.*

We shall use the following finite set of parameters of the matrix $DU(C)$:

(i) The bound d on the G-degree of the structures in C.
(ii) The rank $DU(C)$.
(iii) Representatives $\mathfrak{B}_j$ for each of the equivalence classes $\mathcal{D}_j$.
(iv) For fixed $m \in \mathbb{N}$ the finitely many values of $d_{\mathcal{E}}(n) = (d_{\mathcal{D}_1}(n), \ldots, d_{\mathcal{D}_s}(n))$.

In the case of, say, Gessel classes of graphs, these parameters are easily available. To compute $d_{\mathcal{E}}(n)$ in general we also need an algorithm for checking whether a finite τ-structure $\mathfrak{A}$ belongs to C. A further algorithm for checking whether two τ-structures are DU-equivalent for C can then be obtained from the parameters and the algorithm above.

2.5 *Orbits*

Given a permutation group G that acts on a set A (and in the natural manner acts on models over the universe A), the *orbit* in G of a model $\mathfrak{A}$ with the universe A is the set $\mathrm{Orb}_G(\mathfrak{A}) = \{\sigma(\mathfrak{A}) : \sigma \in G\}$.

For $A' \subset A$ we denote by $S_{A'}$ the group of all permutations for which $\sigma(u) = u$ for every $u \notin A'$. The following lemma is useful for showing linear congruences modulo m.

Lemma 1 *Given $\mathfrak{A}$, if a vertex $v \in A - A'$ has exactly d neighbors in A', then $|\mathrm{Orb}_{S_{A'}}(\mathfrak{A})|$ is divisible by $\binom{|A'|}{d}$.*

Proof Let N be the set of all neighbors of v which are in A', and let $G \subset S_{A'}$ be the subgroup $\{\sigma_1\sigma_2 : \sigma_1 \in S_N \wedge \sigma_2 \in S_{A'-N}\}$; in other words, G is the subgroup of the permutations in $S_{A'}$ that in addition send all members of N to members of N. It is not hard to see that $|\mathrm{Orb}_{S_{A'}}(\mathfrak{A})| = \binom{|A'|}{|N|}|\mathrm{Orb}_G(\mathfrak{A})|$. □

The following simple observation is used to enable us to require in advance that all structure in C have a degree bounded by d.

Observation 6 *We denote by C_d the class of all members of C that in addition have bounded degree d. If C has a finite DU-index, then so does C_d.*

2.6 *Other Parameters of the Recurrence Relation*

In order to show that $d_C^{(d)}(n)$ is ultimately periodic modulo m, we show a linear recurrence relation modulo m on the vector function $(d_{\mathcal{E}}(n))_{\mathcal{E}}$ where $\mathcal{E}$ ranges over all other equivalence classes with respect to C_d.

The following parameters will be used in the recurrence relation:

(i) $c = m \cdot (d!)$ which depends only on the modulus m and the G-degree d. We note that for every $t \in \mathbb{N}$ and $0 < d' \leq d$, m divides $\binom{tc}{d'}$.
(ii) $t = t(n) = \lfloor \frac{n-1}{c} \rfloor$.

This, together with Lemma 1, allows us to prove the following.

Lemma 2 *Let $\mathcal{D} \neq N_\phi$ be an equivalence class for ϕ, that includes the requirement of the maximum degree not being larger than d. Then*

$$d_{\mathcal{D}}(n) \equiv \sum_{\mathcal{E}} a_{\mathcal{D},\mathcal{E},m,(n \bmod C)} d_{\mathcal{E}}(C\lfloor\frac{n-1}{C}\rfloor) \pmod{m},$$

for some fixed appropriate $a_{\mathcal{D},\mathcal{E},m,(n \bmod C)}$.

Proof Let $t = \lfloor \frac{n-1}{C} \rfloor$. We look at the set of structures in $\mathcal{D}$ with the universe $[n]$, and look at their orbits with respect to $S_{[tC]}$. If a model $\mathfrak{A}$ has a vertex $v \in [n] - [tC]$ with neighbors in $[tC]$, let us denote the number of its neighbors by d'. Clearly $0 < d' \leq d$, and by Lemma 1 the size of $\text{Orb}_{S_{[tC]}}(\mathfrak{A})$ is divisible by $\binom{tC}{d'}$, and therefore it is divisible by m. Therefore, $d_{\mathcal{D}}(n)$ is equivalent modulo m to the number of structures in $\mathcal{D}$ with the universe $[n]$ that in addition have no vertices in $[n] - [tC]$ with neighbors in $[tC]$.

We now note that any such structure can be uniquely written as $\mathfrak{B} \sqcup \mathfrak{C}$ where $\mathfrak{B}$ is any structure with the universe $[n - tC]$, and $\mathfrak{C}$ is any structure over the universe $[tC]$. We also note using Observation 4(i) that the question as to whether $\mathfrak{A}$ is in $\mathcal{D}$ depends only on the equivalence class of $\mathfrak{C}$ and on $\mathfrak{B}$ (whose universe size is bounded by the constant C). By summing over all possible $\mathfrak{B}$ we get the required linear recurrence relation (cases where $\mathfrak{C} \in N_C^{(d)}$ do not enter this sum because that would necessarily imply $\mathfrak{A} \in N_C^{(d)} \neq \mathcal{D}$). □

2.7 Proof of the Main Theorem

Proof of Theorem 3: (i) We use Lemma 2: Since there is only a finite number of possible values modulo m to the finite dimensional vector $(d_{\mathcal{E}}(n))_{\mathcal{E}}$, the linear recurrence relation in Lemma 2 implies ultimate periodicity for n's which are multiples of C. From this the ultimate periodicity for other values of n follows, since the value of $(d_{\mathcal{E}}(n))_{\mathcal{E}}$ for an n which is not a multiple of C is linearly related modulo m to the value at the nearest multiple of C.

(ii) Finally, if all structures are connected we use Lemma 1. Given $\mathfrak{A}$, connectedness implies that there exists a vertex $v \in A'$ that has neighbors in $A - A'$. Denoting the number of such neighbors by d_v, we note that $|\text{Orb}_{S'_A}(\mathfrak{A})|$ is divisible by $\binom{|A'|}{d_v}$, and since $1 \leq d_v \leq d$ (using $|A'| = tC$) it is also divisible by m. This makes the total number of models divisible by m (remember that the set of all models with $A = [n]$ is a disjoint union of such orbits), so $f_C^{(d)}(n)$ ultimately vanishes modulo m. □

2.8 Unbounded G-Degree

In the case of density functions for classes C of unbounded G-degree the situation is quite different. In 2023, [8], the following is shown:

Theorem 7 *There is a class $\mathcal{D}$ of τ-structures where τ consists of one ternary relation such that*

(i) $\mathcal{D}$ has unbounded G-degree,
(ii) $DU(\mathcal{D})$ has finite DU-rank, and
(iii) $d_{\mathcal{D}}(n)$ is not MC-finite.

Already in 2003, E. Fischer [4] constructed a class of τ'-structures $\mathcal{D}'$ with the same properties as $\mathcal{D}$ where τ consists of one quaternary relation.

Remark 1 The original Theorem 7 actually proved that $\mathcal{D}$ is definable in First Order Logic **FOL**. Already in [1] it was noticed that **FOL**-definability implies finite DU-rank. One way to see this is by using the Feferman-Vaught Theorem for **FOL**, see [5, Theorem 14.4].

However, in the original paper [1], C. Blatter and E. Specker did not use the DU-rank but the substitution rank $Subst(C)$ of classes of τ-structures C where τ only contains finitely many relations of arity at most 2. They called it the *index* of C. Their main result is:

Theorem 8 (C. Blatter and E. Specker, 1981) *Let C be a class of τ-structures where τ only contains finitely many relations of arity at most 2. Assume that C has finite substitution rank. Then $d_C(n)$ is MC-finite.*

For τ-structures $\mathfrak{A}$ where τ only contains finitely many relations we define

(i) A pointed τ-structure $(\mathfrak{A}, a_0)$ is a τ-structure with a distinguished element a_0. We also allow $a_0 \notin A$.
(ii) Let $\mathfrak{B}$ another τ-structure. $Subst((\mathfrak{A}, a_0), \mathfrak{B}))$ is the structure obtained by replacing a_0 by $\mathfrak{B}$ and connecting all elements $a \in A$ with $R(a, a_0)$ for some binary $R \in \tau$ with all the elements $b \in B$.
(iii) If $a_0 \notin A$, the structure $Subst((\mathfrak{A}, a_0), \mathfrak{B}))$ is just the disjoint union of $\mathfrak{A}$ and $\mathfrak{B}$.
(iv) The matrix $Subst(C)$ has columns labeled by τ-structures and rows labeled by pointed τ-structures. The entry for $(\mathfrak{A}, a_0), \mathfrak{B})$ is 1 if $Subst((\mathfrak{A}, a_0), \mathfrak{B})) \in C$ and 0 otherwise.
(v) The substitution rank $rk(Subst(C)$ is the rank of $Subst(C)$ over $\mathbb{Z}_2$.

Observations 9 *(i) There is no obvious definition of $Subst((\mathfrak{A}, a_0), \mathfrak{B})$ for ternary relations which works to obtain an analogue of Theorem 8.*
(ii) If τ only contains finitely many relations of arity at most 2, we have that $rk(DU(C)) \leq rk(Subst(C))$, because $DU(C)$ is a submatrix of $Subst(C)$.

The proof of Theorem 8 also uses orbits but is much more complicated. A transparent proof may be found in E. Specker's later exposition in [13] or in [5, Sect. 16]. A *ternary structure* is given by the pair $\mathcal{A} = (A, B)$ where A is a set and $B \subseteq A^3$. A *ternary property* is a class of ternary structures closed under isomorphism. Theorem 3 holds for ternary properties of bounded G-degree. However, the combinatorial part of the Specker-Blatter theorem Theorem 8 does not hold for ternary properties of unbounded G-degree due to Theorem 7. One immediate reason for this is the fact that the substitution rank for ternary properties is not well defined. However, this does not preclude the possibility that another general condition on classes of ternary structures $\mathcal{T}$ exists in order to establish MC-finiteness.

2.9 Examples of Ternary Structures and Properties

Natural finite ternary structures occur when a binary operation is represented by a ternary relation. Examples are groups and semi-groups. The Gaifman graph of a binary operation as a ternary operation is a complete graph, therefore such a ternary property is of unbounded G-degree, provided it contains arbitrarily large ternary structures. The number of labeled ternary structures on $[n]$ is bounded by 2^{n^3}. In the literature the number of finite groups is usually counted up to isomorphisms.

Problem 1 What is the number of ternary structures on $[n]$ which are finite groups? Is it MC-finite?

The most studied classes of ternary structures are motiviated by the study of the *betweenness relation* in elementary geometry. There are many axiomatizations of betweenness relations having various geometric applications in mind. For an encyclopedic survey the reader may consult [11]. In the case where we study *finite* ternary structures the betweenness relations usually comes from partial orders, or more specifically, from lattices. In [12] the betweenness relations derived from lattices are extensively studied.

Another class of ternary structures are the *ternary spaces* studied in [9]. A ternary structure $\mathcal{A}$ is a *ternary space* if it satisfies the following axioms $\mathcal{T}$.

T_1: $\forall a, b, c(B(a, b, c) \rightarrow B(c, b, a))$
T_2: $\forall a, b, c(B(a, b, c) \wedge B(a, c, b) \rightarrow b = c)$
T_3: $\forall a, b, c, d(B(a, b, c) \wedge B(a, c, d) \rightarrow B(b, c, d))$
T_4: $\forall a, b, c, d(B(a, b, c) \wedge B(a, c, d) \rightarrow B(a, b, d))$

Problem 2 What is the number of labeled ternary spaces? Is it MC-finite?

2.10 Showing MC-Finitness for Ternary Properties

What do we have to know about a ternary property $\mathcal{P}$ in order to show that $d_{\mathcal{P}}(n)$ is MC-finite?

Here are some applicable methods:

Bounded degree: In [6] it is shown that if all the structures in C are of G-degree at most d, then definability of C in **CMSOL** implies MC-finiteness.

PRS: In [3] it is shown that if $d_C(n)$ satisfies a polynomial recurrence sequence, then it is MC-finite. It may be difficult to find such a recurrence. The method was briefly discussed in Sect. 1.

Reducing to binary: In some cases there is a bijection of the ternary relations on $[n]$ of a ternary structure in $\mathcal{T}$ with the binary relations on $[n]$ of structures on $[n]$ in a class of structures C. In such a case (and similar cases), the density function $d_{\mathcal{T}}(n)$ is the same as $d_C(n)$ where $DU(C)$ is of finite DU-rank. This is the case for the betweenness relation of certain classes of partial orders.

Problem 3 Formulate and prove an analogue of Theorem 8 for ternary properties.

3 Conclusions

Theorem 8 uses the finite rank of $Subst(C)$ to prove MC-finiteness of $d_C(n)$. In [1, 13] it is shown that if C is definable in Monadic Second Order Logic **MSOL**, then C has finite substitution rank. This still holds if C is definable in **CMSOL**, the extension of **MSOL** by modular counting quantifiers, [6]. The proof of this uses the Feferman-Vaught Theorem for **MSOL**, respectively for **CMSOL**.

In its simplest form for disjoint unions and for **FOL** it says:

Theorem 10 (Feferman-Vaught Theorem for Disjoint Unions) *If $\mathfrak{A}$ and $\mathfrak{B}$ satisfy the same τ-sentences of* **FOL** *of quantifier rank at most q, then $\mathfrak{A} \sqcup \mathfrak{C}$ and $\mathfrak{B} \sqcup \mathfrak{C}$ satisfy the same first-order sentences of quantifier rank q.*

The corresponding theorems for **MSOL** and **CMSOL** also hold, see [10].

From this one can derive that the DU-rank is finite. However, such a proof gives very bad upper bounds for the DU-rank. If C is defined by a τ-formula of **MSOL** of quantifier rank q, the bound is roughly speaking determined by the number of non-equivalent τ-formulas of **MSOL** of quantifier rank q.

Another formulation of Theorem 10 is as follows:

Theorem 11 *Let $\mathfrak{A}, \mathfrak{B}$ be two τ-structures. Assume ϕ is an* **FOL**(τ) *sentence of quantifier rank q. Then there are* **FOL** *sentences $\psi_i : i \in I$, each of quantifier rank q, and there is a boolean function B in variables $x_i, y_i : i \in I$ such that the following holds:*

$$\mathfrak{A} \sqcup \mathfrak{B} \models \phi \textit{ iff } B(a_i, b_i : i \in I) = 1$$

where a_i, b_i are the truth value of $\mathfrak{A} \models \psi_i$ respectively $\mathfrak{B} \models \psi_i$.

We call $\psi_i : i \in I$ a reduction sequence of ϕ and $|I| = I(\phi)$ its *FV-width*. In this case the bound obtained for the DU-rank of $DU(\phi)$ is related to the FV-width of ϕ.

In fact, in [2, Theorem 6.1] it is suggested that no better bounds can be obtained using the Feferman-Vaught Theorem. They show a related result for the length of the formulas rather than the quantifier rank.

Theorem 12 *There is no elementary function g such that the following holds for all trees* $\mathfrak{A}, \mathfrak{B}, \mathfrak{C}$: *If* $\mathfrak{A}$ *and* $\mathfrak{B}$ *satisfy the same first-order sentences of length at most* $g(\lambda)$, *then* $\mathfrak{A} \sqcup \mathfrak{C}$ *and* $\mathfrak{B} \sqcup \mathfrak{C}$ *satisfy the same first-order sentences of length at most* λ.

Problem 4 Can one prove a theorem similar to Theorem 12 which estimates the minimal FV-width of ϕ?

They also note that if all the structures of C are of G-degree bounded by d, the situation is much better. In this case there is an elementary function g' and g' is bounded by an iterated exponential function of at most 5 iterations.

In this paper we discussed effective MC-finiteness in purely combinatorial terms. In Examples 1 we have seen that in many cases the DU-rank of a class C is very low and can be determined easily, although using the Feferman-Vaught Theorem would only give a very large upper bound for it. Similarly, one can construct examples with very low substitution rank. For example the class of finite connected graphs is definable in **MSOL** by a formula of quantifier rank 3, hence has finite substitution rank. But it is easily seen that its substitution rank is 2. We leave it to the reader to find more examples.

Problem 5 Find more combinatorial properties of C which allow to determine the DU-rank or the substitution rank efficiently.

Acknowledgements The proof of Theorem 3 was taken verbatim from [6, 7]. It also appears in [5].

References

1. Blatter, C., Specker, E.: Recurrence relations for the number of labeled structures on a finite set. In: Börger, E., Hasenjaeger, G., Rödding, D. (eds.) In Logic and Machines: Decision Problems and Complexity, Lecture Notes in Computer Science, vol. 171, pp. 43–61. Springer (1984)
2. Dawar, A., Grohe, M., Kreutzer, S., Schweikardt, N.: Model theory makes formulas large. In: Automata, Languages and Programming: 34th International Colloquium, ICALP 2007, Wrocław, Poland, July 9-13, 2007. Proceedings 34, pp. 913–924. Springer (2007)
3. Filmus, Y., Fischer, E., Makowsky, J.A., Rakita, V.: MC-finiteness of restricted set partition functions. J. Integer Sequences **26**(2), 3 (2023)
4. Fischer, E.: The Specker-Blatter theorem does not hold for quaternary relations. J. Comb. Theory, Ser. A **103**, 121–136 (2003)
5. Fischer, E., Kotek, T., Makowsky, J.A.: Application of logic to combinatorial sequences and their recurrence relations. Model Theor. Methods Finite Comb. **558**, 1–42 (2011)

6. Fischer, E., Makowsky, J.A.: The Specker-Blatter theorem revisited. In: Computing and Combinatorics: 9th Annual International Conference, COCOON 2003, Lecture notes in computer science, vol. 2697, pp. 90–101 (2003)
7. Fischer, E., Makowsky, J.A.: The Specker-Blatter theorem revisited: generating fuctions for definable classes of structures (2003). https://doi.org/10.3929/ethz-a-004518887. FIM preprints, ETZ, Zurich
8. Fischer, E., Makowsky, J.A.: Extensions and limits of the Specker–Blatter theorem. J. Symb. Logic 1–29 (2024)
9. Hedlíková, J.: Ternary spaces, media, and Chebyshev sets. Czechoslov. Math. J. **33**(3), 373–389 (1983)
10. Makowsky, J.A.: Algorithmic uses of the Feferman-Vaught theorem. Ann. Pure Appl. Logic **126**(1–3), 159–213 (2004)
11. Pambuccian, V.: The axiomatics of ordered geometry: I. ordered incidence spaces. Expositiones mathematicae **29**(1), 24–66 (2011)
12. Pitcher, E., Smiley, M.: Transitivities of betweenness. Trans. Am. Math. Soc. **52**(1), 95–114 (1942)
13. Specker, E.: Application of logic and combinatorics to enumeration problems. In: Ernst Specker Selecta, pp. 324–350. Springer (1990)

An Overview of Property B

Karl Grill and Daniel Linzmayer

Abstract Following Erdős and Hajnal [12], a hypergraph (V, E) is said to have property B if it admits a proper two-coloring. Let $m(n)$ denote the smallest number of edges in an n-uniform hypergraph that does not have property B, and $m_v(n)$ the analogous number for an n-uniform hypergraph with $|V| = v$ vertices. We give an overview of classical and recent bounds on $m(n)$ and $m_v(n)$, respectively.

1 Introduction

Let $H = (V, E)$ be a n-uniform hypergraph, i.e., a hypergraph with

$$E \subseteq \binom{V}{n} = \{A \subset V : |A| = n\}$$

H is said to have **property B**, if it can be 2-colored, i.e., if its vertices can be colored with two colors (traditionally called "red" and "blue"), such that none of its edges is monochromatic. Let **m(n)** be the lowest number of edges a n-uniform hypergraph (V, E) can have, such that it is **not** possible to find a proper coloring.

Furthermore, we define $m_v(n)$ in same way, but for hypergraphs with a fixed number v of vertices.

We only consider the question of two-colorability here, and we are particularly interested in the case where n and v are small. For a more general overview, we refer the reader to Raigorodskii and Cherkashin [21] or Raigorodskii and Shabanov [22].

K. Grill (✉) · D. Linzmayer
Institute of Statistics and Mathematical Methods in Economics, TU Wien, Vienna, Austria
e-mail: karl.grill@tuwien.ac.at

D. Linzmayer
e-mail: daniel@linzmayer.at

G. O. H. Katona et al. (eds.), *Sum(m)it280*, Bolyai Society Mathematical Studies 32,
https://doi.org/10.1007/978-3-032-18810-6_9

2 Exact Results

For $v < 2n - 1$, every balanced coloring is proper, so the question does not really make sense in this case, or we could let $m_v(n) = \infty$. Erdős observed that, in general, the inequality $m_{v+1}(n) \le m_v(n)$ holds, and that $m_{2n-1}(n) = m_{2n}(n) = \binom{2n-1}{n}$, which together imply the inequality $m(n) \le \binom{2n-1}{n}$, which was already found by Erdős and Hajnal [12].

Erdős [9] states $m(2) = 3$ and $m(3) = 7$, which imply $m_v(2) = 3$ for all $v \ge 3$ and $m_v(3) = 7$ for all $v \ge 7$.

For $n = 4$, Abbott and Hanson [1] showed $m(4) \le 24$, which was improved by Seymour [23] or Toft [25] (independently) to $m(4) \le 23$. Östergård [18] conducted an exhaustive computer search and found $m_9(4) = 26$, $m_{10}(4) = 25$, and $m_{11}(4) = m(4) = 23$.

deVries [8] proves that the pigeon-hole lower bound

$$m_{2n+1}(n) \ge \frac{1}{n}\binom{2n+1}{n-1}$$

holds with equality if and only if there is a Steiner system $S(n-1, n, 2n+1)$. The only known Steiner systems of this form are $S(2, 3, 7)$ and $S(4, 5, 11)$. The latter yields $m_{11}(5) = m_{12}(5) = 66$.

3 Constructive Bounds

Bounding $m(n)$ or $m_v(n)$ above is conceptually simple: one only has to provide (or at least prove the existence) of a non-colorable hypergraph of the appropriate dimensions. Unfortunately, putting this into practice is not so simple. In this section, we review a few explicit constructions that have emerged over the years. These generally use one or more smaller non-colorable hypergraphs around which the new hypergraph is built. These constructions yield upper bounds that are asymptotically weaker than the probabilistic bound that we will discuss later, but they yield the best known upper bounds for some small values of n.

The first of these constructions was obtained by Abbott and Moser [3] (here and below, an (n, v, m)-hypergraph is meant to be a non-colorable n-uniform hypergraph with v vertices and m edges): given an (n_1, v_1, m_1) and an (n_2, v_2, m_2)-hypergraph which are non-2-colorable, an $(n_1 n_2, v_1 v_2, m_1 m_2^{n_1})$-hypergraph without property B is constructed. To this end, v_1 copies H_i of hypergraph 2 with disjoint edge sets are used. Then, every edge of hypergraph 1 is replaced by the $m_2^{n_1}$ edges obtained by replacing node i by any of the from H_i.

This construction yields the current best bounds $m(6) \le 147$, $m(9) \le 2401$, $m(10) \le 7801$, $m(12) \le 55223$, and $m(15) \le 857157$. In more detail, these are actually upper bounds for $m_{21}(6)$, $m_{49}(9)$, $m_{51}(10)$, $m_{77}(12)$, and $m_{119}(15)$.

The constructions by Abbott and Hanson [1], Toft [25], and Mathews et al. [17] have a common theme in that a $(n-2, v, m)$-hypergraph is used to construct a $(n, v+2n, m')$-hypergraph. There are slight differences in the construction method and, of course, the value of m'. The construction of Abbott and Hanson [1] works for odd n with $m' = 2^{n-1} + nm$, Toft [25] uses a similar method for even n, yielding $m' = 2^{n-1} + nm + \binom{n}{n/2}$. Mathews et al. [17] provide two different constructions, with the first one yielding

$$m' = (n+1)2^{n-2} + (n-1)m + \frac{1}{2}\binom{n}{n/2} + (n-1)\binom{n-2}{(n-2)/2},$$

and the second one

$$m' = (n+4)2^{n-3} + (n-2)m + n\binom{n-2}{(n-2)/2} + \frac{1}{2}\binom{n}{n/2}.$$

In both formulas, we avoid the case distinction for odd/even n by using the convention than $\binom{n}{k} = 0$ for non-integer k.

Aglave et al. [4] present the following constructions:

- A generalized Abbot-Hanson-Toft type construction, using a $(n-2k, v, m)$-hypergraph to construct a $(n, v+2(n+k-1), m')$-hypergraph with

$$m' = \binom{n+k-1}{k}m + \binom{n+k-1}{k-1}\left(2^{n-1} + \binom{n}{n/2}/2\right),$$

- A multi-core construction: for $k < n$, let $w = \lfloor n/k \rfloor$, $x = n \pmod{k}$, $y = \lfloor k/x \rfloor$ and $z = k \pmod{x}$. The construction uses a $(n-k, v_c, m_c)$-hypergraph, a (k, v_k, m_k)-hypergraph, and (given $x > 0$) a (x, v_x, m_x)-hypergraph to construct a (n, v, m)-hypergraph, where

  – if $x > 0$ and $z > 0$, $v = v_c + wv_k + yv_x + x + z - 1$, and

$$m = wm_cm_k + ym_xm_k^w + \binom{x+z-1}{z}m_cm_x^y + \binom{x+z-1}{x}m_k^w,$$

  – if $x > 0$ and $z = 0$, $v = v_c + wv_k + yv_x$ and

$$m = wm_cm_k + ym_xm_k^w + m_cm_x^y,$$

  – if $x = 0$, $v = v_c + wv_k$ and

$$m = wm_cm_k + m_k^w.$$

- A block construction, using a $(n-2k, v_c, m_c)$-hypergraph and a collection of t hypergraphs with parameters (n_i, v_i, m_i), $i = 1, \ldots, t$, satisfying $n_i \geq k$ and $\sum_{i=1}^{t} n_i \geq n$, to construct a (n, v, m)-hypergraph with $v = v_c + 2\sum_{i=1}^{t} v_i$ and

$$m \leq m_c \sum_{i=1}^{t} m_i^2 + \left(2^{t-1} + \frac{1}{2}\binom{t}{t/2}\right) \prod_{i=1}^{t} m_i.$$

- A modification of this block construction for $n = 3k+1$ that gives the recurrence relation

$$m(3k+1) \leq (m(k-1) + 2^{k-1})m(k+1)^2 + 2\tilde{m}(k+1)m(k)^2 + 4m(k+1)m(k)^2,$$

where $\tilde{m}(k+1)$ is the number of edges in a non-colorable $k+1$-uniform hypergraph obtained by the Abbott-Hanson-Toft construction.

Gebauer [13] gives a deterministic construction of a non-colorable hypergraph on $2^{\Theta(n^{1/3})}$ vertices with $2^{n+\Theta(n^{2/3})}$ edges. Kozik [15] improves this to $2^{n+\Theta(\sqrt{n\log(n)})}$ edges on $2^{\Theta(\sqrt{n/\log(n)})}$ vertices.

One common feature of these constructions is that they use a relatively large number of vertices. In particular, there don't seem to be any useful constructions for v close to $2n$, especially since the gap between the probabilistic lower and upper bounds is quite narrow there. One tempting approach is to try and find some upper bounds by heuristic methods. Naturally, this is only feasible for small values of n and v. We (Linzmayer and Grill [16]) used a few variants of heuristic searches (mostly simulated annealing) to obtain upper bounds on $m_v(n)$. In particular, we were able to reproduce the best results for $n \leq 4$, and also $m_{11}(5) = 66$. Also, the best known results $m(5) \leq 51$ and $m(6) \leq 147$ could be found in this way. It is worth mentioning that all hypergraphs that we found are isomorphic to those obtained by constructive methods. This seems to indicate that there could actually hold equality in these inequalities. A summary of these heuristic upper bounds is given in Table 1.

Table 1 Upper bounds for $m_v(n)$ cf. [16] Entry (6, 19, 189) comes from a modified Moser-Toft construction with two (3, 7, 7) Fano planes and a (3, 5, 10) complete hypergraph. The others have been (re-)obtained by various heuristic search algorithms

	v										
n	11	12	13	14	15	16	17	18	19	20	21
5	66	66	65	65	63	57	51	—	—	—	—
6	—	—	242	242	225	—	—	—	189	—	147
7	—	—	—	—	849	—	832	—	—	—	—
8	—	—	—	—	—	—	3400	3311	2850	—	—
9	—	—	—	—	—	—	—	—	12111	—	—

4 Probabilistic Bounds

The asymptotically most successful bounds, both above and below, have been found by probabilistic methods.

4.1 Upper Bounds

The best known upper bounds on $m(n)$ and $m_v(n)$ are still those obtained by Erdős [10, 11] by the probabilistic method. The central idea is to pick m edges at random and to estimate the probability that there is a proper coloring. For a given coloring with b blue and $v - b$ red vertices, the probability that a randomly picked edge is properly colored is

$$1 - \frac{\binom{b}{n} + \binom{v-b}{n}}{\binom{v}{n}}.$$

This takes its maximum for $b = \lfloor v/2 \rfloor$. Thus, the chance that m edges picked independently are properly colored is bounded above by

$$\left(1 - \frac{\binom{\lfloor v/2 \rfloor}{n} + \binom{\lceil v/2 \rceil}{n}}{\binom{v}{n}}\right)^m.$$

As there are 2^v possible colorings, we arrive at the upper bound

$$2^v \left(1 - \frac{\binom{\lfloor v/2 \rfloor}{n} + \binom{\lceil v/2 \rceil}{n}}{\binom{v}{n}}\right)^m \tag{1}$$

for the probability that there exists a proper coloring. Choosing m so large that this upper bound becomes less than 1 yields a positive probability that no coloring is proper and thus the existence of a non-colorable hypergraph with m edges. This gives us the upper bound

$$m_v(n) \leq (v \log(2) + \epsilon) \frac{\binom{v}{n}}{\binom{\lfloor v/2 \rfloor}{n} + \binom{\lceil v/2 \rceil}{n}}. \tag{2}$$

Choosing $v = n^2$ and $v = \lfloor n^2/2 \rfloor$, this implies the results from Erdős [10]:

$$m(n) < n^2 2^{n+1}$$

and

$$m(n) \leq (1 + o(1))e \log(2) n^2 2^{n-2}. \tag{3}$$

For even v, (2) is the statement of theorem 2 from Erdős [11]. Abbott and Liu [2], for the special case $v = 2n + 1$, improve this to

$$m_{2n+1}(n) \leq (1 + o(1))\binom{2n+1}{n}\frac{\log(n)}{n}.$$

The lemma they use is equivalent to considering (1) as the expectation of the number of proper colorings of the randomly selected hypergraph, and adding one monochromatic edge for each of those colorings to obtain a non-colorable hypergraph with

$$m + 2^v\left(1 - \frac{\binom{\lfloor v/2 \rfloor}{n} + \binom{\lceil v/2 \rceil}{n}}{\binom{v}{n}}\right)^m$$

edges. Choosing m to minimize this expression yields the desired result.

This idea can be applied for any v. For $v = 2n + c$, the gap between the upper bound and the pigeon-hole lower bound is of order $\log(n)$, as in [2]. For $v = O(n)$, the case considered in [11], this improves the constant in the asymptotic bound. For v bigger than that, in particular for $v \approx n^2$ that gave us (3), this only affects lower-order terms and doesn't change the first-order asymptotics.

4.2 Lower Bounds

A very simple instance of the probabilistic methods, equivalent to the pigeon-hole principle, yields the lower bounds (Erdős [9, 11])

$$m(n) \geq 2^{n-1}$$

and

$$m_v(n) \geq \frac{\binom{v}{n}}{\binom{\lfloor v/2 \rfloor}{n} + \binom{\lceil v/2 \rceil}{n}}.$$

We prove the second statement, the proof of the first one runs along similar lines or can be obtained by letting $v \to \infty$ in the second one.

Choose a balanced coloring with $b = \lfloor v/2 \rfloor$ blue and $r = v - b = \lceil v/2 \rceil$ red vertices at random. There are $\binom{v}{b}$ such colorings, and $\binom{v-n}{b}$ of them make a given edge monochromatic red, $\binom{v-n}{b-n}$ blue. That gives a probability

$$\frac{\binom{v-n}{b-n} + \binom{v-n}{b}}{\binom{v}{b}} = \frac{\binom{b}{n} + \binom{r}{n}}{\binom{v}{n}}$$

that this edge is monochromatic. For m edges, the probability that any of them is monochromatic is at most

$$m\frac{\binom{v-n}{b-n}+\binom{v-n}{b}}{\binom{v}{b}}=\frac{\binom{b}{n}+\binom{r}{n}}{\binom{v}{n}},$$

and when this is less than 1, there is a positive probability that no edge is monochromatic, so there exists a proper coloring, and $m(n) \geq m$.

A natural idea is to start out from a random coloring and, if it fails to be proper, try to repair its blemishes by recoloring some of the vertices contained in a monochromatic edge. This is actually a good idea and has been exercised with varying degrees of sophistication.

Beck [5, 6] used this idea to arrive at

$$m(n) \geq n^{1/3-\delta}2^n$$

eventually for any $\delta > 0$. Spencer [24] gave a simplified proof of this result. Radhakrishna and Srinivasan [20] used this approach to arrive at the current best asymptotic lower bound

$$m(n) = \Omega(\sqrt{\frac{n}{\log(n)}}2^n.$$

A different approach was developed by Pluhár [19], He uses a random greedy coloring: the vertices are put in random order and successively colored: a vertex is colored blue, unless it is the last one in an otherwise all-blue edge, in which case it is colored red. Pluhár's analysis only gets to $m(n) = \Omega(n^{1/4}2^n)$, but Cherkashin and Kozik [7] refine it and thus give another (simpler) proof of Radhakrishnan and Srinivasan's result. They assign random weights picked independently uniformly in $[0, 1]$ to the vertices and arrange the vertices in increasing order of their weights, thus removing the dependence on the number of vertices. We (Grill and Linzmayer [14]) re-introduce this dependence, resulting in some improvements over the Cherkashin-Kozik bound for small values of n and v. In the same paper, some additional improvement is obtained by putting up to three low-degree vertices at a fixed position in the middle of the sequence. Thus, we arrive at the lower bounds in Table 2, in particular, we get $m(5) \geq 32$, improving on the previous best $m(5) \geq 29$ obtained by Aglave et al. [4]. For comparison, Table 3 gives the currently best-known upper bounds on $m(n)$ for small n.

Table 2 Lower bounds for $m(n)$ Obtained by variations of the greedy coloring algorithm [14]

5	6	7	8	9
32	64	128	263	538

Table 3 Upper bounds for $m(n)$

n	Upper bound	References
2	3	Erdős [9]
3	7	Erdős [9]
4	23	Östergård [18]
5	51	Toft [25]
6	147	Abbott and Moser [3]

References

1. Abbott, H.L., Hanson, D.: On a combinatorial problem of Erdős. Canad. Math. Bull. **12**, 823–829 (1969)
2. Abbott, H.L., Liu, A.C.: On property B of families of sets. Canad. Math. Bull. **23**(4), 429–435 (1980)
3. Abbott, H.L., Moser, L.: On a combinatorial problem of Erdős and Hajnal. Canad. Math. Bull. **7**, 177–181 (1964)
4. Aglave, S., Amarnath, V.A., Shannigrahi, S., Singh, S.: Improved bounds for uniform hypergraphs without property B. Australasian J. Combinatorics **76**(1), 73–86 (2020)
5. Beck, J.: On a combinatorial problem of P. Erdős and L. Lovász. Discret. Math. **17**, 127–131 (1977)
6. Beck, J.: On 3-chromatic hypergraphs. Discret. Math. **24**(2), 127–137 (1978)
7. Cherkashin, D., Kozik, J.: A note on random greedy coloring of uniform hypergraphs. Random Struct. Algorithms **47**(3), 407–413 (2015)
8. de Vries, H.L.: On property B and on Steiner systems. Math. Z. **153**(2), 153–159 (1977). https://doi.org/10.1007/BF01179788
9. Erdős, P.: On a combinatorial problem. Nordisk Mat. Tidskrift **11**, 5–10 (1963)
10. Erdős, P.: On a combinatorial problem II. Acta. Math. Acad. Sci. Hung. **15**, 445–447 (1964)
11. Erdős, P.: On a combinatorial problem III. Canad. Math. Bull. **12**(4), 413–416 (1969)
12. Erdős, P., Hajnal, A.: On a property of families of sets. Acta. Math. Acad. Sci. Hung. **12**, 87–123 (1961)
13. Gebauer H.: On the construction of 3-chromatic hypergraphs with few edges. J. Comb. Theory Ser. A **120(7)**, 1483–1490 (2013)
14. Grill, K., Linzmayer, D.: Improved lower bounds for property B (2024). arXiv.org, http://arxiv.org/abs/2403.05674
15. Kozik J.: Improving Gebauer's Construction of 3-chromatic hypergraphs with few edges. In: 48th International Colloquium on Automata, Languages, and Programming (ICALP 2021). Leibniz International Proceedings in Informatics (LIPIcs), vol. 198, pp. 89:1–89:9, Schloss Dagstuhl – Leibniz-Zentrum für Informatik (2021). https://doi.org/10.4230/LIPIcs.ICALP.2021.89
16. Linzmayer, D., Grill, K.: Some upper bounds for property B for small values of n by heuristic search (2024)
17. Mathews, J., Panda, M.K., Shannigrahi, S.: On the construction of non-2-colorable uniform hypergraphs. Discret. Appl. Math. **180**, 181–187 (2015)
18. Östergård, P.R.J.: On the minimum size of 4-uniform hypergraphs without property B. Discret. Appl. Math. **163**(2), 199–204 (2014)
19. Pluhár, A.: Greedy colorings of uniform hypergraphs. Random Struct. Algorithms **35**, 216–221 (2009). https://doi.org/10.1002/rsa.20267
20. Radhakrishnan, J., Srinivasan, A.: Improved bounds and algorithms for hypergraph 2-coloring. Random Struct. Algorithms **16**(1), 4–32 (2000)

21. Raigorodskii, A.M., Cherkashin, D.D.: Extremal problems in hypergraph colourings. Russian Math. Surv. **75**(1), 89–146 (2020)
22. Raigorodskii, A.M., Shabanov, D.A.: The Erdős-Hajnal problem of hypergraph colouring, its generalizations, and related problems. Russian Math. Surv. **66**(5), 933–1002 (2011)
23. Seymour P.D.: A note on a combinatorial problem of Erdős and Hajnal. J. Lond. Math. Soc. **(2)8**, 681–682 (1974)
24. Spencer, J.: Coloring n-sets red and blue. J. Comb. Theory Ser. A **30**, 112–113 (1981)
25. Toft, B.: On colour-critical hypergraphs, In: Hajnal, A., Rado, R., Sós, V.T. (eds.), Infinite and Finite Sets, vol. 3, Colloquia Mathematica Societatis János Bolyai, vol. 10, pp. 1445—1457. North-Holland, Amsterdam (1975)

A Survey of the Holroyd-Talbot Conjecture

Glenn Hurlbert

Abstract A family of sets is *intersecting* if every pair of its members has an element in common. Such a family of sets is called a *star* if some element is in every set of the family. Given a graph G, let $\mu(G)$ denote the size of the smallest maximal independent set of G. In 2005, Holroyd and Talbot conjectured the following generalization of the Erdős-Ko-Rado Theorem: for $1 \leq r \leq \mu(G)/2$, there is a maximum size intersecting family of independent r-sets that is a star. In this paper we present the history of this conjecture and survey the results that have supported it over the last 20 years.

1 Introduction

For $0 \leq r \leq n$, set $[n] = \{1, 2, \ldots, n\}$, $2^{[n]} = \{S \subseteq [n]\}$, and $\binom{[n]}{r} = \{S \subseteq [n] \mid |S| = r\}$. For $\mathcal{F} \subseteq 2^{[n]}$, define $\cap\mathcal{F} = \cap_{S\in\mathcal{F}} S$, $\mathcal{F}^r = \{S \in \mathcal{F} \mid |S| = r\} = \mathcal{F} \cap \binom{[n]}{r}$, and $\mathcal{F}_X = \{S \in \mathcal{F} \mid X \subseteq S\}$. We use the notational conventions $A + x = A \cup \{x\}$ and $A - x = A \setminus \{x\}$. For $1 \leq t \leq r$, we call $\mathcal{F}$ *t-intersecting* if $|A \cap B| \geq t$ for every $A, B \in \mathcal{F}$. For convenience, we write $\mathcal{F}_x$ instead of $\mathcal{F}_{\{x\}}$, and say *intersecting* in place of 1-intersecting. If $\cap\mathcal{F} \neq \emptyset$, we say that $\mathcal{F}$ is a *star* (an *r-star* if $\mathcal{F} \subseteq \binom{[n]}{r}$); in this case, any $x \in \cap\mathcal{F}$ is called a *center*. If x is the center of a maximum size star of $\mathcal{F}$, we call x a *maximum* center. If $| \cap \mathcal{F}| = k$ we say that $\mathcal{F}$ is *k-centered*. Erdős, Ko, and Rado [24] proved the following foundational result in extremal set theory, that, for large enough n, the biggest t-intersecting families of r-subsets of $[n]$ are t-centered r-stars.

Theorem 1 (Erdős et al. [24]) *For all $1 \leq t \leq r$, there is an integer $M(r, t)$ such that if $n > M(r, t)$ and $\mathcal{F} \subseteq \binom{[n]}{r}$ is t-intersecting then $|\mathcal{F}| \leq \binom{n-t}{r-t}$, with equality if and only if $\mathcal{F} = \binom{[n]}{r}_X$ for some $X \in \binom{[n]}{t}$.*

G. Hurlbert (✉)
Department of Mathematics and Applied Mathematics, Virginia Commonwealth University, Richmond, VA, USA
e-mail: ghurlbert@vcu.edu

G. O. H. Katona et al. (eds.), *Sum(m)it280*, Bolyai Society Mathematical Studies 32,
https://doi.org/10.1007/978-3-032-18810-6_10

In 1984, Wilson [68] used Delsarte's linear programming bound [20] to prove that the optimal value of $M(r,t) = (r-t+1)(t+1)$. When $n = (r-t+1)(t+1)$, the Erdős-Ko-Rado bound still holds, but there are non-stars of maximum size as well. For example, when $t = 1$ we have $M(r, 1) = 2r$ and any family that includes exactly one of each complementary pair $\{S, \overline{S}\}$ is intersecting and of maximum size. When $t > 1$, the Complete Intersection Theorem of Ahlswede-Khachatrian [1] identifies the maximum size and structure of t-intersecting families in the range of $2r - t < n < 2r$. In this paper we will restrict our attention to the case $t = 1$ only.

Since the publication of Theorem 1, in which the authors used shifting to achieve the result, a number of different methods have been used to prove it: bounding the size of the largest intersecting non-star family via shifting [38]; finding local bounds that extend to global bounds via the cycle method [52]; applying the Kruskal-Katona Theorem to shadows of complements [18, 51]; providing an upper bound on the Shannon capacity of the Kneser graph [57]; using Delsarte's linear programming bound [68]; applying the Hoffman-Delsarte ratio bound to the independence number of the Kneser graph [36]; using the method of linearly independent polynomials [35]; displaying an injection from any intersecting family into a star [31, 48]; and using the Borel Fixed Point Theorem from algebraic group theory [70].

For a graph G, we write $\mathcal{I}(G)$ for the family of all independent sets of G, and use $\mathcal{I}^r(G)$ in place of $\mathcal{I}(G)^r$ and $\mathcal{I}_v(G)$ in place of $\mathcal{I}(G)_v$, etc. Suppose that $\mathcal{F} \subseteq \mathcal{I}^r(G)$ is an intersecting subfamily of maximum size. Define G to be r-EKR if some $v \in V(G)$ satisfies $|\mathcal{I}^r_v(G)| = |\mathcal{F}|$, and *strictly* r-EKR if every such $\mathcal{F}$ equals $\mathcal{I}^r_v(G)$ for some v. Such a v is called an r-*witness* of G (either strictly so or not). Let $\alpha(G)$ denote the independence number of G, and $\mu(G)$ denote the size of a smallest maximal independent set of G; i.e., the size of the smallest independent dominating set of G. Holroyd and Talbot [45] made the following conjecture to generalize the $t = 1$ case of Theorem 1, which is the empty graph case of this conjecture.

Conjecture 2 (Holroyd and Talbot [45]) *For any graph G, if $1 \le r \le \mu(G)/2$, then G is r-EKR, and strictly so if $1 < r < \mu(G)/2$.*

The necessity of the $\mu(G)/2$ condition in this conjecture stems from their example of two disjointcopies of the complete bipartite graph $K_{3,3}$: $G = 2K_{3,3}$, with $E(G) = (A \times B) \cup (C \times D)$. Here we have $\mu(G) = 6$ and a maximum star in $\mathcal{I}^4(G)$ has size 20, while the family consisting of every independent 4-set with at least two elements of A is an intersecting non-star of size 24. Hence G is not 4-EKR. More generally, two disjoint copies of $K_{2k-1,2k-1}$ has $\mu/2 = 2k - 1$ and is not $2k$-EKR.

Because every graph G is technically r-EKR (vacuously) when $r \le 0$ or $r > \alpha(G)$, we will drop the obvious condition "$1 \le r \le \alpha(G)$" that would normally follow "for all r", only writing constraints when they are more restrictive.

2 Precursors to the Holroyd-Talbot Conjecture

In 1972, Berge [3] considered intersecting subfamilies of the collection of k-cliques in $K_{n_1,\ldots,n_k}$, and proved the following.

Theorem 3 *For $k \geq 2$, if $\mathcal{F}$ is an intersecting family of k-cliques in $G = K_{n_1,\ldots,n_k}$, with $n_1 \leq n_2 \leq \cdots \leq n_k$, then $|\mathcal{F}| \leq n_2 \cdots n_k$.*

Of course, if $\mathcal{F}$ is a maximum intersecting family and $f(G)$ is the maximum number of k-cliques containing the same vertex of G, then $n_2 \cdots n_k = f(G) \leq |\mathcal{F}| \leq \chi'(G)$, where χ' denotes the *chromatic index*. Berge achieved the result by proving that $\chi'(G) = f(G)$. The paper was also published in [4]. The result can be considered the first precursor to the Holroyd-Talbot Conjecture because it shows that the complement $\overline{G}$ is k-EKR. For the case that each $n_i = n > 2$, Livingston [56] proved that $\overline{G}$ is strictly k-EKR. Other proofs of this result were given later in [37, 59].

Deza and Frankl [21] considered the following scenario. An *ℓ-signed set* is a pair (A, f), where $A \subseteq [n]$ and $f : A \rightarrow [\ell]$, and its *size* equals $|A|$. For two such pairs (A, f) and (B, g) define $(A, f) \cap (B, g) = (C, h)$, where $C = \{x \in A \cap B \mid f(x) = g(x)\}$ and $h : C \rightarrow [\ell]$ such that $h(x) = f(x)$. Say that a family of ℓ-signed sets is *t-intersecting* if the size of the intersection of any two of its members is at least t. They proved the following, first stated by Meyer in [58] (stated in the complementary form of Berge/Livingston).

Theorem 4 *Let $r \leq n$, $\ell \geq 2$, and $\mathcal{F} \subseteq \binom{[n]}{r}$ be a t-intersecting family of ℓ-signed sets. Then $|\mathcal{F}| \leq \binom{n-1}{r-1}\ell^{r-1}$.*

When $t = 1$, $\ell = 2$, and $r \leq n/2$, the result follows from Theorem 1 and the following Fact.

Fact 5 *If $\mathcal{F} \subseteq 2^{[n]}$ is intersecting, then $|\mathcal{F}| \leq 2^{n-1}$.*

Fact 5 holds because at most one of each pair $\{F, \overline{F}\}$ can be a member of $\mathcal{F}$, and is best possible because of $2^{[n]}_x$. Thus, if $\mathcal{F}$ is an intersecting family of 2-signed sets, then there are at most $\binom{n-1}{r-1}$ r-sets A in some pair of $\mathcal{F}$ when $r \leq n/2$ by Theorem 1. In addition, for each such A, Fact 5 implies that there are at most 2^{r-1} functions f such that $(A, f) \in \mathcal{F}$, since the set of those f is isomorphic to the set of characteristic vectors of an intersecting family of subsets of A. That is, for fixed $A \in \binom{[n]}{r}$, the set of pairs (A, f), where $f : A \rightarrow [2]$ is isomorphic to 2^A by the correspondence $B(A, f) = \{a \in A \mid f(a) = 2\}$; clearly $(A, f) \cap (A', f') \neq \emptyset$ if and only if $B(A, f) \cap B(A', f') \neq \emptyset$. But Theorem 4 needs to be verified for the wider range $r \leq n$, as well as for $t > 1$ and for $\ell > 2$. Deza and Frankl [21] prove this by using the shifting technique introduced in [24], which we define now.

For $1 \leq i < j \leq n$, and for any family $\mathcal{F} \subseteq 2^{[n]}$, define the shifting function $\sigma_{j,i} : \mathcal{F} \rightarrow 2^{[n]}$ by

$$\sigma_{j,i}(A) = \begin{cases} A + i - j & \text{if } j \in A,\ i \notin A, \text{ and } A + i - j \notin \mathcal{F}, \text{ and} \\ \sigma_{j,i}(A) = A & \text{otherwise.} \end{cases}$$

We extend the shifting function to families of sets by $\sigma_{j,i}(\mathcal{F}) = \{\sigma_{j,i}(A) \mid A \in \mathcal{F}\}$. It is well known that $|\sigma_{j,i}(\mathcal{F})| = |\mathcal{F}|$ and that $\sigma_{j,i}(\mathcal{F})$ is t-intersecting whenever $\mathcal{F}$ is. Thus, when proving EKR-type results, one may always assume that an intersecting family $\mathcal{F}$ is *shifted*; that is, $\sigma_{j,i}(\mathcal{F}) = \mathcal{F}$ for all $1 \leq i < j \leq n$. (See [29] for the many other wonderful properties and implications of shifting.) The shifting technique as been applied to many varieties of intersection theorems, as well as to other types of problems in extremal set theory and other areas of combinatorics and graph theory. Its power and utility makes it a primary tool in tackling Conjecture 2.

The reason that Theorem 4 is a precursor to Conjecture 2 is that the family of all ℓ-signed r-subsets of $[n]$ corresponds to $\mathcal{I}^r(G)$, where G is the disjoint union of n copies of the complete graph K_ℓ, which we denote nK_ℓ. With this correspondence, we may think of the $t = 1$ case of Theorem 4 as being the second verification of Conjecture 2. In fact, this result goes farther than Holroyd-Talbot because it holds true for $r \leq n = \mu(nK_\ell)$. Deza and Frankl [21] also showed that nK_ℓ is strictly r-EKR unless $\ell = 2$ and $r = n \geq 3$.

Bollobás and Leader [7] gave an alternate proof of Theorem 4 by using the idea underlying Katona's *cycle method* that proved Theorem 1. The essence of the method is to define, for any permutation π of $[n]$, written cyclically, the family $\mathcal{F}_\pi$ of all sets of $\mathcal{F}$ that appear consecutively on the cycle. For any $A \in \mathcal{F}_\pi$, one can cut the cycle between any of the $r - 1$ consecutive pairs of A, creating a pair of disjoint, consecutive sets on the cycle that begin at the adjacent elements on opposite sides of the cut and progress in opposite directions. Since $\mathcal{F}$ is intersecting, at most one of each pair is in $\mathcal{F}_\pi$, implying that $|\mathcal{F}_\pi| \leq r$. Because each $A \in \mathcal{F}$ is in a uniform number of $\mathcal{F}_\pi$, double-counting the pairs $(A, \mathcal{F}_\pi)$ with $A \in \mathcal{F}_\pi$ leads to the result.

In the language of independent sets in nK_ℓ instead of signed sets, a the jth vertex (i, j) in the ith (maximum) clique of nK_ℓ represents the element x with sign $f(x) = i$, What Bollobás and Leader were able to do is make the double-counting work by only considering what they called *good* permutations of $[n\ell]$: permutations π in which, for each clique, the vertices of that clique are equally spaced around the cycle. This guarantees that consecutive r-sets in π are independent in nK_ℓ, and that the necessary uniformity condition holds.

Recently, Feghali [26] gave an injective proof of the $t = 1$ and $r \leq n/2$ instance of Theorem 4, along the lines of [31, 48].

The above results are for uniform families; in [5], Bey generalized the $t = 1$ instance to antichains in general, proving the following theorem. For an antichain $\mathcal{F} \subseteq 2^{[n]}$, define is f-*vector* $\mathbf{f}(\mathcal{F}) = (\mathbf{f}(r))_{r=0}^n$ by $\mathbf{f}(r) = |\mathcal{F}^r|$. For $0 \leq i \leq n$, let $\mathbf{f}_i$ denote the f-vector with $\mathbf{f}_i(i) = \binom{n-1}{i-1}\ell^{i-1}$ and $\mathbf{f}_i(r) = 0$ otherwise; that is, the f-vector of a star in $\mathcal{I}^i(\overline{G})$, where $G = K_{n_1,\ldots,n_k}$. Finally, set $\mathbf{0} = (0)_{r=0}^n$ and write $\mathbf{F}$ for the convex hull of $\{\mathbf{0}, \mathbf{f}_1, \ldots, \mathbf{f}_n\}$.

Theorem 6 *If $\mathcal{F} \subseteq \mathcal{I}(\overline{G})$ is an intersecting antichain, for $G = K_{n_1,\ldots,n_k}$ with $k \geq 2$, then $\mathbf{f}(\mathcal{F}) \in \mathbf{F}$.*

For the instance of $t = 1$ and $\ell = 2$, Fuentes and Kamat [34] proved the following common generalization of Theorems 1 and 4. Let $\mathcal{H}^{(p,s)}(n)$ denote the family of vertex sets of subgraphs of nK_2 that have p edges and s isolated vertices.

Theorem 7 *Let $n \geq 2p + s$. Then*

1. *if $n \geq 2p + 2s$, then $\mathcal{H}^{(p,s)}(n)$ is $(2p + s)$-*EKR*, and strictly so when $n > 2p + 2s$,*
2. *$\mathcal{H}^{(p,s)}(2p + s)$ is $(2p + s)$-*EKR*, but not strictly so, and*
3. *if $n > 2p + 1$, then $\mathcal{H}^{(p,1)}(n)$ is strictly $(2p + 1)$-*EKR*.*

They conjecture that $\mathcal{H}^{(p,s)}(n)$ is $(2p + s)$-EKR for all $n \geq 2p + s$.

Another scenario involves families of subsets of integers, in which no subset contains consecutive elements. Such subsets of $[n]$ correspond to independent sets in the path P_n. Disallowing the additional pair $\{1, n\}$ corresponds to independent sets in the cycle C_n; such subsets are known as *separated* sets. More generally, for a graph G, denote by $G^{(k)}$ the kth *power of* G: the graph on $V(G)$ having an edge xy precisely when the distance in G between x and y is positive and at most k. Then a *k-separated* set is any set in $\mathcal{I}(C_n^{(k)})$. Talbot [67] proved the following result, stated in terms of k-separated sets, which had been conjectured by Holroyd and Johnson in [43] (originally posed by Holroyd [42] in terms of sorties of King Arthur's knights sitting far apart from one another at the Round Table).

Theorem 8 *The cycle power $C_n^{(k)}$ is r-*EKR* for all r, and strictly so unless $k = 1$ and $n = 2r + 2$.*

Talbot's proof uses induction, a simplified shift operator, a very careful partition of $\mathcal{F}$ into $k + 3$ parts, and seven pages of detailed analysis. When $k = 1$ and $n = 2r + 2$, a 1-centered r-star in $C_n^{(k)}$ has size $|\mathcal{I}^{r-1}(P_{2r-1})| = \binom{r+1}{r-1} = \binom{r+1}{2}$, and an example of a non-star intersecting family of that size is given by the following construction. Call a vertex of $V = V(C_{2r+2})$ *even* (*odd*) if its index is even (resp. odd). For any $1 \leq i \leq r$, any set S of i consecutive even vertices, there is a unique set $T = T(S)$ of $r - i$ odd vertices such that $F_S = S \cup T \in \mathcal{I}^r(C_{2r+2})$. Let $\mathcal{S}$ denote the family of such S for $i > (r + 1)/2$; if r is odd, also include half of the sets S for $i = (r + 1)/2$ by choosing one from each pair $\{S, \overline{S}\}$, where the complement is over even vertices. Then $\mathcal{S}$ is intersecting of size $\binom{r+1}{2}$; hence $\mathcal{F} = \{F_S \mid S \in \mathcal{S}\}$ is intersecting of that size.

The *Kneser graph* $K(n, r)$ consists of vertex set $\binom{[n]}{r}$, with edges between disjoint pairs of vertices. Thus, intersecting families of r-subsets of $[n]$ correspond to independent sets in $K(n, r)$, and Theorem 1 is equivalent to the fact that $\alpha(K(n, r)) = \binom{n-1}{r-1}$, established by Godsil and Meagher [36]. For an interesting connection, Lovász [57] proved that the chromatic number of $K(n, r)$ is $n - 2r + 2$ and, subsequently, Schrijver [63] showed that the subgraph of $K(n, r)$ induced by the separated sets is a minimal subgraph having this chromatic number.

Building on Talbot's ideas, Holroyd, Spencer, and Talbot [44] proved the following three analogous results for disjoint unions of complete graphs, for powers of paths, and for certain other disjoint unions of graphs. This was the first paper to introduce the "r-EKR" terminology.

Theorem 9 *If G is a disjoint union of $n \geq r$ complete graphs, each of size at least two, then G is r-EKR.*

Note that Theorem 9 is a common generalization of Theorem 3 (to $r < k$) and the $t = 1$ case of Theorem 4 (to non-uniform part sizes).

Theorem 10 *The path power $P_n^{(k)}$ is r-EKR for all r, and strictly so unless $k = 1$ and $n = 2r + 1$.*

Theorem 10 follows from Theorem 8 by a straightforward lemma that they prove: If G is r-EKR with center v then, for any set $S \subseteq N_G(v)$ of neighbors of of v, $G - S$ is also r-EKR with center v. The trick in applying the lemma is to remove, not all neighbors of v but, rather, just those on the clockwise side of v.

Theorem 11 *If G is a disjoint union of $n \geq 2r$ complete graphs, cycles, and paths, including an isolated vertex (K_1 component), then G is r-EKR.*

Note that all three theorems verify Conjecture 2 for their given instances, while only Theorem 11 is restricted to the range $r \leq \mu(G)/2$; it was later proven for the complete range of r by Borg and Holroyd in [14].

Talbot's proof of Theorem 8 used a shift operator that is equivalent to contracting an edge of a graph. This connection was made explicit in [44], where Holroyd, Spencer, and Talbot denoted by G/e the graph obtained from G by contracting the edge e. Additionally, they defined $G{\downarrow}e$ to be the graph obtained from G by deleting the vertices of e and their neighbors; i.e., $G - N(u) - N(v)$, where $e = uv$ and $N(x)$ denotes the neighbors of a vertex x. They were able to show that if $\mathcal{F} \subseteq \mathcal{I}^r(G)$ is intersecting, $e = uv$ is an edge of G, and $x \in V(G{\downarrow}e)$, then there are families $\mathcal{A}$, $\mathcal{B}$, $\mathcal{C}$, and $\mathcal{D}$ such that

- $|\mathcal{F}| = |\mathcal{A}| + |\mathcal{B}| + |\mathcal{C}| + |\mathcal{D}|$,
- $\mathcal{A} \subseteq \mathcal{I}^r(G/e)$ is intersecting,
- $\mathcal{B} \subseteq \mathcal{I}^{r-1}(G{\downarrow}e)$ is intersecting,
- $\mathcal{C} = \{F \in \mathcal{F}_u \mid N(v) \cap (F - \{u\}) \neq \emptyset\}$,
- $\mathcal{D} = \{F \in \mathcal{F}_v \mid N(u) \cap (F - \{v\}) \neq \emptyset\}$,
- $B \in \mathcal{B}$ and $E \in \mathcal{C} \cup \mathcal{D}$ implies that $B \cap E \cap V(G{\downarrow}e) \neq \emptyset$,
- $C \in \mathcal{C}$ and $D \in \mathcal{D}$ implies that $C \cap D \cap V(G{\downarrow}e) \neq \emptyset$, and
- $|\mathcal{I}_x^r(G)| = |\mathcal{I}_x^r(G/e)| + |\mathcal{I}_x^{r-1}(G{\downarrow}e)| + |\mathcal{C}_x| + |\mathcal{D}_x|$.

These properties are then used to prove the above results by induction on both r and n. For example, in the case of Theorem 9, one chooses u and v from the same component to obtain that $\mathcal{C} = \mathcal{D} = \emptyset$, so that

$$|\mathcal{F}| = |\mathcal{A}| + |\mathcal{B}| \leq |\mathcal{I}_x^r(G/e)| + |\mathcal{I}_x^{r-1}(G{\downarrow}e)| = |\mathcal{I}_x^r(G)|.$$

Variations on this theme produce the other results, in some cases by choosing an appropriate vertex x. In the case of Theorem 11, x is an *isolated* vertex (a vertex with no neighbors), which makes sense by virtue of it being a member every maximal independent set. Thus, importantly, knowing which vertices are maximum centers is sometimes necessary to show that a particular graph is EKR.

3 Further Support for the Holroyd-Talbot Conjecture

Conjecture 2 was made in [45]. In that paper, Holroyd and Talbot proved the following theorem for the $r = 2$ case.

Theorem 12 *Let G be an n-vertex graph of minimum degree δ.*

1. *If $\alpha(G) = 2$, then G is strictly* 2-EKR.
2. *If $\alpha(G) \geq 3$, then G is* 2-EKR *if and only if $\delta \leq n - 4$, and strictly so if and only if $\delta \leq n - 5$, with any vertex of minimum degree as a maximum center.*

They also noted an interesting set of natural examples: If G is the graph of vertices and edges of a platonic solid S then G is $\alpha(G)$-EKR if and only if S is not the dodecahedron.

Additionally, they proved an EKR result involving graph products. For graphs G and H, define the *lexicographic product* $G[H]$ to have vertex set $V(G) \times V(H)$, with edges $(u, x)(v, y)$ if uv is an edge of G or if $u = v$ and xy is an edge of H. For example, if $G = K_2$ and $H = mK_1$ (the empty graph on m vertices) then $G[H] = K_{m,m}$ (the m-regular complete bipartite graph), while $H[G] = mK_2$ (the perfect matching on $2m$ vertices).

Theorem 13 *If G is r-*EKR *and $m \geq 1$ then $G[K_m]$ is r-*EKR.

They achieved this result by (a) exhibiting conditions that produce an r-witness and (b) using those conditions to show that if v is an r-witness of G then (v, x) is an r-witness of $G[K_m]$ for all $x \in V(K_m)$.

Furthermore, they generalized the lexicographic product as follows. If G has n vertices $v_1, \ldots, v_n$ and $\mathcal{H} = (H_1, \ldots, H_n)$ is a sequences of n graphs then the graph $G[\mathcal{H}]$ has vertices $\cup_{i=1}^{n}(\{v_i\} \times V(H_i))$, with edges $(v_i, x)(v_j, y)$ if $v_i v_j$ is an edge of G or if $i = j$ and xy is an edge of H_i. Then $G[H]$ is the case in which each $H_i = H$. They used this general construction to provide several informative examples, such as (1) that if $\mathcal{H}$ is a sequence of complete graphs then the conclusion of Theorem 13 does not necessarily hold if the complete graphs are not all of the same size, and (2) that, for any $r \geq 3$, there are graphs of arbitrarily large independence number that are not r-EKR.

Finally, they used shifting to prove the following theorem.

Theorem 14 *Let G be a disjoint union of two complete multipartite graphs. Then G is r-*EKR *for all $r \leq \mu(G)/2$ and strictly so unless $r = \mu(G)/2$.*

Borg and Holroyd [15] then proved much more.

Theorem 15 *Let G be a disjoint union of complete multipartite graphs, at least one of which is a single vertex. Then G is* r-EKR *for all* $r \leq \mu(G)/2$ *and strictly so unless* $r = \mu(G)/2$.

Their work was phrased in the form of *double partitions*: a collection of disjoint *large* sets A_i, each of which is partitioned into *small* sets $A_{i,j}$, with every intersecting family of sets chosen from those that intersect each large set in some portion of exactly one of its small sets, if at all.

In 2009, Borg [9] proved one of the more significant results to date, roughly that every graph G is r-EKR for every r up to about $(\frac{2}{9}\mu(G))^{1/3}$. More precisely, he proved this for all hereditary families, not just families of all independent sets of a graph. A family is *hereditary* if all subsets of its members are members. Call a family (r, t)-EKR if no t-intersecting subfamily of its r-sets is larger than some t-centered r-star. Define $m(r, t) = (r - t)\binom{3r-2t-1}{t+1} + r$.

Theorem 16 *Let* $\mathcal{F}$ *be any hereditary family of sets and let* $t \leq r$ *be such that* $m(r, t) \leq \mu(\mathcal{F})$. *Then, for any* $\emptyset \neq S \subseteq \{t, \ldots, r\}$, *the family* $\cup_{s \in S} \mathcal{F}^s$ *is* (r, t)-EKR.

The rough statement just mentioned is the instance $t = 1$, $S = \{r\}$, and $(r - 1)\binom{3(r-1)}{2} + r \leq \mu(\mathcal{F})$. The theorem is not proven with the aid of shifting, but instead by the use of a version of the so-called Local LYM Inequality (see [6], p.12) of Sperner [65] on sizes of shadows, generalized to hereditary families.

In [8], Borg had already generalized Conjecture 2 to include all hereditary families. In [9], he generalized it further, conjecturing the strengthening of Theorem 16 that it holds with $m(r, t) = (t + 1)(r - t + 1)$, adding also that the conclusion holds strictly unless $\mu(\mathcal{F}) = m(r, t)$ and $S = \{r\}$. He proved that this strengthening is true for shifted families $\mathcal{F}$.

Also in 2009, Borg and Holroyd [14] expanded upon Theorem 11 significantly, as follows. They call a graph a *modified* power of a cycle if it can be obtained from some $C_n^{(k)}$ by adding a set of consecutive edges of distance $k + 1$ in C_n.

Theorem 17 *Let G be a disjoint union of complete graphs, copies of mnd graphs, powers of cycles, modified powers of cycles, and at least one singleton. Then G is* r-EKR *for all* $r \leq \alpha(G)/2$.

The list includes what they call *mnd graphs*, which are a generalization of path powers. While vertices of a path are adjacent to all vertices to its right at distance at most some fixed k to form a path power, the fixed k condition is relaxed to a non-decreasing sequence of distances to form an mnd graph. As graphs by themselves, Borg [11] later proved that mnd graphs satisfy Conjecture 2. Such graphs, of course, are chordal. Observing that several verifications of Conjecture 2 involved chordal graphs and graphs with isolated vertices, the authors of [47] proved the following theorem in 2011. Of course, as noted above, the role of an isolated vertex in these theorems is that such a vertex is a maximum center.

Theorem 18 *If G is a chordal graph with at least one singleton, then G is r-*EKR *for all $r \leq \alpha(G)/2$.*

Disjoint unions of chordal graphs are chordal, and so it is not necessary to state the theorem with disjoint union language. While stated only for chordal graphs, their proof actually shows that the statement is true for the larger class of dismantlable graphs. By Dirac's theorem [23], a graph is chordal if and only if it has a simplicial elimination ordering; that is, one can repeatedly remove a simplicial vertex—one whose neighbors form a clique—until only a complete graph remains. Analogously, a graph is *dismantlable* if one can repeatedly remove a *corner* vertex—one whose neighbors are neighbors of one of its neighbors—until only a complete graph remains.

More carefully, v is a corner vertex if $N[v] \subseteq N[u]$ for some vertex $u \neq v$, where $N[x] = N(x) \cup \{x\}$ is the *closed* neighborhood of x; any such vertex u is called a *witness*. The proof of Theorem 18 uses an induction idea for vertices similar to Talbot's edge contraction operator used to prove Theorem 8. Here they define $G{\downarrow}v = G - N[v]$ and note that both $G - v$ and $G{\downarrow}v$ preserve dismantlability for all v. Additionally, they observe that, if v is a corner of G and u is a witness, then $\mu(G-u) \geq \mu(G)$ and $\mu(G{\downarrow}u) + 1 \geq \mu(G)$. These imply that, if $r \leq \mu(G)/2$, then $r \leq \mu(G-u)/2$ and $r - 1 \leq \mu(G{\downarrow}u)/2$, which allows for induction to be applied. From a dismantling order $(v_1, \ldots, v_n)$ on $V(G)$, they define a shift from any v_i to v_1 and proceed with a simplified version of the partition ideas mentioned above, making use of the identity $|\mathcal{I}_x^r(G)| = |\mathcal{I}_x^r(G-v)| + |\mathcal{I}_x^{r-1}(G{\downarrow}v)|$ (for isolated x and $v \neq x$) that drives the induction.

Soon after, Woodroofe [69] proved a more general result for hereditary families via *(exterior) algebraic* shifting, as opposed to the *combinatorial* shifting discussed above. Algebraic shifting was introduced by Kalai [50] and takes advantage of the geometry that arises from viewing a hereditary family as an abstract simplicial complex. For example, the concepts of a simplicial elimination (or dismantling) ordering for chordal (or dismantlable) graphs generalize to shellability and the even more general sequentially Cohen-Macauley property for simplicial complexes, which we will not define here. A graph G is called *sequentially Cohen-Macaulay* if $\mathcal{I}(G)$ is sequentially Cohen-Macaulay. Among the graphs in this class are chordal graphs, graphs with no induced cycles of length other than 3 or 5, incomparability graphs of shellable posets, and others (see [69]). Woodroofe proved the following.

Theorem 19 *If G is a sequentially Cohen-Macauley graph with an isolated vertex, then G is r-*EKR *for all $r \leq \mu(G)$.*

Additionally, Woodroofe uses these tools to prove the most general version to date along the lines of Theorem 17.

Theorem 20 *Let G be the disjoint union of k graphs, including a singleton. Then G is r-*EKR *for all $r \leq k/2$.*

Regarding graphs without singletons, in [47] we also find the verification of Conjecture 2 for a connected sequence of complete graphs called *chains*. For a sequence of complete graphs $K_{a_1}, \dots, K_{a_m}$, we form the chain graph $G = G(a_1, \dots, a_m)$ by allowing two consecutive K_{a_i} and $K_{a_{i+1}}$ to share a single, distinct vertex, which requires of course that each $a_i \geq 2$. The chain G is called *special* if $m \leq 2$ or if $m > 2$, $a_i > a_{i-1}$ for each $2 \leq i < m$, and $a_m \geq a_{m-1}$. They proved the following.

Theorem 21 *If G is a disjoint union of at most two special chains, then G is r-*EKR *for all r.*

Disjoint unions G of powers of cycles were considered by Hilton, Holroyd, and Spencer [39]. A *simple* cycle is just the cycle $C_n = C_n^{(1)}$. For $r = \alpha(G)$, they proved that such unions are r-EKR, provided that some component is simple, and were also able to identify which vertices are r-witnesses. Included as "cycles" are the degenerate cases of a vertex ($C_1^{(1)}$) and an edge ($C_2^{(1)}$). The problems for other values of r remain open. An (a, b)-*coloring* of a graph G is a function $c : V(G) \to \binom{[a]}{b}$ such that adjacent vertices receive disjoint subsets. The *fractional chromatic number* of G, denoted $\hat{\chi}$, is the minimum of a/b over all (a, b)-colorings of G.

Theorem 22 *Let G be a disjoint union of powers of cycles, at least one of which is simple. Then G is $\alpha(G)$-*EKR, *and all $\alpha(G)$-witnesses belong to the component having least fractional chromatic number.*

Important ingredients in the proof of this result include the partitioning and shifting from Talbot's proof of Theorem 8, as well as the fact that if $\hat{\chi}(C_n^{(k)}) \leq \hat{\chi}(C_{n'}^{(k')})$ then there is a homomorphism from $C_n^{(k)}$ to $C_{n'}^{(k')}$. For the case of two simple cycles, they prove more.

Theorem 23 *If G is a disjoint union of two cycles then G is r-*EKR *for all r; furthermore, all of the vertices of any of its even cycles are r-witnesses.*

Hilton and Spencer [40, 41] then expanded on Theorem 22, handling all values of r.

Theorem 24 *Let G be a disjoint union of powers of cycles and a power of H, where H is either a path or a cycle. Write $G = H^k \cup \left(\cup_{i=1}^m C_{n_i}^{k_i}\right)$ and suppose that $k \leq min_i k_i$. Then G is $\alpha(G)$-*EKR *for all r, and some r-witness has minimum degree.*

Borg and Feghali [13] then prove the strictness of Theorem 24 when $k < \min_i k_i$. Additionally, they give a simpler proof that relies on Katona's Shadow Intersection Theorem [51].

In addition to studying the lexicographic product, as in Theorem 13, it is natural to consider Cartesian products as well. For graphs G and H, the *Cartesian product* $G \square H$ has vertices $V(G) \times V(H)$ and edges $(u, x)(v, y)$ whenever $u = v$ and $xy \in E(H)$, or $uv \in E(G)$ and $x = y$. For example, $K_2 \square K_2 = C_4 = Q^2$, and the *$d$-dimensional cube* Q^d can be defined by $Q^d = K_2 \square Q^{d-1}$. In [46], we find that the *ladder* $L_k = K_2 \square P_k$ is 3-EKR. Subsequently, Li and Zhang [55] proved much more, using the traditional shifting and partitioning techniques outlined above.

Theorem 25 *The ladder L_k is r-EKR for all r, and is strictly so unless $r = k - 1$. Additionally, the maximum centers are the minimum degree vertices of L_k.*

One of the simplest open cases for disjoint unions of graphs was the union of k length-2 paths, denoted kP_3. Generalizing Theorem 4 (on $kK_2 = kP_2$) in this manner, this case was partially resolved in 2020 [27] by also generalizing Bollobás and Leader's use of the idea behind Katona's cycle method.

Theorem 26 *The graph $G = kP_3$ is r-EKR for all $r \leq \mu(G)/2 = k/2$.*

Note that this also similar to Theorem 11 without the isolated vertex. The authors conjectured that kP_3 is r-EKR for all r. The main hurdle in generalizing Bollobás and Leader's proof is that, unlike kP_2, the graph kP_3 is not vertex transitive: The k degree-2 *middle* vertices live in one orbit and the $2k$ leaves live in another. This causes the usual counting arguments to not line up properly, and the way around it is to state the cycle argument in a well-known probabilistic form, so that objects from the resulting probability space can be sampled in a controlled, but non-uniform manner that compensates for the imbalance.

A second hurdle arises in each component of kP_3 having the ability to contribute up to two vertices (leaf *siblings*) to an independent set, instead of up to one for kP_2. For this, the authors still use "good" permutations of the leaves, but pair "main" intervals with opposite sub-intervals to distinguish leaves that appear alone or with their siblings in an independent set. Further, the "main" intervals have an adjacent "invisible" interval that picks up middle vertices. So that enumeration works well, this structure is reflected as well before rotating as in the original cycle method.

The obvious next generalization to attempt would be kP_4. However, the authors suggest that unions of 3-claws also continue this sequence of graphs and might more easily succumb to a cycle-method-like symmetry argument—a *k-claw* is the complete bipartite graph $K_{1,k}$, so $K_{1,1} = P_2$ and $K_{1,2} = P_3$. Thus, an EKR theorem for unions of claws would be similar to the relevant portion of Theorem 11 without an isolated vertex.

An alternative to studying specifically structured families of graphs is to consider density; i.e. minimum or average degree. For vertices u and v in a graph G, we use the notations $\text{deg}_G(u)$, $\Delta(G)$, $\overline{d}(G)$, and $\text{dist}_G(u, v)$ for the degree of vertex u, the maximum degree of G, the average degree of G, and the distance between u and v in G, respectively; we may omit the subscript if the context is clear. In [32] we find the following four theorems for sparse graphs.

Theorem 27 *If G is a graph on n vertices with $\Delta(G) \leq d$, then G is r-EKR for $r < (8n/27d)^{1/2}$.*

Note that, for bounded-degree graphs, this improves the range of r from the $O(n^{1/3})$ of Theorem 16 to $O(n^{1/2})$.

Theorem 28 *If G is a graph on n vertices with $\overline{d}(G) \leq c$, for any $c \geq e/36$, then G is r-EKR for $r < (n/18c)^{1/3}$.*

Here, bounded average graphs yield a range of the same order of magnitude as in the more general Theorem 16, but with an increased constant.

A *split* vertex is a vertex of degree at least 3. A *spider* is a tree with exactly one split vertex.

Theorem 29 *If G is a spider on n vertices, then G is r-*EKR *for $r \le (n \ln 2)^{1/2} - ((\ln 2)/2)^{1/2}$.*

This returns the range of r for spiders to $O(n^{1/2})$ and increases the constant found in Theorem 27.

Theorem 30 *If G is a tree on n vertices, with exactly $s > 1$ split vertices, then G is r-*EKR *for $2s < r \le (n \ln c)^{1/2} - ((\ln c)/2)^{1/2}$, where $c = 2 - 2s/r$.*

There is curiosity about there being a lower bound on r here; the authors don't believe there should be one but didn't find a way to avoid it.

The key to these theorems is the method of *diversity* (originally called *unbalance* in [54]), which for a family $\mathcal{F}$ is defined to be $\gamma(\mathcal{F}) = |\mathcal{F}| - |\mathcal{F}_z|$ for any maximum center z. The origins of the concept are found in a result of [22] that states all but at most $O(\binom{n-2}{k-2})$ members of any intersecting $\mathcal{F} \subseteq \binom{[n]}{r}$ contain a common element. The key lemma in [32] is a result from [30] that every intersecting $\mathcal{F} \subseteq \binom{[n]}{r}$, with $n > 72r$, has $\gamma(\mathcal{F}) \le \binom{n-3}{r-2}$. How this is used to prove Theorem 27, for example, is to let z be a maximum center of $\mathcal{F}$ and choose some $E \in \mathcal{F} - \mathcal{F}_z$. Because G has bounded degree d, one can compute a simple lower bound $f(n, r, d)$ on the number of sets $F \in \mathcal{I}^r_z(G)$ that are disjoint from E and so can't be in $\mathcal{F}$. Then $|\mathcal{F}| = |\mathcal{F}_z| + \gamma(\mathcal{F}) \le |\mathcal{I}^r_z(G)| - f(n, r, d) + \binom{n-3}{r-2}$, which is at most $|\mathcal{I}^r_z(G)|$ when $f(n, r, d) \ge \binom{n-3}{r-2}$. Hence the bound on r in the theorem comes from making the last inequality true. The specific properties of the graphs in the other theorems are used to modify their bounds on r similarly.

Another theorem on trees was given by Feghali, Hurlbert, and Kamat [28]. A spider is *ℓ-uniform* if every leaf has distance ℓ from the split vertex; it has also been called a *depth-ℓ claw* in the literature. For example, the 1-uniform spider with k leaves is $K_{1,k}$, which is r-EKR for all $r \le k/2$ by Theorem 1.

Theorem 31 *Every 2-uniform spider G is r-*EKR *for all $r \le (\mu(G) + 1)/2$. In addition, if G has k leaves then G is not k-*EKR.

Finally, Holroyd and Talbot [45] opined that, if a counterexample G to Conjecture 2 exists, then one exists with $\mu(G) = \alpha(G)$. This motivates studying such graphs, which are equivalent to the well-known class of well-covered graphs: a *vertex cover* is a set of vertices that touch all edges, and a graph is *well-covered* if all its minimal vertex covers have the same size. Examples of such graphs already studied above include disjoint unions of complete graphs, balanced complete bipartite graphs, and graphs of independence number 2 with no dominating vertex (e.g., complements of cycles). A *pendant vertex* has degree 1. Another construction of a well-covered graph is a *fully pendant graph* G^*, which is formed from a graph G by adding to each of its vertices a pendant neighbor. De Silva, Dionne, Dunkelberg, and Harris [19] recently proved the following.

Theorem 32 *The fully pendant graph K_n^* is r-EKR for all $r \leq n/2$, and strictly so when $r < n/2$; all r-witnesses are pendant vertices.*

4 HK Graphs

As in the case of isolated vertices, it can be part of a useful technique to know where the maximum r-centers are located in a graph, especially within inductive or injective methods. In studying the EKR property for connected graphs, it is reasonable to start with the fewest possible edges; i.e. trees. Theorems 29, 30, and 31 are important results in this direction. As has been noted regarding the above results, among known witnesses are minimum degree vertices. It makes sense, therefore, to guess that every tree has some leaf as a maximum center, as was conjectured in [47], and verified for small r.

Theorem 33 *If T is a tree and $r \leq 4$, then some maximum r-center of T is a leaf.*

However, numerous counterexamples to the conjecture were soon found for $r \geq 5$ [2, 12, 28]. The maximum star centers in [2, 12] were of degree 2. However, Feghali, Johnson, and Thomas [28] showed that r-centers can be far from leaves and have any degree.

Theorem 34 *For all $\ell \geq 3$ and $d \geq 2$ there is a tree T and $r \leq \alpha(T)$ with a maximum r-center of degree d at distance ℓ from every leaf.*

Let $S^{t,\ell}$ be the ℓ-uniform spider with split vertex of degree t. Then define the tree $T = T^{d,t,\ell}$ by joining a vertex v to the split vertices of d copies of $S^{t,\ell-1}$. Through very careful analysis, they are able to show that $|\mathcal{I}_v(T)| > |\mathcal{I}_u(T)|$ for every leaf u when t is large enough. It follows, then, that $|\mathcal{I}_v^r(T)| > |\mathcal{I}_u^r(T)|$ for some value of r. They complete the proof by showing that, for every remaining vertex x, $|\mathcal{I}_v^r(T)| \geq |\mathcal{I}_x^r(T)|$ for all r.

Based on these results, Estrugo and Pastine [25] define a tree T to be r-HK if some leaf of T is a maximum r-center, and simply HK if it is r-HK for all r. More generally, one can define a graph G to be r-HK if some minimum-degree vertex of G is a maximum r-center, and simply HK if it is r-HK for all r. The problem then arises to discover which graphs are (r-)HK.

One of the first such results is found in [49], that spiders are HK. They not only show where the maximum r-center is, but also describe a partial order on the sizes of all r-centers. Let $S = S(\ell_1, \ldots, \ell_k)$ denote the spider with k leaves $v_1, \ldots, v_k$, with v_i at distance ℓ_i from the split vertex v. Say that S is in *spider order* if the following three conditions hold for all $i < j$:

- if ℓ_i and ℓ_j are both odd then $\ell_i \le \ell_j$;
- if ℓ_i and ℓ_j are both even then $\ell_i \ge \ell_j$; and
- otherwise ℓ_i is odd and ℓ_j is even.

Theorem 35 *Suppose that $S = S(\ell_1, \ldots, \ell_k)$ is in spider order, with split vertex v and leaves $v_1, \ldots, v_k$ as defined above. Then S is* HK*. Moreover, for every r, $1 \le i \le j \le k$, and degree-2 vertex y on the vv_i-path in S we have*

1. $|\mathcal{I}^r_y(S)| \le |\mathcal{I}^r_{v_i}(S)|$,
2. $|\mathcal{I}^r_v(S)| \le |\mathcal{I}^r_{v_k}(S)|$, *and*
3. $|\mathcal{I}^r_{v_j}(S)| \le |\mathcal{I}^r_{v_i}(S)|$.

In particular, v_1 is a maximum r-center.

They proved each part of the theorem injectively, using a *flipping* operation on paths. For example, if one flips the yv_i-path by mapping it onto itself in reverse order, independent r-sets containing y get mapped to independent r-sets containing v_i, thereby injecting $\mathcal{I}^r_y(S)$ into $\mathcal{I}^r_{v_i}(S)$. The same can be done for v instead of y. The interesting case is statement 3, of course, where the flipping operation is trickier, leading to an injection that works only for spider order.

In [25], Estrugo and Pastine generalize the flip to certain paths in graphs. A *thread* in a graph G is a path P such that every interior vertex of P has degree 2 in G. Vacuously, an edge is a thread, for example. They define a path $P = v_1 \cdots v_h$ to be an *escape path* for v_1 if $v_1 \cdots v_{h-1}$ is a thread and v_h is pendant. By flipping on P they prove that $|\mathcal{I}^r_{v_1}(G)| \le |\mathcal{I}^r_{v_h}(G)|$, which yields the following theorem.

Theorem 36 *If every non-pendant vertex of a graph G has an escape path, then G is* HK.

Hence caterpillars, for example, are HK. A graph G is *pendant* if every split vertex of G has a pendant neighbor. Another example is that pendant graphs are HK. A pendant spider has also been called *short* spider. For a vertex x in G one can *join a spider S at x* by identifying the split vertex of S with x. Then, for a set D of vertices of G, a *short D-spidering of G* joins to each vertex of D a short spider. It follows that if D is a dominating set of G then any short D-spidering of G is HK. Define a set of vertices D to be a *thread dominating set* of G if every vertex not in D has a thread to some vertex in D. Estrugo and Pastine then observe the following, more general consequence of Theorem 36.

Theorem 37 *If D is a thread dominating set of G, then any short D-spidering of G is* HK.

Hence all pendant graphs are HK.

The authors of [49] conjectured that trees with no degree-2 vertices are HK. As a test case, the authors of [32] propose showing that the following family of graphs is r-EKR. For $1 \le i \le 3$, let $T_i(h)$ be a complete binary tree of depth h, having unique degree-2 vertex v_i. Now join a new vertex v to each v_i to create the tree $T(h)$, in which every interior vertex has degree 3.

Finally, Zhang, Ran, and Huang [71] prove the following result, similar to Theorem 33. A *unicyclic graph* contains exactly one cycle.

Theorem 38 *If G is a connected, unicyclic graph with at least one pendant vertex then G is r-*HK *for $r \leq 4$. There exist such graphs that are not r-*HK *for $r \geq 5$.*

They also construct a caterpillar $P^{k,t}$ by attaching t pendant vertices to each vertex of $P_k = v_1 \cdots v_k$; thus $P^{k,1} = P_k^*$, the fully pendant contruction for well-covered graphs. They prove, for all r and t, that every pendant neighbor of v_2 or v_{k-1} is a maximum r-center (the $t = 1$ case had been done in [19]).

Zhang, et al. finish by offering the following conjecture.

Conjecture 39 *If T is a caterpillar, $r \geq 1$, and z is a maximum r-center, then the neighbor of z has maximum degree.*

5 Chvátal's Conjecture

One cannot avoid Chvátal's Conjecture [16] when discussing this topic. It is the elephant in the room. We say that a family $\mathcal{F}$ of sets is EKR if some element x satisfies $|\mathcal{F}_x| \geq |\mathcal{H}|$ for every intersecting family $\mathcal{H} \subseteq \mathcal{F}$.

Conjecture 40 (Chvátal [16]) *Every hereditary family $\mathcal{F}$ is* EKR.

It is a monument to the difficulty of the conjecture that it is so well-known and yet so few papers have addressed it in the half century since its birth. There were some early papers. Kleitman and Magnanti [53] solved the case in which $\mathcal{F}$ is the union of two stars. Sterboul [66] solved the rank at most 3 case, where the *rank* of $\mathcal{F}$ is the size of its largest set. In 1976, Schönheim [62] solved case in which the maximal elements of $\mathcal{F}$ intersect. Then there was a long gap until Snevily [64] solved the case in which $\mathcal{F}$ is 1-shifted in 1992, and quite another long period before Borg [10] generalized the conjecture to weighted families in 2011, solving the 1-shifted case as well. Then in 2017, the paper [17] reproved Sterboul's result, once with a relatively similar approach, and once with the Sunflower Lemma, in hopes that the method might yield more fruit. Subsequently, Olarte, Santos, and Spreer [61] reproved both Kleitman-Magnanti and Sterboul with shorter proofs. Most recently, Frankl and Kupavskii [33] solved the case in which $\mathcal{F}$ has cover number 2.

In the meantime, some of the results above actually validate Conjecture 40 when $\mathcal{F} = \mathcal{I}(G)$ for various graphs G that are r-EKR for all r. For example, in Theorem 8, because a cycle power $G = C_n^{(k)}$ is vertex transitive, every vertex is a maximum r-center for every r. This means that, for any fixed vertex x, if $\mathcal{H} \subseteq \mathcal{I}(G)$, then $|\mathcal{H}| = \sum_r |\mathcal{H}^r| \leq \sum_r |\mathcal{I}_x^r(G)| = |\mathcal{I}_x(G)|$. This not only works for vertex transitive graphs, but for graphs for which some vertex is an r-witness for all r: e.g., path powers $P_n^{(k)}$ (Theorem 10, minimum degree vertex), lexicographic products $P_n^{(k)}[K_m]$ (Theorem 13, minimum degree vertex), disjoint union of two cycles, one of which is even (Theorem 23, even cycle vertex), and ladders (Theorem 25, minimum degree vertex). This highlights the importance of tracking maximum r-centers.

6 Questions and Remarks

This subject is clearly rich and vibrant, teeming with tantalizing questions, problems, variations, and conjectures. Surely the reader will think of lovely problems to attack beyond what is included below, but here are a few thoughts to try in hopes of making progress.

1. EKR problems.
   a. What Cartesian (or other) products of various graphs are r-EKR; e.g., products of paths, cycles, or complete graphs? For some graph product $\times$, could it be that, if G and H are r-EKR for all r, then is $G \times H$ r-EKR for all r?
   b. Are other well-covered graphs r-EKR? The *rook graph* $K_m \square K_m$ is particularly attractive because it is both well-covered and a Cartesian product.
   c. Might the techniques applied to sparse graphs be tweaked somewhat so as to work for dense graphs?
   d. Are all spiders r-EKR? Is the ℓ-uniform spider with k leaves r-EKR if and only if $kP_{\ell+1}$ is r-EKR?
2. HK problems.
   a. Is the internally 3-regular tree $T(h)$ r-HK for all r?
   b. Where are the maximum r-centers for caterpillars? (See [25].)
   c. What other graphs G (besides cycles) with $\delta(G) = 2$ are r-HK?

Acknowledgements This paper arose from a talk I presented at the July, 2024, Summit280 Conference in Budapest in celebration of the 70th birthdays of Péter Frankl, Zoltán Füredi, Ervin Győri, and János Pach, two of whom were my professors in graduate school and had a strong influence in my pursuit of combinatorial mathematics. The author extends his gratitude to the organizers of the conference for inviting me to write this survey.

References

1. Ahlswede, R., Khachatrian, L.H.: The complete intersection theorem for systems of finite sets. European J. Combin. **18**(2), 125–136 (1997)
2. Baber, R.: Some results in extremal combinatorics. Ph.D. thesis, Department of Mathematics, University of London, London, UK (2011)

3. Berge, C.N.: de coloration de l'hypergraphe h-partie complet. In: Hypergraph Seminar, Columbus, Ohio,: vol. 141 of Lecture Notes in Mathematics, vol. 1974, pp. 13–20. Springer, Berlin (1972)
4. Berge, C.: Nombres de coloration de l'hypergraphe h-partie complet. Ann. Mat. Pura Appl. **103**(4), 3–9 (1975)
5. Bey, C.: An intersection theorem for weighted sets. Discret. Math. **235**(1–3), 145–150 (2001)
6. Bollobás, B.: Combinatorics. Cambridge University Press, Cambridge (1986). Set systems, hypergraphs, families of vectors and combinatorial probability
7. Bollobás, B., Leader, I.: An Erdős–Ko–Rado theorem for signed sets. Comput. Math. Appl. **34**, 11, 9–13 (1997)
8. Borg, P.: Some set intersection theorems of extremal type. Ph.D. thesis, The Open University, Milton Keynes, UK (2007)
9. Borg, P.: Extremal t-intersecting sub-families of hereditary families. J. London Math. Soc. **(2)** **79**, 3, 167–185 (2009)
10. Borg, P.: On Chvátal's conjecture and a conjecture on families of signed sets. European J. Combin. **32**(1), 140–145 (2011)
11. Borg, P.: Erdős-Ko-Rado with separation conditions. Australas. J. Combin. **59**, 39–63 (2014)
12. Borg, P.: Stars on trees. Discret. Math. **340**(5), 1046–1049 (2017)
13. Borg, P., Feghali, C.: The Hilton-Spencer cycle theorems via Katona's shadow intersection theorem. Discuss. Math. Graph Theory **43**(1), 277–286 (2023)
14. Borg, P., Holroyd, F.: The Erdős-Ko-Rado properties of various graphs containing singletons. Discret. Math. **309**(9), 2877–2885 (2009)
15. Borg, P., Holroyd, F.: The Erdős-Ko-Rado properties of set systems defined by double partitions. Discret. Math. **309**(9), 4754–4761 (2009)
16. Chvátal, V.I.: families of edges in hypergraphs having the hereditary property. In: Hypergraph Seminar, Columbus, Ohio,: vol. 141 of Lecture Notes in Mathematics, vol. 1974, pp. 13–20. Springer, Berlin (1972)
17. Czabarka, E., Hurlbert, G., Kamat, V.: Chvátal's conjecture for downsets of small rank (2017). arxiv.org/abs/1703.00494
18. Daykin, D.E.: Erdős-Ko-Rado from Kruskal-Katona. J. Combin. Theory Ser. A **17**, 254–255 (1974)
19. De Silva, J., Dionne, A.B., Dunkelberg, A., Harris, P.E.: Very well-covered graphs with the Erdős-Ko-Rado property. Involve **16**(1), 35–47 (2023)
20. Delsarte, P.: An algebraic approach to the association schemes of coding theory. Philips Res. Rep. Suppl. **10**, vi+97 (1973)
21. Deza, M.-M., Frankl, P.: Erdős-Ko-Rado theorem—22 years later. SIAM J. Algebraic Discret. Methods **4**(4), 419–431 (1983)
22. Dinur, I., Friedgut, E.: Intersecting families are essentially contained in juntas. Combin. Probab. Comput. **18**(1–2), 107–122 (2009)
23. Dirac, G.A.: On rigid circuit graphs. Abh. Math. Sem. Univ. Hamburg **25**, 71–76 (1961)
24. Erdős, P., Ko, C., Rado, R.: Intersection theorems for systems of finite sets. Quart. J. Math. Oxford Ser. **(2)** **12**, 313–320 (1961)
25. Estrugo, E., Pastine, A.: On stars in caterpillars and lobsters. Discret. Appl. Math. **298**, 50–55 (2021)
26. Feghali, C.: Intersecting families, signed sets, and injection. Australas. J. Combin. **76**, 226–231 (2020)
27. Feghali, C., Hurlbert, G., Kamat, V.: An Erdős–Ko–Rado theorem for unions of length 2 paths. Discret. Math. **343**, 12, 112121, 6pp (2020)
28. Feghali, C., Johnson, M., Thomas, D.: Erdős-Ko-Rado theorems for a family of trees. Discret. Appl. Math. **236**, 464–471 (2018)
29. Frankl, P.T.: Shifting technique in extremal set theory. In Surveys in combinatorics,: (New Cross, 1987), vol. 123 of London Mathematical Society Lecture Note Series, vol. 1987, pp. 81–110. Cambridge University Press, Cambridge (1987)

30. Frankl, P.: Maximum degree and diversity in intersecting hypergraphs. J. Combin. Theory Ser. A **144**, 81–94 (2020)
31. Frankl, P., Füredi, Z.: A new, short proof of the EKR theorem. J. Combin. Theory Ser. A **119**(6), 1388–1390 (2012)
32. Frankl, P., Hurlbert, G.: On the Holroyd-Talbot conjecture for sparse graphs. Discret. Math. **347**, 1, Paper No. 113742, 8 (2024)
33. Frankl, P., Kupavskii, A.: Perfect matchings in down-sets. Discret. Math. **346**, 5, Paper No. 113323, 7 (2023)
34. Fuentes, M., Kamat, V.: On intersecting families of subgraphs of perfect matchings (2024). arxiv.org/abs/2407.12289
35. Füredi, Z., Hwang, K.-W., Weischel, P.: A new proof and generalizations of the Erdős–Ko–Rado theorem using the method of linearly independent polynomials. In: Topics in Discrete Mathematics, vol. 26 of Algorithms Combinatorial, pp. 215–224. Springer, Berlin (2006)
36. Godsil, C., Meagher, K.: Erdős-Ko-Rado Theorems: Algebraic Approaches. Cambridge Studies in Advanced Mathematics, vol. 149. Cambridge University Press, Cambridge (2015)
37. Gronau, H.-D.O.F.: More on the Erdős-Ko-Rado theorem for integer sequences. J. Combin. Theory Ser. A **35**(3), 279–288 (1983)
38. Hilton, A.J., Milner, E.C.: Some intersection theorems for systems of finite sets. Quart. J. Math. Oxford Ser. **(2)18**, 369–384 (1967)
39. Hilton, A.J.W., Holroyd, F.C., Spencer, C.L.: King Arthur and his knights with two round tables. Q. J. Math. **62**(3), 625–635 (2011)
40. Hilton, A.J.W., Spencer, C.L.: A graph-theoretical generalization of Berge's analogue of the Erdős-Ko-Rado theorem. In: Graph theory in Paris, Trends Mathematics Birkhäuser, Basel, pp. 225–242 (2007)
41. Hilton, A.J.W., Spencer, C.L.: A generalization of Talbot's theorem about King Arthur and his Knights of the Round Table. J. Combin. Theory Ser. A **116**(5), 1023–1033 (2009)
42. Holroyd, F.C.: Problem 338 (BCC16.25), Erdős–Ko–Rado at the court of King Arthur. Discret. Math. **197/198**, 812 (1999)
43. Holroyd, F.C., Johnson, A.: Problem 25. In: Cameron, P.J. (ed.), Problems from the Sixteenth British Combinatorial Conference (1997)
44. Holroyd, F.C., Spencer, C., Talbot, J.: Compression and Erdős–Ko–Rado graphs. Discret. Math. **293**(1–3), 155–164 (2005)
45. Holroyd, F.C., Talbot, J.: Graphs with the Erdős–Ko–Rado property. Discret. Math. **293**(1–3), 165–176 (2005)
46. Hurlbert, G., Kamat, V.: Erdős-Ko-Rado theorems for chordal and bipartite graphs (2009). arxiv.org/abs/0903.4203
47. Hurlbert, G., Kamat, V.: Erdős–Ko–Rado theorems for chordal graphs and trees. J. Combin. Theory Ser. A **118**(3), 829–841 (2011)
48. Hurlbert, G., Kamat, V.: New injective proofs of the Erdős–Ko–Rado and Hilton-Milner theorems. Discret. Math. **341**(6), 1749–1754 (2018)
49. Hurlbert, G., Kamat, V.: On intersecting families of independent sets in trees. Discret. Appl. Math. **321**, 4–9 (2022)
50. Kalai, G.A.: Shifting. In: Computational Commutative Algebra and Combinatorics (Osaka): vol. 33 of Advanced Studies in Pure Mathematics Mathematical Society of Japan, Tokyo vol. 2002, pp. 121–163 (1999)
51. Katona, G.O.H.: Intersection theorems for systems of finite sets. Acta Math. Acad. Sci. Hungar. **15**, 329–337 (1964)
52. Katona, G.O.H.: A simple proof of the Erdős–Ko–Rado theorem. J. Combin. Theory (B) **13**, 183–184 (1972)
53. Kleitman, D.J., Magnanti, T.L.: On the number of latent subsets of intersecting collections. J. Combin. Theory Ser. A **16**, 215–220 (1974)
54. Lemons, N., Palmer, C.: The unbalance of set systems. Graphs Combin. **24**(4), 361–365 (2008)
55. Li, Y.-S., Zhang, H.-J.: Erdös-Ko-Rado theorem for ladder graphs. Acta Math. Appl. Sin. Engl. Ser. **30**(3), 583–588 (2014)

56. Livingston, M.L.: An ordered version of the Erdős-Ko-Radó theorem. J. Combin. Theory Ser. A **26**(2), 162–165 (1979)
57. Lovász, László, J.: Kneser's conjecture, chromatic number, and homotopy. J. Combin. Theory Ser. A **25**, 319–324 (1978)
58. Meyer, J.-C.: Quelques problèmes concernant les cliques des hypergraphes h-complets et q-parti h-complets. In: Hypergraph Seminar, pp. 127–139. Springer (1974)
59. Moon, A.: An analogue of the Erdős-Ko-Rado theorem for the Hamming schemes $H(n, q)$. J. Combin. Theory Ser. A **32**(3), 386–390 (1982)
60. Nowakowski, R., Winkler, P.: Vertex-to-vertex pursuit in a graph. Discret. Math. **43**(2–3), 235–239 (1983)
61. Olarte, J.A., Santos, F., Spreer, J.: Short proof of two cases of Chvátal's conjecture. Discret. Math. **342**(8), 2192–2194 (2019)
62. Schönheim, J.: Hereditary systems and Chvátal's conjecture. In: Proceedings of the Fifth British Combinatorial Conference (University of Aberdeen, Aberdeen, 1975), vol. No. XV of Congress. Numer., Utilitas Mathematica, Winnipeg, MB, pp. 537–539 (1976)
63. Schrijver, A.: Vertex-critical subgraphs of kneser graphs. Nieuw Arch. Wisk. **26**(3), 454–461 (1978)
64. Snevily, H.: A new result on Chvátal's conjecture. J. Combin. Theory Ser. A **61**(1), 137–141 (1992)
65. Sperner, E.: Ein Satz über Untermengen einer endlichen Menge. Math. Z. **27**(1), 544–548 (1928)
66. Sterboul, F.: Sur une conjecture de V. Chvátal. In: Hypergraph Seminar (Proc. First Working Sem., Ohio State Univ., Columbus, Ohio, 1972; dedicated to Arnold Ross), vol. 411 of Lecture Notes in Mathematics, pp. 152–164. Springer, Berlin-New York (1974)
67. Talbot, J.: Intersecting families of separated sets. J. London Math. Soc. **(2)68**, 1, 37–51 (2003)
68. Wilson, R.M.: The exact bound in the Erdős-Ko-Rado theorem. Combinatorica **4**(2–3), 247–257 (1984)
69. Woodroofe, R.: Erdős-Ko-Rado theorems for simplicial complexes. J. Combin. Theory Ser. A **118**(4), 1218–1227 (2011)
70. Woodroofe, R.: An algebraic groups perspective on Erdős-Ko-Rado. Linear Multilinear Algebra **70**(22), 7825–7835 (2022)
71. Zhang, H., Ran, Y., Huang, J.: On the star of the family of independent sets in a graph. Discret. Appl. Math. **344**, 170–178 (2024)

A Survey of Cameron-Liebler Sets and Low Degree Boolean Functions in Grassmann Graphs

Ferdinand Ihringer

Abstract We survey results for Cameron-Liebler sets and low degree Boolean functions for Hamming graphs, Johnson graphs and Grassmann graphs from the point of view of association schemes. This survey covers selected results in finite geometry, Boolean function analysis, design theory, coding theory, and cryptography.

1 Introduction

We survey *Cameron-Liebler sets* and *low degree Boolean functions* in hypercube, Johnson graph and Grassmann graph. Going back to work by Cameron and Liebler in 1982 on permutation groups [17], the study of Cameron-Liebler sets and *antidesigns* in finite geometries has been a particularly flourishing topic in recent years. The seemingly unrelated *Fourier analytic study* of Boolean functions on the hypercube $\{0, 1\}^n$ is a fundamental topic in theoretical computer science which emerged in the 1990s. Recently, Boolean function analysis expanded from the study of the hypercube towards the investigation of more algebraic structures such as the subspace lattice of a finite vector space or bilinear forms over finite fields. Most notably, the remarkable proof of the 2-to-2 Games Conjecture by Khot, Muli and Safra [79] has been obtained by the study of expansion properties of Grassmann graphs. Thus, both fields converge, but due to different traditions the exchange of ideas remains low. One reason for this is the lack of mutual intelligibility of publications in all the involved areas. We hope that this survey helps with amending this situation.[1]

Our aim is a survey of selected results written from a perspective of association schemes and Delsarte theory. This provides a reasonably uniform framework which most researchers in finite geometry, spectral graph theory, finite group theory, coding theory and design theory are familiar with. For readers coming from Boolean function

[1] And if not, then at least it summarizes all the aspects of the area which the author enjoys.

F. Ihringer (✉)
Southern University of Science and Technology, Shenzhen, Guangdong Province, People's Republic of China
e-mail: ihringer@sustech.edu.cn

G. O. H. Katona et al. (eds.), *Sum(m)it280*, Bolyai Society Mathematical Studies 32,
https://doi.org/10.1007/978-3-032-18810-6_11

analysis and closely related areas in theoretical computer science, the advantage is that most things are phrased in terms of linear algebra and we can mostly avoid the vast specific terminology of areas such as finite geometry or Delsarte-style coding theory. We will also mention some parallel developments in cryptography to clarify some precedence.

This survey focuses on three structures: the hypercube $\{0, 1\}^n$, the Johnson graph of m-sets of $\{1, \ldots, n\}$ (or slice of the hypercube), and the Grassmann graph of m-spaces in $V(n, q)$ (where $V(n, q)$ denotes the n-dimensional vector space over $\mathrm{GF}(q)$, the field with q elements). At the end we will also mention some other recent developments in permutation groups as well as other finite geometries such as bilinear forms and affine spaces.

2 Preliminaries

The main difficulty of the surveyed subject is that it spans several communities, each of them using their own notation. Thus, we start by surveying several perspectives on the topic.

2.1 *Spectral Graph Theory*

A graph $\Gamma = (X, \sim)$ has vertex set X and an adjacency relation $\sim$. For us, all graphs are finite and simple.

Let $v = |X|$. Say that Γ is k-regular if each vertex is adjacent to exactly k vertices. Denote the *adjacency matrix* of Γ by A. Write $\lambda_1 \geq \lambda_2 \geq \ldots \geq \lambda_v$ for the spectrum of a symmetric matrix A. Let $u_1, \ldots, u_v$ be a corresponding orthogonal basis of eigenvectors. Let $V(\lambda)$ denote the eigenspace of λ. We denote the idempotent projection onto $V(\lambda)$ by $E(\lambda)$, so $E(\lambda) = \sum_{\lambda_j = \lambda} u_j u_j^T$. Clearly, $\mathrm{rank}(E(\lambda)) = \dim(V(\lambda))$ and

$$A = \sum_{\lambda \in \{\lambda_1, \ldots, \lambda_v\}} \lambda E(\lambda).$$

For a subset $Y \subseteq X$, we write f_Y for the *characteristic function* (or: *characteristic vector*) of Y. That is, $f_Y(S) = 1$ if $S \in Y$ and $f_Y(S) = 0$ otherwise. There is some easy, but useful arithmetic. Consider a subspace W of $\mathbb{C}^v$ with the all-ones vector j in W.

(i) If $f_Y \in W$, then $f_{X \setminus Y} \in W$.
(ii) If Y and Z disjoint and $f_Y, f_Z \in W$, then $f_{Y \cup Z} = f_Y + f_Z \in W$.
(iii) If $Z \subseteq Y$ and $f_Y, f_Z \in W$, then $f_{Y \setminus Z} = f_Y - f_Z \in W$.

In particular, if one intends to classify sets Y with $f_Y \in W$ (as we want to), then it suffices to execute this classification up to these three operations.

We will survey the following graphs.

Example 1 (Hamming Graph) Let $X = \{0, \ldots, q-1\}^n$. Say that $x \sim y$ if x and y have Hamming distance 1. Then $(X, \sim)$ is the Hamming graph $H(n, q)$. The case $q = 2$ is known as the *hypercube*. It has $n+1$ distinct eigenvalues $\theta_j = q(n-j) - d$ with $\dim(V(\theta_j)) = \binom{d}{j}(q-1)^j$, where $0 \leq j \leq n$, see [12, Th. 9.2.1].

Example 2 (Johnson Graph) Write $[n]$ for $\{1, \ldots, n\}$ and $\binom{[n]}{m}$ for all m-sets in $[n]$. Let $X = \binom{[n]}{m}$. Say that $x \sim y$ if $|x \setminus y| = 1$. Then $(X, \sim)$ is the *Johnson graph* $J(n, m)$ or the *slice (of the hypercube)*. For $n \geq 2m$, it has $m+1$ distinct eigenvalues $\theta_j = (m-j)(n-m-j) - j$ with $\dim(V(\theta_j)) = \binom{n}{j} - \binom{n}{j-1}$, where $0 \leq j \leq n$, see [12, Th. 9.1.2].

Example 3 (Grassmann Graph) For q a prime power, put $V = V(n, q)$. Write $\left[{V \atop m}\right]$ for the set of all m-spaces in V and define the Gaussian coefficient (or: q-binomial coefficient) by $\left[{n \atop m}\right]_q = \left[{n \atop m}\right] = \left|\left[{V \atop m}\right]\right|$. Note that

$$\begin{bmatrix} n \\ m \end{bmatrix} = \frac{(q^n - 1) \cdots (q^{n-m+1} - 1)}{(q^m - 1) \cdots (q - 1)}.$$

Let $X = \binom{V}{m}$. Say that $x \sim y$ if $\dim(x \cap y) = m - 1$. Then $(X, \sim)$ is the *Grassmann graph* $J_q(n, m)$ or *q-Johnson graph*. For $n \geq 2m$, it has $m+1$ distinct eigenvalues $\theta_j = q^{j+1}\left[{m-j \atop 1}\right]\left[{n-m-j \atop 1}\right] - \left[{j \atop 1}\right]$ with $\dim(V(\theta_j)) = \binom{n}{j} - \binom{n}{j-1}$, where $0 \leq j \leq n$, see [12, Th. 9.1.2].

2.2 Association Schemes

Let X be a finite set. A symmetric m-class association scheme is a pair $(X, \mathcal{R})$ such that

(i) $\mathcal{R} = \{R_0, R_1, \ldots, R_m\}$ is a partition of $X \times X$;
(ii) $R_0 = \{(x, x) : x \in X\}$;
(iii) for each i, $0 \leq i \leq m$, $R_i = R_i^T$.
(iv) there are numbers p_{ij}^k such that for any $(x, y) \in R_k$ the number of $z \in X$ with $(x, z) \in R_i$ and $(z, y) \in R_j$ is p_{ij}^k.

Note that (iii) implies $p_{ij}^k = p_{ji}^k$. Delsarte's PhD thesis is a good introduction to the topic [32].

Let A_i denote the adjacency matrix of the graph $\Gamma_i = (X, R_i)$. Observe that Γ_i is regular of degree $k_i := p_{ii}^0$. Let J denote the all-ones matrix and I the identity matrix. The properties (i)–(iv) above translate to

$$\sum_{i=0}^{m} A_i = J, \qquad A_0 = I, \qquad A_i = A_i^T, \qquad A_i A_j = \sum_{k=0}^{m} p_{ij}^k A_k.$$

As $p_{ij}^k = p_{ji}^k$, $A_i A_j = A_j A_i$. Thus, the A_i generate a $(m+1)$-dimensional commutative algebra $\mathcal{A}$. Hence, we can diagonalize the A_i simultaneously (over $\mathbb{C}$). We obtain a decomposition of $\mathbb{C}^v$ into $m+1$ (maximal common) eigenspaces $V_0, V_1, \ldots, V_n$ of dimensions $f_0, f_1, \ldots, f_m$. As $J \in \mathcal{A}$, one eigenspace has dimension 1 and is spanned by the all-ones vector $\mathbf{j}$. By convention, $V_0 = \langle \mathbf{j} \rangle$ and $f_0 = 1$. Let E_j be the idempotent projection onto the jth eigenspace V_j. This is a basis of minimal idempotents of $\mathcal{A}$. We have

$$\sum_{j=0}^{m} E_j = I, \qquad E_0 = v^{-1} J.$$

There exist constants P_{ji} and Q_{ij} such that

$$A_i = \sum_{j=0}^{m} P_{ji} E_j, \qquad E_j = v^{-1} \sum_{i=1}^{m} Q_{ij} A_i.$$

Note that the P_{ji} are the eigenvalues of A_i. A useful equality is

$$f_i P_{ij} = k_j Q_{ji}.$$

For a non-empty subset $Y \subseteq X$ of size y with characteristic vector f_Y, define the *inner distribution* a of Y by

$$a_i = \frac{1}{y} f_Y^T A_i f_Y.$$

Delsarte's linear programming (LP) bound states that

$$(aQ)_j = \frac{v}{y} f_Y^T E_j f_Y \geq 0 \tag{1}$$

for all $j \in \{0, 1, \ldots, m\}$, see Proposition 2.3.2 in [12]. The inequality follows from the fact that E_j is positive semidefinite. It is known as *MacWilliams transform* of the vector a. It implies that we can check if f_Y is orthogonal to an eigenspace V_j by looking at the $(m+1)\times(m+1)$-matrix Q (instead of the usually much larger $(v \times v)$-matrix E_j).

Define $\|f\| = \left(\sum f_i^2\right)^{1/2}$ as the 2-norm for vectors, not the (normalized) 2-norm $\|f\|_2 = \left(\frac{1}{v} \sum f_i^2\right)^{1/2}$ as used in Boolean function analysis. We distinguish both by writing $\|\cdot\|$ and $\|\cdot\|_2$, respectively. Then

$$|Y| \cdot a_i = \sum_{j=0}^{m} P_{ji} \|E_j f_Y\|^2. \tag{2}$$

Hence, we can investigate expansion (as in *expander-mixing lemma* or *expander graph*) for all graphs in an association scheme using the $(m+1) \times (m+1)$-matrices.

Two important classes of association schemes are *metric* and *cometric* association schemes. An association scheme is metric if there exists an ordering of the A_i and polynomials p_i of degree i such that $A_i = p_i(A_1)$. An association scheme is cometric if there exists an ordering of the E_j and polynomials q_j of degree j such that $E_j = q_j(E_1)$ (here multiplication is $\circ$, the Hadamard product). Metric association schemes are also known as *distance-regular graphs* or *P-polynomial association schemes*. Cometric association schemes are also known as *Q-polynomial association schemes*. All our three main examples belong to association schemes which are metric and cometric.

Example 4 (Hamming Scheme) Let $X = \{0, \ldots, q-1\}^n$. Say that $(x, y) \in R_i$ if x and y have Hamming distance i. This defines the n-class Hamming scheme, see [12, §9.2]. Particularly, (X, R_1) is the Hamming graph $H(n, q)$. Formulas for the P_{ji}'s are given by the Krawtchouk polynomials, see [32, p. 39].

Example 5 (Johnson Scheme) Let $n - m \geq m$. Let $X = \binom{[n]}{m}$. Say that $(x, y) \in R_i$ if $|x \setminus y| = i$. This defines the Johnson scheme $J(n, m)$, an m-class association scheme, see [12, §9.1]. In particular, (X, R_1) is the Johnson graph $J(n, m)$ and (X, R_m) is the *Kneser graph*. Formulas for the P_{ji}'s are given by the Eberlein polynomials, see Delsarte [32, p. 48].

Example 6 (Grassmann Scheme) Let $n \geq m$ and $V = V(n, q)$. Let $X = \binom{V}{m}$. Say that $(x, y) \in R_i$ if $\dim(x \cap y) = m - i$. This defines the Grassmann scheme $J_q(n, m)$ (also: q-Johnson scheme), an m-class association scheme, see [12, §9.3]. In particular, (X, R_1) is the Grassmann graph $J_q(n, m)$. Formulas for the P_{ji}'s are given by the q-Eberlein polynomials, see Delsarte [33].

All formulas for the P_{ji} for these examples which are known to the author can be found in [11].

Another useful way of determining if $f_Y \in \sum_{j \in J} V_j$ for some subset J of $\{0, \ldots, m\}$ is *design-orthogonality*. Let Y be a subset of X. Suppose that $\mathcal{Z}$ is a family of subsets of X such that $|Y \cap Z| = \frac{|Y| \cdot |Z|}{|X|}$ for all $Z \in \mathcal{Z}$. Note that $f_Y^T f_Z = |Y \cap Z|$. Then for all $1 \leq j \leq m$, $E_j f_Y = 0$ or $E_j f_Z = 0$. For example, take the Johnson scheme $J(n, m)$. There $X = \binom{[n]}{m}$. Recall that a d-(n, m, λ) design Z is a subset of X such that any d-set of $[n]$ lies in precisely λ elements of Z. Suppose that $\mathcal{Z}$ is a family of d-(n, m, λ) designs and that $|Y \cap Z| = \frac{|Y| \cdot |Z|}{|X|}$ for all $Z \in \mathcal{Z}$. Then Y and Z are design-orthogonal. In particular, Y is an *antidesign*. Indeed, if a non-trivial d-(n, m, λ) design Z exists, then for all images of Z under the automorphism group, the f_Z span $V_0 + V_{d+1} + \cdots + V_m$ in the canonical ordering of the eigenspaces of $J(n, m)$. Hence, $f_Y \in V_0 + V_1 + \cdots + V_d$. More generally, we can replace f_Y and f_Z by functions $f \in V_0 + V_1 + \cdots + V_d$ and $g \in V_0 + V_{d+1} + \cdots + V_m$ and everything said remains true. In particular,

$$f^T g = \frac{f^T \mathbf{j} \cdot g^T \mathbf{j}}{|X|}. \tag{3}$$

If m divides n, then there exists a partition of $[n]$, respectively, $V(n, q)$ into m-sets, respectively, m-spaces. In the vector space case, this is a *spread*. For a fixed $J(n, m)$ or $J_q(n, m)$, let $\mathcal{Z}$ denote the set of all such partitions. Then the f_Z with $Z \in \mathcal{Z}$ span $V_0 + V_2 + \cdots + V_m$. Cameron-Liebler sets are often defined as those Y such that $|Y \cap Z| = |Y| \cdot |Z|/|X|$ for all $Z \in \mathcal{Z}$. See [31] for $J(n, m)$ and [10] for $J_q(n, m)$.

Also see Ito [76] for a treatment of the topic of design-orthogonality from the point of view of *coset geometries*.

2.3 Graded Posets

Let $(\mathcal{X}, \subseteq)$ be a graded poset of rank n. That is, a partially ordered set with a rank function. Let X_i denote all elements of $\mathcal{X}$ of rank i. Let D be a subset of $\mathcal{X}$. For $D \in X_i$, define $x_D \colon X \to \{0, 1\}$ as the (Boolean) function with $x_D(S) = 1$ if $D \subseteq S$ and $x_D(S) = 0$ otherwise.

Now only consider x_P with $P \in X_1$. Then for any $f : \mathcal{X} \to \mathbb{C}$, f can be written as a multivariate polynomial with x_P as variables and degree at most n. For instance,

$$\sum_{S \in \mathcal{X}} f(S) \left(\prod_{P \in X_1 : P \subseteq S} x_P \right) \left(\prod_{P \in X_1 : P \not\subseteq S} 1 - x_P \right).$$

We say that f has degree d if d is the minimum degree of such a polynomial. If $D \in \mathcal{X}_d$, then x_D has degree d as

$$x_D = \prod_{P \in X_1 : P \not\subseteq D} x_P.$$

If we consider $f : x_m \to \mathbb{C}$, then $\deg(f) \leq m$. This can be seen by considering

$$\sum_{S \in X_m} f(S) \prod_{P \in X_1 : P \not\subseteq S} x_P.$$

In several examples of metric association schemes, there is a metric ordering of relations such that $(x, y) \in R_1$ if and only if x and y have minimal distance in $\mathcal{X}$ restricted to X (usually, either distance 1 if $X = \mathcal{X}$ or distance 2 if X is one layer of $\mathcal{X}$). Similarly, there is often a cometric ordering such that f is a degree d function if and only if $f \in V_0 + V_1 + \ldots + V_d$, but not $f \in V_0 + V_1 + \ldots + V_{d-1}$. See also [33, 96].

In the literature on cometric association schemes, what we call *degree* is usually referred to as *dual degree*. For instance, see [13]. For sets $Y, Z \subseteq X$, let f_Y and f_Z be the corresponding Boolean functions, that is, from now on we identify characteristic vectors and Boolean functions.

(i) We have $\deg(f_Y) = \deg(1 - f_Y)$.
(ii) If Y and Z disjoint, then $f_Y + f_Z$ is Boolean with degree at most $\max(\deg(f_Y), \deg(f_Z))$.
(iii) If $Z \subseteq Y$, then $f_Y - f_Z$ is Boolean with degree at most $\max(\deg(f_Y), \deg(f_Z))$.

Example 7 (Hamming Graph) There are (at least) two natural choices for a poset associated with the Hamming graph. The first one is very natural for $q = 2$ if we identify $\{0, 1\}^n$ with the subset poset of $[n]$. Namely, $\mathcal{X} = \{0, \ldots, q-1\}^n$ and we say that $x \subseteq y$ if $x_i = y_i$ for all i with $x_i \neq 0$. Here, for the Hamming scheme, $X = \mathcal{X}$.

The second choice is that $\mathcal{X}$ is the set of all words of length n out of $\{0, \ldots, q-1\} \cup \{\cdot\}$ and we say that $x \subseteq y$ if $x_i = y_i$ for all i with $x_i \neq \cdot$. This is called the *Hamming lattice*. For the Hamming scheme, $X = X_n$.

Example 8 (Johnson Graph) Let $\mathcal{X}$ be all subsets of $[n]$. Then $X = X_m$ for the Johnson scheme $J(n, m)$.

Example 9 (Grassmann Graph) Let $\mathcal{X}$ be all subspaces of $V(n, q)$, the *subspace lattice* of $V(n, q)$. Then $X = X_m$ for the Grassmann scheme $J_q(n, m)$.

Suppose that Y is a subset of vertices in any of the three examples. Then $\deg(f_Y) \leq d$ if and only if $f_Y \in V_0 + V_1 + \ldots + V_d$ in the usual cometric ordering. Similarly, distance in the poset corresponds naturally to distance in the graph.

Suppose that we are in the Johnson graph or the Grassmann graph. If $\deg(f) \leq d$, then we can write f as

$$f = \sum_{D \in X_d} c_D x_D$$

for uniquely determined constants c_D. Hence, we only need to consider $X_d \times X_m$ in this case.

2.4 Equitable Partitions

The concept of *equitable partition* is also related. For instance, they often correspond to the extremal examples in spectral bounds, see [72] and our discussion in Sect. 2.6. Let Γ be a graph of order v with vertex set X. For A, B sets of vertices of Γ, let $E(A, B)$ denote the set of edges of Γ in $A \times B$. Now let $\mathcal{X} = \{X_1, X_2, \ldots, X_\nu\}$ be a partition of the vertex set X. Let e_{ij} be the average number of edges from X_i to X_j, that is $e_{ij} = |E(X_i, X_j)|/|X_i|$. The matrix $E = (e_{ij})$ of order ν is the *quotient matrix* of the partition. Let $\mu_1 \geq \mu_2 \geq \ldots \geq \mu_\nu$ be the eigenvalues of E.

Theorem 1 *We have* $\lambda_j \geq \mu_j \geq \lambda_{v-\nu+j}$ *for all* $1 \leq j \leq \nu$.

If for all i, j each vertex $x \in X_i$ has exactly e_{ij} neighbors in X_j, then the partition is an equitable partition. Other words for *equitable partition* include *perfect coloring*, *regular partition*, and, for $\nu = 2$, (often) *regular set* or (in parts of finite geometry) *intriguing set*. We have the following characterization of equality:

Theorem 2 *Suppose that* $\mathcal{X}$ *is equitable. Then there exist* $1 \leq i_1 \leq \ldots \leq i_\nu \leq v$ *such that* $\lambda_{i_j} = \mu_j$. *Furthermore,* f_{X_i} *lies in* $V(\lambda_{i_1}) + \cdots + V(\lambda_{i_\nu})$.

The most interesting case for us is that of equitable bipartitions, that is, $\nu = 2$. For example, take (the distance-1-graph on) the hypercube. As μ_1 is the degree for a regular graph, $\mu_1 = n$. The eigenvalues of the hypercube are $n, n-2, \ldots, -n+2, -n$. If $\mu_2 \geq n - 2d$, then f_{X_i} has degree at most d.

Special cases of equitable partitions often get rediscovered and are investigated from various points of views, so many names for relatively similar things exist. For instance, finite geometry alone uses *Cameron-Liebler set*, *tight set*, *m-ovoid*, *hemisystem* and *intriguing set* for particular equitable partitions. Design theory has *design* and *antidesign* [96]. Coding theory has *completely regular codes*, see [32, 66, 85]. In our three families, Boolean degree 1 functions correspond to *completely regular strength* 0 *codes of covering radius* 1 (but there is no such correspondence for larger degree). Latin squares have *k-plexes*. A more recent example is the notion of *graphical design* and *extremal graphical design*, for instance, see [68, 98, 110]. Finally, let us mention that there is vast literature on equitable partitions themselves. For instance, equitable partitions of graphs of degree at most 5 are classified [24].

2.5 *Tactical Decompositions*

Let X and Y be finite sets and let $\mathbf{I} \in X \times Y$. Call the elements of X *points*, the elements of Y *blocks*, and $\mathbf{I}$ an *incidence relation*. If $(x, y) \in \mathbf{I}$, then x and y are *incident*. We write $x \, \mathbf{I} \, y$.

Let $\mathcal{X} = \{X_1, X_2, \ldots, X_s\}$ be a partition of X and let $\mathcal{Y} = \{Y_1, Y_2, \ldots, Y_t\}$ be a partition of Y. Then $(\mathcal{X}, \mathcal{Y})$ is a *tactical decomposition* (or: *generalized orbit*) if the following holds:

1. For $x \in X_i$, the number of $y \in Y_j$ with $x \, \mathbf{I} \, y$ only depends on j.
2. For $y \in Y_i$, the number of $x \in X_j$ with $x \, \mathbf{I} \, y$ only depends on i.

Define a (bipartite) graph $\Gamma = (X \cup Y, \sim)$ where $x \sim y$ if $x \, \mathbf{I} \, y$. Hence, a tactical decomposition is an equitable partition of a bipartite incidence graph.

Many examples for tactical decomposition come from finite groups. In this context Block's lemma [9] is important as it shows (under certain assumptions) for the group case that $s \leq t$. We use the formulation given by Vanhove [107].

Lemma 1 (Block's Lemma) *Let G be a group acting on two finite sets X and Y, with respective sizes n and m. Let $O_1, \dots, O_s$, respectively $P_1, \dots, P_t$ be the orbits of the action on X, respectively Y. Suppose that $R \subseteq X \times Y$ is a G-invariant relation and call $B = (b_{ij})$ the $n \times m$ matrix of this relation, i.e. $b_{ij} = 1$ if and only if $x_i R y_j$ and $b_{ij} = 0$ otherwise.*

(i) The vectors $B^T f_{O_i}$, $i = 1, \dots, s$, are linear combinations of the vectors f_{P_j}.
(ii) If B has full row rank, then $s \leq t$. If $s = t$, then all vectors f_{P_j} are linear combinations of the vectors $B^T f_{O_i}$, hence $f_{P_j} \in \mathrm{Im}(B^T)$.

Under the conditions of Lemma 1(ii), groups can be used to construct low degree Boolean functions. In particular, it applies to Grassmann graphs due to a result by Kantor [77]. More generally, most of the examples in Sect. 5.1 were found by assuming a group action and investigating the corresponding tactical decomposition with computational tools such as integer linear programming (ILP) solvers and group theoretic methods.

2.6 *Erdős-Ko-Rado Theorems & Hoffman's Ratio Bound*

Let us briefly mention the Erdős-Ko-Rado (EKR) theorem [44]. The EKR theorem for sets states that a family Y of pairwise intersecting elements of $\binom{[n]}{m}$, $2 \leq m \leq n/2$, has size at most $\binom{n-1}{m-1}$. For $m < n/2$ equality occurs if and only if Y consists of all m-sets containing some fixed element. Essentially the same holds for $\begin{bmatrix} V \\ m \end{bmatrix}$, $V = V(n, q)$. This can be seen using Delsarte's LP bound or, more famously, the well-known Hoffman's Ratio Bound which bounds the size of a *coclique* (or: *independent set*, *stable set*) of a graph.

Theorem 3 (Hoffman's Ratio Bound) *Let A be the adjacency matrix of a graph Γ of order v with degree k and smallest eigenvalue λ. Suppose that Y is a coclique of Γ. Then $|Y| \leq \frac{v}{1-k/\lambda}$. If equality holds, then $f_Y \in \langle \mathbf{j} \rangle + V(\lambda)$.*

It has been observed by Lovász [84] that this implies $|Y| \leq \binom{n-1}{m-1}$ for an intersecting family in $J(n, m)$ (i.e., a coclique of Γ_m). For $n > 2m$, equality implies that $f_Y \in V_0 + V_1$, so $\deg(f_Y) \leq 1$. Similarly, Frankl and Wilson did show that $|Y| \leq \begin{bmatrix} n-1 \\ m-1 \end{bmatrix}$ for an intersecting family in $J_q(n, m)$ [58]. Similarly, for $n \geq 2m$, equality implies $f_Y \in V_0 + V_1$, that is, $\deg(f_Y) \leq 1$. See [67] for a unified treatment. In general, there has been a fruitful interaction between the theory of intersecting families and low degree Boolean functions going back to the work by Friedgut [59], and Dinur and Friedgut [38].

3 The Hypercube

The most basic question for any graded poset under consideration is the classification of the degree 1 Boolean functions.[2] For the hypercube, this is trivial and one sees that the only examples are 0, 1, x_i, and $1 - x_i$: Say, $f : \{0, 1\}^n \to \{0, 1\}$ is not the constant function. Write $f(x) = c + \sum c_i x_i$. Clearly, $c, c_1, \ldots, c_n \in \{0, 1\}$ by considering the elements x of $\{0, 1\}^n$ with weight at most 1. As we can always replace f by $1 - f$, assume that $f(0) = 0$ or, equivalently, $c = 0$. By considering the elements of $\{0, 1\}^n$ of weight 2, we see that at most one c_i can be equal to 1. In general, the spectral analysis on the hypercube is easy as it possesses an orthogonal basis corresponding to the well-known *Walsh-Hadamard transform*.

The approximate case is more interesting for degree 1. It is known as the Friedgut-Kalai-Naor (FKN) Theorem [60]. In line with the usual notation in Boolean function analysis, we write $\|f\|_2^2 = \frac{1}{|X|} \sum_{x \in X} f(x)^2$. We say that f is *ε-close* to g if $\|f - g\|_2^2 < \varepsilon$. One possible formulation is as follows, see also [93, §2.6].

Theorem 4 (FKN Theorem) *Suppose that $f : \{0, 1\}^n \to \{0, 1\}$ is ε-close to g for some degree* 1 *function $g : \{0, 1\}^n \to \mathbb{R}$. Then f is $O(\varepsilon)$-close to* 0, 1, x_i, *or* $1 - x_i$ *for some $i \in [n]$.*

Here and in the following asymptotics with big-O-notation are always with respect to n. For instance, read $\|f - g\|_2^2 = O(\varepsilon)$ as $\limsup_{n \to \infty} \|f - g\|_2^2 \leq C\varepsilon$ for some constant C. Note that $\|f - g\|_2^2 = \Pr(f \neq g)$ if f and g are Boolean.

This survey aims to state everything in the theory of association schemes, so let us rephrase the FKN theorem accordingly. This will give a better idea about how an "FKN theorem for an association scheme" should look like. Recall that E_0 and E_1 are idempotent matrices which project onto $V_0 = \langle j \rangle$ and V_1, respectively.

Theorem 5 (FKN Theorem, alternative) *Suppose that $Y \subseteq \{0, 1\}^n$ has $\|(E_0 + E_1) f_Y\|_2^2 \geq 1 - \varepsilon$. Then either $|Y|/2^n = O(\varepsilon)$ or $1 - |Y|/2^n = O(\varepsilon)$ holds, or there exists an $i \in \{1, \ldots, n\}$ such that $|Y \Delta \{z \in \{0, 1\}^n : i \in z\}|/2^n = O(\varepsilon)$ or $|Y \Delta \{z \in \{0, 1\}^n : i \notin z\}|/2^n = O(\varepsilon)$.*

A set Y is called a *ℓ-junta* if f_Y depends on at most ℓ coordinates of its input. That is, there exists a set $L = \{c_1, \ldots, c_\ell\} \subseteq [n]$ and a function $g \colon \{0, 1\}^\ell \to \{0, 1\}$ such that $f_Y(x_1, \ldots, x_n) = g(x_{c_1}, \ldots, x_{c_\ell})$. Theorem 5 states that an almost degree 1 Boolean function is close to a 1-junta (also called a *dictator*). The term *essential variable* is also used [102]. Boolean degree d functions have been first characterized as juntas by Nisan and Szegedy in 1994 [92]. The same result has been rediscovered by Tarannikov, Korolev, Botev in 2001 [104, Theorem 6].

Theorem 6 (Nisan-Szegedy Theorem) *Suppose that $Y \subseteq \{0, 1\}^n$ has $\deg(f_Y) \leq d$. Then f_Y depends on at most $d2^{d-1}$ coordinates, i.e., is a $d2^{d-1}$-junta.*

---

[2] The term *affine Boolean function* is maybe less clumsy. Here we avoid it due to the potential confusion with affine spaces.

Is this bound tight? For $d = 2$ it is. A complete classification of Boolean degree 2 functions was to our knowledge first given by Carlet et al. in [18]. The following description is due to Yuval Filmus. We do not give complements:

$$0, \ x, \ x \text{ AND } y = xy, \ x \text{ XOR } y = x + y - xy, \ xy + (1 - x)z,$$
$$\text{Ind}(x{=}y{=}z) = xy + xz + yx - x - y - z + 1,$$
$$\text{Ind}(x{\leq}y{\leq}z{\leq}w \text{ OR } x{\geq}y{\geq}z{\geq}w).$$

Here Ind is the indicator function.

For $d = 3$, the correct answer is 10, not 12 as suggested by Theorem 6. The degree 3 case has been classified by Kirienko (computationally) in [103] and Zverev (humanly) in [111]. The extremal examples belong to a general family of Boolean degree d functions due to Tarannikov from 2000 [102] which depends on $3 \cdot 2^{d-1} - 2$ coordinates. The same construction has been rediscovered in 2020 by Chiarelli, Hatami, and Saks [19]. Put

$$H_2(x_1, x_2, x_3, x_4) = x_1 \oplus x_2 \oplus (x_1 \oplus x_3)(x_2 \oplus x_4).$$

This is an example for $d = 2$. Now define H_d recursively by

$$H_d(x_1, x_2, y, z) = x_1 \oplus (x_1 \oplus x_2 \oplus 1)H_{d-1}(y) \oplus (x_1 \oplus x_2)H_{d-1}(z).$$

Tarannikov gives the construction in terms of *resilient functions*. This is equivalent with the notation of essential variables, see the remark after Proposition 6.23 in [93] or the introduction of [83]. Another relevant terms is that of a *correlation-immune function*.

For $d \geq 4$, exact values are unknown.

For the upper bound, a breakthrough occured with the recent result by Chiarelli, Hatami, and Saks who did show an upper bound of $O(2^d)$ for the number of essential variables [19]. The current best upper bound is due to Wellens [109]. We obtain the following state of the art:

Theorem 7 *Suppose that $Y \subseteq \{0, 1\}^n$ has $\deg(f_Y) = d$ and f_Y depends on the maximum number of coordinates (for that d). Then f_Y depends on at least $3 \cdot 2^{d-1} - 2$ and at most $8.788 \cdot 2^{d-1}$ coordinates.*

Lastly, one can ask what happens when a function is close to degree d. This question got answered by Kindler and Safra in [81].

3.1 The p-Biased Hypercube

The p-biased hypercube is a hypercube $\{0, 1\}^n$ with a distribution where each coordinate is 1 with probability p. Hence, for $x \in \{0, 1\}^n$, we have a measure μ_p defined by

$\mu_p(x) = p^{\sum x_i}(1-p)^{\sum(1-x_i)}$. For $p = \frac{1}{2}$ we have the classical hypercube. Expanding Boolean function analysis to the p-biased setting has been very fruitful in the last years, but it is outside the scope of this document. There are an FKN theorem [49, 52] and a Kindler-Safra-type theorem are known [37]. The utility of the p-biased hypercube is that it can give a good model for more complex structures. For instance, for $p = \min(\frac{m}{n}, 1 - \frac{m}{n})$, the p-biased hypercube and $J(n, m)$ behave similarly.

Recently, Tanaka and Tokushige suggested a p-biased measure for $V(n, q)$ in the context of Erdős-Ko-Rado type results for vector spaces [101].

3.2 *The General Hamming Scheme*

Consider the Hamming graph $H(n, q)$ with $q > 2$. A FKN theorem due to Alon, Dinur, Friedgut and Sudakov can be found in Lemma 2.4 in [2]. One can show that Theorem 7 implies that a Boolean degree d function on $H(n, q)$ depends on at most $4.394 \cdot 2^{\lceil \log_2 q \rceil d}$ variables. Recently, Valyuzhenich did show an upper bound $\frac{dq^{d+1}}{4(q-1)}$ for $q \neq 2^s$ [106].

4 The Johnson Scheme

A Boolean degree d function of the hypercube is, restricted to words of weight m, also a Boolean degree d function of the Johnson scheme $J(n, m)$. Any Boolean function on $J(n, m)$ has degree at most $\min(m, n - m)$. Thus, one requires at least $\min(m, n - m) > d$ for a classification. For $d = 1$, this has been done many times. For instance, it is a special case of more general results by Meyerowitz [90]. Short proofs limited to the degree 1 case can be found in [49, 54].

Theorem 8 *Let $m, n - m \geq 2$. Suppose that $Y \subseteq \binom{[n]}{m}$ has $\deg(f_Y) \leq 1$. Then Y or its complement is one of the following:*

1. *the empty set (that is, $f = 0$),*
2. *the set of all m-sets which contain a fixed element i (that is, $f = x_i$).*

Recall that the FKN theorem for the hypercube says that if a Boolean function is ε-close to degree 1, then it is $O(\varepsilon)$-close to a Boolean degree 1 function. For the Johnson scheme, we cannot hope for such a result. Say, for $m = 2$ and n large, $x_1 + x_2$ is close to degree 1 on the Johnson scheme, while it is not on the hypercube. Filmus gives an FKN Theorem for the Johnson scheme [49]. Again, we will state the theorem in two formulations.

Theorem 9 (FKN Theorem for the slice) *Let $m, n - m \geq 2$ and put $p = \min(m/n, 1 - m/n)$. Suppose that $f : \binom{[n]}{m} \to \{0, 1\}$ is ε-close to g for some degree 1 function $g : \binom{[n]}{m} \to \mathbb{R}$. Then either f or $1 - f$ is $O(\varepsilon)$-close to*

$$x_{i_1} + \cdots + x_{i_t} \text{ or } 1 - (x_{i_1} + \cdots + x_{i_t})$$

for some set $I = \{i_1, \ldots, i_t\}$ *of size at most* $\max(1, O(\sqrt{\varepsilon}/p))$.

It can be checked that the two functions in the statement are $O(\varepsilon)$-close to being Boolean. Thus, the theorem is tight. Let us also state a version of the theorem using sets.

Theorem 10 (FKN Theorem for the Johnson scheme) *Let* $m, n-m \geq 2$ *and put* $p = \min(m/n, 1-m/n)$. *Suppose that* $Y \subseteq \binom{[n]}{m}$ *has* $\|f_Y - (E_0 + E_1)f_Y\|_2^2 \geq 1-\varepsilon$. *Then there exists an* $I \subseteq \{1, \ldots, n\}$ *with* $|I| = \max(1, O(\sqrt{\varepsilon}/p))$ *such that either* $|Y \Delta \{z \in \binom{[n]}{m} : |z \cap I| \geq 1\}|/\binom{n}{m} = O(\varepsilon)$ *or* $|Y \Delta \{z \in \binom{[n]}{m} : |z \cap I| = 0\}|/\binom{n}{m} = O(\varepsilon)$.

For $m = 3$, equitable partitions of degree 2 are classified in [45, 61].

Let $\gamma(d)$ denote the maximum number of relevant coordinates for a Boolean degree d function on the hypercube. The main results of [53, 55, 57] show the following.

Theorem 11 *There exist a constant* C *and functions* $\xi, n_0 : [d] \to \mathbb{R}$ *such that the following holds. Suppose that* $Y \subseteq \binom{[n]}{m}$ *with* $\deg(f_Y) \leq d$.

1. *If* $C^d \leq m \leq n - C^d$, *then* f_Y *depends on at most* $\gamma(d)$ *coordinates.*
2. *If* $n - m \geq n_0(d)$ *and* $m \geq 2d$, *then* f_Y *depends on at most* $\gamma(d)$ *coordinates.*
3. *If* $n \geq 2m$ *and* $m \geq 2d$, *then* f_Y *depends on at most* $\xi(d)$ *coordinates.*

It is also shown in [53] that for every positive integer m, where $d \leq m \leq 2d-1$, and any positive integer ℓ, there exists a $n \geq 2m$ and a $Y \subseteq \binom{[n]}{m}$ such that f_Y depends on ℓ coordinates. To see this, take some integer e such that $\ell \leq de$. Consider the function g defined by

$$g = \sum_{i=1}^{e} \prod_{j=1}^{d} x_{(d-1)i+j}.$$

It is clear that $\deg(g) \leq d$ and that it depends on at least de coordinates for $n \geq 2de$. The condition $m \leq 2d-1$ guarantees that g is Boolean. Written differently, we have $g = f_Y$ for

$$Y = \bigcup_{i=1}^{e} \left\{ y \in \binom{[n]}{m} : \{(d-1)i+1, \ldots, (d-1)i+d\} \subseteq y \right\}.$$

In Theorem 11(2), $\gamma(d)$ and $\xi(d)$ are not the same. We have $\gamma(2) \leq 4$ by the Nisan-Szegedy theorem (and, indeed, equality), but [28, §3] gives an examples for $(n, m) = (8, 4)$ which depends on 5 coordinates, so $\xi(2) \geq 5 > 4 = \gamma(2)$: identify $J(8, 4)$ with the elements of $\{0, 1\}^8$ with Hamming weight 4 and take all elements which start with one of

$$11000, 01100, 00110, 00011, 10001,$$
$$11100, 01110, 00111, 10011, 11001.$$

This construction generalizes to $J(2m, m)$ for all $m \geq 4$, see Construction 3 in [108]. A variant of the Kindler-Safra theorem for $J(n, m)$ due to Keller and Klein can be found in [78].

In [80] it is shown by Kiermaier, Mannaert and Wassermann that

$$\gcd\left(\binom{n}{m}, \binom{n-1}{m-1}, \ldots, \binom{n-d}{m-d}\right)$$

divides $|Y|$.

The spectral analysis on the Johnson scheme $J(n, m)$ is much helped by the existence of a "nice" orthogonal basis. This basis has been described by Srinivasan in [97] as well as by Filmus in [50]. See also [56].

5 The Grassmann Scheme

In the Grassmann scheme $J_q(n, m)$, Boolean degree 1 functions are traditionally considered in the projective space $\mathrm{PG}(n-1, q)$ and are called *Cameron-Liebler sets* (or *Cameron-Liebler* $(m-1)$*-space classes* and *Cameron-Liebler line classes* for $m = 2$). Considering the very easy classification of Boolean degree 1 functions on the hypercube and in the Johnson scheme $J(n, m)$, it is tempting to assume that for $n \geq 2m \geq 4$, the situation in the Grassmann scheme is similar. That is, for $V = V(n, q)$, if $Y \subseteq \begin{bmatrix} V \\ m \end{bmatrix}_q$ with $\deg(f_Y) \leq 1$, then Y is one of the obvious, say, *trivial examples*:

(I) The empty set.
(II) All m-spaces through a fixed 1-space P.
(III) All m-spaces in a fixed hyperplane H of V.
(IV) The union of the previous two examples when $P \not\subseteq H$.
(V) The complement of any of the examples (I) to (IV).

Surprisingly, the classification question is one of the oldest ones in this survey as it goes back to work by Cameron and Liebler from 1982 [17]. Consider a subgroup G of $\mathrm{P\Gamma L}(n, q)$ acting on the 1-spaces and 2-spaces of $V(n, q)$. Then Lemma 1 shows that G has at least as many orbits on 2-spaces as on 1-spaces. Examples (I)–(V) arise as orbits of such groups G. Are there any more examples which arise as orbits in this way? In 2008, Bamberg and Penttila did confirm the conjecture, that is, there are no additional examples for such groups [4].

Cameron and Liebler also relaxed their group theoretic question about G to a combinatorial question, limited to $J_q(4, 2)$. Here one asks, in the language of this survey, about the Boolean degree 1 functions of $J_q(4, 2)$. They suggested that the list

(I)–(V) is complete. This turned out to be false. We will survey various constructions in the next section.

A relevant portion of the literature on $J_q(4, 2)$ uses projective notation and the Klein correspondence between buildings of type A_3 and D_3. We refer the reader to Sect. 4.6 of [8] for a geometric point of view of the Klein correspondence and to Chap. 12 of [105] for an algebraic one. The recent survey by Gavrilyuk and Zinoviev on completely regular codes in $J(n, m)$ and $J_q(n, m)$ covers the degree 1 case [66].

5.1 Exceptional Degree 1 Examples

Only in $J_q(4, 2)$ non-trivial example for Boolean degree 1 function/Cameron-Liebler sets are known. As all the descriptions are in terms of projective geometry, we will call 1-spaces *points* and 2-spaces *lines*. There are $\left[{4 \atop 2}\right] = (q^2 + 1)(q^2 + q + 1)$ lines. Examples necessarily have size $x(q^2 + q + 1)$ for some integer x, where $1 \leq x \leq q^2 + 1$, see Eq. (4) in Sect. 5.3. By taking complements, we can always guarantee $x \leq \frac{q^2+1}{2}$. There are no established naming conventions for the various families, so we make some up for the convenience of the reader.[3] The group in the table is a known subgroup of the the automorphism group of the example. Usually, some modularity condition on q is necessary.

q	x	group	name	ref
odd	$\frac{q^2+1}{2}$	$O^-(4, q)$	Bruen-Drudge	Sect. 5.1
odd	$\frac{q^2+1}{2}$	pt.-stab. in $O^-(4, q)$	derived Bruen-Drudge	Sect. 5.1
odd	$\frac{q^2+1}{2}$	$q^2(q + 1)$	derived Bruen-Drudge	Sect. 5.1
$1 \pmod 4$	$\frac{q^2+1}{2}$	$PGL(2, q)$	Cossidente-Pavese	Sect. 5.1
4	7	stab. of cone over hyperoval	Govaerts-Penttila	Sect. 5.1
5	10	stab. of a sp. cap	Gavrilyuk-Metsch	Sect. 5.1
$5, 9 \pmod{12}$	$\frac{q^2-1}{2}$	$(C(q^2 + q + 1) \rtimes C(\frac{q-1}{4})) \rtimes C(3)$	Affine I	Sect. 5.1
$2 \pmod 3$	$\frac{(q+1)^2}{3}$	$C(q^2 + q + 1) \rtimes C(3)$	Affine II	Sect. 5.1
27	336	$(C(q^2 + q + 1) \rtimes C(2)) \rtimes C(9)$	Rodgers' Sporadic I	Sect. 5.1
32	495	$C(q^2 + q + 1) \rtimes C(15)$	Rodgers' Sporadic II	Sect. 5.1

The Bruen-Drudge Family

This family requires q odd and has $x = \frac{q^2+1}{2}$. It has been described by Drudge for $q = 3$ [39] and later in general by Bruen and Drudge [15]. The stabilizer of the example is the finite simple group of type $O^+(4, q)$.

[3] If a construction can be clearly attributed to three or fewer authors, then we name them after them. Otherwise, we tried to pick a descriptive name.

The example has a concise description: Recall that the finite field GF(q) of odd order q has $(q-1)/2$ non-zero squares and $(q-1)/2$ nonsquares. Let $Q(x) = \xi x_1^2 + x_2^2 + x_3^2 + x^4$ where ξ is some non-square. Then the discriminant disc(Q) of Q equals ξ. It can be checked that there are q^2+1 points $\langle x \rangle$ with $Q(x) = 0$, $q(q^2+1)/2$ with $Q(x)$ a non-zero square, and $q(q^2+1)/2$ with $Q(x)$ a non-square. For each subspace S, we can consider $\mathrm{disc}_{|S}(Q)$, the discriminant of Q restricted to S. Then we have the following census of lines:

(i) There $\binom{q^2+1}{2}$ lines L with $\mathrm{disc}_{|L}(Q)$ a non-zero square.
(ii) There are $\binom{q^2+1}{2}$ lines L with $\mathrm{disc}_{|L}(Q)$ a non-square.
(iii) There are $\frac{(q^2+1)(q+1)}{2}$ lines with $\mathrm{disc}_{|L}(Q) = 0$ and $Q(x)$ either 0 or a non-zero square for all points $\langle x \rangle$ on L.
(iv) There are $\frac{(q^2+1)(q+1)}{2}$ lines with $\mathrm{disc}_{|L}(Q) = 0$ and $Q(x)$ either 0 or a non-square for all points $\langle x \rangle$ on L.

Any union of size $\frac{q^2+1}{2}(q^2+q+1)$ gives an example, that is, (i) and (iii), (i) and (iv), (ii) and (iii), (ii) and (iv).

Derived Bruen-Drudge Families

There are several more families with $x = \frac{q^2+1}{2}$ which are obtained by modifying the Bruen-Drudge family locally. All of them require q odd. One family has been described by Cossidente and Pavese [20] and, independently, by Gavrilyuk, Matkin and Penttila [62] admitting the stabilizer of a point in $O^-(4,q)$, one other family by Cossidente and Pavese [21] admitting a subgroup of order $q^2(q+1)$. Furthermore, Remark 3.16 in [21] notes that there are $1, 3, 5$ additional examples for $q = 7, 9, 11$, respectively.

The Cossidente-Pavese Family

One last family with $x = \frac{q^2+1}{2}$. It requires $q \equiv 1 \pmod 4$, admits $PGL(2,q)$ as a stabilizer, and has been described by Cossidente and Pavese [22]. Francesco Pavese also informed the author that there are sporadic examples for $q = 13, 25, 41$ known which admit $C_{(q-1)/2} \times PGL(2,q)$ as group.

The Govaerts-Penttila Example

This example requires $q = 4$ and has $x = 7$. It has been described by Govaerts and Penttila [69]. This is the only known example for which q is an even power of 2.

It has a very concise description: Fix a projective plane π (that is, a 3-space). In π, take a hyperoval $\mathcal{H}$, that is, a set of $q+2$ points with no three collinear. Pick a point P not in π. Let C denote the cone with vertex P and base $\mathcal{H}$. Then the set consists of the following is an Boolean degree 1 function with $x = 7$:

(i) The $q + 2 = 6$ lines in C.
(ii) The $\frac{(q+2)(q-1)\cdot(q+1)(q-1)}{2} = 135$ lines which meet C in precisely two points and are disjoint from $\mathcal{H}$.
(iii) The $\frac{q(q-1)}{2} = 6$ lines in $V(4, q)$ which are disjoint from $\mathcal{H}$.

The Gavrilyuk-Metsch Example

This example requires $q = 5$ and has $x = 10$. It is described by Gavrilyuk and Metsch [64]. It uses a cap of 20 points which has been constructed in [1]. For $q = 5$, this is the unique example with $x = 10$.

Affine Family I

This family requires $q \equiv 5 \pmod{12}$ or $q \equiv 9 \pmod{12}$ and has $x = \frac{q^2-1}{2}$. Small examples for this family have been first discovered by Rodgers in his PhD thesis [94]. Descriptions of the infinite family are due to De Beule, Demeyer, Metsch and Rodgers [25] as well as Feng, Momihara and Xiang [48].

We call a family *affine* if it has the property that all their elements are not contained in a fixed (hyper)plane π. In particular, for $x = \frac{q^2-1}{2}$, we can add all the lines in π to obtain a family with $x = \frac{q^2+1}{2}$. This also makes them non-trivial examples for Cameron-Liebler sets in affine spaces, see [34, 36] and Sect. 6.2.

For $q \equiv 9 \pmod{12}$, the family belongs to an affine two-intersection set with parameters $(\frac{1}{2}(3^{2e} - 3^e), \frac{1}{2}(3^{2e} + 3^e))$. That is, there exists a set of points $\mathcal{P}$ not in π such that any line not in π intersects $\mathcal{P}$ in one of these two numbers. The one type of lines corresponds to the lines in the family.

Affine Family II

This family requires $q \equiv 2 \pmod{3}$ and has $x = \frac{(q+1)^2}{3}$. Small examples for this family have been first discovered by Rodgers [94]. The infinite family has been described [47] by Feng, Momihara, Rodgers, Xiang and Zou. The family includes the only infinite family for q an odd power of 2.

Rodgers' Sporadic Affine Examples

In §4.3 of [94], Rodgers describes one sporadic example with $(q, x) = (27, 336)$ and two sporadic examples with $(q, x) = (32, 495)$.

5.2 Classification Results for Degree 1

We have seen that in $J_q(4, 2)$ a classification of Boolean degree 1 functions is hopeless. The general question of the classification of Boolean degree 1 function for $J_q(n, m)$ is surely (at least for $m = 2$) implicit in the work by Cameron and Liebler from 1982. The case for general n and $m = 2$ has been intensively investigated by Drudge in his PhD thesis [39] in 1998, while the case for general m found a more explicit treatment in the late 2010s [10, 54, 95].

The crucial ingredient for several of the classification results is the use of design-orthogonality together with the following weighted design in $J_q(4, 2)$ due to Gavrilyuk and Mogilnykh [65]. For a point P and a hyperplane H, define $g_{P,H} : \begin{bmatrix} V(4,q) \\ 2 \end{bmatrix} \to \{-(q-1), 0, 1\}$ by

$$g_{P,H}(L) = \begin{cases} 1 & \text{if } P \subseteq L \not\subseteq H, \\ 1 & \text{if } P \not\subseteq L \subseteq H, \\ -(q-1) & \text{if } P \subseteq L \subseteq H, \\ 0 & \text{otherwise.} \end{cases}$$

One verifies that $g_{P,H}^T \mathbf{j} = q^2 + 1$. A line L contains $q + 1$ points $P_1, \ldots, P_{q+1}$ and lies on $q + 1$ planes $H_1, \ldots, H_{q+1}$. By Equation (3), $f_Y^T g_{P_i,H_j} = x$ for a Boolean degree 1 function f_Y of size $x(q^2 + q + 1)$. Hence,

$$\begin{aligned} x =& |\{L \in Y : P \subseteq L\}| + |\{L \in Y : L \subseteq H\}| \\ &- (q+1)|\{L \in Y : P \subseteq L \subseteq H\}|. \end{aligned}$$

For a fixed line L, let t_{ij} denote the number of elements of Y in P_i and H_j (distinct from L). Together with the fact that $\{Y, X \setminus Y\}$ is an equitable bipartition, the possible values of t_{ij} (called pattern) are very restricted, see [64].

The combined work by Drudge [39], by Gavrilyuk, Matkin and Mogilnykh [63, 65, 86], and by Filmus and the author [54] gives a complete classification result for small q:

Theorem 12 *Let $q \in \{2, 3, 4, 5\}$. Suppose that Y is a family of m-spaces of $V(n, q)$ with* $\deg(f_Y) \leq 1$. *If Y is not trivial, then $(n, m) = (4, 2)$.*

This result is based on a complete classification of the non-trivial Boolean degree 1 functions in $J_q(4, 2)$ and then showing that these do not extend to $J_q(5, 2)$. Recently, the author showed in [73] that for $|n - 2m|$ sufficiently large, the same classification results holds.

Theorem 13 *Let* $\min(n - m, m) \geq 2$. *Then there exists a function $c(q)$ such that the following holds: Suppose that Y is a family of m-spaces of $V(n, q)$ with* $\deg(f_Y) \leq 1$ *and $|n - 2m| \geq c(q)$. Then Y is trivial.*

The constant $c(q)$ depends on vector space Ramsey numbers (which exist, see [70]). For most parameters these implicitly depends on the repeated application of the Hales-Jewett theorem. Thus, the $c(q)$ given by the proof is very large. By Theorem 12, we know $c(2) = 0$ and $c(3) = c(4) = c(5) = 1$. For small n, there is a plenitude of results which restrict the possible sizes of non-trivial examples.

Let us now summarize what is known for small n. The best investigated case is $J_q(4, 2)$. Recall that here any Y with $\deg(f_Y) \leq 1$ has $|Y| = x(q^2 + q + 1)$ for some integer x. If Y is one of the trivial examples, then $x \in \{0, 1, 2, q^2 - 1, q^2, q^2 + 1\}$ and we can without loss of generality assume that $x \leq \frac{q^2+1}{2}$ (as we can always consider the complement of Y). Here results by Metsch [87, 88] and Metsch and Gavrilyuk [64] show the that non-trivial examples are very restricted in terms of their possible sizes. We summarize these in the following statement.

Theorem 14 *Suppose that Y is a family of 2-spaces of $V(4, q)$ of size $x(q^2 + q + 1)$ with $\deg(f_Y) \leq 1$. If $2 < x$, then the following holds:*

1. *We have $x \geq q + 1$.*
2. *We have $x > q\sqrt[3]{q/2} - \frac{2}{3}q$.*
3. *The equation*

$$\binom{x}{2} + \ell(\ell - x) \equiv 0 \pmod{q + 1}$$

has an integer solution in ℓ.

In the general case, the first bounds on x were found by Rodgers, Storme, and Vansweevelt [95] for $n = 2m$ and by Blokhuis, De Boeck, and D'haeseleer [10] for general n. The currently best known conditions on the sizes, which the author is aware of, are summarized in the following theorem:

Theorem 15 *Suppose that Y is a family of m-spaces of $V(n, q)$ of size $x\begin{bmatrix} n-1 \\ m-1 \end{bmatrix}$ with $\deg(f_Y) \leq 1$. If Y is non-trivial, then the following holds:*

1. *If $(n, m) = (6, 3)$, then $x \geq \frac{1}{3}q$.*
2. *If $n = 2m$ and $m \geq 3$, then $x > \sqrt[3]{q/2}$.*
3. *If $n = 2m$, $m \geq 3$, and $q \geq q_0$ for some universal constant q_0, then $x \geq \frac{1}{5}q$.*
4. *If $n \geq 3m - 1$ and $m \geq 2$, then $x > \frac{1}{8}q^{n-3m+2}$.*
5. *If $n \geq 3m + 1$ and $m \geq 2$, then*

$$x \geq \left(\frac{1}{\sqrt[8]{2}} + o(1)\right)q^{n-\frac{5}{2}m+\frac{1}{2}}.$$

Proof The first claim is Theorem 1.3 in [89]. The second claim is Theorem 6.7 in [95]. The third claim is Theorem 1.4 in [89] (with the minor improvement from Theorem 1.8 in [74]). The fourth claim is Theorem 5.1 in [73]. The fifth claim is Theorem 3.6 in [29]. □

There are also general modular conditions which can be found in [26, 29]. Note that these are less strong than the results for $(n, m) = (4, 2)$.

A particular special case are *two-intersection sets*. A two-intersection set is a family of 1-spaces $\mathcal{P}$ such that any m-space contains either precisely α or precisely β elements of $\mathcal{P}$. There are many examples for such sets for $(n, m) = (3, 2)$, but already for $(n, m) = (4, 2)$ only trivial constructions are known: $\mathcal{P}$ is the empty set $(\alpha = \beta = 0)$, all the 1-spaces $(\alpha = \beta = q + 1)$, a hyperplane $(\alpha = 1, \beta = q + 1)$, or the complement of a hyperplane $(\alpha = 0, \beta = q)$. If there is a non-trivial example for $n \geq 4$ and $m = 2$, then q must be an odd square. The first open case is $(n, m, q) = (4, 2, 9)$. The work by Tallini Scafati from the 1970s discusses this in detail [99, 100].[4] There is a close connection to so-called *two-weight codes*, see Theorem 12.5 in [16].

If Y is the set of all m-spaces which contain precisely α elements of $\mathcal{P}$, then $\deg(f_Y) \leq 1$. Here Theorem 13 implies that there are no non-trivial examples if $|n - 2m| \geq c(q)$. See [73, §4].

5.3 *Results Beyond Degree* 1

The results beyond the degree 1 case are sparse.

The Khot-Minzer-Safra Theorem

Let us start with the most prominent result, the recent proof of the 2-to-2 Games Conjecture by Khot, Minzer, and Safra [79]. Loosely speaking, the Unique Games Conjecture states that it is hard to approximate a certain type of NP-complete problem and is an important conjecture in complexity theory. The 2-to-2 Games Conjecture is a weaker version and its proof has been a great breakthrough in theoretical computer science. We refer to Boaz Barak's blog post [6] for an excellent summary.

Put $V = V(n, 2)$ and $X = \begin{bmatrix} V \\ m \end{bmatrix}$. For $Y \subseteq X$, an r-space R and an s-space S, let $Y(R, S)$ denote the set of all elements y of Y with $R \subseteq y \subseteq S$. Informally, the main result by Khot, Minzer, and Safra is as follows: If Y has significant weight on low degree, then there exist R, S such that $Y(R, S)$ is a significant proportion of $X(R, S)$. Formally, define $\Phi(Y)$ by $\Phi(Y) = \frac{|E(Y, \overline{Y})|}{k \cdot |Y|}$, where k is the degree of $J_q(n, m)$.

Theorem 16 (Khot-Minzer-Safra) *For all $\alpha \in (0, 1)$ there exists a $\varepsilon > 0$ and an integer $t \geq 0$ such that for all $m \geq m_0(\alpha)$ and all $n \geq n_0(m, \alpha)$, the following holds. Suppose that Y is a family of m-spaces in $V(n, 2)$ with $\Phi(Y) \leq \alpha$. Then there exist subspace R, S with* $\dim(R) + \operatorname{codim}(S) \leq t$ *and*

---

[4] In particular, the problem of the existence of a two-intersecting with respect to lines in PG(3, 9) has been open for more than five decades. This deserves a footnote.

$$\frac{|Y(R, S)|}{|X(R, S)|} \geq \varepsilon.$$

Recall that the distinct eigenvalues θ_j of $J_2(n, m)$ are:

$$\theta_j = P_{j1} = 2^{j+1}(2^{m-j} - 1)(2^{n-m-j} - 1) - (2^j - 1) \approx 2^{n-j} \text{ for } j \ll n.$$

Using these eigenvalues in Equation (2) we see that $\Phi(Y) \leq \alpha$ for some constant α is equivalent with $\left\| \sum_{j=0}^{d(\alpha)} E_j f_Y \right\| / \|f_Y\| \geq \beta(\alpha)$. For instance, for $\alpha = \frac{1}{2}$, we can take $d(\alpha) = 1$ and $\beta(\alpha) = 1 + o(1)$. Hence, their result does indeed describe the structure of families Y with a significant weight on low degree.

Degree 2

The degree 2 in $J_q(n, m)$ has been investigated for small dimension n in [28]. In contrast to Theorem 11(2), but similar to the behavior for degree 1 in $J_q(4, 2)$, $n - m, m \geq 2d$ does not seem to suffice for an FKN-type result. The authors provide the following example in Sect. 5.2 for $(n, m) = (8, 4)$ which is reminiscent of the Bruen-Drudge example for degree 1. The construction goes as follows. Define a quadratic form Q by

$$Q(x) = x_1^2 + \alpha x_1 x_2 + \beta x_2^2 + x_3^2 + x_4^2 + x_5^2 + x_6^2 + x_7^2 + x_8^2$$

such that $1 + \alpha x + \beta x^2$ is irreducible over GF(q). There are six types of 4-spaces with respect to Q. Take for Y a union of the three types with a radical of dimension at least 2. Then $\deg(f_Y) = 2$.

Many non-trivial degree 2 examples for $(n, m) = (6, 3)$ can be found in [28]. As for $J(n, m)$, a Nisan-Szegedy type theorem (in a narrow sense) in $J_q(n, m)$ is impossible for $m < 2d$.

Divisibility Conditions

Kiermaier, Mannaert and Wassermann recently investigated divisibility conditions for the general case. Let Y be a family of m-spaces of $V(n, q)$ such that $\deg(f_Y) \leq d$. Then Theorem 4.7 in [80] shows that

$$\gcd\left(\begin{bmatrix} n \\ m \end{bmatrix}, \begin{bmatrix} n-1 \\ m-1 \end{bmatrix}, \ldots, \begin{bmatrix} n-d \\ m-d \end{bmatrix}\right) \tag{4}$$

divides $|Y|$. For n even and $d = 2$, this implies that $|Y| = x\begin{bmatrix} n-1 \\ m-1 \end{bmatrix}$ for some integer x. For instance, Lemma 4.11 in [80] shows that $|Y| = x\begin{bmatrix} n-d \\ m-d \end{bmatrix}$ for some integer x if $m - i \mid n - i$ for all $i \in \{0, \ldots, d-1\}$.

6 Other Structures

Let us mention some further results in some other structures.

6.1 Permutation Groups

Affine Boolean functions in the symmetric group $Sym(n)$ have been classified in [40]. FKN theorems for $Sym(n)$ can be found in [41, 42, 51]. While a Nisan-Szegedy type theorem does not hold in the sense that there is a junta, it has been shown in [23] that a *constant depth decision tree* suffices to decide the value of Boolean degree d function on $Sym(n)$. An approximate version of this result holds too. The different *complexity measures* discussed in [23] also give inspiration for how classification results might need to be phrased.

The more general setting of Cameron-Liebler sets (or Boolean degree 1 functions) of permutation groups has been recently investigated in [35].

6.2 Bilinear Forms, Affine Spaces, Polar Spaces

Bilinear forms [12, §9.5], affine spaces, and polar spaces [12, §9.4] all share the property that they are closely related to Grassmannians.

Bilinear Forms

The bilinear forms graph Γ can be seen as an induced subgraph of the Grassmann graph $J_q(n, m)$: Fix a $(n-m)$-space S. The induced subgraph of $J_q(n, m)$ on all m-spaces disjoint from S is the bilinear forms graph, see [12, §9.5]. Barak, Kothari, and Steuer give a proof of the 2-to-2 games conjecture using Γ with $q = 2$ instead of $J_2(n, m)$ [7]. In [54] it is shown that the restriction of the Bruen-Drudge example (see 5.1) to Γ is also non-trivial (in some sense). In [3] equitable partitions for $(q, m) = (2, 2)$ are studied. Bounds on the possible sizes of degree 1 examples are given in [71, 75]. One central important property of the hypercube and the Johnson graph, which helps with the investigation of low degree Boolean functions, is *hypercontractivity*. Unlike for the Grassmann scheme, hypercontractivity results exists for the bilinear forms scheme, see [43, 46].

Affine Spaces

The affine case of Cameron-Liebler sets/Boolean degree 1 functions has been investigated in [26, 27, 34, 36] and an asymptotic classification is implied by [73]. Classification results in the affine setting are strictly easier as any Boolean degree d function on the affine $V(n-1, q)$ is a Boolean degree d function on $V(n, q)$. The modular condition in Theorem 14 generalizes to affine spaces for $m = 2$ [26]. Here it reads

$$x(x-1)\frac{q^{n-3}-1}{q-1} \equiv 0 \pmod{2(q+1)},$$

where x only depends on the size of Y. The affine case also has applications in property testing. For instance, Theorem 2.4 in [82] is a weak type of FKN theorem which is used for testing Reed-Muller codes. A similar application for degree d functions can be found in Theorem 1.1 in [91].

Note that there are two natural ways of defining degree in the affine lattice by considering the dual. We are not aware of any investigations of low degree Boolean functions for the dual.

Polar Spaces

Let σ be a non-degenerate, reflexive sesquilinear form on $V(n, q)$. A *classical polar space* consists of the subspaces of $V(n, q)$ which vanish on σ, that is, the isotropic subspaces with respect to σ. For $m \geq 2$, this gives us all the classification problems as for bilinear forms and affine spaces on an induced subgraph of $J_q(n, m)$. See [30] for an investigation for maximal isotropic subspaces and [54] for $2 \leq m \ll n$. Note that at least in the case of maximal isotropic subspaces the term *Cameron-Liebler class* does not correspond to a degree 1 function in this context: the former is defined using design-orthogonality with so-called *spreads*, while degree 1 is defined as in this survey.

As for affine spaces, there are two choices for the degree: Either 1-spaces or maximal isotropic subspaces have rank 1. The above discusses the former. For the latter and $m = 1$, the corresponding families are called *tight sets*. FKN-type and Nisan-Szegedy-type theorems are surely a very challenging problem due to the amount of examples known. See [5] as starting point. An up-to date survey can be found in Chap. 2 in [14].

Acknowledgements The author thanks Yuval Filmus, Alexander Gavrilyuk, Jonathan Mannaert, Dor Minzer, Francesco Pavese, Morgan Rodgers, and Yuriy Tarannikov for discussing topics related to this survey. We also thank the anonymous reviewer for carefully reading this manuscript.

Competing Interests The authors have no conflicts of interest to declare that are relevant to the content of this chapter.

References

1. Abatangelo, V., Korchmaros, G., Larato, B.: Classification of maximal caps in PG(3, 5) different from elliptic quadrics. J. Geom. **57**(1–2), 9–19 (1996)
2. Alon, N., Dinur, I., Friedgut, E., Sudakov, B.: Graph products, fourier analysis and spectral techniques. Geom. Funct. Anal. **14**, 913–940 (2004)
3. Bailey, R.A., Cameron, P.J., Gavrilyuk, A.L., Goryainov, S.V.: Equitable partitions of Latin-square graphs. J. Combin. Des. **27**(3), 142–160 (2018)
4. Bamberg, J., Penttila, T.: Overgroups of cyclic sylow subgroups of linear groups. Comm. Algebra **36**(7), 2503–2543 (2008)
5. Bamberg, J., Kelly, S., Law, M., Penttila, T.: Tight sets and m -ovoids of finite polar spaces. J. Combin. Theory Ser. A **114**(7), 1293–1314 (2007)
6. Barak, B.: Unique Games Conjecture – halfway there? (2024). https://windowsontheory.org/2018/01/10/unique-games-conjecture-halfway-there/, Accessed from 31 October 2024
7. Barak, B., Kothari, P.K., Steurer, D.: Small-set expansion in shortcode graph and the 2-to-2 conjecture. In: 10th Innovations in Theoretical Computer Science Conference (ITCS 2019). Leibniz International Proceedings in Informatics (LIPIcs), vol. 124, pp. 9:1-9:12, Schloss Dagstuhl – Leibniz-Zentrum für Informatik (2019)
8. Beutelspacher, A., Rosenbaum, U.: Projective Geometry: From Foundations to Applications, Cambridge University Press (1998)
9. Block, R.E.: On the orbits of collineation groups. Math. Z. **96**, 33–49 (1967)
10. Blokhuis, A., De Boeck, M., D'haeseleer, J.: Cameron-Liebler sets of k-spaces. Des. Codes Cryptogr. **87**, 1839–1856 (2019)
11. Brouwer, A.E., Cioabă, S.M., Ihringer, F., McGinnis, M.: The smallest eigenvalues of Hamming graphs, Johnson graphs and other distance-regular-graphs with classical parameters. J. Combin. Theory Ser. B **133**, 88–121 (2018)
12. Brouwer, A.E., Cohen, A.M., Neumaier, A.: Distance-regular graphs. Springer, Heidelberg (1989)
13. Brouwer, A.E., Godsil, C.D., Koolen, J.H., Martin, W.J.: Width and dual width of subsets in polynomial association schemes. J. Combin. Th. (A) **102**(2), 255–271 (2003)
14. Brouwer, A.E., Van Maldeghem, H.: Strongly Regular Graphs. Cambridge Univ, Press (2022)
15. Bruen, A.A., Drudge, K.: The construction of Cameron-Liebler line classes in PG$(3, q)$. Finite Fields Appl. **5**(1), 35–45 (1999)
16. Calderbank, R., Kantor, W.M.: The geometry of two-weight codes. Bull. Lond. Math. Soc. **18**, 97–122 (1986)
17. Cameron, P.J., Liebler, R.A.: Tactical decompositions and orbits of projective groups. Linear Algebra Appl. **46**, 91–102 (1982)
18. Camion, P., Carlet, C., Charpin, P., Sendrier, N.: On Correlation-immune functions. In: Feigenbaum, J. (eds), Advances in Cryptology – CRYPTO '91. CRYPTO 1991. Lecture Notes in Computer Science, vol. 576. Springer, Berlin, Heidelberg
19. Chiarelli, J., Hatami, P., Saks, M.: An asymptotically tight bound on the number of relevant variables in a bounded degree boolean function. Combinatorica **40**(2), 237–244 (2020)
20. Cossidente, A., Pavese, F.: Intriguing sets of quadrics in PG$(5, q)$. Adv. Geom. **17**, 339–345 (2017)
21. Cossidente, A., Pavese, F.: New Cameron-Liebler line-classes with parameter $\frac{q^2+1}{2}$. J. Algebr. Comb. **49**, 193–208 (2018)
22. Cossidente, A., Pavese, F.: Cameron-Liebler line classes of PG$(3, q)$ admitting $PGL(2, q)$. J. Combin. Theory Ser. A **167**, 104–120 (2019)
23. Dafni, N., Filmus, Y., Lifshitz, N., Lindzey, N., Vinyals, M.: Complexity measures on symmetric group and beyond. In: 12th Innovations in Theoretical Computer Science Conference (ITCS 2021) (2021)
24. Damasco, J.R.C., Frettlöh, D.: Perfect colourings of regular graphs. Contemp. Math. **764**, 43–56 (2021)

25. De Beule, J., Demeyer, J., Metsch, K., Rodgers, M.: A new family of tight sets in $Q^+(5, q)$. Des. Codes Cryptogr. **78**(3), 655–678 (2016)
26. De Beule, J., Mannaert, J.: A modular equality for Cameron-Liebler line classes in projective and affine spaces of odd dimension. Finite Fields Appl. **82**, 102047 (2022)
27. De Beule, J., Mannaert, J., Storme, L.: Cameron-Liebler k -sets in subspaces and non-existence conditions. Des. Codes Cryptogr. **90**, 633–651 (2022)
28. De Beule, J., D'haeseleer, J., Ihringer, F., Mannaert, J.: Degree 2 boolean functions on grassmann graphs. Electron. J. Combin. **30**(1), P1.31 (2023)
29. De Beule, J., Mannaert, J., Storme, L.: On two non-existence results for Cameron-Liebler k -sets in $\mathrm{PG}(n, q)$ (2024). arXiv:2403.00519 [math.CO]
30. De Boeck, M., D'haeseleer, J.: Equivalent definitions for (degree one) Cameron–Liebler classes of generators in finite classical polar spaces. Disc. Math. **343**(1), 111642 (2020)
31. De Boeck, M., Storme, L., Svob, A.: The Cameron-Liebler problem for sets. Disc. Math. **339**(2), 470–474 (2016)
32. Delsarte, P.: An Algebraic Approach to the Association Schemes of Coding Theory. Supplements. N.V. Philips' Gloeilampenfabrieken, Philips research reports (1973)
33. Delsarte, P.: Association schemes and t -designs in regular semilattices. J. Combin. Th. (A) **20**, 230–243 (1976)
34. D'haeseleer, J., Ihringer, F., Mannaert, J., Storme, L.: Cameron-Liebler k-sets in $\mathrm{AG}(n, q)$. Electron. J. Combin. **28**(4), P4.11 (2021)
35. D'haeseleer, J., Meagher, K., Pantangi, V.R.T.: Cameron–Liebler sets in permutation groups, Algebr. Comb. **7**(4), 1157–1182 (2024)
36. D'haeseleer, J., Mannaert, J., Storme, L., Švob, A.: Cameron-Liebler line classes in $\mathrm{AG}(3, q)$. Finite Fields Appl. **67**, 101706 (2020)
37. Dinur, I., Filmus, Y., Harsha, P.: Sparse juntas on the biased hypercube, TheoretiCS **3** (2024)
38. Dinur, I., Friedgut, E.: Intersecting families are essentially contained in juntas. Combin. Probab. Comput. **18**(1–2), 107–122 (2009)
39. Drudge, K.W.: Extremal sets in projective and polar spaces, Ph.D. thesis, The University of Western Ontario (1998)
40. Ellis, D., Friedgut, E., Pilpel, H.: Intersecting families of permutations. J. Amer. Math. Soc. **24**, 649–682 (2011)
41. Ellis, D., Filmus, Y., Friedgut, E.: A quasi-stability result for dictatorships in S_n. Combinatorica **35**(5), 573–618 (2015)
42. Ellis, D., Filmus, Y., Friedgut, E.: A stability result for balanced dictatorships in S_n. Random Struct. Algorithms **46**(3), 494–530 (2015)
43. Ellis, D., Kindler , G., Lifshitz, N.: An analogue of Bonami's Lemma for functions on spaces of linear maps, and 2-2 Games (2022). arXiv:2209.04243v1 [math.CO]
44. Erdős, P., Ko, C., Rado, R.: Intersection theorems for finite sets. Quart. J. Math. Oxford Ser. (2) **12**, 313–320 (1961)
45. Evans, R.J., Gavrilyuk, A.L., Goryainov, S., Vorob'ev, K.: Equitable 2-partitions of the Johnson graphs $J(n, 3)$ (2022), arXiv:2206.15341 [math.CO]
46. Evra, S., Kindler, G., Lifshitz, N.: Polynomial Bogolyubov for special linear groups via tensor rank (2024). arXiv:2404.00641v1 [math.CO]
47. Feng, T., Momihara, K., Rodgers, M., Xiang, Q., Zou, H.: Cameron-Liebler line classes with parameter $x = \frac{(q+1)^2}{3}$. Adv. Math. **385**, 107780 (2021)
48. Feng, T., Momihara, K., Xiang, Q.: Cameron-Liebler line classes with parameter $x = \frac{q^2-1}{2}$. J. Combin. Theory Ser. A **133**, 307–338 (2015)
49. Filmus, Y.: Friedgut-Kalai-Naor theorem for slices of the Boolean cube. Chic. J. Theoret. Comput. Sci. **14**(1–14), 17 (2016)
50. Filmus, Y.: Orthogonal basis for functions over a slice of the Boolean hypercube. Electron. J. Combin. **23**(1), P1.23 (2016)
51. Filmus, Y.: Boolean functions on S_n which are nearly linear. Discrete Anal. **25**(1–25), 27 (2021)

52. Filmus, Y.: FKN theorems for the biased cube and the slice, Manuscript (2021). https://yuvalfilmus.cs.technion.ac.il/Papers/FKN-Exposition.pdf
53. Filmus, Y.: Junta threshold for low degree boolean functions on the slice. Electron. J. Combin. **30**(1), P1.55 (2023)
54. Filmus, Y., Ihringer, F.: Boolean degree 1 functions on some classical association schemes. J. Combin. Theory Ser. A **162**, 241–270 (2019)
55. Filmus, Y., Ihringer, F.: Boolean constant degree functions on the slice are juntas. Disc. Math. **342**(12), 111614 (2019)
56. Filmus, Y., Lindzey, N.: Murali's basis for the slice (2020), https://yuvalfilmus.cs.technion.ac.il/Manuscripts/Murali.pdf, Accessed: 27 Oct 2024
57. Filmus, Y., Vinciguerra, A.: On the restriction threshold for Boolean functions on the slice, Unpublished Note (2023)
58. Frankl, P., Wilson, R.M.: The Erdős-Ko-Rado theorem for vector spaces. J. Combin. Theory Ser. A **43**(2), 228–236 (1986)
59. Friedgut, E.: On the measure of intersecting families, uniqueness and stability. Combinatorica **28**(5), 503–508 (2008)
60. Friedgut, E., Kalai, G., Naor, A.: Boolean functions whose Fourier transform is concentrated on the first two levels. Adv. Appl. Math. **29**(3), 427–437 (2002)
61. Gavrilyuk, A.L., Goryainov, S.V.: On perfect 2-colorings of Johnson graphs $J(v, 3)$. J. Combin. Des. **21**(6), 232–252 (2013)
62. Gavrilyuk, A., Matkin, I., Penttila, T.: Derivation of Cameron-Liebler line classes. Des. Codes Cryptogr. **86**(1), 231–236 (2018)
63. Gavrilyuk, A.L., Matkin, I.: Cameron-Liebler line classes in PG(3, 5). J. Combin. Des. **26**(12), 563–580 (2018)
64. Gavrilyuk, A., Metsch, K.: A modular equality for Cameron-Liebler line classes. J. Comb. Theory A **127**, 224–242 (2014)
65. Gavrilyuk, A.L., Mogilnykh, I.Y.: Cameron-Liebler line classes in PG(n, 4). Des. Codes Cryptogr. **73**(3), 969–982 (2014)
66. Gavrilyuk , A., Zinoviev, V.A.: Completely Regular Codes in the Johnson Graph, In: Completely Regular Codes in Distance Regular Graphs, CRC Press (2025)
67. Godsil, C., Meagher, K.: Erdős-Ko-Rado Theorems: Algebraic Approaches. Cambridge Studies in Advances Mathematics. Cambridge University Press (2016)
68. Golubev, K.: Graphical designs and extremal combinatorics. Linear Algebra Appl. **604**, 490–506 (2020)
69. Govaerts, P., Penttila, T.: Cameron-Liebler line classes in PG(3, 4). Bull. Belg. Math. Soc. Simon Stevin **12**(5), 793–804 (2005)
70. Graham, R., Leeb, K., Rothschild, B.: Ramsey's theorem for a class of categories. Adv. Math. **8**, 417–433 (1972)
71. Guo, J.: Cameron-Liebler sets in bilinear forms graphs. Des. Codes Cryptogr. **89**, 1159–1180 (2021)
72. Haemers, W.H.: Interlacing eigenvalues and graphs. Linear Algebra Appl. **226–228**, 593–616 (1995)
73. Ihringer, F.: The classification of Boolean degree functions in high-dimensional finite vector spaces. Proc. Am. Math. Soc. **152**(12), 5355–5365 (2024)
74. Ihringer, F.: The chromatic number of the q-Kneser graph for large q. Electron. J. Combin. **26**(1), P1.50 (2019)
75. Ihringer, F.: Remarks on the Erdős matching conjecture for vector spaces. Eur. J. Comb. **94**, 103306 (2021)
76. Ito, T.: Designs in a coset geometry: Delsarte theory revisited. Europ. J. Comb. **25**(2), 229–238 (2004)
77. Kantor, W.M.: On incidence matrices of finite projective and affine spaces. Math. Z. **124**, 315–318 (1972)
78. Keller, N., Klein, O.: A structure theorem for almost low-degree functions on the slice. Israel J. Math. **240**, 179–221 (2020)

79. Khot, S., Minzer, D., Safra, M.: Pseudorandom sets in Grassmann graph have near-perfect expansion. Ann. of Math. (2) **198**, 1–92 (2023)
80. Kiermaier, M., Mannaert, J., Wassermann, A.: The degree of functions in the Johnson and q-Johnson schemes (2024). arXiv:2405.07572v1 [math.CO]
81. Kindler, G., Safra, S.: Noise-resistant Boolean functions are juntas, Manuscript (2002)
82. Kaufman, T., Minzer, D.: Improved Optimal Testing Results from Global Hypercontractivity (2022). arXiv:2202.08688v1 [cs.CC]
83. Krotov, D.S., Valyuzhenich, A.A.: On degree-3 and $(n-4)$ -correlation-immune perfect colorings of n -cubes. Discrete Math. **347**, 114138 (2024)
84. Lovász, L.: On the Shannon capacity of a graph. IEEE Trans. Inform. Theory **25**(1), 1–7 (1979)
85. Martin, W.J.: Completely regular designs of strength one. J. Algebraic Combin. **3**(2), 177–185 (1994)
86. Matkin, I.: Cameron-Liebler line classes in $PG(n, 5)$, Trudy Inst. Mat. i Mekh. UrO RAN **24**(2), 158–172 (2018)
87. Metsch, K.: The non-existence of Cameron-Liebler line classes with parameter $2<x \leq q$. Bull. Lond. Math. Soc. **6**, 991–996 (2010)
88. Metsch, K.: An improved bound on the existence of Cameron-Liebler line classes. J. Comb. Theory A **121**, 89–93 (2014)
89. Metsch, K.: A gap result for Cameron-Liebler k -classes. Discrete Math. **340**(6), 1311–1318 (2017)
90. Meyerowitz, A.D.: Cycle-balanced partitions in distance-regular graphs. J. Combin. Inform. System Sci. **17**(1–2), 39–42 (1992)
91. Minzer, D., Zheng, K.: Optimal Testing of Generalized Reed-Muller Codes in Fewer Queries (2023). arXiv:2304.05598v1 [cs.CC]
92. Nisan, N., Szegedy, M.: On the degree of Boolean functions as real polynomials. Comput. Complex. **4**(4), 301–313 (1994)
93. O'Donnell, R.: Analysis of Boolean Functions, Cambridge University Press (2014)
94. Rodgers, M.: On Some New Examples of Cameron–Liebler Line Classes, Ph.D. thesis, University of Colorado, Denver (2012)
95. Rodgers, M., Storme, L., Vansweevelt, A.: Cameron-Liebler k -classes in $\mathrm{PG}(2k+1, q)$. Combinatorica **38**(3), 739–757 (2018)
96. Roos, C.: On antidesigns and designs in association schemes. Delft Prog. Rep. **7**(2), 98–109 (1982)
97. Srinivasan, M.K.: Symmetric chains, Gelfand-Tsetlin chains, and the Terwilliger algebra of the binary Hamming scheme. J. Algebraic Combin. **34**, 301–322 (2011)
98. Steinerberger, S.: Generalized designs on graphs: sampling, spectra, symmetries. J. Graph Theory **93**(2), 253–267 (2020)
99. Tallini Scafati, M.: Calotte di tipo (m, n) in uno spazio di Galois $S_{r,q}$, Atti Accad. Peloritana Pericolanti Cl. Sci. Fis. Mat. Natur., Rendiconti, Serie 8 (**53**)(1–2), 71–81 (1972)
100. Tallini Scafati, M.: Sui k–insiemi di uno spazio di Galois $S_{r,q}$ a due soli caratteri nella dimensione d, Atti Accad. Peloritana Pericolanti Cl. Sci. Fis. Mat. Natur., Rendiconti, Serie 8 **60**(6), 782–788 (1976)
101. Tanaka, H., Tokushige, N.: Extremal problems for intersecting families of subspaces with a measure (2024). arXiv:2404.17385v1 [math.CO]
102. Tarannikov, Y.: On resilient Boolean functions with maximal possible nonlinearity. In: Proceedings of Indocrypt'2000, Lecture Notes in Computer Science, vol. 1977, pp. 19–30 (2000)
103. Tarannikov, Y.V., Kirienko, D.P.: Spectral analysis of high order correlation immune function. In: Proceedings of International School-seminar "Synthesis and Complexity of Controlling Systems", Moscow (2001)
104. Tarannikov, Y., Korolev, P., Botev, A., Coefficients, A., Immunity, C.: of Boolean Functions, Advances in Cryptology - ASIACRYPT,: Lecture Notes in Computer Science, vol. 2248. Springer, Berlin, Heidelberg (2001)

105. Taylor, D.E.: The Geometry of the Classical Groups. Heldermann, Berlin (1992)
106. Valyuzhenich, A.: An upper bound on the number of relevant variables for Boolean functions on the Hamming graph (2024). arXiv:2404.10418v1 [math.CO]
107. Vanhove, F.: Incidence geometry from an algebraic graph theory point of view. Ph.D. thesis, University of Ghent (2011)
108. Vorob'ev, K.: Equitable 2-partitions of Johnson graphs with the second eigenvalue (2020). https://arxiv.org/pdf/2003.10956 [math.CO]
109. Wellens, J.: Relationships between the number of inputs and other complexity measures of Boolean functions. Discrete Anal. **19** (2022)
110. Zhu, Y.: Optimal and extremal graphical designs on regular graphs associated with classical parameters. Des. Codes Cryptogr. **91**(8), 2737–2754 (2023)
111. Zverev, A.: On the structure of the spectrum support of boolean functions, boolean functions. In: Cryptology and Information Security, NATO Sci. Peace Secur. Ser. D Inf. Commun. Secur., vol. 18 (2008)

Results and Problems on Equitable Coloring of Graphs

H. A. Kierstead, Alexandr Kostochka, and Zimu Xiang

Abstract A proper coloring of vertices of a graph is *equitable* if the sizes of any two color classes differ by at most 1. Such colorings have many applications and are interesting by themselves. In this paper, we discuss the state of art and unsolved problems on equitable coloring and its list versions.

Mathematics Subject Classification 05C07, 05C15, 05C35

1 Introduction

Consider the following idealized problem. A service company employs a set W of k equivalent workers. Each day the company must complete a set V of n orders. The parameters W and V may vary from day-to-day. For various reasons (confidentiality, custom, location, etc.) some pairs of orders should not or cannot be handled by the same worker. In this case the pairs are *incompatible*. Let $G = (V, E)$ be the *incompatibility* graph on V, where $xy \in E$ iff the orders x and y are incompatible. The company would like to assign orders to workers so that no two incompatible orders are assigned to the same worker. In graph theoretic terms, this assignment should be a *proper coloring* $f : V \to W$. But the company may have the additional

Alexandr Kostochka. Research is supported in part by NSF Grant DMS-2153507 and by NSF RTG Grant DMS-1937241.
Zimu Xiang. Research is supported in part by the Campus Research Board Award RB24000 of the University of Illinois Urbana-Champaign.

H. A. Kierstead
Arizona State University Tempe, Tempe, USA
e-mail: kierstead@asu.edu

A. Kostochka (✉) · Z. Xiang
University of Illinois at Urbana–Champaign, Urbana, USA
e-mail: kostochk@illinois.edu

Z. Xiang
e-mail: zimux2@illinois.edu

G. O. H. Katona et al. (eds.), *Sum(m)it280*, Bolyai Society Mathematical Studies 32,
https://doi.org/10.1007/978-3-032-18810-6_12

requirement that the numbers of orders assigned to any two workers should differ by at most one. In other words, the proper coloring f should be *equitable*. Suppose $|V| = n$ and $|W| = k$.

Now the problem is whether G has an equitable k-coloring, that is, a (proper) k-coloring f such that $\lfloor \frac{n}{k} \rfloor \leq |f^{-1}(w)| \leq \lceil \frac{n}{k} \rceil =: m$ for all colors/workers $w \in W$. Notice that this is a decision problem, not an optimization problem; the company already employs k workers. Being equitable means that any two of the k workers handle the same number of jobs (± 1). However there are other interesting questions. For instance, suppose m is a limit on the number of orders that a worker can handle. Then on days that $n \geq km$ the company might hire additional workers. If G has an equitable k-coloring, does it have an equitable k'-coloring for $k' > k$? Maybe the company employs a fixed set of $k = 4$ workers, but takes as long as is needed to complete all orders. When can orders be equitably assigned to these four workers?

This example hints on possible applications of equitable coloring and similar notions in graph theory. Among such applications are scheduling in communication systems, construction timetables, mutual exclusion scheduling problem, and round-a-clock scheduling; see, e.g. [5, 22, 23, 48, 54, 62, 63]. For example, in [63] equitable coloring modeled scheduling weekly routes of garbage trucks.

This concept is also useful in studying extremal combinatorial and probabilistic problems. Alon and Füredi [1] used it to study existence of some spanning subgraphs in random graphs. Alon and Yuster [2] applied results on equitable coloring to the problem of the existence of H-factors in dense graphs. Janson and Ruciński [25], Pemmaraju [60] and Janson [24] used equitable colorings to derive deviation bounds for sums of dependent random variables with limited dependency. Rödl and Ruciński [61] used results on equitable coloring to give a simpler proof of the Blow-Up Lemma. Kaul and Jacobson [26] used equitable coloring of a dependency graph of random variables to generalize a classic bound of Arnold and Groenveld [3] on order statistics of independent random variables.

1.1 Framing the Theory

Here are two informative negative examples.

Let S be the star on n vertices. It is 2-colorable, but the maximum size of a color class containing its root is one. Now by the definition of an equitable coloring, every color class has size at most two. Thus, for S to have an equitable k-coloring, we need $k > n/2 > \Delta(S)/2$. More generally, if v is a vertex of a graph G on n vertices such that the maximum size of an independent set containing v is m then every equitable coloring of G uses more than $\frac{n}{m+1}$ colors.

Let $K = K_{t,t}$. Now K has an equitable 2-coloring. If K has an equitable t-coloring, then every class must have size two and be contained in one of the two parts of K. Thus t is even. So when t is odd, increasing the number of required colors from 2 to t makes it impossible to equitably color K.

Early positive results followed from the complementary version of the problem. Let $\overline{G}$ be the complement of G. Now a proper coloring of G is a partition of $V(\overline{G})$ into cliques. Suppose for simplicity that $n = km$. Then

$$\Delta(G) < k \Leftrightarrow \Delta(G) + 1 \leq k \Leftrightarrow \delta(\overline{G}) \geq n - \Delta(G) - 1 = (1 - \frac{1}{m})n.$$

If $m = 2$ and $\Delta(G) < k$ then $\delta(\overline{G}) \geq \frac{n}{2}$, so Dirac's Theorem [16] implies that $V(\overline{G})$ has a k-partition into 2-cliques. Thus G has an equitable k-coloring. Corradi and Hajnal [10] proved the case $m = 3$: if $\delta(\overline{G}) = \frac{2}{3}n$ then $V(\overline{G})$ has a k-partition into 3-cliques. Then Erdős conjectured:

Conjecture (Erdős [18] *1964*) If H is a graph on $n = km$ vertices with $\delta(H) \geq n(1 - \frac{1}{m})$, then $V(H)$ has a k-partition into m-cliques.

Hajnal offered 1000 cups of coffee for a proof of the conjecture. Grünbaum [20] proved the case $m = 4$, and reformulated the conjecture as a statement about equitable coloring.

Conjecture (*Grünbaum* [20], *1968*) Every graph G with maximum degree $\Delta(G) < k$ has a k-equitable coloring.

The Grünbaum's version includes the case that k does not divide n; this case was handled by Larman [50] (see the second paragraph of Subsect. 2.1). By the end of the sixties, the state of the art was as follows.

Theorem 1 *A graph G on n vertices with maximum degree Δ is equitably k-colorable if $\Delta < k$ and either*

(a) (Dirac *[16]*, 1952) $n/2 \leq k$,
(b) (Corradi and Hajnal *[10]*, 1963) $n/3 \leq k$, *or*
(c) (Grünbaum *[20]*, 1968) $n/4 \leq k$.
(d) (Zelinka *[69]*, 1966) $k \leq 3$.

Finally, in 1970 Hajnal and Szemerédi proved Erdős' Conjecture (in Grünbaum's form).

Theorem 2 (Hajnal and Szemerédi [21], 1970) *Every graph with $\Delta < k$ is equitably k-colorable.*

In fact, most of the mathematical applications of equitable coloring that we know use the Hajnal-Szemerédi Theorem (mostly in the complementary form).

Another line of research considers equitable k-coloring of sparse graphs G with $k < \Delta(G)$. Early results in this area include the theorem of Meyer [54] from 1973 on equitable k-coloring of trees T for $k > \Delta(T)/2$ and the following result of Bollobás and Guy [6] on equitably 3-coloring trees.

Theorem 3 (Bollobás and Guy [6], 1983) *Let T be a tree on n vertices. If $n \geq |V(T)| \geq 3\Delta(T) - 8$ or $n = 3\Delta(T) - 10$, then T has an equitable 3-coloring.*

Results in this spirit are discussed in Sect. 5.

1.2 Organization

In Sect. 2, our focus is on results for equitable k-coloring graphs G with k close to maximum degree. We provide a setup for proving results related to the Hajnal-Szemerédi Theorem, give a complete proof of the theorem, and show that there is a polynomial time algorithm for constructing the coloring. We generalize Theorem 2 by showing that if the Ore-degree $\Theta(G) := \max\{d(x) + d(y) : xy \in E(G)\}$ is less than $2k$, then G has an equitable k-coloring, and conjecture that there is a polynomial algorithm for finding such a coloring. Finally we discuss a strengthening of Theorem 2 to directed graphs.

In Sect. 3, we discuss the Chen-Lih-Wu Conjecture and partial results towards it. The conjecture is wide open, but is proved for some interesting classes of graphs. We also discuss an equivalent conjecture and some harder conjectures, including the Ore-version of the conjecture.

In Sect. 4, we consider problems on equitable list coloring including list versions of the Hajnal-Szemerédi Theorem and Chen-Lih-Wu Conjecture. We also introduce strongly equitable (SE) list coloring and compare results and problems on equitable list coloring and SE list coloring.

In Sect. 5, we discuss equitable coloring (and list equitable coloring) of "sparse" graphs, when we need significantly fewer colors than the maximum degree of the graph. Finally, in Sect. 6, we list known results on some variations of equitable coloring.

1.3 Notation

We view a graph G as a pair (V, E) where $V = V(G)$ denotes the set of its vertices and $E = E(G)$ denotes the set of its edges. We let $|G| = |V(G)|$ and $\|G\| = |E(G)|$. Further, $\Delta(G)$ is the maximum degree of G, and $\delta(G)$ is the minimum degree of G.

For disjoint $X, Y \subseteq V(G)$, $G[X]$ denotes the subgraph of G induced by X, $E(X, Y)$ denotes the set of edges in G connecting X with Y and $\|X, Y\| = \|X, Y\|_G := |E(X, Y)|$. If $X \cap Y \neq \emptyset$, then the edges of $G[X \cap Y]$ belong to $E(X, Y)$ and count twice in $\|X, Y\|$. If $X = \{x\}$ we may simply write $\|x, Y\|$ for $\|X, Y\|$. The degree of x in G is noted by $d(x) = d_G(x) := \|x, V(G) - x\|_G$.

We use Greek letters for colors and Roman letters for vertices. For a coloring $f : V \to C$ and $\gamma \in C$, set $\widetilde{\gamma} := f^{-1}(\gamma)$. For $v \in \widetilde{\gamma}$, set $\widetilde{v} := \widetilde{\gamma}$. Let $f|_A$ be the function obtained from f by restricting its domain to A.

For a positive integer n, set $[n] := \{1, \ldots, n\}$. For a set X and element y, set $X + y := X \cup \{y\}$ and $X - y := X \setminus \{y\}$.

2 Theorems Related to Bounded Maximum Degree

We begin this section with a short proof of Theorem 2. This proof is the basis for proofs of several stronger theorems, including the existence of an efficient polynomial-time algorithm for equitably k-coloring graphs when $k > \Delta$.

2.1 Setup for Equitable Coloring in Terms of Maximum Degree

In this subsection, we organize a proof of Theorem 2 and many of its generalizations. The notation is different, but the setup matches the original proof until Case 2, where it takes a major short cut.

Suppose $G = (V, E)$ is a graph with $\Delta := \Delta(G)$, k is an integer with $k > \Delta$, and we want to equitably color G with exactly k colors. If $|G| = mk - p$, where $1 \leq p \leq k - 1$, then set $G' := G + K_p$. Now $|G'|$ is divisible by k, $\Delta(G') \leq \max\{\Delta, k - 2\}$, and the restriction of any equitable k-coloring of G' to G is an equitable k-coloring of G. So we may assume $n := |G| = km$. Thus every color class of an equitable k-coloring will have size m.

Argue by induction on $\|G\|$. If G has no edges, we are done. Else let $v \in V$ be non-isolated, and set $G' := G - E(v, V)$. By induction there is an equitable k-coloring f_0 of G'. If f_0 is a proper coloring of G, we are done. Else some neighbors of v are in $\widetilde{v}$. As $d(v) < k$, there is a class $\widetilde{\beta}$ with no neighbors of v. Shift v from $\widetilde{v}$ to $\widetilde{\beta}$, obtaining a new coloring f with one *small* class $\widetilde{\alpha}_0 := \widetilde{v} - v$, one *large* class $\widetilde{\beta}_0 := \widetilde{\beta} + v$, and all other classes with size m. Such a coloring is *nearly equitable*.

To finish coloring G, we will do a sequence of *shift* operations that improve our nearly equitable coloring, until finally it becomes equitable. First define an auxiliary directed graph $\mathcal{H} = \mathcal{H}(f)$ on the colors of f so that $\sigma\tau$ is a directed edge if and only if $\|y, \widetilde{\tau}\| = 0$ for some vertex $y \in \widetilde{\sigma}$.[1] In this case we say that y *witnesses* $\sigma\tau$. For edge $\sigma\tau \in \mathcal{H}$ with witness w, let shift$(\widetilde{\sigma}\widetilde{\tau}, w)$ be the operation that produces a new proper coloring by *shifting* w from $\widetilde{\sigma}$ to $\widetilde{\tau}$. We also write shift$(\widetilde{\sigma}\widetilde{\tau}, w)$ for this coloring. When the particular witness is unimportant, we may simply write shift$(\widetilde{\sigma}\widetilde{\tau})$. We say that color σ_0 *reaches* color σ_k and that σ_k is *reached by* σ_0 in $\mathcal{H}$, if there is a directed σ_0,σ_k-path $\mathcal{P} = \sigma_0 \ldots \sigma_k \subseteq \mathcal{H}$. In this case, put shift$(\mathcal{P}) :=$ shift$(\widetilde{\sigma}_{k-1}\widetilde{\sigma}_k) \circ \cdots \circ$ shift$(\widetilde{\sigma}_0\widetilde{\sigma}_1)$. This reduces the size of $\widetilde{\sigma}_0$ by one, increases the size of $\widetilde{\sigma}_k$ by one, and maintains all other class sizes.

[1] As the vertices of $\mathcal{H}$ are colors, not color classes, it is possible that different colorings have the same $\mathcal{H}$.

Let $\mathcal{A}$ be the set of colors that reach α_0 in $\mathcal{H}$, and set $\mathcal{B} := V(\mathcal{H}) \smallsetminus \mathcal{A}$, $a := |\mathcal{A}|$, $b := |\mathcal{B}|$, $A := \bigcup\{\widetilde{\alpha} : \alpha \in \mathcal{A}\}$, and $B := V \smallsetminus A$. So $k = a + b$ and $|B| = bm + 1$.

Argue by a secondary induction on b, i.e., reverse induction on a.

CASE 0: $\beta_0 \in \mathcal{A}$; this includes the case $b = 0$. Let $\mathcal{P}$ be a β_0, α_0-path in $\mathcal{H}[\mathcal{A}]$. Now shift$(\mathcal{P})$ is an equitable k-coloring.

Otherwise, $\beta_0 \in \mathcal{B}$. A color $\alpha \in \mathcal{A}$ is *terminal*, if every color $\gamma \in \mathcal{A} - \alpha$ reaches α_0 in $\mathcal{H} - \alpha$. Let $\mathcal{A}'$ be the set of terminal colors, $a' := \left|\mathcal{A}'\right|$, and $A' := \bigcup\{\widetilde{\alpha} : \alpha \in \mathcal{A}'\}$. An edge $wz \in E(A, B)$ is *solo* if $f(w) \in \mathcal{A}'$ and $\|z, \widetilde{w}\| = 1$. In this case, w is a *solo root* of z, and z is a *solo leaf* of w. For $z \in B$, let s_z be the number of solo roots (in A') of z. Then

$$\|z, A\| \geq a + a' - s_z \geq a. \tag{1}$$

CASE 1: Some solo root $w \in \widetilde{\omega}$ witnesses an edge $\omega\alpha$ of $\mathcal{H}[\mathcal{A}]$. As $\omega \in \mathcal{A}'$, some α, α_0-path $\mathcal{P}$ avoids ω. Let y be a solo leaf of w. Set $f :=$ shift$(\widetilde{y}\widetilde{\omega}, y) \circ$ shift$(\mathcal{P}) \circ$ shift$(\widetilde{\omega}\widetilde{\alpha}, w)$. Now $f|_{A+y}$ is equitable. Set $G^* := B - y$. By (1), $\Delta(G^*) < b$. By primary induction, G^* has an equitable b-coloring g. Thus $f|_{A+y} \cup g$ is an equitable k-coloring of G.

Now assume no solo root witnesses an edge of $\mathcal{H}[\mathcal{A}]$.

CASE 2: Some solo root $w \in \widetilde{\alpha}$ has two nonadjacent solo leaves $y, z \in B$; say $z \in \widetilde{\beta}$. Set $B^* := B - y$. By (1), $\Delta(B^*) < b$. Using induction, equitably b-color B^*. Now $\|w, B^*\| < b$, since w witnesses no edge of $\mathcal{H}[\mathcal{A}]$ and $wy \in E$. Shift w to some class $\widetilde{\gamma} \subseteq B^*$, and shift y to $\widetilde{\alpha} - w$. This results in a new nearly equitable k-coloring f'. Now z witnesses that $\beta\alpha \in \mathcal{H}(f')$, so $\beta \in \mathcal{A}(f') \supset \mathcal{A}$, and we are done by secondary induction.

2.2 Finishing the Proof

In this subsection we finish the proof of Theorem 2 and discuss how it differs from the original proof. Using Sect. 2.1, it suffices to show that one of Cases 0–2 holds.

Proof of Theorem 2 Assume that Case 0 and Case 1 fail. Then $b \geq 1$ and no solo root witnesses an edge of $\mathcal{H}[\mathcal{A}]$. So for all solo roots x,

$$\text{(a) } \|x, A\| \geq a - 1 \text{ and (b) } \|x, B\| \leq d(x) - a + 1 \leq b. \tag{2}$$

Order $\mathcal{A}$ as $\alpha_0, \ldots, \alpha_{a-1}$ so that for all $i \in [a-1]$, there is $j < i$ with $\alpha_i\alpha_j \in \mathcal{H}$; then α_{a-1} is terminal. Let l be the least index such that α_j is terminal whenever $l < j < a$. Now $a' \geq a - l - 1$, and there is $t > l$ such that all nontrivial α_t, α_0-paths meet α_l. As α_t has no out-neighbors before α_l,

$$\text{(a) } \|w, A\| \geq l - 1 \geq a - a' - 1 \text{ and (b) } \|w, B\| \leq b + a' \tag{3}$$

for all $w \in \widetilde{\alpha}_t$. Let S be the set of solo roots in $\widetilde{\alpha}_t$, and set $N := N(S) \cap B$. Then $\|z, S\| \geq 1$ for all $z \in N$ and $\|z, \widetilde{\alpha}_t\| \geq 2$ for all $z \in B \smallsetminus N$. By (2), $|N| \leq b|S|$, so

$$\begin{aligned} |N| + 2|B \smallsetminus N| \leq \|S \cup (\widetilde{\alpha} \smallsetminus S), N \cup (B \smallsetminus N)\| &\leq_{(2,3)} b|S| + (b + a')(m - |S|) \\ 2bm + 2 - b|S| &\leq_{(2)} b|S| + bm - b|S| + a'(m - |S|) \\ b(m - |S|) &< a'(m - |S|) \\ b &< a'. \end{aligned} \tag{4}$$

Let $z \in B$. As $a + b - 1 \geq \Delta \geq d(z) \geq_{(1)} a + a' - s_z + \|z, B\|$,

$$s_z \geq a' - b + 1 + \|z, B\|. \tag{5}$$

Pick a maximum independent set $I \subseteq B$. By maximality, $|I| \geq m$. Also $|I| + \sum_{z \in I} \|z, B\| \geq |B| = bm + 1$. Now

$$\sum_{z \in I} s_z \geq_{(5)} \sum_{z \in I} (a' - b + \|z, B\| + 1) > m(a' - b) + bm =_{(4)} a'm = |A'|. \tag{6}$$

Thus some vertex in A' has two (nonadjacent) solo leaves in I, and so Case 2 holds. □

This proof uses the failure of Case 1 to quickly derive $a' > b$. From this we easily establish that Case 2 holds—some solo root has two nonadjacent solo neighbors—and quickly finish. In the original proof, a much more complicated argument used the failure of Case 1 to derive $a' > 2b$. From this another complicated argument proved that Case 2 failed—no solo vertex has nonadjacent solo vertices. Finally, a nontrivial effort established a contradiction.

2.3 A Polynomial-Time Algorithm

In 2007, Szemerédi mentioned to Kostochka that the original proof of Theorem 2 does not provide a polynomial-time algorithm for finding the equitable coloring, but that he had recently found one. Thinking about this led the first two authors to a new proof of the theorem, another polynomial-time algorithm, and other new results [28, 29]. Combining our ideas with those of Szemerédi and his student, Mydlarz, we developed a simple, efficient algorithm for finding such a coloring.

Theorem 4 (Kierstead, Kostochka, Mydlarz and Szemerédi [38], 2010) *Every graph on n vertices with maximum degree less than k can be equitably k-colored in $O(kn^2)$ time using at most $2kn$ shifts.*

Proof We will prove the bound on the number of shifts; the full result follows from standard data-structure methods. For this we modify the proof of Theorem 2 in two ways. First, let $\mathcal{B}' \subseteq \mathcal{B}$ be the set of colors that can be reached from β_0, $B' := \bigcup\{\widetilde{\beta} : \beta \in \mathcal{B}'\}$, and $b' := |\mathcal{B}'|$. Now choosing $I \subseteq B'$ and using $\|z, B\| \geq \|z, B'\| + b - b'$ for $z \in B' \subseteq B$ in (6) yields

$$\sum_{z\in I} s_z \geq_{(5)} \sum_{z\in I}(a' - b + \|z, B'\| + b - b' + 1) > m(a' - b') + b'm =_{(4)} a'm = |A'|,$$

so some solo root in A' has two (nonadjacent) solo leaves in $I \subseteq B'$. Thus we can obtain the first recoloring of $\mathcal{B}$ in Case 2 with at most b' shifts. Secondly, we prove a stronger technical statement that will allow us to avoid the secondary induction on b. Instead we show by induction on k that a nearly equitable k-coloring can be transformed to an equitable k-coloring using $2k - 1$ shifts.

Lemma 5 *Let G be a graph on $n = km$ vertices with a nearly equitable k-coloring f. If*

$$d(x) < k \text{ for all } x \in A' \cup B, \tag{*}$$

then G can be equitably k-colored in $2k - 1$ shifts.

Proof Argue as in the previous proof, but by induction on k. In Case 0, shift($\mathcal{P}$) is an equitable k-coloring G that uses $k - 1$ shifts. In Case 1, we use a shifts to equitably a-color $G[A + y]$ and produce a nearly equitable b-coloring of $G[B - y]$. Using induction, we equitably b-color $G[B - y]$ in $2b - 1$ shifts. So consider Case 2. As Case 0 fails, (4) implies $1 \leq b < a' \leq a$, so $a \geq 2$. Now α_0 is nonterminal, so $b < a' < a$. By integrality, $b + 2 \leq a$.

Let w be a solo root with nonadjacent solo leaves $y, z \in B'$; say $f(w) = \alpha$. Then there is a $\beta_0, f(y)$-path $\mathcal{P} \subseteq \mathcal{H}$. As Case 1 fails, some $x \in \widetilde{\alpha} - w$ is the first witness of an α, α_0-path $Q \subseteq \mathcal{H}$. Now $f' :=$ shift$(\widetilde{y}(\widetilde{\alpha} - w), y) \circ$ shift$(Q, x) \circ$ shift$(\mathcal{P})$ is an equitable coloring of $G - w$ with a small class $\widetilde{\alpha}(f') = \widetilde{w} - w - x + y$. We have used at most $a - 1 + b \leq 2a - 3$ shifts to obtain f'. Set $B^* := B \cup \widetilde{\alpha}(f')$, $G^* := G[B^*]$, and $A^* = V \setminus B^*$. Every vertex $v \in B^* \setminus \widetilde{\alpha}(f')$ satisfies $\|v, A^*\| \geq a - 1$ (some vertices of $\widetilde{\alpha}(f')$ might not). So we can shift w to some class $\widetilde{\beta}$ to obtain a nearly equitable $(b + 1)$-coloring f^* of $B^* + w$ with big class $\widetilde{\beta}$. Now z witnesses $\beta\alpha \in \mathcal{H}(f')$, so α is not terminal in f^*, and (*) holds for G^*. By induction, we can extend f^* to an equitable $(b + 1)$-coloring of B^*, using at most $2b + 1$ shifts. In total we have used at most $(2a - 3) + 1 + 2b + 1 = 2k - 1$ shifts to equitably k-color G. □

This completes the proof of Theorem 4. □

2.4 *Equitable Coloring in Terms of Maximum Ore-Degree*

The Ore-degree of an edge xy is the sum $\Theta(xy) := d(x) + d(y)$. The maximum Ore-degree of a graph G is $\Theta(G) := \max\{\Theta(xy) : xy \in E\}$. Kostochka and Yu [45, 46] conjectured that every graph G with $\Theta(G) + 1 < 2k$ is equitably k-colorable; this would strengthen Theorem 2. We proved a stronger version.

Theorem 6 *([29],* **2008)** *If $\Theta(G) < 2k$ then G has an equitable k-coloring.*

Proof Arguing by induction, we continue to use the same setup. In particular, we can assume $|G| = km$ and there is a nearly equitable coloring f of G'. For the latter, let $G' = G - E(v, V)$, where $d(v) \leq k$. Also $\mathcal{H}, \mathcal{A}, \mathcal{A}', \mathcal{B}$, etc. are defined as before. These definitions depend on f, not any hypothesis about maximum degree, so (1) and (2)(a) still hold. By (1), $\Theta(G[B]) < 2b$. We also need the subset $\mathcal{A}'' \subseteq \mathcal{A}'$ of colors that are separated from α_0 in $\mathcal{H}$ by the last nonterminal color α_l. Set $a'' := |\mathcal{A}''|$ and $A'' := \bigcup\{\widetilde{\alpha} : \alpha \in \mathcal{A}''\}$. As in (3)(a), for all $x \in A''$,

$$\|x, A\| \geq a - a'' - 1. \tag{3*}$$

For $y \in B$ let s^*_y be the number of solo neighbors of y in A''. As in (1),

$$\|y, A\| \geq a + a'' - s^*_y. \tag{1*}$$

Using the secondary induction, it is enough to find a nearly equitable k-coloring f' with $a(f') > a$. If Case 0 or Case 1 holds, we finish as before, so assume both fail. The following lemma allows us to deal with Case 2 and to make progress, even when all cases fail.

Lemma 7 *Suppose Cases 0 and 1 fail and wy is a solo edge with $f(w) := \alpha \in \mathcal{A}''$. Then G' has a nearly equitable k-coloring f' with $y \in \widetilde{\alpha}(f') - w$, $f'(x) = f(x)$ for all $x \in A - w$, $\mathcal{A} \subseteq \mathcal{A}(f')$, and $\mathcal{A}''(f') \subseteq \mathcal{A}''$. Moreover, if y' is another solo leaf of w with $yy' \notin E$, then Case 0 holds for f' or $a(f') > a$.*

Proof Shift y to $\widetilde{\alpha} - w$. By induction, $B - y$ has an equitable b-coloring g. By the failure of Case 1, (2), and (1), $\|wy, A + y\| \geq 2a$, so $\|w, B - y\| \leq 2b - 1$. If $\|w, \widetilde{\beta}(g)\| = 0$ for some $\beta \in \mathcal{B}$ then shift w to $\widetilde{\beta}(g)$; else $d(w) \geq a + b$, and $N(w) \cap \widetilde{\beta}(g) = \{z\}$ for some $\beta \in \mathcal{B}$ and $z \in \widetilde{\beta}(g)$. Then $d(z) < a + b$ and $\|z, A\| \geq a$, so $\|z, B \cup \widetilde{\alpha} - w\| \leq b - 1$. Thus $\|z, \widetilde{\gamma}(g)\| = 0$ for some $\gamma \in \mathcal{B} + \alpha$ where (i) $\widetilde{\gamma}(g) = \widetilde{\alpha}(g) = \widetilde{\alpha} + y - w$ or (ii) $\gamma \in \mathcal{B}$. Shift z to $\widetilde{\gamma}(g)$ and w to $\widetilde{\beta}(g)$. Let f' be the final coloring obtained from f and g. If (i) holds then Case 0 holds for f'. Else if wy' is solo and $yy' \notin E$, then y' witnesses $f(y')\alpha \in \mathcal{H}(f')$, so $a(f') > a$.

Both w and y witness no edges of $\mathcal{H}$ which are directed toward $\mathcal{A}$, so $\mathcal{A} \subseteq \mathcal{A}(f')$ and $E_{\mathcal{H}}(\mathcal{A}'', \mathcal{A} \smallsetminus \mathcal{A}'') = E_{\mathcal{H}(f')}(\mathcal{A}'', \mathcal{A} \smallsetminus \mathcal{A}'')$. It follows that α_l separates $\mathcal{A}''$ from α_0 in $\mathcal{H}(f')$. But some terminal colors of $\mathcal{H}$ could be nonterminal colors in $\mathcal{H}(f')$. Anyway, $\mathcal{A}''(f') \subseteq \mathcal{A}''$. □

We call the operation described in Lemma 7 a wy-swap and write $f' :=$ swap$(wy) :=$ swap$_f(wy)$. Assume Case 2 fails for f. In the remainder of the proof we show that some sequence of swaps ends with Case 0, 1 or 2. Define a function $\mu : E(A'', B) \to \mathbb{Q}$ by $xz \mapsto \frac{b}{\|x,B\|}$, and for $z \in B$, let $\mu(z) = \sum_{x \in N(z) \cap A''} \mu(xz)$. Now

$$a''mb = |A''|b = \sum_{x \in A''} b = \sum_{xz \in E(A'',B)} \mu(xz) = \sum_{z \in B} \mu(z).$$

By the Pigeonhole principle and the failure of Case 0, there is $y \in B$ with $\mu(y) < a''bm/|B| < a''$. Pick $x \in A'' \cap N(y)$ with $\|x, B\|$ maximum, and if $s^*_y \geq 1$, let $w \in A''$ be a solo neighbor of y with $\|w, B\|$ maximum. So

$$a'' > \mu(y) \geq \frac{s^*_y b}{\|w, B\|} + \frac{2(a'' - s^*_y)b}{\|x, B\|}, \text{ where } \|w, B\| := 1 \text{ if } s^*_y = 0. \tag{7}$$

Now (i) $\|w, B\| \geq b + 1$ or (ii) $\|x, B\| \geq 2b + 1$: else $\mu(y) \geq s^*_y + (a'' - s^*_y) = a''$, a contradiction.

Lemma 8 *Vertex y is incident to a solo edge from A''.*

Proof Suppose not. Then $s^*_y = 0$, and by (7), $\|x, B\| \geq 2b + 1$. This yields the contradiction

$$\Theta(xy) = d(x) + d(y) \geq_{(3*)(1*)} (a - a'' - 1 + 2b + 1) + (a + a'') = 2a + 2b = 2k. \qquad \square$$

First suppose $a'' > b$. Set $f' :=$ swap(wy). We claim that one of the cases holds for f' or (iii) $\|A''(f'), B(f')\| < \|A'', B\|$. Suppose (iii) fails. Then $\|w, B\| + \|y, A''\| \leq \|w, A''\| + \|y, B\|$. Counting wy twice yields,

$$2k \geq \|wy, (A \setminus A'') \cup A'' \cup B\| \geq 2(a - a'') + \|w, B\| + \|y, A''\| + \|w, A''\| + \|y, B\|$$
$$a'' + b \geq \|w, B\| + \|y, A''\|.$$

Now $\|y, A''\| \geq a''$, so (iv) $\|w, B\| \leq b$. Thus (v) $s^*_y \geq 2a'' - \|y, A''\| \geq a'' - b + \|w, B\|$. So

$$\mu(y) \geq_{(7)} \frac{s^*_y b}{\|w, B\|} \geq_{(v)} \frac{a'' - b + \|w, B\|}{\|w, B\|} b \geq (a'' - b)\frac{b}{\|w, B\|} + b \geq_{(iv)} a'',$$

a contradiction. So each unsuccessful swap decreases $\|A'', B\|$. Eventually, one of the cases holds or $a'' \leq b$.

Now suppose $a'' \leq b$. Then using (7), $\|w, B\| > s^*_y$. Set $f' :=$ swap(wy). We claim that $a < a(f')$ or one of the cases holds for f' or $\|B\| < \|B(f')\|$. Suppose not; then $\|B\| \geq \|B(f')\|$. Now $\|y, B\| \geq \|w, B\| - 1 \geq s_y^*$. Now we have the contradiction:

$$\Theta(G) \geq \begin{cases} \Theta(wy) \geq_{(2,1)} (a - 1 + \|w, B\|) + (a + \|y, B\|) \geq 2k & \text{if (i) holds,} \\ \Theta(xy) \geq_{(3*,2)} (a - a'' - 1 + \|x, B\|) + (a + a'' - s_y^* + \|y, B\|) \geq 2k & \text{if (ii) holds.} \end{cases}$$

By Lemma 7, $a''(f') \leq a'' \leq b = b(f')$. So each unsuccessful swap increases $\|B\|$. □

Because of the repeated use of Lemma 7, this proof only yields an exponential-time algorithm.

Conjecture 1 There is a polynomial-time algorithm for finding an equitable k-coloring of every graph with maximum Ore-degree $\Theta(G) < 2k$.

2.5 *Equitable Coloring of Directed Graphs*

Next we consider the problem of extending the Hajnal-Szemerèdi Theorem to directed graphs. It is not immediately clear what this should mean. Let $\overrightarrow{G}$ be a *simple* directed graph—a loopless directed graph such that for every ordered pair (x, y) of vertices there is at most one edge from x to y. Let $d^-(v)$ and $d^+(v)$ denote the in- and out-degree of $\overrightarrow{G}$. The *maximum* and *minimum* degree of $\overrightarrow{G}$ are $\Delta(\overrightarrow{G}) := \max\{d^-(v) + d^+(v)\}$ and $\delta(\overrightarrow{G}) := \min\{d^-(v) + d^+(v)\}$. A directed graph is *acyclic* if it contains no directed cycle. In the context of directed graphs, an *acyclic coloring* is a coloring whose classes are acyclic (instead of independent).

Theorem 9 (Czygrinow, DeBiasio, Kierstead and Molla [12], 2015) *Every directed graph $\overrightarrow{G}$ with maximum degree $\Delta(\overrightarrow{G}) < 2k$ has an equitable acyclic k-coloring.*

The theorem implies the Hajnal-Szemerédi Theorem. Suppose $G = (V, E)$ is an undirected graph with $\Delta(G) < k$. Let $\overrightarrow{G}$ be the directed graph obtained from G by replacing every edge $uv \in E$ by directed edges uv and vu, and observe (*) vuv is a cycle in $\overrightarrow{G}$. Now $\Delta(\overrightarrow{G}) \leq 2k - 2 < 2k - 1$, so $\overrightarrow{G}$ has an equitable acyclic k-coloring f by Theorem 9. Using (*), f is an equitable k-coloring of G.

Here is another consequence of Theorem 9. A simple directed graph $\overrightarrow{G}$ is *complete* if it contains every possible edge. If $\overrightarrow{G}$ is complete and $|\overrightarrow{G}| = m$, we write $\overrightarrow{G} = \overrightarrow{K}_m$. Set $\overrightarrow{G}^c := \overrightarrow{K}_{|\overrightarrow{G}|} - E(\overrightarrow{G})$. When $|\overrightarrow{G}| = km$, an *equitable k-tiling* of $\overrightarrow{G}$ by transitive tournaments is a k-partition of $V(\overrightarrow{G})$ such that each part induces a transitive tournament.

Corollary 10 *Every directed graph $\overrightarrow{H}$ on $n = km$ vertices with $\delta(\overrightarrow{H}) \geq 2n(1 - \frac{1}{m}) - 1$ has an equitable k-tiling by transitive tournaments.*

Proof Set $\overrightarrow{G} := \overrightarrow{H}^c$. Then $\Delta(\overrightarrow{G}) = 2n - \delta(\overrightarrow{H}) - 1 \leq 2k - 1 < 2k$. By Theorem 9, $\overrightarrow{G}$ has an equitable acyclic coloring f, so $\overrightarrow{G}[\widetilde{\gamma}]$ is acyclic for every class $\widetilde{\gamma}$ of f. Thus $\overrightarrow{G}[\widetilde{\gamma}]$ is a spanning subgraph of a transitive tournament $\overrightarrow{Q}$. So $\overrightarrow{Q}^c$ is a transitive tournament spanning $\overrightarrow{G}^c[\widetilde{\gamma}] = \overrightarrow{H}[\widetilde{\gamma}]$. □

The proof of Theorem 9 is very similar to that of Theorem 2, and it even provides a polynomial-time algorithm. One significant difference is in the definition of the auxiliary graph $\mathcal{H}$. Now $\gamma\alpha \in \mathcal{H}$ if and only if for some $x \in \widetilde{\gamma}$, there is no cycle in $\widetilde{\alpha} + x$. In particular, if $\gamma\alpha \notin \mathcal{H}$ then for all $x \in \widetilde{\gamma}$ there are edges from x to $\widetilde{\alpha}$ and from $\widetilde{\alpha}$ to x.

One might ask why, in the statement of Corollary 10, attention is restricted to transitive tournaments. Of course one answer is that they are easy to handle. The special (and exceptional) case $m = n/k = 3$ provides some direction. There are two tournaments on three vertices: one $\overrightarrow{T}$ is transitive, the other $\overrightarrow{C}$ is cyclic. The next theorem gives a degree condition for tiling $\overrightarrow{G}$ with various mixtures of transitive and cyclic tournaments.

Theorem 11 (Czygrinow, Kierstead and Molla [14], 2014) *Suppose $\overrightarrow{G}$ is a directed graph on $n = 3k$ vertices with $\delta(\overrightarrow{G}) \geq \frac{4}{3}n - 1$. For all $t \in [k]$, $\overrightarrow{G}$ has a k-tiling by t transitive and $k - t$ cyclic tournaments.*

Notice that the theorem does not assert the existence of a tiling entirely by cyclic tournaments since $t \geq 1$ is assumed. This is necessary. Indeed, in 2000, Wang [64] observed that even $\delta(\overrightarrow{G}) \geq \frac{3|\overrightarrow{G}|-5}{2}$ is not enough to force a tiling with cyclic tournaments. Let 3 divide $2s + 1$, and set $G := \overrightarrow{K}_s \cup \overrightarrow{K}_{s+1} \cup \overrightarrow{E}(K_s, K_{s+1})$, where $\overrightarrow{K}_s$ and $\overrightarrow{K}_{s+1}$ are disjoint and $\overrightarrow{E}(K_s, K_{s+1})$ is the set of all edges from K_s to K_{s+1}. Then $\overrightarrow{G}$ satisfies the degree condition. As one of s and $s + 1$ is not divisible by 3, any 3-tiling has a 3-tournament that meets both $\overrightarrow{K}_s$ and $\overrightarrow{K}_{s+1}$. Any such tournament must be transitive because its two crossing edges are oriented in the same direction. Wang also proved that this example is tight. But it is not strongly 2-connected.

Conjecture 2 (*Czygrinow, DeBiasio, Kierstead and Molla* [12], *2015*) If $\overrightarrow{G}$ is a strongly 2-connected, simple, directed graph on $3k$ vertices such that $\delta(\overrightarrow{G}) \geq 4k - 1$, then $\overrightarrow{G}$ has a cyclic triangle factor.

Molla, and two students, Eion Mulrenin and Panith Thiruvenkatasmy, are currently preparing a manuscript to show that Conjecture 2 holds for sufficiently large graphs.

Standard Multigraphs

Removing the orientation from the edges of a simple directed graph $\overrightarrow{G}$ leaves a loopless multigraph $M(\overrightarrow{G})$ such that every edge has multiplicity 1 or 2. Call such a multigraph *standard*. The edges of a standard multigraph are light if their multiplicity is 1; else they are heavy. Now $\delta(M(\overrightarrow{G})) = \delta(\overrightarrow{G})$ and $\Delta(M(\overrightarrow{G})) = \Delta(\overrightarrow{G})$. The multigraph M is *acyclic* if it contains no cycles, including 2-cycles; a coloring of M is *acyclic* if its color classes contain no cycles. Thus every acyclic coloring of $M(\overrightarrow{G})$ is an acyclic coloring of $\overrightarrow{G}$, but not the converse. Now the following conjecture implies Theorem 9.

Conjecture 3 (*Czygrinow, DeBiasio, Kierstead and Molla* [12], 2015) Every standard multigraph M with $\Delta(M) \leq 2k - 1$ has an equitable acyclic k-coloring.

An *s-sided triangle* is a standard multigraph on three vertices that contains K_3 and has s edges. Any 5-sided triangle in $M(\overrightarrow{G})$ spans both a cyclic and a transitive tournament in $\overrightarrow{G}$, and a 4-sided triangle in $M(\overrightarrow{G})$ spans a transitive tournament in $\overrightarrow{G}$. Thus Theorem 11 follows from:

Theorem 12 (Czygrinow, Kierstead and Molla [14], 2014) *Suppose G is a standard multigraph on* $n = 3k$ *vertices with* $\delta(G) \geq \frac{4}{3}n - 1$. *Then G has a k-tiling by triangles such that all but one of the triangles are* 5*-sided, and the remaining triangle is* 4*-sided.*

Call a directed graph $\overrightarrow{H}$ *universal* if $\overrightarrow{Q} \subseteq \overrightarrow{H}$ for every tournament $\overrightarrow{Q}$ with $|\overrightarrow{Q}| = |\overrightarrow{H}|$. All simple orientations of 5-sided triangles are universal, but 4-sided triangles can be oriented so that they do not contain a cyclic triangle. Call a standard multigraph M *universal* if all simple directed graphs $\overrightarrow{G}$ with $M(\overrightarrow{G}) = M$ are universal. Note that 5-sided triangles are universal, but 4-sided triangles are not. It is easy to check that a standard multigraph M on m vertices is universal if and only if it contains K_m and every simple orientation of M contains a copy of every orientation of $L(M)$, where $L(M)$ is the subgraph of M induced by its light edges. In particular, $L(M)$ is acyclic when M is universal.

Conjecture 4 (*Czygrinow, DeBiasio, Kierstead and Molla* [12], *2015*) Suppose $m \neq 3$ and $\overrightarrow{G}$ is a simple directed graph on $n = km$ vertices with $\delta(\overrightarrow{G}) \geq 2(1 - \frac{1}{m})n - 1$. Then $\overrightarrow{G}$ has a k-tiling by universal directed graphs on m vertices.

Czygrinow et al. proved the following stronger statement, but only for sufficiently large graphs.

Theorem 13 (Czygrinow, DeBiasio, Molla and Treglown [13], 2018) *For all* $m \geq 4$, *there exist* n_0 *such that every standard multigraph M on* $n = km \geq n_0$ *vertices with* $\delta(M) \geq 2(1 - \frac{1}{m})n - 1$, *has a k-tiling with universal standard multigraphs on m vertices.*

3 Chen-Lih-Wu Conjecture

Brooks' Theorem states that every graph G with $\Delta(G) \leq k$ is k-colorable except when $K_{k+1} \subseteq G$ or $k = 2$ and G contains an odd cycle. Could we also require that the k-coloring be equitable? When k is odd, $G = K_{k,k}$ is a 2-colorable graph with $\Delta(G) = k$ and $|G| = 2k$. Thus any equitable k-coloring of G consists of k classes of size 2, but each color class is contained in one part of G, so G has at most $2\lfloor \frac{k}{2} \rfloor = k - 1$ such classes. So, $K_{k,k}$ is not equitably k-colorable when k is odd. The following conjecture, asserting that this is the only counterexample, is the most important open question on equitable coloring.

Conjecture 5 (*Chen, Lih and Wu* [9], *1994*) If G is a k-colorable graph with $\Delta(G) \leq k$, then either G is equitably k-colorable or k is odd and G contains a $K_{k,k}$.

By Theorem 2, we only need to consider the case $k = \Delta(G)$. Lih and Wu [52] proved the conjecture for bipartite graphs. For general graphs, the conjecture is only known to hold for special cases of k. The case $k \leq 2$ is trivial. Chen, Lih and Wu [9] themselves proved the cases $k = 3$ and $k \geq |G|/2$. The first two authors proved the following.

Theorem 14 *Suppose G is a k-colorable graph with $\Delta(G) = k$ such that $K_{k,k} \nsubseteq G$ if k is odd. Then G has an equitable coloring if $k \leq 4$ ([30],* 2012) *or $k \geq |G|/4$ ([36],* 2014).

The arguments used in [30] and [36] are based on the approach described in Sect. 2, but every step is more difficult. For example, the existence of a nearly equitable coloring in the proof of Hajnal-Szemerédi Theorem is obvious, but to show this existence in the proof of Theorem 14 we used a sharpening of Brooks' Theorem due to Kostochka and Nakprasit [42]. They showed that if $\Delta(G), \chi(G) \leq k$ then for every k-coloring f of $G - v$ there is a k-coloring f' of G such that for all colors γ, except one, $|\widetilde{\gamma}(f)| = |\widetilde{\gamma}(f')|$. Another difficulty is that while using induction in the proof of Hajnal-Szemerédi Theorem, we automatically have an equitable b-coloring of $G[B - y]$, but in the proof of Theorem 14 we need to check that when b is odd, $G[B - y]$ contains no complete $K_{b,b}$. We will have more to say about this in the next subsection.

It is notable that the state of knowledge (Theorem 14) concerning Conjecture 5 is analogous to that (Theorem 1) concerning Erdős' conjecture before the proof of Theorem 2.

For sparse graphs, more is known. Yap and Zhang [68] proved that the conjecture holds for all outerplanar graphs. Yap and Zhang [72] also proved that the conjecture holds for planar graphs when $k \geq 13$, Nakprasit [57, 58] proved it for planar graphs when $9 \leq k \leq 12$, and Kostochka, Lin and Xiang [47] settled the case $k = 8$. Thus, for planar graphs Conjecture 5 is not verified only for graphs with maximum degree $5, 6$ and 7. For these degrees, Conjecture 5 was proved for planar graphs with extra restrictions, mainly with restrictions on cycle structure. In 2008, Zhu and Bu [73] proved that the conjecture holds for C_3-free planar graphs with maximum degree

$\Delta \geq 8$. It also holds for C_4, C_5-free planar graphs with maximum degree $\Delta \geq 7$. In 2009, Li and Bu [51] proved that the conjecture holds for C_4, C_6-free planar graph with maximum degree $\Delta \geq 6$. In 2012, Nakprasit and Nakprasit [56] proved that the conjecture holds for C_3-free planar graphs with maximum degree $\Delta \geq 6$, C_4-free planar graphs with maximum degree $\Delta \geq 7$, and planar graphs with maximum degree $\Delta \geq 5$ and girth at least 6.

Zhang [70] proved the conjecture for the wider class of 1-planar graphs, that is, graphs that admit a drawing in the plane such that each edge crosses at most one other edge but with the stronger restriction on k: for $k \geq 17$. The restriction $k \geq 17$ was relaxed to $k \geq 13$ for 1-planar graphs by Cranston and Mahmoud [11]. In fact, Zhang proved his bound 17 for the more general class $\mathcal{Z}$ of graphs G such that $\|G[A]\| \leq 4(|A| - 2)$ for every $A \subseteq V(G)$ with $|A| \geq 3$.

3.1 An Equivalent Conjecture

Note that for an odd k, Conjecture 5 does not describe all graphs G with $\Delta(G) \leq k$ that are not equitably k-colorable. For example, for each odd $k \geq 3$, the graph consisting of two disjoint copies of $K_{k,k}$ has an equitable k-coloring, but the graph consisting of a copy of $K_{k,k}$ and a copy of K_k does not. This construction can be generalized. Say that a graph H is *k-equitable* if $|H|$ is divisible by k, H is k-colorable and every k-coloring of H is equitable. For an odd k, if G contains $K_{k,k}$ and $G - K_{k,k}$ is k-equitable, then G does not have an equitable k-coloring. This motivates the study of equitable graphs, i.e., graphs that are k-equitable for some k. It was proved in [37] that there is a good description of the family of all k-equitable graphs; they can all be built from simple examples in a simple way. Note first that if a spanning subgraph of a graph G is the disjoint union of k-equitable graphs, then G is k-equitable. Clearly, K_k is k-equitable. Figures 1 and 2 show 10 other equitable 2-connected graphs.

Together with K_k, the k-equitable graphs in Figs. 1 and 2 are the *k-basic* graphs. A *k-decomposition* of G is a partition on $V(G)$ into subsets $V_1, \ldots, V_t$ such that each $G[V_i]$ is k-basic.

Theorem 15 (*[37]*, **2010**) *Let $k \geq 3$. Let G be a k-colorable graph with $\Delta(G) \leq k$ and $|V(G)|$ divisible by k. Then the following are equivalent:*

(a) G has a k-decomposition;
(b) G is k-equitable;
(c) G has an equitable k-coloring but does not have a nearly equitable k-coloring.

Theorem 15 is used in the proof of Theorem 14 to get b-equitable colorings of $B - y$. It also inspired the following conjecture which (if true) would describe all graphs G with $\Delta(G) \leq k$ that are not equitably k-colorable.

Conjecture 6 ([37], 2010) Suppose that $k \geq 3$ and G is a k-colorable graph with $\Delta(G) \leq k$. Then G has no equitable k-coloring if and only if k is odd and there exists $H \subseteq G$ such that $H = K_{k,k}$ and $G - H$ has a k-decomposition.

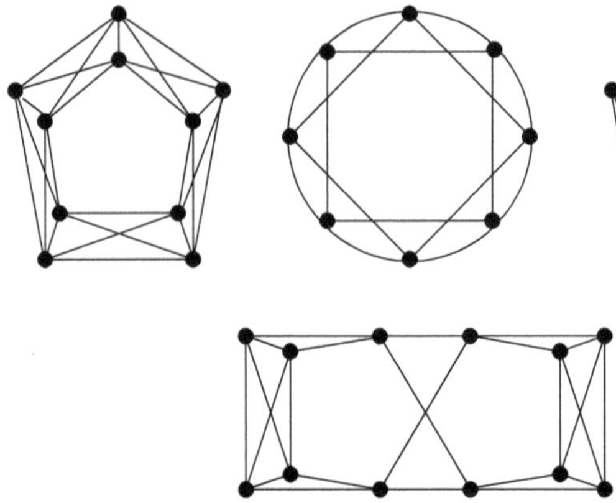

Fig. 1 One 5-equitable and three 4-equitable basic graphs

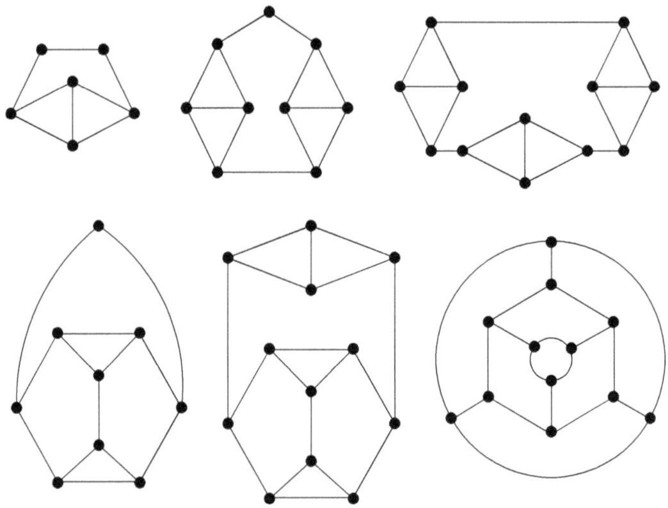

Fig. 2 Six 3-equitable basic graphs

Theorem 15 yields the following implications.

Corollary 16 *For all positive integers k and $n > k$, Conjecture 6 holds for all k-colorable graphs G with $\Delta(G) \leq k$ and at most n vertices if and only if Conjecture 5 holds for all such graphs.*

Corollary 17 *Every k-equitable k-colorable graph G with $\Delta(G) \leq k$ has a unique k-decomposition.*

Corollary 18 *There exists a polynomial time algorithm for deciding whether a graph G with $\Delta(G) \leq k$ is k-equitable.*

3.2 An Ore Version of the Chen-Lih-Wu Conjecture

Recall that the Ore-degree of a graph G is $\Theta(G) := \max\{d(x) + d(y) : xy \in E\}$. The analog of the trivial bound $\chi(G) \leq \Delta(G) + 1$ in terms of the Ore-degree is the claim that if $2k > \Theta(G)$, then G is k-colorable. This claim is true and is easy to prove. The bound is achieved for complete graphs and odd cycles. Moreover, there are infinitely many graphs G with $\Theta(G) = 7$ and $\chi(G) = 4$. Also the 5-Ore graph with 9 vertices has Ore-degree 9 and chromatic number 5. However, the following analog of Brooks' Theorem for the Ore-degree holds.

Theorem 19 ([35], 2009) *If $k \geq 7$ and G is a connected graph with $\Theta(G) \leq 2k + 1$ distinct from K_{k+1}, then $\chi(G) \leq k$.*

We think that the claim of Theorem 19 holds also for $k = 6$ but cannot prove this.

Theorem 6 in Sect. 2.4 proves an analog of Theorem 2 in terms of the Ore-degree. So, in view of Theorems 19 and 6, the following analog of Conjecture 5 was posed in [28] and [35].

Conjecture 7 ([28, 35], 2009) Let G be a connected graph with $\Theta(G) \leq 2k$ and $k \geq 3$. If G is distinct from K_{k+1} and from $K_{m,2k-m}$ for odd m, then G has an equitable k-coloring.

This conjecture was proved for $k = 3$. In [35], the following strengthening of it was conjectured.

Conjecture 8 ([32, 35], 2009) Suppose that $k \geq 3$. A k-colorable, n-vertex graph G with $\Theta(G) \leq 2k$ has no equitable k-coloring if and only if n is divisible by k, and there exists $W \subset V(G)$ such that $G[W] = K_{m,2k-m}$ for some odd m and $G - W$ is k-decomposable.

It was proved in [35] that Conjectures 7 and 8 are equivalent even for graphs with restricted number of vertices and restricted values of Ore-degree. This implies that Conjecture 8 holds for $k = 3$. This stronger claim may help to prove Conjecture 7 by induction, because we would need to prove the formally weaker conjecture but use the fact that the stronger conjecture holds for smaller graphs.

4 Equitable List Coloring

Returning to the initial idealized example of this paper about equitable distribution of jobs among workers, we consider the more complicated situation when not all workers have the same skills. Maybe each job J has a list $L(J)$ of workers who can perform this job. Then it might be that some workers are able to perform only few jobs. In this case, it is not realistic to insist on almost the same number of jobs being performed by each worker. It is more natural to ask for an upper bound on the number of jobs performed by each worker. Among the models for this more complicated situation there are different kinds of *equitable list coloring*.

A *list assignment* L for a graph G assigns to each vertex v a set $L(v)$ of colors. An L*-coloring* f is a proper coloring of G such that $f(v) \in L(v)$ for each $v \in V(G)$. If $|L(v)| = k$ for each $v \in V(G)$, then L is a k*-list assignment*.

Kostochka, Pelsmajer and West [44] proposed a list version of equitable coloring as follows. For a k-list assignment L, a graph G is *equitably* L*-colorable* if it has an L-coloring such that every color class has at most $\lceil |G|/k \rceil$ vertices. Now G is *equitably* k*-choosable* if it is equitably L-colorable for every k-list assignment L.

In contrast to ordinary equitable coloring, in this setting it is possible that the sizes of two color classes differ significantly, as it could be that some color appears only in the lists of few vertices.

The following analog of Theorem 2 was conjectured in [44].

Conjecture 9 (*Kostochka, Pelsmajer and West, Conjecture 1.1* [44], *2003*) Every graph G is equitably k-choosable for each $k > \Delta(G)$.

In [44] Conjecture 9 was confirmed for $k \geq \max\{\Delta(G) + 1, |G|/2\}$. Pelsmajer [59] and Wang and Lih [65] independently proved Conjecture 9 for $k \leq 3$.

In [31] the conjecture was proved for a larger class of graphs; several relaxations of the conjecture were also proved.

Theorem 20 ([31], 2013) *Let* G *be a graph with* $\Delta(G) = \Delta$.

(a) *If* $\Delta \leq 7$ *and* $k \geq \Delta + 1$, *then* G *is equitably* k*-choosable.*
(b) *If* $|G| \geq \Delta^3$ *and* $k \geq \Delta + 2$, *then* G *is equitably* k*-choosable.*
(c) *If* $\omega(G) \leq k$ *and* $|G| \geq 3(k + 1)k^8$, *then* G *is equitably* $(k + 1)$*-choosable.*

For each of the conditions in Theorem 20, there is a polynomial-time (in terms of n) algorithm, that finds an equitable L-coloring of G for every graph G satisfying the condition, and every k-list assignment L on G. The method used to prove Theorem 20 is based on ideas of the proof of the Hajnal-Szemerédi Theorem and its variations, such as Theorem 14. But several new tricks were applied to handle distinct lists. In particular, the definition of the auxiliary digraph $\mathcal{H}$ needed to be modified so that $\sigma\tau$ is a directed edge of $\mathcal{H}$ if and only if there is $w \in \widetilde{\sigma}$ with $\tau \in L(w)$ and $\|w, \widetilde{\tau}\| = 0$. Still, these tricks were insufficient to prove the conjecture in full.

An Ore-type version of Conjecture 9 (which also can be viewed as the list version of Theorem 6) was proposed in [32]:

Conjecture 10 (*Conjecture 6.3* [32], *2009*) Every graph G is equitably $(1 + 0.5\Theta(G))$-choosable.

In support of this conjecture, it was proved in [32] that every graph G with $\Theta(G) \leq 6$ is equitably 4-choosable. A Chen-Lih-Wu version of Conjecture 9 (in other words, the list version of the Chen-Lih-Wu Conjecture) was also proposed in [44] in the following form.

Conjecture 11 (*Conjecture 1.2, Kostochka, Pelsmajer and West* [44], 2003) If G is a k-colorable graph with $3 \leq \Delta(G) \leq k$, then either G is equitably k-choosable or k is odd and G contains a $K_{k,k}$.

Conjecture 11 was confirmed for $k \geq \max\{\Delta(G), |G|/2\}$, connected interval graphs and 2-degenerate graphs G with $k \geq \max\{\Delta(G), 5\}$ in [44]. Zhang and Wu [71] proved the Conjecture for series-parallel graphs. Dong and Zhang [17] studied list equitable coloring of graphs G with low *maximum average degree*, $\mathrm{mad}(G) := \max_{H \subseteq G} \frac{2\|H\|}{|H|}$. They proved that if G is a graph with $\mathrm{mad}(G) < 3$ then G is equitably k-choosable for $k \geq \max\{\Delta(G), 4\}$, and if $\mathrm{mad}(G) < \frac{12}{5}$ then G is equitably k-choosable for $k \geq \max\{\Delta(G), 3\}$. Zhu and Bu [74] proved that if G is a triangle-free planar graph, then G is equitably k-choosable for $k \geq \max\{\Delta(G), 8\}$ and if G is a planar graph without 4- and 5-cycles, then G is equitably k-choosable for $k \geq \max\{\Delta(G), 7\}$.

It is worth mentioning that for every k, there are graphs that are equitably k-colorable but not equitably k-choosable, and vice versa. One of the examples is the star $K_{1,2k}$.

Claim For each integer $k \geq 3$, the star $K_{1,2k}$ is equitably k-choosable but not equitably k-colorable.

Proof Let $G = K_{1,2k}$ and let v be the unique vertex of degree $2k$ in G.

Assume that every vertex is assigned a list of size k. Note that regardless of what color is used for v, there are at least $k-1$ colors to choose from for each neighbor of v. By the definition of equitable list coloring, each color can be used up to $\lceil (2k+1)/k \rceil = 3$ times. Since $3(k-1) \geq 2k$, we may color the neighbors of v greedily so that no color is used more than 3 times.

If $K_{1,2k}$ had an equitable k-coloring, then the color class of v cannot be used for any other vertex. Also in this case, each of the remaining $k-1$ color classes would have at most two vertices. Thus $|G| \leq 1 + 2(k-1)$, a contradiction. □

The next example of an equitably k-colorable but not equitably k-choosable graph is a bit more sophisticated.

Example 21 Consider any integer $k \geq 3$ and disjoint sets $V_0, U_0, V_1, U_1, \ldots, V_k, U_k$ with $|V_i| = k-1$ and $|U_i| = k^3 - 2k^2 + 1$. Define the graph $G = G_k = (V, E)$ on $V := \bigcup_{i=0}^{k}(V_i \cup U_i)$ by

$$E := \bigcup_{i=1}^{k} \{vw : (v, w) \in ((V_0 \times V_i) \cup (V_i \times U_i))\} \cup \{vw : v \in V_0, w \in V_0 \cup U_0\}.$$

Claim For each integer $k \geq 3$, G_k is equitably k-colorable but not equitably k-choosable.

Proof Notice that

(a) $|G| = (k+1)|V_0 \cup U_0| = (k+1)k(k-1)^2$,
(b) $|N(V_0)| = k|V_1| + |U_0| = k^3 - 2k^2 + 1 + k(k-1) = k^3 - k^2 - k + 1 = (k+1)(k-1)^2$,
(c) $|V \setminus N[V_0]| = (k-1)(k+1)(k-1)^2$,
(d) V_0 is a $(k-1)$-clique, and
(e) $N(V_0)$ and $V \setminus N[V_0]$ are independent.

Now G is equitably k-colorable, since we can color V_0 with $k-1$ distinct colors, use the same $k-1$ colors to equitably color $V \setminus N(V_0)$, and color $N(V_0)$ with the remaining color.

To see that G is not equitably k-choosable, we assign a particular k-list L to V and show that G does not have an equitable L-coloring. Set $A := [2k-1] \setminus [k]$ and $V_i := \{v_{i,j} : j \in [k-1]\}$ for $i \in [k]$. For $v \in V_0 \cup U_0$, let $L(v) = [k]$. For each $v_{i,j}$, let $L(v_{i,j}) = ([k] - i) \cup \{k+j\}$. For $v \in U_i$ with $i \in [k]$, let $L(v) = A + i$.

Since V_0 is a clique, we must use $k-1$ distinct colors from $[k]$ on V_0. Without loss of generality, we do not use the color 1; then all vertices in U_0 must be colored 1. By the definition of L, $v_{1,j}$ must get color $k+j$ for $j \in [k-1]$. Now every color in A is used on V_1, so vertices in U_1 must get color 1. Now there are $|U_0| + |U_1| = 2k^3 - 4k^2 + 2$ vertices that must be colored with 1, which already exceeds $|V|/k = (k-1)^2(k+1)$ when $k \geq 3$. □

The authors [33] defined the *strongly equitable* (SE) L-coloring which has the property that every SE k-choosable graph is also equitably k-choosable and equitably k-colorable. An L-coloring of a graph G is SE if it has at most $|G| \bmod^* k$ color classes of size $\left\lceil \frac{|G|}{k} \right\rceil$, where $n \bmod^* k$ is the unique integer $m \in [k]$ with $n - m$ divisible by k. So, if n is divisible by k then $n \bmod^* k = k$. A multigraph is SE k-choosable if it is SE L-colorable for every k-list assignment L. If a multigraph is SE k-choosable, then it is equitably k-colorable and equitably k-choosable.

Let $\mathcal{B}$ denote the class of graphs such that

$$\|X, Y\| \leq 2|X \cup Y| - 4 \text{ for all disjoint } X, Y \subseteq V(G) \text{ with } |X \cup Y| \geq 3. \quad (8)$$

Observe that class $\mathcal{B}$ is topologically much broader than the class of planar graphs. For example, for any graph H, the graph B_H obtained from H by subdividing each edge once is in $\mathcal{B}$. Therefore, every graph is a minor of a graph in $\mathcal{B}$. In particular, the acyclic chromatic number of graphs in $\mathcal{B}$ can be arbitrarily large, while it is at most 5 for any planar graph. Note that the class $\mathcal{Z}$ of graphs G such that $||G[A]|| \leq 4(|A| - 2)$ for every $A \subseteq V(G)$ with $|A| \geq 3$ considered by Zhang [70] also has these properties.

Adapting the method described in Sect. 2 to SE list coloring and analysing sparsity of graphs in $\mathcal{B}$, we recently proved the following bound.

Theorem 22 ([33], 2024) Every graph $G \in \mathcal{B}$ is SE k-choosable, if $k \geq \max\{9, \Delta(G)\}$.

Theorem 22 combined with Theorem 20(a) implies the following fact.

Corollary 23 *Conjecture 9 holds for every graph in $\mathcal{B}$, in particular, for every planar graph.*

We think that it is natural to extend Conjectures 9, 10 and 11 to SE list coloring as follows.

Conjecture 12 Every graph G is equitably SE k-choosable for each $k > \Delta(G)$.

Conjecture 13 If $2k > \Theta(G)$, then G is SE k-choosable.

Conjecture 14 If G is a k-colorable graph with $3 \leq \Delta(G) \leq k$, then either G is SE k-choosable or k is odd and G contains a $K_{k,k}$.

5 Coloring Sparse Graphs with Fewer Colors

One of the directions of study of equitable coloring and list equitable coloring is trying to color "sparse" graphs G with fewer than $\Delta(G)$ colors. By "sparse" one can mean "small average degree" or "small maximum average degree" or a topologically simple class of graphs, like outerplanar or planar graphs.

Berge and Sterboul [4] studied equitable coloring of sparse graphs and more generally uniform hypergraphs. They considered $m_0(n, r, k)$—the minimum number of edges in an n-vertex r-uniform hypergraph that has no equitable k-coloring. They have found the exact values of $m_0(n, r, k)$ for many triples (n, r, k). In particular, for graphs ($r = 2$), they have found the exact values for all pairs (n, k). They proved that

$$m_0(n, 2, k) = \begin{cases} \min\{\binom{k+1}{2}, \binom{n}{2} - \binom{2(n-k)-1}{2}\} & \text{when } k \leq n \leq 2k; \\ \min\{\binom{k+1}{2}, n - \lfloor \frac{n}{k} \rfloor + 1\} & \text{when } n \leq 2k. \end{cases}$$

The extremal example for small k with respect to n is simply K_{k+1} that has no proper k-coloring at all together with $n - k - 1$ isolated vertices. For "larger" k, if $n \geq 2k$ then the extremal example is the star $K_{1,n-\lfloor \frac{n}{k} \rfloor+1}$ together with some isolated vertices, and if $k \leq n \leq 2k$, then it is the graph obtained from K_n by deleting the edges of a $K_{2(n-k)-1}$.

Meyer [54] proved that every tree with maximum degree Δ has an equitable k-coloring for $k = 1 + \lfloor \frac{\Delta}{2} \rfloor$ and it was observed later that the result holds for every $k \geq 1 + \lfloor \frac{\Delta}{2} \rfloor$. The extremal example is the star $K_{1,\Delta}$.

Recall that Theorem 3 by Bollobás and Guy [6] gives for trees a bound that is not a function of maximum degree. For convenience, we restate it here.

Theorem 24 (Bollobás and Guy [6], 1983) *A tree T is equitably 3-colorable if $|T| \geq 3\Delta(T) - 8$ or $|T| = 3\Delta(T) - 10$.*

They also provided an algorithm for producing the coloring. An extremal example is any n-vertex tree containing a vertex v of degree exactly $\frac{n+9}{3}$ and a perfect matching on the set of the nonneighbors of v. This result was extended to all $k \geq 2$ and to all forests by Chen and Lih [8] and Miyata, Tokunaga and Kaneko [55]. Given a graph G and a vertex $v \in V(G)$, let $\alpha_v = \alpha_v(G)$ denote the size of the maximum independent set in G containing v. If a graph G has an equitable k-coloring, then by definition, $\alpha_v(G) \geq \lfloor n/k \rfloor$ for every $v \in V(G)$. Chen and Lih [8] and independently Miyata, Tokunaga and Kaneko [55] proved (and Chang [7] gave a shorter and nicer proof) that this necessary condition is sufficient for forests.

Theorem 25 (Chen and Lih [8], 1994, Miyata, Tokunaga and Kaneko [55], 1994) *For a forest T of order n and integer $k \geq 3$, T is equitably k-colorable if and only if $\alpha_v \geq \lfloor n/k \rfloor$ for every vertex $v \in T$.*

The "only if" part is obvious. We present the proof of the "if" part due to Chang from [7].

Proof Let $n = |V(G)|$. We choose a bipartition (A, B) of $V(G)$ so that $|A| = a \geq |B| = b$, and modulo this, a is minimum. By this choice,

$$\text{either } |A| - |B| \leq 1 \text{ or } A \text{ does not contain isolated vertices.} \tag{9}$$

For $1 \leq i \leq k$, let $s_i = \lfloor \frac{n+i-1}{k} \rfloor$. Choose the minimum j such that $b \leq \sum_{i=1}^{j} s_i$. If $b = \sum_{i=1}^{j} s_i$, then we partition B into sets of sizes $s_1, \ldots, s_j$ and A into sets of sizes $s_{j+1}, \ldots, s_k$. So we may assume $\sum_{i=1}^{j-1} s_i < b < \sum_{i=1}^{j} s_i$.

Case 1: $j > 1$. Take S be the set of $s := b - \sum_{i=1}^{j-1} s_i$ vertices of lowest degrees in B. By this choice, $|N(S)| \leq \frac{s(n-1)}{b}$. Thus using $a \geq b$ and $b - s \geq s_1$, we have

$$|S \cup (A - N(S))| > s + a - \frac{sn}{b} = \frac{(b-s)a}{b} \geq b - s \geq s_1.$$

So, $|S \cup (A - N(S))| \geq 1 + s_1 \geq s_j$, and $S \cup (A - N(S))$ contains a subset S' with $|S'| = s_j$ containing S. Then we can partition $A - S'$ into (independent) sets of sizes $s_{j+1}, \ldots, s_k$.

Case 2: $j = 1$. In this case, it is enough to prove that

$$\text{there are disjoint independent sets } I_1, I_2 \text{ with } |I_1| \geq s_1, |I_2| \geq s_k \text{ and } B \subset I_1 \cup I_2, \tag{10}$$

since in this case $V(G) - I_1 - I_2$ is independent. Thus we assume that (10) does not hold.

Since $k \geq 3$, (9) implies that A does not contain isolated vertices. Let L be the set of all leaves in A. Then $|L| + 2|A - L| \leq |E(G)| \leq n - 1$ and so $|L| \geq 2a - (n - 1) = a - b + 1$. In particular,

$$|L| + b \geq a + 1 \geq n - b + 1. \tag{11}$$

Call a set $Q \subseteq B$ *good* if $|Q \cup (L - N(Q))| \geq s_k$. Since (10) does not hold, B is not good. By (11), $\emptyset$ is good. Let Q be a largest good subset Q of B. Since B is not good, there is $v \in B - Q$. By the maximality of Q,

$$|Q + v| + |L - N(Q) - N(v)| \leq s_k - 1. \tag{12}$$

If $|(N(Q) \cap L) \cup (B - Q)| \geq s_1$, then we are done by (10). So, suppose $|N(Q) \cap L| + |B - Q| \leq s_1 - 1$. Summing this with (12), we get $|L| - |N(v) \cap L| + b + 1 \leq s_1 + s_k - 2$. Thus, using (11),

$$|N(v) \cap L| + (b - 1) \geq |L| + b + 1 - s_1 - s_k + 2 + (b - 1) \geq n + 2 - s_1 - s_k > s_k. \tag{13}$$

Since $\alpha_v \geq s_1$, there is an independent set R containing v with $|R| \geq s_1$. Choose such R with the minimum $|R \cap B|$. If $R \cap B = \{v\}$, then by (13), there is an independent set $I_2 \subseteq (B - v) \cup (N(v) \cap L)$ containing $B - v$ with $|I_2| = s_k$ disjoint from R. In this case, we are done by (10).

Otherwise, $|R \cap B| \geq 2$, say $v' \in (R - v) \cap B$. If there is $x \in L - R$ with no neighbors in R, then the set $R - v' + x$ has fewer vertices in B than R, contradicting the choice of R. So each $x \in L - R$ has a neighbor in R. Then the set $I_2 := (B \cup L) - R$ is independent, and by (11), $|I_2| \geq n - b + 1 - s_1 \geq s_k$. Thus (10) holds with $I_1 = R$, a contradiction. □

Chang [7] asked about broader classes $\mathcal{F}$ of graphs for which the conclusion of Theorem 25 holds.

Question 1 *Does the claim of Theorem 25 hold for SE list coloring? In other words, is it true that for each $k \geq 3$, every n-vertex forest T is equitably SE k-choosable if and only if $\alpha_v \geq \lfloor n/k \rfloor$ for every vertex $v \in T$?*

When Yap and Zhang [68] proved the CLW-Conjecture for outerplanar graphs, they also conjectured that every outerplanar graph with maximum degree $\Delta \geq 3$ is equitably k-colorable for every $k \geq 1 + \Delta/2$. The conjecture was proved in [40].

Question 2 *Does the claim of Theorem 25 hold for all outerplanar graphs?*

When Zhang and Wu [71] proved the CLW-Conjecture and its list version for the broader class of series-parallel graphs, they also conjectured that every series-parallel graph with maximum degree $\Delta \geq 3$ is equitably k-colorable for every integer $k \geq (\Delta + 3)/2$. Kostochka and Nakprasit [41] proved a more general claim for d-degenerate graphs G with $\Delta(G) \geq 27d$:

Theorem 26 (Kostochka and Nakprasit [41], 2003) *Let $d \geq 2$, $\Delta \geq 27d$, $k \geq (d + \Delta + 1)/2$. Then every d-degenerate graph with maximum degree at most Δ is equitably k-colorable.*

The restriction on k is tight. Consider the graph $G(d, \Delta) = K_d \vee \overline{K}_{\Delta-d+1}$, obtained from the complete graph K_d by adding $\Delta - d + 1$ vertices each of which is adjacent to vertices of our K_d and only to them. This graph is d-degenerate, has maximum degree Δ and needs $\lceil \frac{\Delta+d+1}{2} \rceil$ colors for its equitable coloring because some d color classes must be of size 1.

In particular, Theorem 26 confirms the conjecture by Zhang and Wu for $\Delta \geq 54$. The restriction $\Delta \geq 27d$ in Theorem 26 maybe is not needed, and Kostochka and Nakprasit [41] conjectured that the claim of Theorem 26 holds for all $2 \leq d \leq \Delta$. Partially supporting this conjecture is the result of Verclos and Sereni [15] who proved the conjecture by Zhang and Wu in full.

Pemmaraju [60] proved an analogue of Theorem 24 for outerplanar graphs.

Theorem 27 (Pemmaraju [60], 2001) *A connected outerplanar n-vertex graph with maximum degree at most $n/6$ has an equitable 6-coloring.*

Kostochka, Nakprasit and Pemmaraju [43] generalized this as follows.

Theorem 28 (Kostochka, Nakprasit and Pemmaraju [43], 2005) *For $d, n \geq 1$, every d-degenerate n-vertex graph G with $\Delta(G) \leq n/15$ is equitably k-colorable for each $k \geq 16d$.*

Again, maybe the restrictions $\Delta(G) \leq n/15$ and $k \geq 16d$ can be weakened (but cannot be dropped completely).

In a similar spirit, Wu and Wang [67] proved in 2008 that every planar graph G with minimum degree $\delta(G) \geq 2$ and girth $g(G) \geq 14$ is equitably k-colorable for each $k \geq 4$. Moreover, if the girth of G is at least 26 then G is equitably 3-colorable. Luo, Sereni, Stephens and Yu [53] strengthened this result with the following theorem.

Theorem 29 (Luo, Sereni, Stephens and Yu [53], 2010) *Every planar graph G with $\delta(G) \geq 2$ and $g(G) \geq 10$ is equitably k-colorable for each $k \geq 4$, and if the $g(G) \geq 14$ then G is equitably 3-colorable.*

Luo et al. [53] mention that their proofs yield the same bounds when they replace planarity with appropriate bounds on maximal average degree: for graphs G with $\mathrm{mad}(G) \leq 7/3$ when $k = 3$ and for graphs with $\mathrm{mad}(G) \leq 5/2$ when $k \geq 4$, but they still need the restrictions on the girth and minimum degree.

Theorem 29 was sharpened, generalized and extended to SE list colorings as follows.

Theorem 30 ([34], 2024) *Let G be a graph with $\delta(G) \geq 2$. If*

$$6\|G[A]\| \leq 7\,A| + 2 \quad \textit{for each nonempty } A \subseteq V(G), \tag{14}$$

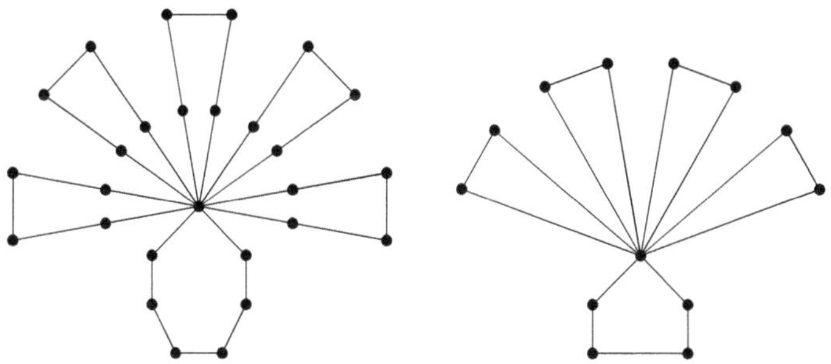

Fig. 3 Graph on the left of the picture has 27 vertices, 32 edges and cannot be equitably 3-colored. Graph on the right of the picture has 13 vertices, 17 edges and cannot be equitably 4-colored

then G SE 3*-choosable, and if*

$$4\|G[A]\| \leq 5\, A| + 2 \quad \textit{for each nonempty } A \subseteq V(G), \tag{15}$$

then G is SE 4*-choosable. Moreover, both results are sharp.*

Sharpness examples for Theorem 30 are shown in Fig. 3. It is interesting that the bounds for ordinary equitable coloring, list equitable coloring and SE list coloring are the same. We think that we can prove an exact bound for SE list coloring with 5 colors, but are not sure what would be the exact bound for more colors.

Krivelevich and Patkos [49] studied equitable colorings of random graphs in the model $G(n, p)$. They showed that if $n^{-1/5+\epsilon} < p < 0.99$ for some $\epsilon > 0$, then almost surely the minimum k such that $G \in G(n, p)$ has an equitable k-coloring is asymptotically the chromatic number of G. They also proved that the maximum k such that if $C/n < p < 0.99$ for a large C, then almost surely the maximum k such that $G \in G(n, p)$ has no equitable k-coloring is asymptotically at most twice the chromatic number of G.

6 Variations of Equitable Coloring

In this section we discuss some variants of equitable coloring. Recall that in equitable list coloring, sizes of color classes could differ as some colors may appear only in few lists. Given a graph G and a k-list assignment L for G, for each color c, denote by $\eta(c)$ the number of vertices $v \in V(G)$ such that $c \in L(v)$. A *proportional L-coloring* of G is a proper L-coloring that uses each color c exactly $\lfloor \eta(c)/k \rfloor$ or $\lceil \eta(c)/k \rceil$ times. A graph G is *proportionally k-choosable* if for every k-list assignment L, G has a proportional L-coloring. The *proportional choice number* $\chi_{pc}(G)$ is the smallest k such that G is proportionally k-choosable.

This notion of proportional coloring was proposed by Kaul, Mudrock, Pelsmajer and Reiniger in [27], and they proved the following upper bound on $\chi_{pc}(G)$ for all graphs G.

Theorem 31 (Kaul, Mudrock, Pelsmajer and Reiniger [27], 2018) *For every graph G,*

$$\chi_{pc}(G) \leq \Delta(G) + \left\lceil \frac{|V(G)|}{2} \right\rceil.$$

They also asked if the analog of Theorem 2 holds for proportional coloring.

Question 3 (Kaul, Mudrock, Pelsmajer and Reiniger *[27]*, **2018)** *For any graph G, is G proportionally k-choosable whenever $k \geq \Delta(G) + 1$?*

Kaul et al. [27] mention that the analog of Conjecture 5 does not hold for proportional coloring. A counterexample is a union of k disjoint $K_{1,k}$ for each integer k.

Coloring of a graph could be viewed as a partition of the vertices into disjoint independent sets. A *k-partition* of a graph is a partition of the vertex set into k disjoint sets. There have been many studies on partitioning the vertex set into sets other than independent sets, for example, sets that induce graphs with bounded degeneracy or maximum degree. Similarly to equitable coloring, an *equitable k-partition* is a partition into k disjoint sets such that the sizes of these sets differ by at most 1.

Kostochka, Nakprasit and Pemmaraju proved the following result on partitioning a d-degenerate graph into $(d-1)$-degenerate graphs equitably.

Theorem 32 (Kostochka, Nakprasit and Pemmaraju [43], 2005) *For $k \geq 3, d \geq 2$, every d-degenerate graph has an equitable k-partition into $(d-1)$-degenerate graphs.*

By this result, each planar graph can be partitioned into 81 forests since every planar graph is 5-degenerate. Esperet, Lemoine, and Maffray found a better upper bound on the number k of forests such that each planar graph could be partitioned into k forests, and asked whether an even better upper bound could be achieved.

Theorem 33 (Esperet, Lemoine, and Maffray [19], 2015) *For each $k \geq 4$, every planar graph has an equitable k-partition into forests.*

Question 4 (Esperet, Lemoine, and Maffray *[19]*, **2015)** *Does every planar graph have an equitable 3-partition into forests?*

Relaxing the degeneracy of the sets in a partition, Kim, Oum and Zhang derived the following results.

Theorem 34 (Kim, Oum and Zhang [39], 2022) *Every planar graph has*

(a) an equitable 2*-partition into* 3*-degenerate graphs,*
(b) an equitable 3*-partition into* 2*-degenerate graphs,*
(c) an equitable 3*-partition into two forests and a graph.*

If a vertex partition $(V_1, \ldots, V_k)$ of a graph G is such that the maximum degree of $G[V_i]$ is at most d for each $1 \leq i \leq k$, then such partition is a *d-defective coloring* of G. Williams, Vandenbussche and Yu proved the following result on equitable defective coloring that is an extension of Theorem 29.

Theorem 35 (Williams, Vandenbussche and Yu [66], 2012) *For every $k \geq 3$, each planar graph G with $\delta(G) \geq 2$ and $g(G) \geq 10$ has an equitable k-partition that is a* 1*-defective coloring.*

Acknowledgements The authors thank Louis DeBiasio, Hemanshu Kaul, and a referee for helpful comments.

References

1. Alon, N., Füredi, Z.: Spanning subgraphs of random graphs. Graphs Combin. **8**(1), 91–94 (1992)
2. Alon, N., Yuster, R.: H-factors in dense graphs. J. Combin. Theory Ser. B **66**(2), 269–282 (1996)
3. Arnold, B.C., Groeneveld, R.A.: Bounds on expectations of linear systematic statistics based on dependent samples. Ann. Statist. **7**(1), 220–223 (1979)
4. Berge, C., Sterboul, F.: Equipartite colorings in graphs and hypergraphs. J. Comb. Theory Ser. B **22**(2), 97–113 (1977)
5. Blazewicz, J., Ecker, K., Pesch, E., Schmidt, G., Weglarz, J.: Scheduling Computer and Manufacturing Processes. Springer Berlin, Heidelberg (2001)
6. Bollobás, B., Guy, R.K.: Equitable and proportional coloring of trees. J. Combin. Theory Ser. B **34**(2), 177–186 (1983)
7. Gerard, Chang, J.: A note on equitable colorings of forests. Eur. J. Combin. **30**(4), 809–812 (2009)
8. Chen, B.-L., Lih, K.-W.: Equitable coloring of trees. J. Combin. Theory Ser. B **61**(1), 83–87 (1994)
9. Chen, B.L., Lih, K-W., Wu, P-L.: Equitable coloring and the maximum degree. Eur. J. Combin. **15**(5), 443–447 (1994)
10. Corradi, K., Hajnal, A.: On the maximal number of independent circuits in a graph. Acta Math. Acad. Sci. Hungar. **14**, 423–439 (1963)
11. Cranston, D.W., Mahmoud, R.: Equitable coloring in 1-planar graphs. Discrete Math. **348**(2):Paper No. 114286 (2025)
12. Czygrinow, A., DeBiasio, L., Kierstead, H.A., Molla, T.: An extension of the Hajnal-Szemerédi theorem to directed graphs. Combin. Probab. Comput. **24**(5), 754–773 (2015)
13. Czygrinow, A., DeBiasio, L., Molla, T., Treglown, A.: Tiling directed graphs with tournaments. Forum Math. Sigma **6**, Paper No. e2, 53 (2018)
14. Czygrinow, A., Kierstead, H.A., Molla, T.: On directed versions of the Corrádi–Hajnal corollary. Eur. J. Combin. **42**, 1–14 (2014)

15. Rémi de Joannis de Verclos and Jean-Sébastien Sereni: Equitable colorings of K_4-minor-free graphs. J. Graph Algorithms Appl. **21**(6), 1091–1105 (2017)
16. Dirac, G.A.: Some theorems on abstract graphs. Proc. Lond. Math. Soc. **3**(2), 69–81 (1952)
17. Dong, A., Zhang, X.: Equitable coloring and equitable choosability of graphs with small maximum average degree. Discuss. Math. Graph Theory **38**(3), 829–839 (2018)
18. Erdős., P.: Theory of Graphs and It's Applications, p. 159. Czeck. Acad. Sci. Publ., Prague (1964). Problem 9
19. Esperet, L., Lemoine, L., Maffray, F.: Equitable partition of graphs into induced forests. Discrete Math. **338**(8), 1481–1483 (2015)
20. Grünbaum, B.: A result on graph-coloring. Michigan Math. J. **15**, 381–383 (1968)
21. Hajnal, A., Szemerédi, E.: Proof of a conjecture of P. Erdős. In: Combinatorial Theory and Its Applications, II (Proc. Colloq., Balatonfüred, 1969), pp. 601–623. North-Holland, Amsterdam (1970)
22. Hansen, P., Hertz, A., Kuplinsky, J.: Bounded vertex colorings of graphs. Discrete Math. **111**(1-3), 305–312 (1993). Graph theory and combinatorics (Marseille-Luminy, 1990)
23. Irani, S., Leung, V.: Scheduling with conflicts, and applications to traffic signal control. In: Proceedings of the Seventh Annual ACM-SIAM Symposium on Discrete Algorithms, pp. 85–94, 01 (1996)
24. Janson, S.: Large deviations for sums of partly dependent random variables. Random Struct. Algorithms **24**(3), 234–248 (2004)
25. Janson, S., Ruciński, A.: The infamous upper tail. Random Struct. Algorithms **20**(3), 317–342 (2002)
26. Kaul, H., Jacobson, S.H.: New global optima results for the Kauffman NK model: handling dependency. Math. Program. **108**(2–3), 475–494 (2006)
27. Kaul, H., Mudrock, J.A., Pelsmajer, M.J., Reiniger, B.: Proportional choosability: a new list analogue of equitable coloring. Discrete Math. **342**(8), 2371–2383 (2019)
28. Kierstead, H.A., Kostochka, A.V.: An Ore-type theorem on equitable coloring. J. Combin. Theory Ser. B **98**(1), 226–234 (2008)
29. Kierstead, H.A., Kostochka, A.V.: A short proof of the Hajnal-Szemerédi theorem on equitable colouring. Combin. Probab. Comput. **17**(2), 265–270 (2008)
30. Kierstead, H.A., Kostochka, A.V.: Every 4-colorable graph with maximum degree 4 has an equitable 4-coloring. J. Graph Theory **71**(1), 31–48 (2012)
31. Kierstead, H.A., Kostochka, A.V.: Equitable list coloring of graphs with bounded degree. J. Graph Theory **74**(3), 309–334 (2013)
32. Kierstead, H.A., Kostochka, A.V., Yu, G.: Extremal graph packing problems: Ore-type versus Dirac-type. In: Surveys in combinatorics 2009, London Mathematical Society Lecture Note Series, vol. 365, pp. 113–135. Cambridge University Press, Cambridge (2009)
33. Kierstead, H.A., Kostochka, A., Xiang, Z.: Equitable list coloring of planar graphs with given maximum degree. J. Graph Theory, n/a(n/a) (2024)
34. Kierstead, H.A., Kostochka, A., Xiang, Z.: Equitable list coloring of sparse graphs (2024)
35. Kierstead, H.A., Kostochka, A.V.: Ore-type versions of brooks' theorem. J. Comb. Theory, Ser. B **99**(2), 298–305 (2009)
36. Kierstead, H.A., Kostochka, A.V.: A refinement of a result of corrádi and hajnal. Combinatorica, pp.1–16 (2014)
37. Kierstead, H.A., Kostochka, A.V.: Equitable versus nearly equitable coloring and the chen-lih-wu conjecture. Combinatorica **30**(2), 201–216 (2010)
38. Kierstead, H.A., Kostochka, A.V., Mydlarz, M., Szemerédi, E.: A fast algorithm for equitable coloring. Combinatorica **30**(2), 217–224 (2010)
39. Kim, R., Oum, S-l., Zhang, X.: Equitable partition of planar graphs. Discrete Math. **344**(6), Paper No. 112351, 6, (2021)
40. Kostochka, A.V.: Equitable colorings of outerplanar graphs. Discrete Math. **258**(1–3), 373–377 (2002)
41. Kostochka, A.V., Nakprasit, K.: Equitable colurings of d-degenerate graphs. Combin. Probab. Comput. **12**(1), 53–60 (2003)

42. Kostochka, A.V., Nakprasit, K.: On equitable Δ-coloring of graphs with low average degree. Theoret. Comput. Sci. **349**(1), 82–91 (2005)
43. Kostochka, A.V., Nakprasit, K., Pemmaraju, S.V.: On equitable coloring of d-degenerate graphs. SIAM J. Discrete Math. **19**(1), 83–95 (2005)
44. Kostochka, A.V., Pelsmajer, M.J., West, D.B.: A list analogue of equitable coloring. J. Graph Theory **44**(3), 166–177 (2003)
45. Kostochka, A.V., Yu, G.: Extremal problems on packing of graphs. Technical Report 1, Oberwolfach report, pp. 55–57 (2006)
46. Kostochka, A.V., Yu, G.: Ore-type graph packing problems. Combin. Probab. Comput. **16**(1), 167–169 (2007)
47. Kostochka, A., Lin, D., Xiang, Z.: Equitable coloring of planar graphs with maximum degree at least eight. Discrete Math. **347**(6), Paper No. 113964, 13 (2024)
48. J. Krarup and D. de Werra. Chromatic optimisation: Limitations, objectives, uses, references. *European Journal of Operational Research*, 11(1):1–19, 1982. Third EURO IV Special Issue
49. Krivelevich, M., Patkós, B.: Equitable coloring of random graphs. Random Structures Algorithms **35**(1), 83–99 (2009)
50. Larman, D.G.: On the number of complete subgraphs and circuits in a graph. Proc. Roy. Soc. London Ser. A **308**, 327–342 (1969)
51. Li, Q., Yuehua, B.: Equitable list coloring of planar graphs without 4- and 6-cycles. Discrete Math. **309**(1), 280–287 (2009)
52. Ko-Wei Lih and Pou-Lin Wu. On equitable coloring of bipartite graphs. *Discrete Math.*, 151(1-3):155–160, 1996. Graph theory and combinatorics (Manila, 1991)
53. Rong Luo, Jean-Sébastien Sereni, D. Christopher Stephens, and Gexin Yu. Equitable coloring of sparse planar graphs. *SIAM J. Discrete Math.*, 24(4):1572–1583, 2010
54. Meyer, W.: Equitable coloring. Amer. Math. Monthly **80**, 920–922 (1973)
55. D. Miyata, S. Tokunaga, and A. Kaneko. Equitable coloring of trees. unpublished
56. Nakprasit, K., Nakprasit, K.: Equitable colorings of planar graphs without short cycles. Theoret. Comput. Sci. **465**, 21–27 (2012)
57. Kittikorn Nakprasit. *Coloring and packing problems for d-degenerate graphs*. ProQuest LLC, Ann Arbor, MI, 2004. Thesis (Ph.D.)–University of Illinois at Urbana-Champaign
58. Nakprasit, K.: Equitable colorings of planar graphs with maximum degree at least nine. Discrete Math. **312**(5), 1019–1024 (2012)
59. Pelsmajer, M.J.: Equitable list-coloring for graphs of maximum degree 3. J. Graph Theory **47**(1), 1–8 (2004)
60. Sriram V. Pemmaraju. Equitable coloring extends Chernoff-Hoeffding bounds. In *Approximation, randomization, and combinatorial optimization (Berkeley, CA, 2001)*, volume 2129 of *Lecture Notes in Comput. Sci.*, pages 285–296. Springer, Berlin, 2001
61. Rödl, V., Ruciński, A.: Perfect matchings in ϵ-regular graphs and the blow-up lemma. Combinatorica **19**(3), 437–452 (1999)
62. Barry F. Smith, Petter E. Bjørstad, and William D. Gropp. *Domain decomposition*. Cambridge University Press, Cambridge, 1996. Parallel multilevel methods for elliptic partial differential equations
63. Tucker, A.: Perfect graphs and an application to optimizing municipal services. SIAM Rev. **15**, 585–590 (1973)
64. Wang, H.: Independent directed triangles in a directed graph. Graphs Combin. **16**(4), 453–462 (2000)
65. Wang, W.-F., Lih, K.-W.: Equitable list coloring of graphs. Taiwanese J. Math. **8**(4), 747–759 (2004)
66. Williams, L., Vandenbussche, J., Gexin, Yu.: Equitable defective coloring of sparse planar graphs. Discret. Math. **312**(5), 957–962 (2012)
67. Jianliang, W., Wang, P.: Equitable coloring planar graphs with large girth. Discrete Math. **308**(5–6), 985–990 (2008)
68. Hian Poh Yap and Yi Zhang: The equitable Δ-colouring conjecture holds for outerplanar graphs. Bull. Inst. Math. Acad. Sinica **25**(2), 143–149 (1997)

69. Zelinka, B.: On the number of independent complete subgraphs. Publ. Math. Debrecen **13**, 95–97 (1966)
70. Zhang, X.: On equitable colorings of sparse graphs. Bull. Malays. Math. Sci. Soc. **39**(1), S257–S268 (2016)
71. Zhang, X., Jian-Liang, W.: On equitable and equitable list colorings of series-parallel graphs. Discrete Math. **311**(10–11), 800–803 (2011)
72. Zhang, Y., Yap, H.-P.: Equitable colorings of planar graphs. J. Combin. Math. Combin. Comput. **27**, 97–105 (1998)
73. Zhu, J., Yuehua, B.: Equitable list colorings of planar graphs without short cycles. Theoret. Comput. Sci. **407**(1–3), 21–28 (2008)
74. Zhu, J., Yuehua, B.: Equitable list colorings of planar graphs without short cycles. Theoret. Comput. Sci. **407**(1), 21–28 (2008)

Delta-System Method: A Survey

Andrey Kupavskii

Abstract A sunflower, or a Δ-system is a collection of sets such that the intersection of any two sets coincides with the intersection of all sets. The topic of this survey is the Δ-system method in extremal set theory. It originated in the 1970s from the work of Erdős, Deza and Frankl and was developed by Frankl and Füredi in the 1980s. Roughly speaking, it comprises several approaches of how to analyze a family of sets by looking at the Δ-systems, or sunflowers, that it contains. In this survey, we cover the method and its various applications.

1 Introduction

In 1960 Erdős and Rado [40] published a paper that, in retrospect, became one of the most influential papers in extremal set theory. (It was also probably the second paper on the subject, after the famous paper of Sperner.) They proved a result of Ramsey theoretic flavour, stating that in any sufficiently large family of sets of bounded size there is a homogeneous substructure, called a Δ-*system* (also known under the name of a *sunflower*). For many qualitative results in Discrete Mathematics and Theoretical Computer Science, this has become a very powerful tool to analyze complex set families. Extremal set theory flourished in the 1970–1980s, and many exciting developments happened then. One of them was the development of the Δ-*system method* in the works of Frankl and Füredi. In this survey, we try to give a concise picture of this method starting from its early stages and to the modern day. We also tried to present the proofs of most of the key results. On top of this, we survey the literature on the problems that the Delta-systems was applied to.

We should also note that two other excellent sources on the Delta-system method are the survey of Füredi [65] and the book of Frankl and Tokushige [60]. The content of this survey turned out to have quite significant intersections with these two sources.

A. Kupavskii (✉)
Moscow Institute of Physics and Technology, St. Petersburg State University, Innopolis University, Innopolis, Russia
e-mail: kupavskii@ya.ru

G. O. H. Katona et al. (eds.), *Sum(m)it280*, Bolyai Society Mathematical Studies 32,
https://doi.org/10.1007/978-3-032-18810-6_13

We decided to leave it as it is, partly because we aimed for completeness, and partly because presenting proofs allowed us to elaborate on some of the aspects of the method.

1.1 Notation

We use notation, standard for extremal set theory: for non-negative integers n, a, b, put $[n] = \{1, \dots, n\}$ and, more generally, $[a, b] = \{a, a+1, \dots, b\}$; for a set X, let 2^X ($\binom{X}{k}$, $\binom{X}{\leqslant k}$) stand for the collection of all subsets of X (all k-element subsets of X, all $\leqslant k$-element subsets of X); a *family* is simply a collection of subsets of the (not necessarily specified) *ground set*; a *matching* is a family of sets that are pairwise disjoint. We typically denote elements of the ground set by lower case letters (x, y, v etc.), sets (subsets of the ground set) by capital letters (A, B, F etc.), and families by calligraphic letters ($\mathcal{A}, \mathcal{B}, \mathcal{F}$ etc.). For a family $\mathcal{F}$, we use $\cup\mathcal{F}$ as a shorthand for $\cup_{F\in\mathcal{F}} F$, and similarly for $\cap\mathcal{F}$. Given a family $\mathcal{F}$ and an element x, the *degree of* x is the number of sets from $\mathcal{F}$ containing x.

We use the following standard notation for families $\mathcal{F}$, $\mathcal{S}$ and a set X.

$$\mathcal{F}(X) = \big\{A \setminus X : A \in \mathcal{F}, X \subset A\big\},$$
$$\mathcal{F}[X] = \big\{A : A \in \mathcal{F}, X \subset A\big\},$$
$$\mathcal{F}[\mathcal{S}] = \bigcup_{X\in\mathcal{S}} \mathcal{F}[X].$$

We use $\mathcal{F}(x)$ as a shorthand for $\mathcal{F}(\{x\})$. We use $O_{k,s}(n)$ to denote a function of the form $g(k,s)\cdot O(n)$ for some $g(k,s)$, and similarly for $\Omega_k(n)$ etc. Notation related to Füredi's structural theorem is given in Sect. 3.1. For a family $\mathcal{G}$ and integer m, the *shadow on level* m is defined as follows: $\partial^{(m)}\mathcal{G} := \cup_{G\in\mathcal{G}}\binom{G}{m}$.

The key extremal functions that we study are $m(n,k,L)$, the largest size of family in $\binom{[n]}{k}$ with pairwise intersections in L (see the beginning of Sect. 2); $f(n,k,\ell,s)$, the largest family in $\binom{[n]}{k}$ with no sunflower with s petals and core of size ℓ (see Sect. 5.1); $sim(n,k,d)$, $ssim(n,k,d)$, $sim_\ell(n,k,d)$, $ssim_\ell(n,k,d)$, the largest sizes of families without different simplices and special simplices (see Sect. 5.3); more generally, $ex_k([n],\mathcal{H})$, the largest family in $\binom{[n]}{k}$ that does not contain a fixed family $\mathcal{H}$. Sometimes we write *avoids* $\mathcal{H}$ or *is* $\mathcal{H}$*-free* to denote the same thing.

1.2 Δ-Systems and the Result of Erdős and Rado

A family $\mathcal{A} = \{A_1, \dots, A_s\}$ of sets is a $\Delta(s)$*-system*, or a *sunflower* if $A_i \cap A_j = \bigcap_{i\in[s]} A_i$ for all $i \neq j \in [s]$. If we want to specify the size, the we say that it is a $\Delta(s)$-system or an s-sunflower. The set $\bigcap_{i\in[s]} A_i$ is called the *core* (or *kernel*) of the Δ-system. Erdős and Rado introduced Δ-systems in 1960 in the paper [40] and proved the following result.

Theorem 1 (Erdős and Rado [40]) *If $\mathcal{F}$ is a family of sets of size $\leqslant k$ and $|\mathcal{F}| > k!(s-1)^k$, then $\mathcal{F}$ contains a $\Delta(s)$-system.*

Proof The proof is by induction on k. The result is evident for $k = 1$. Assuming that the result holds for $k-1$, let us prove it for k. Take the largest matching $M_1, \ldots, M_\ell \in \mathcal{F}$. If $\ell \geqslant s$, then $\{M_1, \ldots, M_s\}$ form a $\Delta(s)$-system, and thus the theorem is proved. If $\ell \leqslant s-1$, then $M := \bigcup_{i\in[\ell]} M_i$ has size at most $(s-1)k$. By maximality of the chosen matching, any set from $\mathcal{F}$ intersects M. Thus, there is an element $x \in M$ such that

$$\mathcal{F}(x) := \{F \setminus \{x\} : x \in F, F \in \mathcal{F}\}$$

has size at least $\frac{|\mathcal{F}|}{(s-1)k} > (k-1)!(s-1)^{k-1}$. The sets in $\mathcal{F}(x)$ have size at most $(k-1)$, and $|\mathcal{F}(x)| > (k-1)!(s-1)^{k-1}$. Thus we may apply the inductive assumption to $\mathcal{F}(x)$ and find a $\Delta(s)$-system in $\mathcal{F}(x)$. If we append x to every set in that $\Delta(s)$-system, we get a $\Delta(s)$-system in $\mathcal{F}$, as claimed. □

Let us denote

$$\phi(k,s) := \max\Big\{|\mathcal{F}| : \mathcal{F} \text{ consists of sets of size } \leqslant k \text{ and } \mathcal{F} \text{ contains no } \Delta(s+1)-\text{system}\Big\}. \quad (1)$$

In these terms, Theorem 1 states that $\phi(k,s) \leqslant k!s^k$. Also note that the maximum is attained on a family of sets of size k: we can always complement sets of size $< k$ by new elements of the ground set that are all pairwise distinct. Clearly, this will not create new Δ-systems. Abbot et al. [1] in 1972, and then Spencer [116] in 1977 improved upper bounds on $\phi(k,s)$. The result of Spencer states that for any fixed s and $\epsilon > 0$ there exists C such that $\phi(k,s) \leqslant Ck!(1+\epsilon)^k$. Kostochka [87] improved this bound for $\Delta(s)$-free systems to $\phi(k,s) \leqslant C(s,\alpha)k!\Big(\frac{(\log\log\log k)^2}{\alpha \log\log k}\Big)^k$ for any $\alpha > 1$ and a certain function $C(s,\alpha)$. For fixed k and large s [93] showed that $\phi(k,s) = (1+o(1))k^s$.

One natural example for a $\Delta(s+1)$-free k-uniform family is a k-partite hypergraph with parts of size s. It has s^k sets. There are, however, examples that are exponentially bigger. The authors of [1] showed that $\phi(2,s) = s(s+1)$ for even s and $s^2 + \frac{s-1}{2}$ for odd s. The following observation allows us to extend the graph bound to higher uniformities.

Observation 2 *We have $\phi(a+b,s) \geqslant \phi(a,s)\phi(b,s)$.*

Proof Let $\mathcal{A}$ be a family of a-element sets with no $\Delta(s)$-system and of size $\phi(a,s)$. Define $\mathcal{B}$, $|\mathcal{B}| = \phi(b,s)$ similarly. Assume that $\mathcal{A}$ and $\mathcal{B}$ live on disjoint ground sets X and Y, respectively. Consider a family $C := \{A \sqcup B : A \in \mathcal{A}, B \in \mathcal{B}\}$. It is easy to see that $|C| = \phi(a,s) + \phi(b,s)$ and that it is a $\Delta(s+1)$-system free. □

Abbott, Hanson and Sauer also showed a lower bound $\phi(k,3) \geqslant 10^{(k/2)-c\log k}$ with some positive constant c, which is exponentially better than 3^k which follows from the construction above. They used a construction of a 3-uniform family of size 10 and with no $\Delta(3)$-system, and then leveraged it to any uniformity using an iterated product construction, which gives a recursion $\psi(ab) \geqslant \psi(a)\psi(b)^a$, where $\psi(a)$ is the size of their iterated construction of uniformity a. The celebrated conjecture of Erdős and Rado, mentioned in their paper [40], suggests that we cannot improve the lower bound given by a complete k-partite hypergraph too much further. It states that the dependence on k is exponential: $\phi(k,s) \leqslant (Cs)^k$ for some absolute constant C.

For a more detailed survey on the Erdős–Rado question and related questions (weak Δ-systems, as well as the Erdős–Szémeredi question on Δ-systems for a restricted ground set), we refer to a survey of Kostochka [88].

1.3 How Δ*-Systems Changed to Sunflowers?*

The original name of the object from the paper of Erdős and Rado is a Δ-*system*, and it was called this way for the next 20 years.

The first occurrence of the name *sunflower* I could trace is in the name of the paper of Deza and Frankl [28] from 1981 "Every large set of equidistant $(0, +1, -1)$-vectors form a sunflower," which can be seen as a follow-up to the paper of Deza [25] that we discuss below. Funnily, the word "sunflower" does not appear in the body of the paper, only in the title. Moreover, the conclusion of the main theorem of that paper guarantees a structure which is not exactly a sunflower, but a (one possible) $(-1, 0, 1)$-generalization of it. Peter Frankl (private communication) confirmed that he came up with the term while working with Michel Deza on the paper [28].

The second occurrence of the term I could trace (and the first one where it is actually used in the paper and gets its present meaning) is the influential paper of Alon et al. [3] from 1985 on the chromatic number of Kneser hypergraphs. Two years later, in 1987, Alon and Boppana [4] write another influential paper "Monotone circuit complexity of Boolean functions", where they use structures akin to sunflowers for proving certain complexity lower bounds. Their result extended a breakthrough result of Razborov [111], who also used similar structures in the argument. My guess is that it is at this point, when the term entered TCS community, that it solidified. Later, it appeared in different other complexity lower bounds papers, in Jukna's book [79], and ultimately took over the Δ-system terminology.

In combinatorics community, Δ-system terminology prevailed for a while. But the object went under different names, notably, it was called an s-*star* in some of the most important papers on the subject of the Δ-system method. This term was introduced by Chung and, in particular, used by Füredi in his paper [63] and by Frankl and Füredi in [54]. Peter Frankl (private communication) pointed out that the s-star term came as a generalization of graph stars, which are 2-uniform Δ-systems with kernel of size 1. In another influential paper [55], however, Frankl and Füredi used the sunflower terminology.

1.4 Recent Progress on the Erdős–Rado Conjecture

A breakthrough in the study of the Erdős–Rado Delta-system question came several years ago in a work of Alweiss et al. [5]. They managed to reduce the dependence on k in Theorem 1 from k^k to essentially $(\log k)^k$. It spurred a lot of research. Several follow-up papers slightly improved upon the bound and presented the proof quite differently. Rao [109] gave a code-theoretic proof with a better bound; Bell et al. [7] improved upon some of the arguments of the original proof, which lead to the bound $\phi(s, k) \leqslant (Cs \log k)^k$; Tao [118] gave a proof based on entropy, which, however, contained a mistake; Lu [75] and then Stoeckl [117] gave another entropy proof, which gave the bound $\phi(s, k) \leqslant (64s \log k)^k$; Mossel et al. [103] gave a second moment proof; finally, Rao [110] gave the same bound with a very short combinatorial proof in the spirit of the original paper [5].

The proof is based on a result, which is called the spread lemma. Given a real number $r > 1$, we say that a family $\mathcal{F}$ of sets is *r-spread* if for each set X we have $|\mathcal{F}(X)| < r^{-|X|}|\mathcal{F}|$. We say that W is a *p-random subset* of $[n]$ if each element of $[n]$ is included in W with probability p and independently of others.

Theorem 3 (The spread lemma, [5], a sharpening due to [117]) *If for some $n, k, r \geqslant 1$ a family $\mathcal{F} \subset \binom{[n]}{\leqslant k}$ is r-spread and W is a $(\beta\delta)$-random subset of $[n]$, then*

$$\Pr[\exists F \in \mathcal{F} \ : \ F \subset W] > 1 - \Big(\frac{2}{\log_2(r\delta)}\Big)^{\beta} k.$$

In order to prove the aforementioned bound on $\phi(k, s)$, one needs, first, to find a spread subfamily $\mathcal{F}(X)$ in a large family $\mathcal{F}$. (Choose X to be a maximal set that violates the r-spreadness condition.) Then, using the spread lemma, one sees that, if the parameters are chosen right (say, $r = 32s \log_2(2k)$, $\delta = 1/(2s \log_2(2k))$, $\beta = \log_2(2k)$), a random $\frac{1}{2s}$-subset contains a set from $\mathcal{F}(X)$ with probability at least $1/2$. Then, in a random $2s$-coloring there are in expectation at least s colors that contain a set from the family. These sets form a matching in $\mathcal{F}(X)$, which gives a $\Delta(s)$-system with kernel X in $\mathcal{F}$.

The paper of Rao [110] gives a broad view on different applications of sunflowers.

1.5 Structure of the Survey

Roughly speaking, the survey consists of three parts. In the first part (Sects. 2–5), we discuss the development of the method in the 1970s and 1980s. In the second part (Sects. 6–8), we discuss aspects of the method and some relevant recent developments. In the third part (Sect. 9), I tried to make a comprehensive survey of the literature that deals with the problems from the earlier parts.

Section 2 contains the early manifestations of the method. It starts with the Erdős–Ko–Rado theorem, then it gives the result of Deza on Δ-systems and families with one intersection size, and then presents the theorem of Deza, Erdős and Frankl on (n, k, L)-systems. The latter is considered to be the starting point of the method. Then we discuss the development of the method by Peter Frankl up until 1983. One thing that I realized while going through the literature is that Peter Frankl developed a (lesser known) approach that can be considered as an 'intermediate' approach between the earlier 'base' variant of the method and the 'homogeneous families' variant.

Section 3 is devoted to Füredi's structural theorem on homogeneous families, which underlies further development of the method, and its early applications. In particular, we present the results of Füredi on families with three intersection sizes.

In Sect. 4, we discuss the 'Forbidding one intersection' paper of Frankl and Füredi, which to a large extent shaped the Δ-system method. The result of Füredi meets the result of Frankl and Katona on families with many intersection conditions, and then they develop the shadow/stability machinery to get exact results.

Section 5 is devoted to the paper 'Exact solutions of Turán-type problems', in which Frankl and Füredi applied the method to a variety of Turán-type questions. Although most of the key Δ-system-related ideas were already present in the previous paper, here they harvest the fruits of their previous findings, but also delineate the scope and the limitations of the method in terms of classes of hypergraphs that it works for.

In Sect. 6, we discuss stability and supersaturation that could be achieved using Delta-systems. Stability via Delta-systems is an aspect that is often overlooked in the literature, and we go through the proof of Frankl and Füredi [54] and derive an efficient stability result. We present some further ideas on stability via Delta-systems for harder problems. We also discuss a recent result of Jiang and Longbrake that gives a supersaturation result for 2-reducible trees via Δ-systems.

In Sect. 7, we discuss different approaches to base constructions. This is a key part of the earlier variant of the method, and can be used in conjunction with other approaches. We go over different approaches that appeared in the literature and present the peeling/simplification procedure, which is an important part of the spread approximation method.

In Sect. 8, we briefly discuss other general methods that were developed recently to tackle Túran-type problems: junta approximations, spread approximations and hypercontractivity. We do not go in depth on the methods, but rather underline the different regimes, in which these methods work. In particular, we show that, as it stands now, each of the Δ-system method, hypercontractivity method and spread approximation method allows to solve Turán-type problems in incompatible regimes of parameters.

In Sect. 9, we survey literature concerning the topics of the previous sections, and especially further studies on the topics of the papers [26, 54, 55]. We cover forbidden one intersection problem; (n, k, L)-systems; forbidden sunflowers and unavoidable hypergraphs; forbidden simplices and clusters; forbidden hypertrees and expanded hypergraphs.

2 Early Days of the Delta-System Method

2.1 (n, k, L)-Systems of Deza, Erdős and Frankl, the Precursor of the Delta-System Method and the Proof of the Erdős–Ko–Rado Theorem for Large n

Δ-system method was introduced in the paper of Deza et al. [26]. They introduced and studied the following question, along with its several generalizations. Given positive integers n, k and a set $L \subset [0, k-1]$, we say that $\mathcal{F} \subset \binom{[n]}{k}$ is an *(n, k, L)-system* if for any $A \neq B \in \mathcal{F}$ we have $|A \cap B| \in L$. The question is as follows:

> what is the largest size $m(n, k, L)$ of an (n, k, L)-system, for different values of n, k, L?[1]

Before discussing the result of Deza, Erdős and Frankl and the Δ-system method, we want to put the result and the method into context. Another cornerstone result (and probably the third paper on the topic of extremal set theory) in extremal set theory is the Erdős–Ko–Rado theorem [38]. It determines for all n, k the size of the largest $(n, k, \{1, 2, \dots, k-1\})$-system, and moreover, the size of the largest $(n, k, \{t, t+1, \dots, k-1\})$-system for any $t \in [1, k-1]$, provided $n \geqslant n_0(k)$. (Such families are called *intersecting* and *t-intersecting*, respectively.) It is instructive to discuss the proof of the EKR theorem.

Theorem 4 (Erdős, Ko and Rado [38]) *Let $n \geqslant k \geqslant t$ be positive integers and suppose that $\mathcal{F} \subset \binom{[n]}{k}$ is t-intersecting, that is, $|A \cap B| \geqslant t$ for any $A, B \in \mathcal{F}$. Then $|\mathcal{F}| \leqslant \binom{n-t}{k-t}$ for $n > n_0(k)$.*

Here, we sketch a variant of their proof. In the proof of Erdős, Ko and Rado, $n_0(k)$ is roughly $\binom{k}{t}^3$.

Proof Take a maximal t-intersecting family $\mathcal{F}$. Maximality implies that there are $A, B \in \mathcal{F}$ with $|A \cap B| = t$. There are two cases to consider. If $|\bigcap_{A \in \mathcal{F}} A| \geqslant t$, then $|\mathcal{F}| \leqslant \binom{n-t}{k-t}$ and we are done. If $|\bigcap_{A \in \mathcal{F}} A| < t$ then take two sets $A, B \in \mathcal{F}$ such that $|A \cap B| = t$ and add a third set C such that $C \not\supset A \cap B$ (we leave the verification of the existence of such A, B, C as an exercise). Then $|A \cap B \cap C| < t$. Any set $F \in \mathcal{F}$ intersects each of A, B, C in $\geqslant t$ elements, and thus it must intersect $A \cup B \cup C$ in at least $t+1$ elements. Therefore, $|\mathcal{F}| \leqslant \binom{|A \cup B \cup C|}{t+1}\binom{n-t-1}{k-t-1} \leqslant \binom{3k}{t+1}\binom{n-t-1}{k-t-1}$. For $n > n_0(k)$ this is smaller than $\binom{n-t}{k-t}$. $\square$

Summarizing, the 'trivial t-intersecting' families are covered by one set of size t (that is, all sets from the family contain a given set of size t), while the non-trivial families are covered by $const(k, t)$ sets of size $t+1$, and for large n the latter gives

---

[1] In the paper [54], Frankl and Füredi attribute the idea of using Delta-systems in this context to Deza.

smaller sizes since the number of ways to extend sets of size $t+1$ to sets of size k is roughly n times smaller than the number of ways to extend sets of size t. The important common point between the Δ-system method and the proof above is that in both we compare contributions of 'cover sets' of different sizes. The Δ-system method, however, finds and deals with these covers more efficiently.

2.2 Families of Sets with Only One Intersection Size

One result that precedes the general question of Deza, Erdős, and Frankl is due to Deza and deals with the case when $L = \{\ell\}$, that is, all pairwise intersections in the family have size ℓ. According to the paper of Deza [24], the following conjecture was posed by Erdős on a seminar in Paris in 1973.

> Let $k > \ell$ be positive integers and consider a family $\mathcal{F}$ of at least $k^2 - k + 2$ sets, each of size k, such that for any $A, B \in \mathcal{F}$ we have $|A \cap B| = \ell$. Then $\mathcal{F}$ is a Δ-system.

The bound $k^2 - k + 2$ is best possible: for $k-1$ being a prime power, one can consider a projective plane of order $k-1$. It contains exactly $k^2 - k + 1$ lines, each two intersecting in exactly 1 point and is obviously not a sunflower.

Deza [24] noticed that this conjecture is true because it immediately reduces to his result from the recently published paper [25], in which he studied one-distance codes. The main result of [25] has an elegant proof with geometric flavour, and we present it here. (For the rest of this subsection, u, v, w will exceptionally stand for vectors.) Recall that, given two vectors v, w in $\{0,1\}^n$, the *Hamming distance* between v and w is the number of coordinates i such that $v_i \neq w_i$. Also recall that there is a natural correspondence between vectors in $\{0,1\}^n$ and subsets of $[n]$: set $\leftrightarrow$ characteristic vector.

Theorem 5 (Deza [25]) *Let m, q be integers and $\mathcal{F}$ be a collection of vectors $v^0, v^1, \ldots v^m \in \{0,1\}^n$ such that the Hamming distance between any two vectors is $2q$. If $m > q^2 + q + 1$ then the code is* trivial*: for each coordinate $i \in [n]$, either $\sum_{j \in [m]} v_i^j \leqslant 1$ or $\sum_{j \in [m]} v_i^j \geqslant m-1$.*

This theorem allows us to deduce the conjecture of Erdős and Lovász.

Corollary 6 (Deza [24]) *Let $k > \ell$ be positive integers and consider a family $\mathcal{F}$ of at least $k^2 - k + 2$ sets, each of size k, such that for any $A, B \in \mathcal{F}$ we have $|A \cap B| = \ell$. Then $\mathcal{F}$ is a Δ-system.*

Proof Consider a collection of m sets, each of size k and with pairwise intersections of size $\ell \geqslant 1$. Note that, seen as vectors $v^1, \ldots, v^m$, they have pairwise Hamming distance $2(k-\ell)$. We want to transform them into vectors with pairwise Hamming distance equal to their Hamming weight. For this, add to each vector $\ell - 1$ new coordinates with ones, such that these $\ell - 1$ coordinates for different vectors are pairwise distinct, as well as $(k - \ell - 1)$ new coordinates where all vectors will have ones. It is easy to check that now each vector has Hamming weight $2q := 2(k-1)$, and pairwise Hamming distances are also $2q$. Now we add the all-zero vector v^0 to the collection and get a family of $m+1$ vectors $v^0, \dots, v^m$ forming a code as in Theorem 5. Thus, if $m > (k-1)^2 + (k-1) + 1$, then the code is trivial. This condition is equivalent to $m > k^2 - k + 1$.

In set terms, what does it mean for the corresponding code to be trivial? If two vectors have a common 1, then all vectors have 1 there, which means that any element common for two sets is common for all sets. This is exactly the definition of a Δ-system. □

Let us prove Theorem 5. The first step can be summarized in the following elegant lemma, which is an interpretation and generalization of [25, Theorem 2.1].

Lemma 7 (Lemma on cross-variance) *Consider two independent random vectors* $\mathbf{X}, \mathbf{Y} \in \mathbb{R}^n$. *Let* $(\mathbf{X_1}, \mathbf{Y_1})$, $(\mathbf{X_2}, \mathbf{Y_2})$ *be independent copies of* $(\mathbf{X}, \mathbf{Y})$. *Then*

$$2\mathsf{E}[(\mathbf{X}-\mathbf{Y})^2] \geqslant \mathsf{E}[(\mathbf{X}_1-\mathbf{X}_2)^2] + \mathsf{E}[(\mathbf{Y}_1-\mathbf{Y}_2)^2], \tag{2}$$

where the norm is the Euclidean norm.

We will need the case when the random vectors are uniformly distributed on sets of vectors in $\mathbb{R}^m$, and the vectors are actually $\{0, 1\}$-vectors.

Proof We have

$$\begin{aligned}
\mathsf{E}[(\mathbf{X}_1-\mathbf{X}_2)^2] &= 2\mathsf{E}[\mathbf{X}^2] - 2\mathsf{E}[\mathbf{X}]^2 \\
\mathsf{E}[(\mathbf{Y}_1-\mathbf{Y}_2)^2] &= 2\mathsf{E}[\mathbf{Y}^2] - 2\mathsf{E}[\mathbf{Y}]^2 \\
\mathsf{E}[(\mathbf{X}-\mathbf{Y})^2] &= \mathsf{E}[\mathbf{X}^2] + \mathsf{E}[\mathbf{Y}^2] - 2\mathsf{E}[\mathbf{X}]\mathsf{E}[\mathbf{Y}].
\end{aligned}$$

The statement now follows from $\mathsf{E}[\mathbf{X}]^2 + \mathsf{E}[\mathbf{Y}]^2 - 2\mathsf{E}[\mathbf{X}]\mathsf{E}[\mathbf{Y}] \geqslant 0$. □

The motivation behind the name *cross-variance of* $\mathbf{X}$ *and* $\mathbf{Y}$ for $\mathsf{E}[\|\mathbf{X}-\mathbf{Y}\|^2]$ is because $\mathsf{E}[\|\mathbf{X}_1-\mathbf{X}_2\|^2]$ is twice the variance of $\mathbf{X}$.

Proof of Theorem 5 We may w.l.o.g. assume that v^0 is the all-zero vector. For any coordinate j, we show that

$$s_j := \sum_{i\in[m]} v^i_j \quad \text{belongs to } [0, q+1] \cup [m-q, m]. \tag{3}$$

Split the set of indices $[0, m]$ into $A \sqcup B$, where $A := \{i \in [0, m] : v^i_j = 1\}$ and $B := [0, m] \setminus A$. For a vector $v^i \in \mathcal{F}$, define the vector $w^i \in \mathbb{R}^{n-1}$ obtained from v^i by deleting the j-th coordinate. Let us define a random vector $\mathbf{X}$ ($\mathbf{Y}$) by selecting a uniformly random $i \in A$ ($i \in B$) and taking the vector w^i. We apply Lemma 7 to $\mathbf{X}$ and $\mathbf{Y}$ and get

$$2\mathsf{E}[\|\mathbf{X}-\mathbf{Y}\|^2] \geqslant \mathsf{E}[\|\mathbf{X}_1-\mathbf{X}_2\|^2]+\mathsf{E}[\|\mathbf{Y}_1-\mathbf{Y}_2\|^2]$$

At the same time, $\|w^i - w^\ell\|^2$ is exactly the Hamming distance between w^i and w^ℓ. It is equal to $2q$ for any two distinct vectors with indices in A, or any two distinct vectors with indices in B. If the vectors correspond to different sets of indices, then their Hamming distance is $2q-1$. Also note that $|A| = s_j$ and $|B| = m - s_j + 1$, and thus $\mathsf{E}[\|\mathbf{X}_1-\mathbf{X}_2\|^2] = \frac{s_j-1}{s_j}\cdot 2q$, and $\mathsf{E}[\|\mathbf{Y}_1-\mathbf{Y}_2\|^2] = \frac{m-s_j}{m-s_j+1}\cdot 2q$. Combining with the displayed inequality, we get

$$\begin{aligned} 2(2q-1) &\geqslant 2q\left(\frac{s_j-1}{s_j}+\frac{m-s_j}{m-s_j+1}\right)\\ \frac{1}{s_j}+\frac{1}{m-s_j+1} &\geqslant \frac{1}{q}\\ (m+1)q &\geqslant s_j(m-s_j+1) \end{aligned}$$

The last inequality and the condition on m from the statement imply that $s_j \leqslant q+1$ or $s_j \geqslant m-q$. Consider the set of coordinates U such that $U := \{j : s_j \geqslant m-q\}$.

Claim 8 *We have* $|U| \leqslant q$.

Proof For any $j \in U$, there are at most q vectors $v \in \{v^1, \ldots, v^m\}$ such that $v_j = 0$. If $|U| \geqslant q+1$, then, taking $U' \subset U$ with $|U'| = q+1$, and by pigeon-hole principle, there are $m-(q+1)q \geqslant 2$ vectors from $\mathcal{F}$ that have 1's on each coordinate from U'. Since v^0 is the all-zero vector, every other vector has exactly $2q$ coordinates equal to 1. Thus, these two vectors have $q-1$ coordinates equal to 1 outside U', and thus their Hamming distance is at most $2q-2$, a contradiction. □

Claim 9 *For any vector* $v^i \in \mathcal{F}$, $i \in [m]$, *the set* $U^i := U \cap \{j : v^i_j = 1\}$ *has size at least* q.

Proof Put $\operatorname{supp}(v^{i'}) := \{j : v^{i'}_j = 1\}$. On the one hand,

$$\sum_{i'=1}^{m} |\operatorname{supp}(v^i) \cap \operatorname{supp}(v^{i'})| = 2q + (m-1)q = (m+1)q.$$

The first term is the sum of coordinates of v^i, and the second term accounts for the fact that each of the $m-1$ non-zero vectors has q common ones with v^i. On the other hand, if $|U^i| \leqslant q-1$ then

$$\sum_{i'=1}^{m} \sum_{j \in \mathrm{supp}(v^i)} v_j^{i'} \leqslant m|U^i| + (q+1)|\mathrm{supp}(v^i) \setminus U| \leqslant m(q-1) + (q+1)^2$$
$$= (m+1)q - m + q^2 + q + 1 < (m+1)q,$$

a contradiction. □

Combining the two claims, we see that $U^i = U$ for all $i \in [m]$, and thus $s_j = m$ for all $j \in U$. We have $|U| = q$, and so in order to assure Hamming distance $2q$, the vectors v_i, $i \in [m]$ must have pairwise disjoint sets of 1's outside U. This completes the proof of the theorem. □

Let us summarize that, as long as $\mathcal{F}$ is an $(n, k, \{\ell\})$-system and $|\mathcal{F}| > k^2 - k + 1$, then $\mathcal{F}$ is a Δ-system with kernel of size ℓ. Since outside the kernel the sets from the family are disjoint, we must have

$$|\mathcal{F}| \leqslant \frac{n-\ell}{k-\ell}.$$

2.3 The Result of Deza, Erdős and Frankl

This result came as a natural generalization of the result of Deza [24] and the result of Deza et al. [27], who studied the $|L| = 2$ case of the problem. In [27] the authors also cite and use a result of a preprint of Deza and Erdős "to appear in Aequationes Math.", which has never actually appeared. (My guess is that it was absorbed by the Deza–Erdős–Frankl paper.)

Theorem 10 (Deza et al. [26]) *Fix positive integers $n, k > 0$ and $r \geqslant 2$, and a set $L = \{\ell_1 < \ldots < \ell_r\} \subset [0, k-1]$. Assume that $\epsilon > 0$, $n \geqslant n_0(k, \epsilon)$, and that $\mathcal{F}$ is an (n, k, L)-system.*

(i) If $|\mathcal{F}| > \left(\epsilon + \max\{k - \ell_1 + 1, \ell_2^2 - \ell_2 + 1\}\right) \prod_{i=2}^{r} \frac{n-\ell_i}{k-\ell_i}$ then there exists a set D of cardinality ℓ_1 such that $D \subset A$ for every $A \in \mathcal{F}$.
(ii) If $|\mathcal{F}| \geqslant k^2 2^{r-1} n^{r-1}$ then

$$(\ell_2 - \ell_1) | (\ell_3 - \ell_2) | \ldots | (\ell_r - \ell_{r-1}) | (k - \ell_r).$$

(iii) We have

$$|\mathcal{F}| \leqslant \prod_{i=1}^{r} \frac{n-\ell_i}{k-\ell_i}.$$

First, remark that we restrict to $r \geqslant 2$ since the case $r = 1$ is dealt with in Corollary 6. Below we sketch the original proof of Deza, Erdős and Frankl and discuss the Δ-system part in detail. The proof of the theorem is by induction on k, the case $k = 1$ being trivial.

We first treat the case $\ell_1 = 0$. For each x, the family $\mathcal{F}(x)$ is an $(n-1, k-1, \{\ell_2 - 1, \dots, \ell_r - 1\})$-system. Apply inductive assumption (iii) to each $\mathcal{F}(x)$ and then do double-counting $k|\mathcal{F}| = \sum_{x\in[n]} |\mathcal{F}(x)|$. It is not hard to check that it gives the bound on $|\mathcal{F}|$ as is claimed in (iii). The divisibility part $(\ell_3 - \ell_2)|\dots|(k - \ell_r)$ is by inductive assumption (ii), applied to any large $\mathcal{F}(x)$. We must have at least one large $\mathcal{F}(x)$ by double counting, provided that $\mathcal{F}$ itself is large.

It remains to show that $\ell_2|(\ell_3 - \ell_2)$ in order to complete the proof of (ii) and the theorem in this case. First, we only leave vertices of high degree. Specifically, we remove elements x with $|\mathcal{F}(x)| \leqslant k^2 2^{r-2} n^{r-2}$ one by one, until we reach the situation with all x having degree more than that expression. (Such deletion is possible thanks to the generous lower bound on $|\mathcal{F}|$ in (ii).) Then, applying inductive statement (i) for $k-1$ to $\mathcal{F}(x)$ with any of the remaining x, we see that there must be a set D_x of size $\ell_2 - 1$ that all sets from $\mathcal{F}(x)$ must contain. Using the lower bound on $|\mathcal{F}(x)|$, it is not difficult to check that sets $D_x \cup \{x\}$ for different x either coincide or are disjoint and that, consequently, each $A \in \mathcal{F}$, if intersecting $D_x \cup \{x\}$, must contain it entirely.[2] Thus all intersections of the sets in $\mathcal{F}$ must be divisible by the size of $D_x \cup \{x\}$, which is ℓ_2. At the same time, intersection ℓ_3 must appear, since otherwise $|\mathcal{F}(x)|$ is too small.

The main challenge is in the case $\ell_1 \geqslant 1$ and, more specifically, in the part (i) of the theorem. Once we managed to show that all sets should contain a set X of size ℓ_1, we can finish the proof by applying inductive statement for the $(n - \ell_1, k - \ell_1, \{0, \ell_2 - \ell_1, \dots, \ell_r - \ell_1\})$ system $\mathcal{F}(X)$. (That is, reduce it to the $\ell_1 = 0$ case with a smaller k.)

Proving (i) for $\ell_1 \geqslant 1$ is where the Δ-system method pops up. The goal is to find an economical family $\mathcal{B}$ (the *base*) that *covers* $\mathcal{F}$ in the sense that, for each $A \in \mathcal{F}$ there is $B \in \mathcal{B}$ such that $B \subset A$, and then to bound the size of $\mathcal{F}$ using the structure of $\mathcal{B}$. The first observation is as follows.

Observation 11 *Let $k \geqslant \ell > 0$ be integers. Let A be set of size k and $\mathcal{F}$ is a $\Delta(k - \ell + 2)$-system with kernel D of size ℓ. Then $|A \cap F| \geqslant \ell$ for every $F \in \mathcal{F}$ implies $|A \cap D| \geqslant \ell$.*

Proof Put $\mathcal{F} = \{F_1, \dots, F_{k-\ell+2}\}$ and assume that $|A \cap D| = \ell - x$ for some $x \geqslant 1$. Then $|A \cap (F_i \setminus D)| \geqslant x$ for each i and, since $F_i \setminus D$ are pairwise disjoint, we get that $|A| \geqslant (\ell - x) + (k - \ell + 2)x = k + 1 + (k - \ell + 1)(x - 1) \geqslant k + 1$, a contradiction. □

[2] If $y \in D_x$ and $D_y \cup \{y\} \neq D_x \cup \{x\}$, then all sets containing x must contain D_x, and thus y. But then they also contain D_y, and have a fixed set of size bigger than ℓ_2, which reduces the number of available distances to $r - 2$. This, in turn, implies by induction that the degree of x is too small.

A very similar observation, that was not explicitly stated in [26], but was de-facto used there in the proof is as follows. We omit the proof, which is akin to the previous proof.

Observation 12 *Let k be an integer. Let A be set of size k and $\mathcal{F}$ is a $\Delta(k+1)$-system with kernel D. Then $|A \cap D| = |A \cap F|$ for some $F \in \mathcal{F}$.*

The following lemma is an important ingredient for the construction of the base of the family, which comes in the lemma after.

Lemma 13 *Let s, k, t be integers such that $t > (s-1)k$. For each $i \in [s]$ let $F_i^1, \dots F_i^t$ form a $\Delta(t)$-system with kernel E_i, where $|F_i^j| = k$ for $i \in [s]$, $j \in [t]$.*

Suppose that $E_1, \dots, E_s$ form a $\Delta(s)$-system with kernel D. Then there are distinct indices $j_1, \dots, j_s \in [t]$, such that the sets $F_1^{j_1}, \dots, F_s^{j_s}$ form a $\Delta(s)$-system with kernel D.

Proof For each $i = 1, \dots, s$ we shall find a set $F_i^{j_i}$ that is disjoint from

$$\Big(\bigcup_{\ell=1}^{i-1} F_\ell^{j_\ell} \cup \bigcup_{\ell=i+1}^{s} E_\ell\Big) \setminus D. \tag{4}$$

Then it should be clear that the sets $F_1^{j_1}, \dots, F_s^{j_s}$ form a $\Delta(s)$-system with kernel D. Why is it possible to find such sets? First, by the choice of $F_\ell^{j_\ell}$, $\ell < i$, we have

$$F_\ell^{j_\ell} \cap F_i^j \setminus D \subset F_i^j \setminus E_i$$

for any $\ell < i$ and any $j \in [t]$. Second, the sets $F_i^j \setminus E_i$, $j \in [t]$ are pairwise disjoint. Third, the set (4) has size at most $k(s-1)$. Concluding, we have $t > (k-1)(s-1)$, and so by the pigeon-hole principle we can find $j_i \in [t]$ such that $F_i^{j_i} \setminus E_i$ is disjoint from the displayed set. □

Using this lemma, we may construct the *base* $\mathcal{B}$ of the family $\mathcal{F}$. For a family $\mathcal{X} \subset 2^{[n]}$, denote by $\mathcal{X}^{(\ell)} := \mathcal{X} \cap \binom{[n]}{\ell}$ the ℓth layer in $\mathcal{X}$.

Lemma 14 *For any (n, k, L)-family $\mathcal{F} \subset \binom{[n]}{k}$ there exists a family $\mathcal{B} \subset \binom{[n]}{\leqslant k}$ such that for any $F \in \mathcal{F}$ there exists $B \in \mathcal{B}$ with $B \subset F$, and that satisfies the following:*

(i) for any $B \in \mathcal{B}$ we have $|B| \in L \cup \{k\}$;
(ii) for any $B \in \mathcal{B}^{(\ell_1)}$, the family $\mathcal{F}$ contains a $\Delta(k - \ell_1 + 2)$-system with center in B. For any $i \in [2, r]$ and $B \in \mathcal{B}^{(\ell_i)}$, the family $\mathcal{F}$ contains a $\Delta(k^i + 1)$-system with kernel B;
(iii) the family $\mathcal{B}^{(\ell_2)}$ does not contain a $\Delta(k - \ell_1 + 2)$-system. For each $i \in [3, r]$, the family $\mathcal{B}^{(\ell_i)}$ does not contain a $\Delta(k^{i-1} + 1)$-system. The family $\mathcal{B}^{(k)}$ does not contain a $\Delta(k^r + 1)$-system.

Note that we treat the first layers of $\mathcal{B}$ slightly differently.

Proof We construct the base $\mathcal{B}$ in a greedy fashion: first, we search for sets of size ℓ_1 that are a kernel of a $\Delta(k-\ell_1+2)$-system. If we find one, we remove the sets forming this Δ-system from the family $\mathcal{F}$, add the kernel to $\mathcal{B}$ and continue our search in the family. Once there are no more kernels of size ℓ_1 of $\Delta(k-\ell_1+2)$-systems, we start searching for sets of size ℓ_2 that are kernels of $\Delta(k^2+1)$-system in the remaining family. Once we ran out of such sets of size ℓ_2, we start searching for sets of size ℓ_3 that are kernels of a $\Delta(k^3+1)$-system in the remaining family etc.

At the last step, we include in $\mathcal{B}$ the remaining sets of size k and note that there is no $\Delta(k^r+1)$-system formed by these sets. Indeed, otherwise its core would have been included into $\mathcal{B}^{(\ell_i)}$ for some $i \in [r]$ and the sets would have been removed.

This verifies the first two parts of the lemma. To verify (iii), we argue indirectly. Assume that for some $i \geqslant 3$ the sets $B_1, \dots, B_{k^{i-1}+1} \in \mathcal{B}^{(\ell_i)}$ form a Δ-system (with core of size in L). Recall that each of them is a core of a $\Delta(k^i+1)$-system of sets from $\mathcal{F}$. We may apply Lemma 13 to these Δ-systems and conclude that the union of these Δ-systems contains a $\Delta(k^{(i-1)}+1)$-system with core of size at most ℓ_{i-1}. However, this contradicts the greedy construction of $\mathcal{B}$: we should have included the kernel of this sunflower in $\mathcal{B}$ before including $B_1, \dots, B_{k^{i-1}+1}$. The argument for $i=2$ is virtually identical. This contradiction proves the lemma. □

In retrospect, this lemma is far from being efficient if one tries to get best bounds on n when applying the Δ-system method. We shall discuss and compare different possible approaches for constructing bases in Sect. 7.

Let us return to the proof of the theorem. Lemma 14 allows us to relate the structure of the base and the size of $\mathcal{F}$. If $\mathcal{F}$ contains a $\Delta(k-\ell_1+2)$-system with core X of size ℓ_1, then by Observation 11, all sets in $\mathcal{F}$ must contain X, and we are done. Thus, we may assume that this is not the case and, in particular, $\mathcal{B}^{(\ell_1)}$ is empty. Next, remark that by Lemma 13 with $s=2$, for any two sets B_1, B_2 in $\mathcal{B}^{(\ell_2)}$ we can find two sets F_1, F_2 from the Δ-systems with cores B_1, B_2, respectively, such that $F_1 \cap F_2 = B_1 \cap B_2$. Since $|B_1 \cap B_2| < |B_1| = \ell_2$ and $|F_1 \cap F_2| \in L$, we get $|B_1 \cap B_2| = \ell_1$. Thus, if $\mathcal{B}^{(\ell_2)}$ contains at least $\max\{k-\ell_1+2, \ell_2^2-\ell_2+2\}$ sets, then $\mathcal{B}^{(\ell_2)}$ must be a $\Delta(\geqslant k-\ell_1+2)$-system with kernel K of size ℓ_1 by Corollary 6. But this contradicts Lemma 14 (iii). Thus, $|\mathcal{B}^{(\ell_2)}| \leqslant \max\{k-\ell_1+1, \ell_2^2-\ell_2+1\}$.

What is left is to count the contribution of each $\mathcal{B}^{(\ell)}$ to the cardinality of $\mathcal{F}$. For each $B \in \mathcal{B}^{(\ell_2)}$, the family $\mathcal{F}(B)$ is an $(n-\ell_2, k-\ell_2, \{0, \ell_3-\ell_2, \ell_4-\ell_2, \dots, \ell_r-\ell_2\})$-system, and so by induction $|\mathcal{F}(B)| \leqslant \prod_{i=2}^r \frac{n-\ell_i}{k-\ell_i}$. We have

$$\sum_{B \in \mathcal{B}^{(\ell_2)}} |\mathcal{F}(B)| \leqslant \max\{k-\ell_1+1, \ell_2^2-\ell_2+1\} \prod_{i=2}^{r} \frac{n-\ell_i}{k-\ell_i}.$$

Next, for each $B \in \mathcal{B} \setminus \mathcal{B}^{(\ell_2)}$, the family $\mathcal{F}(B)$ has only $\leqslant r-2$ distances and thus $|\mathcal{F}(B)| \leqslant n^{r-2}$. At the same time, using Theorem 1, there are at most $\sum_{i=2}^r \varphi(k, k^i+1)$ sets in $\mathcal{B}$, and thus

$$\sum_{B \in \mathcal{B} \setminus \mathcal{B}^{(\ell_2)}} |\mathcal{F}(B)| \leqslant \sum_{i=1}^{r} \varphi(k, k^i + 1) n^{r-2} < \epsilon \prod_{i=2}^{r} \frac{n - \ell_i}{k - \ell_i},$$

provided n is large enough. Thus, if there is no set D of size ℓ_1 contained in all sets of $\mathcal{F}$, then $|\mathcal{F}| < (\max\{k - \ell_1 + 1, \ell_2^2 - \ell_2 + 1\} + \epsilon) \prod_{i=2}^{r} \frac{n-\ell_i}{k-\ell_i}$, contradicting the assumption. This concludes the proof.

Remark. For a different proof of Theorem 10 we refer to the book [60]. That proof uses Observation 11, but avoids the construction of the base for the family.

2.4 Further Developments of the Early Δ-System Method

The development of the method for the next several years came mostly thanks to the work of Peter Frankl. He pushed forward the original approach and proposed another point of view on how to decompose a family using Δ-systems. This work mostly revolved around families with prescribed intersections.

In Sect. 2.2, we discussed families with only one intersection size. Erdős and Sós (see the collection of problems by Erdős [36, p. 18]) suggested the *opposite* question on families $\mathcal{F} \subset \binom{[n]}{k}$ that are allowed to have all intersection but one. Specifically, the conjecture of Erdős and Sós was that if $\mathcal{F} \subset \binom{[n]}{k}$ satisfies $|A \cap B| \neq 1$ for each $A, B \in \mathcal{F}$, then $|\mathcal{F}| \leqslant \binom{n-2}{k-2}$ for $n > n_0(k)$. That is, forbidding intersection 1 gives the same bound as forbidding intersections 0 and 1, by the Erdős–Ko–Rado theorem. In terms of Sect. 2.1, it asks for the value $m(n, k, \{0, 2, 3, \dots, k-1\})$ of the size of the largest $(n, k, \{0, 2, 3, \dots, k-1\})$-system.

In the paper [42], Frankl proved this conjecture using the Δ-system method.[3] He employs a variant of the Δ-system construction described in the previous section, applies it to families $\mathcal{F}(x)$ and, importantly, notices the following:

Observation 15 *If $\mathcal{F}$ is an $(n, \{0, 2, 3, \dots, k-1\}, k)$ system and $\mathcal{B}_x$, $\mathcal{B}_y$ are the bases of $\mathcal{F}(x), \mathcal{F}(y)$, respectively, constructed as in the previous section, then for any $C_x \in \mathcal{B}_x$ and $C_y \in \mathcal{B}_y$, we have $|(\{C_x \cup \{x\}) \cap (C_y \cup \{y\})| \neq 1$.*

That is, he explicitly states the fact that the property of the family is transferred to the base. (It is somewhat implicit in the statement of Lemma 14, although it is being used in the proof.) He then again analyzes the structure of sets of small uniformity from the base and their contributions to the size.

---

[3] This paper was published in 1977, which is earlier than the Deza–Erdős–Frankl paper [26], and does not refer to [26], but the latter was received in the journal earlier.

The Second Approach to Base Construction

In 1980, Peter published two papers on restricted intersections, in which he developed another way to decompose the family using sunflowers. Both use a certain proposition (Proposition 19) on Δ-systems that was proved in another of his papers [43] from 1977, which suggests that he employed and developed aspects of the Δ-system method in that paper. Unfortunately, I could not find an electronic copy of that paper.

In the paper [47], Frankl managed to determine the order of magnitude of $m(n, k, L)$ for all $k \leqslant 7$ and all possible L, with the exceptions of $k = 7$, $L = \{0, 2, 3, 5\}$ and $L = \{0, 2, 3, 5, 6\}$. That is, in this short paper, he dealt with an impressive list of more than 100 possible cases. The methods combined the Δ-system method, linear-algebraic method and algebraic constructions. Below, we present the Δ-system part of his arguments. In what follows, we assume that $L = \{\ell_1 \dots, \ell_r\}$.

Peter first uses Lemma 14 to conclude that the Deza–Erdős–Frankl base has size at most $c(k)$,[4] and thus one of the bases B is contained in $|\mathcal{F}|/c(k)$ sets. Looking only at $\mathcal{F}(B)$ allows to reduce the problem to the case when the smallest element in L is 0. We can formalize this as follows.

Proposition 16 *We have* $m(n, k, \{\ell_1, \dots, \ell_r\}) = \Theta_k(m(n - \ell_1, k - \ell_1, \{0, \ell_2 - \ell_1, \dots, \ell_r - \ell_1\}))$.

Then he proposes the following greedy approach to construct another base. Take the *largest* set K such that K is a kernel of a $\Delta(k+1)$-system in $\mathcal{F}$, add $\{(K, F) : K \subset F \in \mathcal{F}\}$ to a separate 'base' family $\mathcal{S}$, and update $\mathcal{F} := \mathcal{F} \setminus \mathcal{F}[K]$. Repeat until no $\Delta(k+1)$-systems are left in $\mathcal{F}$. At the end of the procedure, we get the family $\mathcal{S}$, which contains both the base and the decomposition of $\mathcal{F}$ by the base sets, and the remainder family of size at most $c(k)$, which we may ignore in this case. This decomposition has several properties, which we list as three propositions. We then illustrate their use.

The first proposition is again about the transference of the property to the base. Moreover, the property is enhanced in some cases.

Proposition 17 *Let* $(K_1, F_1), (K_2, F_2) \in \mathcal{S}$. *Then* $|K_1 \cap K_2| \in L = \{\ell_1, \dots, \ell_r\}$. *Moreover, if* $F_1 \setminus K_1 = F_2 \setminus K_2$, *then* $|K_1 \cap K_2| \in \{\ell_1 - |F_1 \setminus K_1|, \dots, \ell_r - |F_1 \setminus K_1|\}$.

The second proposition is as follows.

Proposition 18 *Suppose that* $\mathcal{F}$ *is an* $(n, k, \{\ell_1, \dots, \ell_r\})$*-system and it does not contain a* $\Delta(k+1)$*-system with kernel of size at least* ℓ_{s+1}, $s \in [r-1]$. *Then there is a family* $\mathcal{F}' \subset \mathcal{F}$ *such that* $|\mathcal{F}'| \geqslant |\mathcal{F}|/c(k)$ *and such that* $\mathcal{F}'$ *is an* $(n, k, \{\ell_1, \dots, \ell_s\})$*-system.*

[4] Here and below, we use $c(k)$ some large constant, depending on k only.

That is, if a certain size does not appear as a kernel of a $\Delta(k+1)$-system, then we can remove it from the allowed intersection list at a cost of reducing the size of the family by a constant.

The proof is akin to a greedy construction of a large independent set in a graph of small maximum degree: include a new set A into the family $\mathcal{F}'$ and then remove all sets that intersect it in at least ℓ_{s+1} elements. The number of these sets is at most $c(k)$ since, for any ℓ_{s+1}-element subset S of A, there is no $\Delta(k+1)$-system with kernel containing S.

Finally, he refers to [43] for the following statement. (It concerns the greedy procedure that we described before Proposition 17.)

Proposition 19 *Each of the families $\mathcal{F}[U]$ removed at each step of the procedure satisfies $|\mathcal{F}[U]| \leqslant c(k)n$. Consequently, the family of base sets $\mathcal{D} := \{K : (K, F) \in \mathcal{S}$ for some $F \in \mathcal{F}\}$ satisfies $|\mathcal{D}| \geqslant |\mathcal{F}|/c(k)n$.*

This is easy to verify since each $\mathcal{F}(U \cup \{x\}), x \notin U$, does not contain a $\Delta(k+1)$-system by the choice of U.

Let us show, how the first two propositions are being applied in order to prove that if $\mathcal{F}$ is an $(n, 5, \{0, 1, 3, 4\})$-system, then $|\mathcal{F}| \leqslant c(5)n^2$. In the base construction process for $\mathcal{F}$, we first select kernels U of size 4. Let s be the last step when we selected a kernel U of size 4. Let $\mathcal{S}_s$ be the pairs added to the base up to the step s, and let $\mathcal{F}_s$ be the remainder family. By the choice of s, the family $\mathcal{F}_s$ does not have $\Delta(k+1)$-systems with kernel of size 4, and thus by Proposition 18 it contains a subfamily $\mathcal{F}'$ of size $|\mathcal{F}'| \geqslant |\mathcal{F}_s|/c(5)$ that is an $(n, 5, \{0, 1, 3\})$-system.

Peter then shows that an $(n, 5, \{0, 1, 3\})$-system must have size at most $c(5)n^2$ with the help of the following proposition (presented in a slightly modified form).

Proposition 20 *Fix integer q and a constant m. Any family $\mathcal{F} \subset 2^{[n]}$ has a subfamily $\mathcal{G} \subset \mathcal{F}$ of size at least $|\mathcal{F}| - m\binom{n}{q}$ such that for any set Q of size q, $Q \subset F \in \mathcal{G}$, it is contained in at least m sets from $\mathcal{G}$.*

The statement and the proof of the proposition are akin to the statement and the proof of the fact that any graph with many edges contains a subgraph with high minimum degree.

Using it, he chooses a subfamily $\mathcal{G} \subset \mathcal{F}'$ as guaranteed by the proposition, and then shows that $\mathcal{G}$ should be empty using a simple, but somewhat tedious, 'covering number' argument which we avoid.

So far, we showed that $|\mathcal{F}_s| \leqslant c(5)n^2$ with the help of Propositions 18 and 20. Next, we bound the size of $\mathcal{S}_s$. Take y such that $\{y\} = F \setminus K$ for at least $|\mathcal{S}_s|/n$ pairs $(K, F) \in \mathcal{S}_s$ (recall here that by construction all K's in $\mathcal{S}_s$ have size 4). Consider the family $\mathcal{F}_y := \{K : (K, F) \in \mathcal{S}_s, F \setminus K = \{y\}\}$. Then, in view of Proposition 17, for $K \neq K' \in \mathcal{F}_y$ we have $|K \cap K'| \in \{0, 3\}$. That is, $\mathcal{F}_y$ is an $(n, \{0, 3\}, 5)$-system. We can apply Theorem 10 (ii) and note that $\ell_2 - \ell_1 \nmid k - \ell_2$, and thus $|\mathcal{F}_y| \leqslant c(5)n$. Therefore, $|\mathcal{S}_s| \leqslant c(5)n^2$. We conclude that $|\mathcal{F}| \leqslant |\mathcal{S}_s| + |\mathcal{F}_s| \leqslant 2c(5)n^2$.

Finally, let us illustrate, how Proposition 19 is being used. Let $\mathcal{F}$ be an $(n, 6, \{0, 1, 3, 4, 5\})$-system. Consider the family $\mathcal{D}$ and take the value of ℓ, for which the family $\mathcal{D}^{(\ell)}$ of ℓ-element sets in $\mathcal{D}$ is the largest. then $\mathcal{D}^{(\ell)}$ is an $(n, \ell, \{0, 1, 3, 4\})$-system (some of the intersection sizes may be redundant if $\ell < 5$). Above, we showed that an $(n, 5, \{0, 1, 3, 4\})$-system should have size at most $c(5)n^2$ (in the cases $\ell < 5$ the same bound holds as well), which implies that an $(n, 6, \{0, 1, 3, 4, 5\})$-system should have size at most $c(6)n^3$.

In the second paper [46], Frankl proved a general result concerning ℓ-avoiding set systems: he showed that if $\mathcal{F} \subset \binom{[n]}{k}$ satisfies $|A \cap B| \neq \ell$ for each $A, B \in \mathcal{F}$, then $|\mathcal{F}| \leqslant c(k)\binom{n-\ell-1}{k-\ell-1}$ for $n > n_0(k)$ and $n \geqslant 3\ell$, and sketched a proof for replacing $c(k)$ by $(1 + o(1))$ for $k > 3\ell + 2$. In order to do that, he used the same machinery as above, combined with a clever induction on k: instead of proving a result for one forbidden intersection, he proves stronger result with a forbidden interval of intersections. The reason behind it is Proposition 17, which may cause the change in the number of forbidden intersections.

Results for t-Intersecting Families

In the paper [44], Frankl obtains various stability results for the intersecting and t-intersecting cases of the Erdős–Ko–Rado theorem. Specifically, he studies the families in which no element is contained in more than c-fraction of sets. Here, he proposes yet another base, which is a variant of the Deza–Erdős–Frankl construction. For a family $\mathcal{F}$, he considers the family $\mathcal{B}$ of all sets that intersect each set in $\mathcal{F}$ in at least t elements. He then restricts his attention to inclusion-minimal sets and, importantly, only to the sets which have many extensions: he fixes the smallest size ℓ of a set in the base and essentially restricts to studying X of size ℓ such that $|\mathcal{F}(X)| \geqslant k\binom{n-\ell-1}{k-\ell-1}$.

In a follow-up paper [62] on the same question, Füredi used a similar notion of a *nucleus*. There, he suggested the following construction: replace sets in a family with their subsets while preserving the property of the family (there, it was the value of the matching number, which obviously included the property of being intersecting). He refers to the work of Erdős and Lovász [39] and Lovász' dissertation for some of the properties of such nucleus. In particular, he gives an efficient bound for the number of sets in such a nucleus. This notion of a base is close to the notion of a *generating set* from the celebrated paper of Ahlswede and Khachatrian [2] (which they claimed to be new). We discuss the construction of Erdős and Lovász and Ahlswede and Khachatrian in Sect. 7.

2.5 A Precursor to Füredi's Structural Result

In the years that followed, Peter Frankl continued to actively work on (n, k, L)-systems. One of these papers [50] developed an approach that, in retrospect, can be seen as a weak version of Füredi's structural result that we discuss in the next section (see Theorem 21 from Sect. 3.1). Frankl studies (n, k, L)-families for $|L| = 3$ and, in particular, gives conditions when $|\mathcal{F}| = O(n)$ and $|\mathcal{F}| = \Omega(n^2)$. We sketch some of the ideas from that paper below.

Using Proposition 16, we assume that $L = \{0, a, b\}$, $b > a$. Using the Δ-system method, he selects a subfamily $\mathcal{F}' \subset \mathcal{F}$ that is much more structured. He defines $\mathcal{A}, \mathcal{B}$ to be the kernels of large Δ-systems in $\mathcal{F}$ of size a and b, respectively. First, he shows that all but $O_k(n)$ sets $F \in \mathcal{F}$ must satisfy the following property: F is equal to the union of kernels from $\mathcal{A} \cup \mathcal{B}$ contained in F. Second, he shows that all but $O_k(n)$ pairs $B_1, B_2 \in \mathcal{B}$ that intersect must intersect in a set $A \in \mathcal{A}$.[5] Every 'bad' pair is contained in at most 1 set from $\mathcal{F}$, and thus we can remove $O_k(n)$ sets to get rid of both situations and get the family $\mathcal{F}'$. Now every $F \in \mathcal{F}'$ is a union of kernels contained in it and, moreover, every two b-kernels in F are either disjoint or intersect in an a-kernel. This property (F equal to the union of kernels) actually gives necessary and sufficient condition for $|\mathcal{F}| \gg n$. He also showed that, if the condition holds, then $m(n, k, L) = \Omega(n^{k/(k-1)})$ and, moreover, showed cases when $\Omega(n^{k/(k-1)}) = m(n, k, L) = O(n^{3/2})$, which was the first manifestation of the fact that determining $m(n, k, L)$ is a hard problem in general.

3 Delta-System Method After Füredi's Structural Theorem

The result of Füredi, discussed below, gave a completely new direction to the Delta-system method. Although it appears to be motivated by Frankl's ideas from [50] and as such was an organic development of the Δ-system method, it would be fair to say that the Delta-system method as is used in the Deza–Erdős–Frankl paper and the one discussed in this section are two completely different methods.

3.1 *Füredi's Theorem*

A key development for the Delta-system method came with the paper of Zoltán Füredi from 1983 [63],[6] who proved a much stronger Ramsey-type statement concerning large families of k-element sets. As Füredi writes in his paper, the following was conjectured by Peter Frankl:

> There is a positive constant $c(k, s)$ such that every family $\mathcal{F} \subset \binom{[n]}{k}$ has a subfamily $\mathcal{F}^*$ of size at least $c(k, s)|\mathcal{F}|$ and such that for any two $F_1, F_2 \in \mathcal{F}^*$ the set $F_1 \cap F_2$ is a kernel of a $\Delta(s)$-system.

[5] The proofs of both claims are similar to, say, the proof of Proposition 19, but slightly more involved.

[6] It is in this paper that $\Delta(s)$-systems are called s-stars, a name that appears in the follow-up papers of Frankl and Füredi.

We can speculate that this conjecture was very natural to state after Frankl's paper [50]. Zoltán Füredi managed to prove it in a stronger form. In order to state the result, we need some preparations.

Let $\mathcal{F} \subset \binom{[n]}{k}$ be a family. This family is *k-partite* if there exists a partition $[n] = X_1 \sqcup X_2 \sqcup \ldots \sqcup X_k$ such that for any $A \in \mathcal{F}$ and $i \in [k]$ we have $|A \cap X_i| = 1$. Further, for each set $B \in \mathcal{F}$ define the *trace* $\mathcal{F}|_B := \{A \cap B : A \in \mathcal{F}, A \neq B\}$. (We exclude B itself from the trace for convenience.) If $\mathcal{F}$ is k-partite with parts $X_1, \ldots, X_k$, then each set in $\mathcal{F}|_B$ intersects each part X_i in at most 1 element. The following notion encodes, which parts are intersected. Define the *projection* $\pi : 2^{[n]} \to [k]$ as follows: $\pi(Y) = \{i \in [k] : Y \cap X_i \neq \emptyset\}$. We can naturally extend this definition to $\pi(\mathcal{F}|_B) \subset 2^{[k]}$.

Theorem 21 (Füredi [63]) *Given integers $s > k \geqslant 2$ and a family $\mathcal{F}$ of k-element sets, there exists $c(k, s) > 0$ and a family $\mathcal{F}^* \subset \mathcal{F}$ such that*

1. $|\mathcal{F}^*| \geqslant c(k, s)|\mathcal{F}|$;
2. *$\mathcal{F}^*$ is k-partite;*
3. *$\pi(\mathcal{F}^*|_A) = \pi(\mathcal{F}^*|_B)$ for any $A, B \in \mathcal{F}$;*
4. *for each distinct $A, B \in \mathcal{F}^*$ their intersection $A \cap B$ is a kernel of a $\Delta(s)$-system contained in $\mathcal{F}^*$, i.e., there exist $F_1, \ldots, F_s \in \mathcal{F}^*$ such that $F_i \cap F_j = A \cap B$ for all $i \neq j \in [s]$.*

Let us call such $\mathcal{F}^*$ a *(k, s)-homogeneous structure*. As we shall see, this lemma is a very powerful tool to discern structure in the family $\mathcal{F}$. Unfortunately, the bound on $c(k, s)$ that comes out of Füredi's proof is doubly-exponential in k: $c(k, s) > (sk2^k)^{-2^k}$. This limits the applicability of Theorem 21 to $n > n_0(k, s)$, where the latter function is doubly-exponential in k. We present a more quantitative variant of this theorem due to Jiang and Longbrake (with a very similar proof) in Sect. 6.2. Recently, Janzer et al. [76] showed that the doubly-exponential dependency of $c(k, s)$ on k is necessary. Moreover, they proved a weaker version of the theorem, sufficient for many applications, with a single-exponential bound on the constant. See Sect. 6.3 for details.

The family $\mathcal{M} := \pi(\mathcal{F}^*|_A)$ for some $A \in \mathcal{F}^*$ is called the *intersection structure* of $\mathcal{F}^*$. Note that it does not depend on A. Let us mention one important implication.

Corollary 22 *Suppose that $\mathcal{F}^* \subset \mathcal{F}$ is a $(k, k+1)$-homogeneous structure and $\mathcal{M}$ is the intersection structure of $\mathcal{F}^*$. Then*

- *for each $A \in \mathcal{F}^*$ and $Y_1, Y_2 \in \mathcal{F}^*|_A$ there are sets $B_1, B_2 \in \mathcal{F}^*$ such that $B_1 \cap B_2 = Y_1 \cap Y_2$;*
- *If $\mathcal{F}$ is an (n, k, L)-system, then all sizes and intersections of sets in $\mathcal{M}$ lie in L;*
- *$\mathcal{M}$ is* closed under intersection, *i.e., if $Y_1, Y_2 \in \mathcal{M}$ then $Y_1 \cap Y_2 \in \mathcal{M}$.*

Proof The proof of the first part is an application of Lemma 13 with $s = 2$ to the $\Delta(k+1)$-systems with kernels Y_1, Y_2. The second part immediately follows from the first. Let us prove the third part. Take $A \in \mathcal{F}^*$ and note that $\mathcal{M} = \pi(\mathcal{F}^*|_A)$. Take sets $A_1, A_2 \in \mathcal{F}^*$ such that $\pi(A \cap A_i) = Y_i$. By the first part, we can find $B_1, B_2 \in \mathcal{F}^*$ such that $B_1 \cap B_2 = A \cap A_1 \cap A_2$. Then $Y_1 \cap Y_2 = \pi(A \cap A_1 \cap A_2) = \pi(B_1 \cap B_2) \in \pi(\mathcal{F}^*|_{B_1}) = \mathcal{M}$. □

In the remainder of the section, we prove Füredi's structural theorem. First, we note that any k-uniform $\mathcal{F}$ contains a large k-partite subfamily.

Lemma 23 (Erdős and Kleitman [32]) *Given* $\mathcal{F} \subset \binom{[n]}{k}$, *it contains a* k*-partite subfamily* $\mathcal{F}'$ *that satisfies* $|\mathcal{F}'| \geqslant \frac{k!}{k^k}|\mathcal{F}|$.

The proof is a simple probabilistic counting: take a random partition by including each element of the ground set in part i with probability $1/k$, and calculate the expected number of edges that 'respect' that partition. In what follows, we assume that $\mathcal{F}$ is k-partite.

We now come to the heart of the proof, which is a delicate refinement argument. Put

$$\mathcal{M}(\mathcal{F}) := \bigcup_{A\in\mathcal{F}} \pi(\mathcal{F}|_A),$$

$$\mathcal{B}(\mathcal{F}) := \{X \subset [k] : \text{there is } Y \text{ s.t. } \pi(Y) = X \text{ and } Y \text{ is a kernel of a } \Delta(s)\text{-system } F_1, \dots, F_s \in \mathcal{F}\}$$

Note that any $X \in \mathcal{B}(\mathcal{F})$ is actually an intersection of any two sets from the corresponding sunflower, and thus $X \in \mathcal{M}(\mathcal{F})$. Thus, $\mathcal{B}(\mathcal{F}) \subset \mathcal{M}(\mathcal{F})$. The proof of the theorem consists of at most 2^k refinement steps, where at each step we want to decrease the size of $|\mathcal{M}(\mathcal{F})|$. Namely, we do the following: at first we refine $\mathcal{F}$ to $\mathcal{F}'$ in order to find a set $I \in \mathcal{M}(\mathcal{F}) \setminus \mathcal{B}(\mathcal{F}')$, and then, if needed, further refine $\mathcal{F}'$ to $\mathcal{F}''$ so that $I \in \mathcal{M}(\mathcal{F}) \setminus \mathcal{M}(\mathcal{F}'')$. The exact statement is as follows.

Lemma 24 *Assume that* $\mathcal{F}$ *does not satisfy the conditions of the theorem. Then we can find* $\mathcal{F}'' \subset \mathcal{F}$ *such that*

$$|\mathcal{F}''| \geqslant \frac{|\mathcal{F}|}{1 + s(k-1)|\mathcal{B}(\mathcal{F})|} \geqslant \frac{|\mathcal{F}|}{1 + s(k-1)2^k}$$

and such that either $\mathcal{F}''$ *satisfies the conditions of the theorem or such that there is a set* $I \in \mathcal{M}(\mathcal{F}) \setminus \mathcal{M}(\mathcal{F}'')$.

Proof Let us define a partition

$$\mathcal{F} = \mathcal{F}_0 \cup \bigcup_{I\subset[k]} \mathcal{F}_I$$

as follows. At each iteration we go over all sets A of $\mathcal{F}$ and check if for every $X \subset A$ such that $\pi(X) \in \mathcal{M}(\mathcal{F})$ the set X is a kernel of a $\Delta(s)$-system. If it is not, then we delete this set A from $\mathcal{F}$ and include it into $\mathcal{F}_{\pi(X)}$. At some point the procedure stops and we are left with a family $\mathcal{F}_0$ in which each such $X \subset A$ is a center of an s-sunflower for every A.

Next, we note two things. First, we claim that $\mathcal{F}_0$ is a (k, s)-homogeneous structure. We see that $\mathcal{F}_0$ satisfies conditions (2) and (4) of the theorem by definition. Let us also check that (3) holds for $\mathcal{F}_0$. Suppose that, for some A and B, there is $I \in \pi(\mathcal{F}_0|_A) \setminus \pi(\mathcal{F}_0|_B)$. Then take the set $X \subset B$ such that $\pi(X) = I$ and note that X is not a kernel of an $\Delta(s)$-system $F_1, \ldots, F_s$: otherwise, by Observation 12, there would have been i such that $F_i \cap B = X$ since $s > k$. Since X is not a kernel of a $\Delta(s)$-system, such B should have then been 'filtered' into $\mathcal{F}_{\pi(X)}$ during the initial procedure.[7] Therefore, $\pi(\mathcal{F}_0|_A) = \pi(\mathcal{F}_0|_B)$ for all A, B and $\mathcal{F}_0$ is as desired.

Second, we claim that, for each $I \subset [k]$, $I \notin \mathcal{B}(\mathcal{F}_I)$. Indeed, take any X such that $\pi(X) = I$ and look at the first set F containing X that was included in $\mathcal{F}_I$. At that point, we know that X is not a center of a $\Delta(s)$-system in the remaining part of $\mathcal{F}$. Only sets from the remainder of $\mathcal{F}$ may end up in $\mathcal{F}_I$, and thus X will not become such a center.

Now, for each $I \subset [k]$ and X such that $\pi(X) = I$ and $\mathcal{F}_I[X]$ is non-empty, we take a maximal Δ-system $F_1, \ldots, F_\ell \in \mathcal{F}_I$ with kernel X, where $\ell < s$. Any other set in $\mathcal{F}_I[X]$ (the sets in $\mathcal{F}_I$ containing X) must intersect one of $F_i \setminus X$, so we may take the element $v_X \in \cup F_i \setminus X$ that is contained in at least a $1/\ell k$-proportion of sets from $\mathcal{F}_I[X]$. Recall that $\mathcal{F}$ is k-partite, and thus $\mathcal{F}[X_1], \mathcal{F}[X_2]$ are disjoint for distinct X_1, X_2 such that $\pi(X_i) = I$. Now put

$$\mathcal{F}'_I := \bigsqcup_{X:\pi(X)=I} \mathcal{F}_I[X \cup \{v_X\}].$$

We note that, first, $|\mathcal{F}'_I| \geqslant \frac{1}{(s-1)k}|\mathcal{F}_I|$ and, second, $I \notin \mathcal{M}(\mathcal{F}'_I)$.

Finally, we take as $\mathcal{F}''$ the largest family out of $\mathcal{F}_0$ and $\{\mathcal{F}'_I\}$. Since $\mathcal{F}_0$ and $\{\mathcal{F}'_I\}$ decompose $\mathcal{F}$, we can see that one of the families in question has size at least $\frac{|\mathcal{F}|}{1+s(k-1)|\mathcal{B}(\mathcal{F})|}$. □

After each application of Lemma 24 we either get a subfamily with desired properties, or pass from $\mathcal{F}$ to $\mathcal{F}''$, for which $|\mathcal{M}(\mathcal{F}'')| < |\mathcal{M}(\mathcal{F})|$. Repeating this at most 2^k times inevitably gives us a family with desired properties. One can easily track the value of $c(k, s)$ given by the proof.

3.2 First Applications: (n, k, L)-Systems

In the paper [63], Zoltán Füredi points out several applications to (n, k, L)-systems, which seems to have been the main motivation for finding and proving Theorem 21. We again assume $L = \{\ell_1, \ldots, \ell_r\}$. Recall that $m(n, k, L)$ is the size of the largest

---

[7] Note here the delicate detail of the initial procedure of 'filtering' sets. We check the 'kernel condition' for all $X \subset A$ such that $\pi(X) \in \mathcal{M}(\mathcal{F})$, and not only, say, $\pi(X) \in \pi(\mathcal{F}|_A)$.

(n, k, L)-system. Remark that the bound $m(n, k, L) = \Omega_k(n)$ is evident for any non-empty L, by taking a $\Delta(n/k)$-system with core of size in L. He states without proof the following two theorems. The first one is as follows.

Theorem 25 *We have $m(n, k, L) = O_k(n)$ if and only if there is no family $\mathcal{M} \subset 2^{[k]}$ that is closed under intersection, and such that $\cup_{M \in \mathcal{M}} M = [k]$ and $|M| \in L$ for all $M \in \mathcal{M}$.*

That is, he generalizes several previous results, including the result of Frankl [50] for $|L| = 3$. In order to see that the 'if' direction of the theorem is valid, take any (n, k, L)-family $\mathcal{F}$ and find a $(k, k+1)$-homogeneous structure $\mathcal{F}^* \subset \mathcal{F}$. Then by Corollary 22, the family $\mathcal{M}(\mathcal{F}^*)$ is closed under intersection and satisfies $|M| \in L$ for all $M \in \mathcal{M}(\mathcal{F}^*)$. Therefore, we conclude that $\cup_{M \in \mathcal{M}(\mathcal{F}^*)} \neq [k]$. That is, there is a part i such that each element x from that part is contained in at most one set from $\mathcal{F}^*$. It implies that there are at most n sets in $\mathcal{F}^*$. The 'only if' direction is a generalization of the construction provided by Frankl [50] and mentioned in Sect. 2.5.

The second theorem is the following reduction theorem.

Theorem 26 *Assume that $\ell_1, \ldots, \ell_r, k$ all have a common divisor d. Then we have $m(n, k, \{\ell_1, \ldots, \ell_r\}) = \Theta_k\big(m\big(\lfloor \frac{n}{d} \rfloor, \frac{k}{d}, \{\frac{\ell_1}{d}, \ldots, \frac{\ell_r}{d}\}\big)\big)$.*

The Ω_k part of the theorem is easy to see: take the largest $\big(\lfloor \frac{n}{d} \rfloor, \frac{k}{d}, \{\frac{\ell_1}{d}, \ldots, \frac{\ell_r}{d}\}\big)$-system $\mathcal{F}$ and then replace each element of the ground set with d elements. We get an $(n, k, \{\ell_1, \ldots, \ell_r\})$-system of the same size. The O_k direction requires Theorem 21. Take the largest $(n, k, \{\ell_1, \ldots, \ell_r\})$-system $\mathcal{F}$ and select a $(k, k+1)$-homogeneous structure $\mathcal{F}^* \subset \mathcal{F}$.

Consider $\mathcal{M} := \mathcal{M}(\mathcal{F}^*)$ and define its *atoms* $P_1, \ldots, P_m$, with some integer m. There are disjoint subsets of $[k]$ defined by the property that the elements of each P_i are indistinguishable w.r.t. $\mathcal{M}$: for each $M \in \mathcal{M}$ and $x \in P_i$, $x \in M$ iff $P_i \subset M$. It is easy to see that all atoms form a partition of $[k]$: $P_1 \sqcup \ldots \sqcup P_m = [k]$, and that each $|P_i|$ is divisible by d. For the latter property, we use the fact that $\mathcal{M}$ is closed under intersection.

Next, we use the intersection structure $\mathcal{M}$ to analyze the structure of the family $\mathcal{F}^*$. Note that if $F \in \mathcal{F}^*$ and $Q \subset F$ is such that $\pi(Q) = P_i$, then any set $F' \in \mathcal{F}^*$ that intersects Q must contain Q. Therefore, the partition $P_1 \sqcup \ldots \sqcup P_m = [k]$ induces a matching $Q' \sqcup Q_1 \sqcup \ldots \sqcup Q_p = [n]$ (with some integer p that is potentially much larger than m) such that, first, d divides $|Q_i|$ for each i and, second, whenever a set F intersects Q_i, it must contain it. In other words, elements in Q_i are 'indistinguishable' by sets in $\mathcal{F}^*$. The set Q' is the set of elements not contained in any set from the family. Now we can define a family $\mathcal{G}$ as follows: replace each Q_i by $|Q_i|/d$ elements and let $\mathcal{G}$ be the image of the family $\mathcal{F}^*$ under this transformation. We note that, by the above discussion, $\mathcal{G}$ is correctly defined and is a $\big(n', \frac{k}{d}, \{\frac{\ell_1}{d}, \ldots, \frac{\ell_r}{d}\}\big)$-system for some $n' \leqslant \frac{n}{d}$. Finally, $|\mathcal{G}| = |\mathcal{F}^*|$.

3.3 Families with 3 Intersection Sizes

In the paper [64], continuing the study of Frankl [50], Füredi applied his structural theorem to the study of (n, k, L)-systems with $|L| = 3$, $L = \{\ell_1, \ell_2, \ell_3\}$. For $|L| = 1$ the situation has a perfect description by Deza's Corollary 6. For $|L| = 2$ we either have $|\mathcal{F}| = \Theta_k(n)$ or $|\mathcal{F}| = \Theta_k(n^2)$ (the lower bound here is valid when Deza–Erdős–Frankl divisibility constraints hold, and is due to Babai and Frankl [8]). For $|L| = 3$, however, the situation becomes much more complicated. Theorem 25 gives a characterization of the case when $|\mathcal{F}| = O(n)$. We also know from Theorem 10 that, unless $\ell_1|\ell_2 - \ell_1|\ell_3 - \ell_2|k - \ell_3$, we have $m(n, k, L) = O_k(n^2)$. We, however, do not seem to know necessary and sufficient conditions for $m(n, k, L) = \Theta_k(n^3)$. Frankl and Füredi explored the range between n and n^2 and showed that different non-integral exponents are possible. Using that $m(n, k, \{\ell_1, \ell_2, \ell_3\}) = \Theta_k(m(n - \ell_1, k - \ell_1, \{0, \ell_2 - \ell_1, \ell_3 - \ell_1\}))$, in what follows we assume for simplicity that $L = \{0, a, b\}$.

In the paper that we discussed in Sect. 2.5, Frankl [50] showed that if we cannot find non-negative integers α, β, such that $\alpha a + \beta(b - \lfloor b/a \rfloor a) = k$ and the divisibility conditions of Theorem 10 do not hold, then $m(n, k, L) = O(n^{3/2})$. At the same time, if we are not in the situation of Theorem 25, then $m(n, k, L) = \Omega(n^{k/(k-1)})$. Frankl [50] also showed that $|\mathcal{F}| = \Theta_k(n^{3/2})$ in some cases. At that time, there was hope that maybe 3/2 is the only exceptional rational exponent. However, Füredi showed that the situation is even more complicated. We need some preparations in order to state the results.

Let $\mathcal{M} \subset 2^{[k]}$ be the prospective intersection structure, $\mathcal{M} = \{A_1, \dots, A_f, B_1, \dots, B_g\}$, where A_i and B_j are the prospective intersections of sizes a and b, respectively. We of course assume that $\mathcal{M}$ is closed under intersection. Let $I = \{i \in [f] : A_i \subset B_j$ for at least two B_j's$\}$. Let $B'_j := \{i \in I : A_i \subset B_j\}$ encode the indices of sets A_i that are contained in B_j. Finally, form a family out of these sets: put $C := \{B'_j : j \in [g]\}$. Consider the following condition.

> There exists $\mathcal{M}$ as above and such that every pair of elements of I is contained in exactly one set in C. That is, C is a 2-design on I.

Note that any pair of elements of I is contained in at most one set in C, otherwise, two sets of size b intersect in more than a elements. This condition was one of the requirements for the (n, k, L)-families of size $\Theta_k(n^{3/2})$, constructed in [50]. Füredi in [64] shows that if this condition is not satisfied, then $|\mathcal{F}| = O(n^{4/3})$ and, moreover, constructs such families of size $\Theta(n^{4/3})$ for some values of parameters.

Let us give some proof ideas. First, he passes to the $(k, k+1)$-homogeneous structure $\mathcal{F}^*$. As we assumed, $\mathcal{M} := \mathcal{M}(\mathcal{F}^*)$ does not satisfy the displayed condition. Next, let $\mathcal{A} := \{F_1 \cap F_2 : F_1, F_2 \in \mathcal{F}^*, |F_1 \cap F_2| = a\}$, and define $\mathcal{B}$ analogously for intersection size b. It is not difficult to check that $|\mathcal{A}|, |\mathcal{B}| = O_k(n)$. E.g., sets in $\mathcal{A}$ are pairwise disjoint. Then, using an argument similar to Proposition 20, he 'cleans' the set system so that each set from the modified families $\mathcal{A}, \mathcal{B}$ is contained in at least $n^{1/3}$ sets from (the modified) $\mathcal{F}^*$. (This is again the argument as in proving that a graph with many edges has a subgraph with high minimum degree.) Abusing notation, we use $\mathcal{A}, \mathcal{B}, \mathcal{F}^*$ for these 'cleaned' families. In this process, we removed $O_k(n^{4/3})$ sets. Consider the family $\mathcal{A}_0 \subset \mathcal{A}$ of sets in $\mathcal{A}$ that are contained in at least $n^{1/3}$ sets from $\mathcal{B}$. A simple double-counting argument, together with the fact that sets in $\mathcal{A}$ are disjoint and $|\mathcal{B}| = O_k(n)$, shows that $|\mathcal{A}_0| = O_k(n^{2/3})$.

Finally, we use the fact that the property is violated. Take an arbitrary $F \in \mathcal{F}^*$ and two sets $A_1, A_2 \subset F$ that satisfy the following: both are contained in at least two sets from $\mathcal{B}$, but there is no set $B \in \mathcal{B}$ such that $A_1 \cup A_2 \subset B$. He then shows that, first, $A_1, A_2 \in \mathcal{A}_0$.[8] Second, he shows that A_1, A_2 uniquely determines the set F. Indeed, the latter is easy to see: $A_1 \cup A_2$ is not contained in any intersection of sets from $\mathcal{F}^*$ (in the terminology of the next section, we may call it an *own subset*). We conclude that $|\mathcal{F}^*| \leqslant \binom{|\mathcal{A}_0|}{2} = O_k(n^{4/3})$.

4 Forbidding Just One Intersection

One of the first and most prominent applications of Theorem 21 is the paper of Frankl and Füredi [54], in which they solve for $n > n_0(k)$ the 'forbidden one intersection problem'. Recall that the question is to determine the largest size of a family $\mathcal{F} \subset \binom{[n]}{k}$ such that $|F_1 \cap F_2| \neq \ell$. See the problem collection of Erdős [36, page 18]. For $\ell = 1$ this problem was proposed by Erdős and Sós. Then Erdős in [36] suggested the generalization of this problem for all values of ℓ. He wrote:

> I just thought of (2) while writing these lines and thus would not be surprised if it would be completely false or at best not completely accurate.

The equation (2) above stands for the two prospective extremal families. One is the family of all sets containing a given $(\ell + 1)$-element set, and the other one is a (partial) Steiner system/design: the largest collection of k-element sets such that any two intersect in at most $\ell - 1$ elements. The bound was indeed not completely accurate, but not completely false, as we shall see in Sect. 4.1 below.

We already mentioned the paper of Frankl [46], in which he managed to show that for $n \geqslant n_0(k), k \geqslant 3\ell$ the largest ℓ-avoiding family has size at most $c(k)\binom{n-\ell-1}{k-\ell-1}$. Using Theorem 21, Frankl and Füredi managed to obtain great progress on this problem. In this section, we discuss the approximate result for all ℓ, and in the next section we discuss the exact result for $k \geqslant 2\ell + 2$.

Theorem 27 (Frankl and Füredi [54]) *If $\mathcal{F} \subset \binom{[n]}{k}$ is an $(n, [0, k-1] \setminus \{\ell\}, k)$-system, then*

$$|\mathcal{F}| = O_k(n^{\max\{\ell, k-\ell-1\}}).$$

---

[8] For this we need the fact that each A_i is contained in at least two B_j's and, ultimately, must be contained in many, since B_j's have high degree.

4.1 Examples

Let us explain why the maximum in the exponent is for a good reason. This problem has very different prospective extremal examples depending on whether $k > 2\ell + 1$ or $k \leqslant 2\ell + 1$.

In the former case, it is of the Erdős–Ko–Rado type: the family of all sets containing a given $(\ell+1)$-element set. That is, it avoids all intersections smaller than ℓ. It has size $\binom{n-\ell-1}{k-\ell-1}$.

In the latter case, it is based on a partial *design* (partial *Steiner system*). Consider the following partial design: a collection of $(2k-\ell-1)$-element sets such that any ℓ-element set is contained in at most one of them. A design is *complete* if each ℓ-element set is contained in exactly one set. Any such family has size at most $\binom{n}{\ell}/\binom{2k-\ell-1}{\ell}$, and Rödl [112] proved that such families of size $(1+o(1))\binom{n}{\ell}/\binom{2k-\ell-1}{\ell}$ can be found using a greedy/random procedure called Rödl Nibble Method. Keevash [81] proved that (complete) designs exist for large n, provided that natural divisibility conditions hold. Back to the construction, take the k-shadow of such family, that is, all k-element sets contained in one of the $(2k-\ell-1)$-element 'blocks'. Note that any two k-sets inside one $(2k-\ell-1)$-element block intersect in at least $\ell+1$ elements, while sets from different blocks intersect in at most $\ell-1$ elements. The size of such construction is $(1+o(1))\binom{n}{\ell}\binom{2k-\ell-1}{k}/\binom{2k-\ell-1}{\ell}$.[9]

Let us compare the two constructions when $\ell = k-\ell-1$, i.e., when $k = 2\ell+1$. The size of the latter one is $(1+o(1))\binom{n}{\ell}\binom{k+\ell}{k}/\binom{k+\ell}{\ell} = (1+o(1))\binom{n}{\ell}$, while the size of the former one is $\binom{n-\ell-1}{\ell}$. That is, the two constructions are roughly of the same size, but the design-based construction is potentially slightly bigger, at least when complete designs exist.

The conjecture of Erdős [36], corrected by Frankl [49], is as follows.

Conjecture 1 (*Erdős* [36],*Frankl* [49]) Assume that $n \geqslant n_0(k)$ and $\mathcal{F} \subset \binom{[n]}{k}$ is avoids intersection ℓ.

- If $k \geqslant 2\ell+2$ then
$$|\mathcal{F}| \leqslant \binom{n-\ell-1}{k-\ell-1}.$$
- If $k \leqslant 2\ell+1$ then
$$|\mathcal{F}| \leqslant (1+o(1))\binom{n}{\ell}\binom{2k-\ell-1}{k}\Big/\binom{2k-\ell-1}{\ell}.$$

[9] We may compare it with the example suggested by Erdős. The latter has size (at most) $\binom{n}{\ell}/\binom{k}{\ell}$, and the ratio of the sizes is $\frac{1}{(k-\ell)!}\prod_{i=1}^{k-\ell}(2k-2\ell-i)$, which is strictly bigger than 1 for $\ell \leqslant k-2$. For $\ell = k-1$ the examples coincide.

Actually, Frankl [49] conjectured that the second part holds without the $n \geqslant n_0(k)$ assumption. He proved it for $k - \ell$ being a prime number. In Sect. 4.3 we will see the proof by Frankl and Füredi of the first part of the conjecture.

4.2 Proof of Theorem 27

In this section, we present the proof of Theorem 27. In the theorem, we assume that k is constant. Thus, we may apply Theorem 21 and select $\mathcal{F}^* \subset \mathcal{F}$ that is a $(k, k+1)$-homogeneous structure. In what follows, we bound the size of $\mathcal{F}^*$, which then implies the bound on $|\mathcal{F}|$. Informally, passing to $\mathcal{F}^*$ allows us to work with the intersection structure $\mathcal{M} := \mathcal{M}(\mathcal{F}^*)$ instead of the family $\mathcal{F}$ itself. Let us say that the *rank* $r(\mathcal{M})$ of $\mathcal{M}$ is the size of the smallest set in $2^{[k]}$ not contained in any set from $\mathcal{M}$. The next lemma contains the information we need concerning the structure of $\mathcal{M}$.

Lemma 28 *We have* $r(\mathcal{M}) \leqslant \max\{\ell, k - \ell - 1\}$.

Given this lemma, the proof of Theorem 27 is easy to complete. Indeed, recall that $\mathcal{M} = \pi(\mathcal{F}^*|_F)$ for any set $F \in \mathcal{F}^*$. By Lemma 28, the set F contains a subset $X(F)$ of size $m := r(\mathcal{M}) \leqslant \max\{\ell, k - \ell - 1\}$ that is not contained in any other $B \in \mathcal{F}^*$. (We call it an *own subset.*) Otherwise, $B \cap F \supset X(F)$, which contradicts the definition of the rank. Therefore, there is an injection from $\mathcal{F}^*$ to the sets of size m, and thus $|\mathcal{F}^*| \leqslant \binom{n}{m}$. Consequently, $|\mathcal{F}| \leqslant c(k, k+1)^{-1}\binom{n}{m} \leqslant c(k, k+1)^{-1} n^m$. We are left to prove Lemma 28. The proof of the lemma relies on a result of Frankl and Katona [56] on 'sets with many intersection conditions'. As Frankl and Katona write in their paper, this problem arose in the context of a problem on database systems.

Theorem 29 (Frankl and Katona [56]) *Let* $A_1, \ldots, A_m$ *be not necessarily distinct subsets of* $[n]$ *and assume that* $|\bigcap_{j=1}^{t} A_{i_j}| \neq t - 1$ *for any distinct* $i_1, \ldots, i_t$. *Then* $m \leqslant n$.

At the first glance, the theorem has an unusual condition for an extremal set theoretic question. But, actually, it is kind of a Hall's condition in disguise. We shall comment on this in the proof.

The theorem is sharp: the bound is attained on $A_1 = \ldots = A_m = [n]$, in which case all intersections are larger than the right hand side of the condition. This is, however, not the only possibility. In [54] Frankl and Füredi discuss well-intersection designs: other families that could attain equality in Theorem 29.

Proof It is convenient for us to think of a (containment) bipartite graph G between sets A_i on one side and elements of the ground set $[n]$ on the other side. For an element $x \in [n]$, let us denote its neighborhood in G by $N(x) = \{i : x \in A_i\}$ and its degree by $d(x) = |N(x)|$. We essentially want to show is there is a matching covering the set part. In order to do that, we may try to verify Hall's conditions. The condition in the theorem for $t = 1$ gives that all sets are non-empty. For $t = 2$, it prevents two sets to intersect in exactly 1 element, which excludes the only possible violation of Hall's condition for 2 sets: that both sets consist of the same singleton. Continuing further this way, however, becomes complicated.

It is convenient to prove the theorem by induction on m. For any $m' < m$, we can apply inductive assumption to any subcollection of m' sets, and it then implies that the union of any m' sets A_i has size at least m'. That is, Hall's conditions for all proper subsets of $\{A_1, \ldots, A_m\}$ are verified. Arguing indirectly, assume that $m > n$. By induction for $m' = m - 1$, the union of the first $m - 1$ sets has size at least $m - 1$, but it is at most n, and thus $m = n + 1$. Then there is a matching covering the element side in G and, say $A_1, \ldots, A_n$. We may w.l.o.g. assume that i is matched to A_i.

Below we show that for each A_i and $x \in A_i$ we have $d(x) \leqslant |A_i|$. This concludes the proof, since, double counting the edges of G, we have $\sum_{i=1}^m |A_i| = \sum_{x=1}^n d(x)$, but $d(i) \leqslant |A_i|$ for all $i \in [n]$. It implies that $|A_m| = 0$, which is impossible.

In order to prove $d(x) \leqslant |A_i|$, we again apply induction, but this time we change the sets to essentially their intersections with A_i. Take distinct $j_1, \ldots, j_t \in N(x) \setminus \{i\}$ and observe that

$$\left|A_i \cap \bigcap_{\ell=1}^{t} (A_i \cap A_{j_\ell})\right| \neq t$$

by assumption. Removing A_i from the list and x from the sets, we get

$$\left|\bigcap_{\ell=1}^{t} (A_i \cap A_{j_\ell} \setminus \{x\})\right| \neq t - 1.$$

We see that the sets $A_i \cap A_j \setminus \{x\}$ for $j \in N(x) \setminus \{i\}$ satisfy the condition of the theorem. Now apply inductive hypothesis to these sets and get that $|N(x) \setminus \{i\}| \leqslant |A_i \setminus \{x\}|$, which is equivalent to $d(x) \leqslant |A_i|$. □

The proof we presented is essentially the proof of Frankl and Katona. Frankl and Füredi [54] give a different proof.

Corollary 30 *Let $r \geqslant 1$ be an integer and $A_1, \ldots, A_m$ be not necessarily distinct subsets of $[n]$. Assume that $|\cap_{j=1}^{t} A_{i_j}| \neq t - 1 - r$ for any distinct $i_1, \ldots, i_t$. Then $m \leqslant n + r$.*

The corollary is again tight: again, take all A_i to be equal to $[n]$. Interestingly, in contrast to $r = 0$ case, for $r \geqslant 1$ this is the only extremal example. We shall need this fact later.

Proof We modify the sets: consider $(\cap_{i=1}^{r} A_i) \cap A_j$ for $j = r + 1, \ldots, m$. These sets satisfy the assumption of Theorem 29, and thus their number is at most n, i.e., $m - r \leqslant n$. □

We are now ready to prove Lemma 28.

Proof of Lemma 28 Take a minimal subset B not contained in any set from $\mathcal{M}$ and assume that $|B| > \ell$ holds. Assume that $B = \{x_1, \ldots, x_b\}$ and for each $i \in [b]$ define $B_i := B \setminus \{x_i\}$. By minimality of B, $B_i = B \cap A_i$ for some $A_i \in \mathcal{M}$. Put $Y = [k] \setminus B$ and define $D_i = Y \cap A_i$.

We claim that for any $i_1 < \cdots < i_t$ we have

$$|D_{i_1} \cap \cdots \cap D_{i_t}| \neq t - (b - \ell). \tag{5}$$

Indeed, otherwise $A = A_{i_1} \cap \cdots \cap A_{i_t}$ satisfies $|A \cap B| = b - t$ and $A \cap Y = t - (b - \ell)$, implying $|A| = \ell$. This is a contradiction, since $\mathcal{F}$ avoids intersection ℓ and thus $\mathcal{M}$ has no sets of size ℓ.

Finally, condition (5) allows us to apply Corollary 30 to sets D_i with $r = b - \ell - 1$, $n = |Y| = k - b$ and $m = b$. (Note that $r \geqslant 0$ because $b > \ell$ by assumption.) We conclude that $b \leqslant (k - b) + b - \ell - 1$, that is, $b \leqslant k - \ell - 1$. This concludes the proof. □

4.3 A Sharp Result in the Case $k > 2\ell + 1$

Frankl and Füredi also managed to prove an exact result in the case when $k > 2\ell + 1$, i.e., when the tentative extremal family consists of all sets containing a given $\ell + 1$-element set.

Theorem 31 (Frankl and Füredi [54]) *If $n > n_0(k)$, $k \geqslant 2\ell + 2$ and $\mathcal{F} \subset \binom{[n]}{k}$ avoids intersection ℓ, then*

$$|\mathcal{F}| \leqslant \binom{n - \ell - 1}{k - \ell - 1}.$$

The proof follows the same logic, but requires several refinements. For some technical reasons, it is easier to leave out the case $k = 2\ell + 2$ and treat the case $k \geqslant 2\ell + 3$. We sketch the proof in that assumption.

The first step is a refinement of Corollary 30, which states that for $r \geqslant 1$ equality $m = n + r$ is only possible if the sets $A_1, \ldots, A_m$ are all equal to the ground set $[n]$. It is essentially done by carefully running the proof of Theorem 29, and we skip it for conciseness.[10] Arguing as in the proof of Lemma 28, it implies the following lemma:

Lemma 32 *In the setting of Lemma 28 and for $k \geqslant 2\ell + 3$, we have $r(\mathcal{M}) \leqslant k - \ell - 2$, unless $\mathcal{M}$ has the following special form: there is a set C, $|C| = \ell + 1$, such that $\mathcal{M} \supset \{F : C \subset F \subset [k]\}$. (In which case, $r(\mathcal{M}) = k - \ell - 1$.)*

[10] This is the part where it is helpful to assume that $k \geqslant 2\ell + 3$. In this regime, we apply the corollary, while for $k = 2\ell + 2$, we apply Theorem 29 itself. Thus, in the proof for $k \geqslant 2\ell + 3$ we can rely on the characterisation of the extremal families. But no such characterisation is valid for Theorem 29, and thus for $k = 2\ell + 2$.

Let us comment on the proof. Looking into the proof of Lemma 28 and using equality characterization of Corollary 30, we actually see that, for $b = k - \ell - 1$ to hold, all D_i's must be equal to Y, and Y plays the role of C from the lemma. We also note that the 'own' subset B is the subset complementary to Y.

Equipped with this lemma, we take the largest ℓ-avoiding family $\mathcal{F} \subset \binom{[n]}{k}$ and do the following iterative decomposition: find a $(k, k+1)$-homogeneous structure $\mathcal{F}_1^* \subset \mathcal{F}$, remove it, then find another such structure $\mathcal{F}_2^*$ in $\mathcal{F} \setminus \mathcal{F}_1$ etc. We stop at step m when the rank of $\mathcal{F}_m^*$ is at most $k - \ell - 2$. As in the proof of Theorem 27, it implies that $|\mathcal{F}_m^*| \leqslant n^{k-\ell-2}$, and thus

$$|\mathcal{F} \setminus (\mathcal{F}_1^* \cup \ldots \cup \mathcal{F}_{m-1}^*)| \leqslant C_k n^{k-\ell-2}, \tag{6}$$

that is, an order of magnitude smaller than $\mathcal{F}$. Moreover, using Lemma 32, we know that $\mathcal{M}(\mathcal{F}_i^*)$ contains all supersets of C_i for some C_i of size $\ell + 1$.

For any $F \in \mathcal{F}_i^*$, recall the definition of an own subset $X(F)$ from Sect. 4.2: $X(F)$ is of size $k - \ell - 1$ and is not contained in $F \cap F'$ for $F' \in \mathcal{F}_i^*$, $F' \neq F$. Let us also define $C(F) = F \setminus X(F)$. Recall that the own subset and the center C from Lemma 32 are complementary, and thus $C(F)$ projects into $C_i \in \mathcal{M}(\mathcal{F}_i^*)$. In particular, for any set A, such that $C(F) \subset A \subsetneq F$, there is a $\Delta(k+1)$-system in $\mathcal{F}_i^*$ with center A. We use this property to prove the following proposition.

Proposition 33 *For any $F \in \mathcal{F}_i^*$, $i \in [m-1]$ and $F' \in \mathcal{F}$ we have $X(F) \not\subset F'$.*

That is, an own subset of F within $\mathcal{F}_i^*$ is 'globally' own.

Proof Arguing indirectly, assume that $X(F) \subset F'$. Then $C(F) \not\subset F'$, and thus $|C(F) \cap F'| \leqslant \ell$. At the same time, $|F' \cap F| \geqslant |X(F)| \geqslant k - \ell - 1 \geqslant \ell + 1$. Thus, we may choose a set A, $C(F) \subset A \subsetneq F$ such that $|A \cap F'| = \ell$. By the definition of a $(k, k+1)$-homogeneous system, A is the center of a $\Delta(k+1)$-system in $\mathcal{F}_i^*$, and thus we may find a set $F'' \in \mathcal{F}_i^*$ such that $F'' \cap F' = A \cap F'$. This contradicts the fact that $\mathcal{F}$ avoids intersection ℓ. □

Actually, using the same argument, we can prove the following proposition, which we shall need a bit later.

Proposition 34 *For any $F \in \mathcal{F}_i^*$, $i \in [m-1]$ and $F' \in \mathcal{F}$ the following holds. If $|F \cap F'| \geqslant \ell + 1$ then $C(F) \subset F'$ holds. If additionally $F' \in \mathcal{F}_j^*$ with $j \in [m-1]$ then we have $C(F) = C(F')$.*

Proof The first part is verbatim the same argument as in Proposition 33. For the second part, we note that if $C(F) \neq C(F')$ then $|C(F) \cap C(F')| \leqslant \ell$, but $|F \cap F'| \geqslant \ell + 1$, and thus we can find sets A, A' such that $C(F) \subset A \subsetneq F$, $C(F') \subset A' \subsetneq F'$, such that $|A \cap A'| = \ell$. Both are cores of $\Delta(k+1)$-systems in $\mathcal{F}$. Then we use the same Δ-system argument and find two sets $G, G' \in \mathcal{F}$ that intersect in ℓ elements, a contradiction. □

Proposition 33 is enough to prove an asymptotic version of Theorem 31. Since $X(F)$ are distinct for all $F \in \mathcal{F}_1^* \cup \ldots \cup \mathcal{F}_{m-1}^*$, we have, using (6)

$$|\mathcal{F}| \leqslant |\mathcal{F}_1^* \cup \ldots \cup \mathcal{F}_{m-1}^*| + O(n^{k-\ell-2}) \leqslant \binom{n}{k-\ell-1} + O(n^{k-\ell-2}). \tag{7}$$

In order to prove the exact result, we need to continue our bootstrapping argument. Slightly abusing notation, let $C_1, \ldots, C_h$ be all sets such that $C_i = C(F)$ for some $F \in \mathcal{F}_1^* \cup \ldots \cup \mathcal{F}_{m-1}^*$. Define

$$\mathcal{G}_i := \{G \in \mathcal{F}_1^* \cup \ldots \cup \mathcal{F}_{m-1}^* : C_i \subset G\}.$$

Also, for a family $\mathcal{G}$ and integer m, recall the notion of the *shadow on level* m: $\partial^{(m)}\mathcal{G} := \cup_{G \in \mathcal{G}} \binom{G}{m}$. Proposition 34 immediately implies the following property:

Proposition 35 *The families $\partial^{(\ell)}(\mathcal{G}_i)$ are disjoint.*

Consider the families $\mathcal{G}_i(C_i) = \{G \setminus C_i : G \in \mathcal{G}_i\}$ of $(k-\ell-1)$-element sets. We note that $\partial^{(\ell)}(\mathcal{G}_i(C_i)) \subset \partial^{(\ell)}(\mathcal{G}_i)$, and thus the last proposition implies

$$\sum_{i \in [h]} |\partial^{(\ell)}(\mathcal{G}_i(C_i))| \leqslant \binom{n}{\ell}. \tag{8}$$

However, since $\mathcal{F}$ is extremal and $\mathcal{G}_i$ decompose a bulk of $\mathcal{F}$, we must have

$$\sum_{i \in [h]} |\mathcal{G}_i(C_i)| \geqslant \binom{n}{k-\ell-1} - O(n^{k-\ell-2}).$$

Recall the Kruskal–Katona theorem.

Theorem 36 (Kruskal [94], Katona [80], in Lovász' form) *Given a family $\mathcal{X} \subset \binom{[n]}{k}$ with $|\mathcal{X}| = \binom{x}{k}$ for some real $x \geqslant k$, we have $|\partial^{(\ell)}\mathcal{X}| \geqslant \binom{x}{\ell}$.*

We then apply the Kruskal–Katona theorem to $\mathcal{G}_i(C_i)$ and conclude that the last two displayed inequalities are compatible only if one of the families, say $\mathcal{G}_1$, contains a bulk of the sets: if $|\mathcal{F} \setminus \mathcal{G}_1| = O(n^{k-\ell-2})$. Thus, we discern an approximate structure of the extremal family: almost all sets in $\mathcal{F}$ must contain a given $(\ell+1)$-element set C_1, i.e., lie in $\mathcal{G}_1$. We present this argument explicitly and in a stronger form in Sect. 6.

In the remainder of the proof we need to show that for extremal $\mathcal{F}$, there are no other sets but those that contain C_1. It actually requires a non-trivial effort. Let us partition $\mathcal{F}$ into three parts $\mathcal{F} = \mathcal{A} \cup \mathcal{B} \cup \mathcal{K}$, where

$$\begin{aligned}\mathcal{K} &:= \{F \in \mathcal{F}[C_1] : \text{ every set } B \text{ s.t. } C_1 \subset B \subsetneq F \\ &\qquad \text{is a kernel of a } \Delta(k+1) - \text{sunflower in } \mathcal{F}\},\\ \mathcal{A} &:= \mathcal{F}[C_1] \setminus \mathcal{K},\\ \mathcal{B} &:= \mathcal{F} \setminus \mathcal{F}[C_1].\end{aligned}$$

Clearly, $\mathcal{G}_1 \subset \mathcal{K}$, and thus $\mathcal{A} \cup \mathcal{B}$ are small. The goal is to show that if $\mathcal{A} \cup \mathcal{B}$ is non-empty then $\mathcal{F}$ is smaller than the (supposedly) extremal example $\mathcal{F}_0$ of all sets containing C_1. There are two cases to consider, and in both cases we shall use the arguments concerning shadows in the spirit of the argument above. The idea is to show that, first, the shadows of $\mathcal{K}$ and one of $\mathcal{A}$, $\mathcal{B}$ must be disjoint and, second, to show that the shadows of $\mathcal{A}$ or $\mathcal{B}$ are proportionally much larger than the corresponding shadow of $\mathcal{K}$.

First, assume that $|\mathcal{B}| \geqslant |\mathcal{A}|$. Using Proposition 34, we conclude that $\partial^{(\ell)}(\mathcal{B})$ and $\partial^{(\ell)}(\mathcal{K})$ are disjoint, and thus $|\partial^{(\ell)}(\mathcal{B})| + |\partial^{(\ell)}(\mathcal{K})| \leqslant \binom{n}{\ell}$. Looking at the extremal example $\mathcal{F}_0$, we note that $|\partial^{(\ell)}\mathcal{F}_0|/|\mathcal{F}_0| = \binom{n}{\ell}/\binom{n-\ell-1}{k-\ell-1}$, and using the Kruskal–Katona theorem, it is not difficult to show $|\partial^{(\ell)}\mathcal{K}|/|\mathcal{K}| \geqslant \binom{n}{\ell}/\binom{n-\ell-1}{k-\ell-1}$. If we manage to show that

$$|\partial^{(\ell)}\mathcal{B}|/|\mathcal{B}| \gg \binom{n}{\ell}\Big/\binom{n-\ell-1}{k-\ell-1}, \tag{9}$$

then we get the following chain of inequalities:

$$|\mathcal{A}| + |\mathcal{B}| + |\mathcal{K}| \leqslant 2|\mathcal{B}| + |\mathcal{K}| < \frac{\binom{n-\ell-1}{k-\ell-1}}{\binom{n}{\ell}} \left(|\partial^{(\ell)}\mathcal{B}| + |\partial^{(\ell)}\mathcal{K}|\right) \leqslant \binom{n-\ell-1}{k-\ell-1}.$$

In order to show (9), we first note that $\mathcal{B}$ is ℓ-avoiding, and so we can use (the proof of) Theorem 27 in order to get $|\mathcal{B}| \leqslant C_k|\partial^{(k-\ell-1)}\mathcal{B}|$ (by a bijection from a homogeneous structure $\mathcal{B}^* \subset \mathcal{B}$ into own subsets of size $k-\ell-1$). Then we remark that $\mathcal{B}$, and consequently $\partial^{(k-\ell-1)}\mathcal{B}$, are small (both have size $O(n^{k-1})$), and, when writing down the size of $|\partial^{(k-\ell-1)}\mathcal{B}|$ as $\binom{x}{k-\ell-1}$, we see that $x < n^{1-1/k}$. Thus, using the Kruskal–Katona theorem, we conclude that the ratio $|\partial^{(\ell)}\mathcal{B}|/|\partial^{(k-\ell-1)}\mathcal{B}|$ is much bigger than $\binom{n}{\ell}/\binom{n-\ell-1}{k-\ell-1}$. Combining both observations, we get (9).

Next, assume that $|\mathcal{A}| \geqslant |\mathcal{B}|$. We start by selecting a $(k, k+1)$-homogeneous system $\mathcal{A}^* \subset \mathcal{A}$ with $|\mathcal{A}^*| \geqslant c_k|\mathcal{A}|$. For each set $F \in \mathcal{A}^*$ we find an own[11] subset $H(F)$ of size $k-1$ such that $H \notin \partial^{(k-1)}\mathcal{K}$.[12] The family $\mathcal{H} := \{H(F) : F \in \mathcal{A}^*\}$ has the same size as $\mathcal{A}^*$ and, moreover, all sets in $\mathcal{H}$ and $\mathcal{K}$ contain C_1. Now we look at the families $\mathcal{H}(C_1)$, $\mathcal{K}(C_1)$ and note that $\mathcal{H}(C_1)$ and $\partial^{(k-\ell-2)}\mathcal{K}(C_1)$ are disjoint

[11] Not contained in other sets from $\mathcal{A}^*$.

[12] To see this, take any H', $C_1 \subset H' \subset F$, that does not belong to $\mathcal{M}(\mathcal{A}^*)$. Then H' and none of its supersets can belong to the shadow of $\mathcal{K}$. We only have to pick a superset of size $k-1$ that is not in $\mathcal{M}(\mathcal{A}^*)$. It is easy to see that there is one.

by the construction of $\mathcal{H}$. Thus, the sum of their sizes is at most $\binom{n-\ell-1}{k-\ell-2}$. But, while $|\partial^{(k-\ell-2)}\mathcal{F}_0(C_1)|/|\mathcal{F}_0| = (k-\ell-1)/(n-k-1)$ and $|\partial^{(k-\ell-2)}\mathcal{K}(C_1)|/|\mathcal{K}| \geqslant (k-\ell-1)/(n-k-1)$, we have $|\partial^{(k-\ell-2)}\mathcal{A}(C_1)|/|\mathcal{A}| \geqslant |\mathcal{H}|/|\mathcal{A}| \geqslant 1/C_k$. Then we can conclude that $|\mathcal{F}| < |\mathcal{F}_0|$ using a chain of inequalities similar to the last display.

5 Exact Solution of Some Turán-Type Problems

The next joint paper of Frankl and Füredi [55] on the subject is a pinnacle of the applications of the Δ-system method. From the Δ-system method point of view, it is in a sense harvesting the fruits of their previous work [54] and delineating the scope of applications of the method. But even besides the applications of the Δ-system method, it contains many influential ideas that affected the course of a big part of extremal set theory. In the paper, they solve (for $n \geqslant n_0(k)$) several hard and general extremal set theoretic (or Turán-type, as the name of the paper suggests) problems. In this section, we go over some of the contributions of the paper that are related to the Δ-system method.

5.1 *Δ-Systems with a Fixed Kernel*

The following question was asked by Duke and Erdős [30]: what is the largest size $f(n,k,\ell,s)$ of the family $\mathcal{F} \subset \binom{[n]}{k}$ that does not contain a $\Delta(s+1)$-system with kernel of size ℓ? For $s=1$ this is the forbidden one intersection problem that we discussed before, and for $\ell = 0$ this is the famous Erdős Matching Conjecture [34] (see [57] for a survey). One general question that they asked is whether it is true if $f(n,k,\ell,s) = O_{k,s}(f(n,k,\ell,1))$.

An observation of Füredi from [63] is that Theorem 21 immediately implies that. Take any family $\mathcal{F}$ that does not contain a $\Delta(s+1)$-system with kernel of size ℓ and select a $(k, s+1)$-homogeneous system $\mathcal{F}^*$ in $\mathcal{F}$. If $|F_1 \cap F_2| = \ell$ for some sets $F_1, F_2 \in \mathcal{F}^*$, then there is a $\Delta(s+1)$-system with kernel $F_1 \cap F_2$, a contradiction with the choice of $\mathcal{F}$. That is, $\mathcal{F}^*$ avoids intersection ℓ. At the same time, as Theorem 21 guarantees, $|\mathcal{F}^*| \geqslant c(k,s+1)|\mathcal{F}|$ for some positive constant $c(k,s+1)$ that depends on k,s only.

That is, in view of Theorem 27, we have that

$$f(n,k,\ell,s) = O_k(\max\{n^{k-\ell-1}, n^{\ell}\}).$$

Frankl and Füredi prove a stronger result for $k \geqslant 2\ell+3$. Recall that $\phi(k,s)$ is the largest number of sets in a family of k-element sets that does not have a $\Delta(s+1)$-system.

Theorem 37 **(Frankl and Füredi** *[55])* *For fixed k, s and $k \geqslant 2\ell + 3$ we have*

$$f(n,k,\ell,s) = (\phi(\ell+1,s) + o(1))\binom{n-\ell-1}{k-\ell-1}.$$

Examples

The examples in this section generalize the examples from Sect. 4.1. The example that shows tightness of the bound in Theorem 37 is as follows. Take a family $\mathcal{H}$ of $(\ell+1)$-element sets that has size $\phi(\ell+1,s)$ and has no $\Delta(s+1)$-system. Let $[m]$ be the ground set of $\mathcal{H}$ and note that it has constant size, assuming k, s are constant. Then consider the family of sets

$$\mathcal{A}_1 := \left\{F \in \binom{[n]}{k} : F \cap [m] \in \mathcal{H}\right\}.$$

First, we have $|\mathcal{A}_1| = |\mathcal{H}|\binom{n-m}{k-\ell-1}$, which has the same asymptotic as the bound in Theorem 37 (thanks to the fact that m is constant). Second, assuming that $F_1, \ldots, F_{s+1}$ form a $\Delta(s+1)$-system with kernel of size ℓ, we get that $F_1 \cap [m], \ldots, F_{s+1} \cap [m]$ are distinct and should form a $\Delta(s+1)$-system. But these are elements of $\mathcal{H}$, and $\mathcal{H}$ has no $\Delta(s+1)$-systems. Therefore, there are no $\Delta(s+1)$-systems with kernel of size ℓ in $\mathcal{A}_1$.

The second example is as follows: Take a partial Steiner system of $((s+1)(k-\ell)+\ell-1)$-element sets such that any ℓ-element set is contained in at most 1 of them, and consider all k-element sets that are contained in one of the sets from the Steiner system. This is the family $\mathcal{A}_2$. It has size

$$\frac{\left(\binom{(s+1)(k-\ell)+\ell-1}{k} + o(1)\right)\binom{n}{\ell}}{\binom{(s+1)(k-\ell)+\ell-1}{\ell}}.$$

To see that it has no $\Delta(s+1)$-systems with kernel of size ℓ, note that, if one exists, then all sets must come from the same $((s+1)(k-\ell)+\ell-1)$-element set of the Steiner system. But such a Δ-system spans $\ell+(s+1)(k-\ell)$-elements, and thus cannot fit inside one $((s+1)(k-\ell)+\ell-1)$-element set.

Frankl and Füredi conjectured that one of these examples is (approximately) best possible.

Conjecture 2 (*Frankl and Füredi* [55]) For k, ℓ, s fixed the following holds.

- If $k \geqslant 2\ell + 1$ then

$$f(n,k,\ell,s) \leqslant (\phi(\ell+1,s) + o(1))\binom{n-\ell-1}{k-\ell-1}.$$

- If $k \leqslant 2\ell$ then

$$f(n,k,\ell,s) \leqslant \frac{\left(\binom{(s+1)(k-\ell)+\ell-1}{k} + o(1)\right)\binom{n}{\ell}}{\binom{(s+1)(k-\ell)+\ell-1}{\ell}}.$$

Proof of Theorem 37

The proof of the theorem is mostly a simplified version of the proof of Theorem 31, and we assume that the reader is familiar with the first page of that proof. We start with the same decomposition of $\mathcal{F} = \mathcal{F}_1^* \sqcup \mathcal{F}_2^* \sqcup \ldots$ into $(k, (s+1)k)$-homogeneous structures.[13] We stop when $\mathcal{F}_m^*$ is either empty or has rank at most $k - \ell - 2$. Since we are only proving an asymptotic result, we may ignore this remainder and w.l.o.g. assume that $\mathcal{F} = \mathcal{F}_1^* \sqcup \ldots \sqcup \mathcal{F}_{m-1}^*$. Now each set $F \in \mathcal{F}_i^*$ has an own subset $X(F)$ of size $k - \ell - 1$. At the same time, the set $C(F) = F \setminus X(F)$ by Lemma 32 has the property that, for any A such that $C(F) \subset A \subsetneq F$, there is a $\Delta((s+1)k)$-system in $\mathcal{F}_i^*$ with center A.

It is in this last part that the proofs are somewhat different. For each set $X \in \binom{[n]}{k-\ell-1}$ define the family $\mathcal{G}_X := \{F \setminus X(F) : F \in \mathcal{F}, X(F) = X\}$ of truncated sets, whose own subset is X. These are sets that have the form $C(F)$ for the corresponding F. Note that each $\mathcal{F}_i^*$ contributes at most one set to $\mathcal{G}_X$. Each set in $\mathcal{F}$ has an own subset, and so we have

$$|\mathcal{F}| = \sum_{X \in \binom{[n]}{k-\ell-1}} |\mathcal{G}_X|.$$

In order to complete the proof of the theorem, it is sufficient to show that $|\mathcal{G}_X| \leqslant \phi(\ell+1, s)$ for each X. Note that $\mathcal{G}_X$ consists of $(\ell+1)$-element sets.

Assume that $|\mathcal{G}_X| > \phi(\ell+1, s)$ for some X. Take a $\Delta(s+1)$-system $G_1, \ldots, G_{s+1} \in \mathcal{G}_X$ and assume that its kernel has size t. Note that $t \leqslant \ell$. Take an arbitrary subset $Y \subset X$ of size $\ell - t$. For each $i \in [s+1]$ we have $G_i = C(F)$ for some $F \in \mathcal{F}$, and thus $G_i \cup Y$ is a kernel of a $\Delta((s+1)k)$-system. Moreover, $|(G_i \cup Y) \cap (G_j \cup Y)| = \ell$. Now, say, using Lemma 13, for each $i \in [s+1]$ we can find $F_i \in \mathcal{F}$, such that $F_i \supset G_i \cup Y$ and such that $F_i \cap F_j = (G_i \cup Y) \cap (G_j \cup Y)$. They form a $\Delta(s+1)$-system with kernel of size ℓ in $\mathcal{F}$, a contradiction.

[13] The parameters of the Füredi structure is the only difference between the first part of the two proofs.

5.2 Expanded Hypergraphs

Let $\mathcal{A}$ be an arbitrary family of k-sets. We call $K(A) := \cap_{A\in\mathcal{A}} A$ the *kernel* of $\mathcal{A}$. Let the center $C(\mathcal{A})$ be the set of all elements that are contained in at least two sets in $\mathcal{A}$. Note that $K(\mathcal{A}) \subset C(\mathcal{A})$. For a set Y, let $\mathcal{A}|_Y$ stand for the *trace* of $\mathcal{A}$ on a set Y, that is, $\mathcal{A}_Y := \{A \cap Y : A \in \mathcal{A}\}$.

In the following papers on the subject, a slightly different perspective was taken on $\mathcal{A}$.

Definition 38 (*Expanded hypergraphs*) Given a set family (a hypergraph) $\mathcal{H} = \{H_1, \dots, H_m\}$ of ($\leqslant k$)-element sets, a family of k-element sets $\mathcal{A} = \{A_1, \dots, A_m\}$ is an *expansion* (or *k-expansion*) of $\mathcal{H}$, if A_i is obtained from H_i by adding several elements so that, first, $|A_i| = k$ and, second, all added elements are pairwise distinct and distinct from $\cup\mathcal{H}$.

That is the family $\mathcal{A}$ above is a k-expansion of the family $\mathcal{A}|_{C(\mathcal{A})}$. The most well-studied case is when $\mathcal{H}$ is a graph, i.e., when all sets have uniformity 2.

Frankl and Füredi studied the general quantity

$$ex(n, \mathcal{A}) = \max\left\{|\mathcal{F}| : \mathcal{F} \subset \binom{[n]}{k} : \mathcal{F} \text{ contains no copy of } \mathcal{A}\right\}.$$

While many papers at the time suggested very general extremal quantities as the one above, rarely there were general results proved concerning them. Frankl and Füredi, however, managed to prove one. The first big issue to resolve before proving the extremal result is to decide on the prospective optimal construction.

Examples

Suppose that $|K(\mathcal{A})| = p$, $|C(\mathcal{A})| = q$. One example of a family avoiding $\mathcal{A}$ is the family of sets containing a fixed $(p+1)$-element set. This is, however, suboptimal in many cases. Here, they introduced the very influential notion of a *t-crosscut*.

Definition 39 (*t-crosscut*) A set Y is a *t-crosscut* of $\mathcal{A}$ if $|A \cap Y| = t$ for all $A \in \mathcal{A}$.

Note the difference between 1-crosscut and a cover. The set $K(\mathcal{A})$ is a p-crosscut. We will, however, be interested in $(p+1)$-crosscuts. Let us now introduce the key quantity. We note that it is not guaranteed that a $(p+1)$-crosscut exists for $\mathcal{A}$, but if $\mathcal{A}$ is k-partite, then it does: we may take the union of any $p+1$ parts. We are discussing Turán-type problems in the sparse regime, and so we can limit ourselves to the k-partite case. Indeed, if the forbidden hypergraph is not k-partite, then we can take a complete k-partite hypergraph as an example of the hypergraph that avoids it. It has $(n/k)^k = \Theta(n^k)$ edges. Another scenario in which a $(p+1)$-crosscut always exists if $\mathcal{A}$ is an expanded hypergraph: if each edge has an 'own' vertex of degree 1. We may then form a $(p+1)$-crosscut by taking the union of $K(\mathcal{A})$ and these 'own' vertices for each edge.

Let $\sigma_{p+1}(\mathcal{A})$ be the maximum size of a family of $(p+1)$-element sets that avoids copies of the trace $\mathcal{A}|_Y$ of any $(p+1)$-crosscut Y of $\mathcal{A}$, if such a maximum exists.[14]

Example. If $k \geqslant p+q$, a k-expanded hypergraph $\mathcal{A}$ has many $(p+1)$-crosscuts. Importantly, any $\Delta(|\mathcal{A}|)$-system of sets of size $p+1$ is a crosscut. This is due to the fact that each edge of $\mathcal{A}$ has at least p 'own' vertices. This implies that $\sigma_{p+1}(\mathcal{A}) \leqslant \phi(p+1, |\mathcal{A}|-1)$.

Equipped with this definition and assuming that $\sigma_{p+1}(\mathcal{A})$ is finite, we can give an example of an $\mathcal{A}$-free family in $\binom{[n]}{k}$, similar to the example for Theorem 37. Let $\mathcal{B} \subset \binom{[n]}{p+1}$ be the family that achieves equality in the definition of $\sigma_{p+1}(\mathcal{A})$, and let $Y = \cup\mathcal{B}$. Note that $|\mathcal{B}|$ is finite, and thus $|Y|$ is independent of n. Put

$$\mathcal{F}_{\mathcal{B}} = \left\{F \in \binom{[n]}{k} : F \cap Y \in \mathcal{B}\right\}.$$

Then, first, $|\mathcal{F}| = (\sigma_{p+1}(\mathcal{A}) + o(1))\binom{n}{k-p-1}$ and, second, $\mathcal{F}_{\mathcal{B}}$ is $\mathcal{A}$-free. Indeed, to see this, assume that $\{F_1, \ldots, F_s\}$ form a copy of $\mathcal{A}$ in $\mathcal{F}_{\mathcal{B}}$. Since $|Y \cap F_i| = p+1$, it implies that Y is a $(p+1)$-crosscut for $\{F_1, \ldots, F_s\}$, and thus a $(p+1)$-crosscut for $\mathcal{A}$, a contradiction with the definition of $\mathcal{B}$.

The Result for $k \geqslant 2p+q+2$

Frankl and Füredi showed that the example above is essentially optimal in *intersection condensed* case: when k is somewhat large w.r.t. $C(\mathcal{A})$. Recall that $|K(\mathcal{A})| = p$, $|C(\mathcal{A})| = q$.

Theorem 40 (Frankl and Füredi [55]) *Suppose that* $k \geqslant 2p+q+2$. *Then* $ex(n, \mathcal{A}) = (\sigma_{p+1}(\mathcal{A}) + o(1))\binom{n-p-1}{k-p-1}$.

The proof of this theorem is very similar to the proof of Theorem 37. They are actually presented in parallel in that paper. One necessary ingredient is a substitute for Lemma 32. In order to get prove this analogue of Lemma 32, Frankl and Füredi need a possibly suboptimal bound $k \geqslant 2p+q+2$. The conclusion of the lemma is the same: for a bulk of the sets $F \in \mathcal{F}$, there is a set $C(F)$ with the property that, for any A such that $C(F) \subset A \subsetneq F$, there is a $\Delta(k^2)$-system in $\mathcal{F}_i^*$ with center A. We then do the same double-counting over possible sets $X(F)$ and show that, if $\mathcal{G}_X$ is larger than $\sigma_{p+1}(\mathcal{A})$ and, consequently, contains a trace of a $(p+1)$-crosscut

[14] Note that $\sigma_1(\mathcal{A})$ is one smaller than the size of the smallest 1-crosscut. That is, the extremal family in this case is just the collection of $\sigma_1(\mathcal{A})$ singletons.

of $\mathcal{A}$, then we can complete it to a copy of $\mathcal{A}$ in $\mathcal{F}$ by first embedding the rest of $C(\mathcal{A})$ into $C(X)$ and then completing with disjoint sets using Δ-systems. For this, $C(X)$ must be somewhat large, but $|C(X)| \geqslant |C(\mathcal{A})|$ is enough, which requires a bound $k \geqslant p+q+1$.

5.3 Simplices

The following definition generalizes the notion of a triangle in a graph.

Definition 41 A *d-simplex* is a collection of $d+1$ sets $F_1, \ldots, F_{d+1}$, such that $\bigcap_{i\in[d+1]} F_i = \emptyset$, but for any proper subset $S \subsetneq [d+1]$ we have $\bigcap_{i\in S} F_i \neq \emptyset$.

By analogy with the graph case, a 2-simplex is called a *triangle*. A natural example of a family avoiding any simplex is a star: a family of all sets containing a fixed element. Erdős [35] conjectured that for $n \geqslant 3k/2$ and $k \geqslant 3$, any family that contains no triangle, has size at most $\binom{n-1}{k-1}$. One year later, Chvátal [20] generalized this as follows.

Conjecture 3 (Chvátal [20]) Suppose that $\mathcal{F} \subset \binom{[n]}{k}$, $\mathcal{F}$ contains no d-dimensional simplex, $k > d$, and $n \geqslant (d+1)k/d$. Then $|\mathcal{F}| \leqslant \binom{n-1}{k-1}$.

Let us denote $sim(n, k, d)$ the largest family as in the conjecture. The bound in the conjecture in these terms states that $sim(n, k, d) \leqslant \binom{n-1}{k-1}$. Chvátal himself proved the conjecture for $d = k-1$. Prior to the paper [55], Frankl [41] proved the conjecture for $n \leqslant dk/(d-1)$, and for $n \geqslant n_0(k)$ in the following cases [48]: $d = 2$, $k \geqslant 5$; $k \geqslant 3d+1$. He also proved the bound $sim(n, k, d) \leqslant \binom{n-1}{k-1} + c_k n^{d-2}$ for any $k > d$.

Using linear algebraic methods, Frankl and Füredi [55] improved the last bound to $sim(n, k, d) \leqslant \binom{n}{k-1}$. They proved the following theorem.

Theorem 42 (Frankl and Füredi [55]) *Conjecture 3 holds for any $k > d$, $n > n_0(k)$. Moreover, the equality is only possible if $\mathcal{F} = \{F \in \binom{[n]}{k} : x \in F\}$ for some $x \in [n]$.*

Erdős–Chvátal simplices are, actually, families of hypergraphs. Frankl and Füredi showed that essentially the same results hold if one restricts to one particular type of simplices.

Definition 43 We say that $\{F_1, \ldots, F_{d+1}\}$ is a *special d-simplex* if there is a $(d+1)$-element set $C = \{x_1, \ldots, x_{d+1}\}$, such that $C \cap F_i = C \setminus \{x_i\}$ and that, moreover, $F_i \setminus C$, $i \in [d+1]$ are pairwise disjoint.

Let us denote $ssim(n, k, d)$ the analogue of $sim(n, k, d)$, but for special simplices.

Theorem 44 (Frankl and Füredi [55]) *Let $n \geqslant n_0(k)$. Then $ssim(n, k, d) \leqslant \binom{n-1}{k-1}$, provided $k > d+2$, or $d = 3$ and $k = 4, 5$. Moreover, the equality is only possible if $\mathcal{F} = \{F \in \binom{[n]}{k} : x \in F\}$ for some $x \in [n]$.*

Frankl and Füredi also discussed the following generalization of the notions of a simplex and special simplex.

Definition 45 A (d, ℓ)*-simplex* is defined as a d-simplex, except we ask all sets to have intersection of size at most $\ell - 1$, but any d sets to have intersection size at least ℓ. A special (d, ℓ)*-simplex* is defined as a special d-simplex, to which we append $\ell - 1$ new elements that are contained in all sets.

That is, a (special) $(d, 1)$-simplex is the same as a (special) d-simplex. We denote the corresponding extremal functions by $sim_\ell(n, k, d)$ and $ssim_\ell(n, k, d)$. It may seem just like a generalization of the previous notions, but actually it brings several very interesting connections.

First, it is easy to see that determining $sim_\ell(n, d + \ell - 1, d)$ is equivalent to a notoriously difficult problem of Brown Erdős and Sós of determining the largest family such that any $\ell + d$ vertices span at most d sets.

Second, Frankl and Füredi showed that $sim_\ell(n, 5, 2) = o(n^4)$, but that $sim_3(n, 5, 2)/n^{4-\epsilon} \to \infty$ for any $\epsilon > 0$. (For these parameters, $sim_3(n, 5, 2) = ssim_3(n, 5, 2)$ since the only such simplex is the special one.) This result anwered a question of Erdős on the existence of a single hypergraph with no exponent in the corresponding Turán function. It is obtained by a reduction to a result of Ruzsa and Szemerédi [114] on linear 3-uniform hypergraphs without a triangle, which, in turn, is connected to subsets of integers without 3-term arithmetic progressions.

Frankl and Füredi also noted that, restricting to the most popular $(k - 5)$-element set, the bound above implies $s(n, k, 2, k - 2) = o(n^{k-1})$, and asked whether $s(n, k, 2, k - 2)/n^{k-1-\epsilon} \to \infty$ for all $k \geqslant 6$ and any $\epsilon > 0$. Füredi and Gerbner proved it in [67].

They proved the following theorem concerning (d, ℓ)-simplices.

Theorem 46 (Frankl and Füredi [55]) *Suppose* $k \geqslant \ell + d$ *and* $n \geqslant n_0(k)$. *Then we have*

(i) $sim_\ell(n, k, d) = O\left(n^{\max\{k-\ell, k-d-1\}}\right)$;
(ii) If $k \geqslant 2\ell + d$, *then* $sim_\ell(n, k, d) = \binom{n-\ell}{k-\ell}$;
(iii) If $k \geqslant 2\ell + d$, *then* $ssim_\ell(n, k, d) = (\sigma_\ell(\mathcal{S}) + o(1))\binom{n-\ell}{k-\ell}$, *where* $\mathcal{S}$ *stands for the special* (d, ℓ)*-simplex.*

Examples

One example for all these theorems is, of course, the family of all sets containing a given element or a set of ℓ elements. The other example for Theorem 46 for larger values of ℓ is, as usual, coming from a packing: consider a packing of k-sets so that any $\ell + d - 2$ elements are contained in at most one set. It gives an example of the size $(1 - o(1))\binom{n}{\ell+d-2}/\binom{k}{\ell+d-2}$.

In the part (i) of Theorem 46 for $k \leqslant 2\ell + d - 2$, the second expression gives the maximum, and in this case there is a gap of 1 in the exponent between the upper bound and the packing example. The case of $s(n, 5, 2, 3)$ discussed above partially explains this phenomenon.

Proofs

Asymptotic bounds in Theorem 44 for $k > d + 2$, as well as Theorem 46 (iii) are actually direct consequences of Theorem 40. (One just needs to calculate the sizes of the kernel and the center of these families.) In order to get exact results in all cases except the case of Theorem 44 for $k = 3, d = 2$, Frankl and Füredi employ a strategy which is very similar to the strategy in the proof of Theorem 31.[15] In general, Delta-system proofs follow the same strategy, with one important detail: in each case, one needs an appropriate analogue of Lemma 32.

The case $k = 3, d = 2$ is treated using a clever weighting (double-counting) argument, in which weights are assigned to the sets in $(k-1)$-shadows.

5.4 *Trees*

We define *hypertrees* recursively as follows.

Definition 47 (*Hypertrees*) A single edge E is a hypertree. A family $\mathcal{T}$ is a *hypertree* if there is an ordering of edges $\mathcal{T} = \{E_1, \dots, E_m\}$ such that for each $i \in [m-1]$ the collection of sets $\{E_1, \dots, E_i\}$ is a hypertree and $E_{i+1} \cap \cup\mathcal{T} = E_{i+1} \cap E_j$ for some $j \leqslant i$. A *k-tree* is a hypertree with all edges of having size k.

Definition 48 (*Tight hypertrees*) In terms of Definition 47, we call a k-tree *tight* if $|E_{i+1} \cap E_i| = k - 1$ for all $i \in [m-1]$.

Generalizing the Erdős–Sós conjecture, which states that any graph on n vertices with at least $\lfloor \frac{(k-1)n}{2} \rfloor + 1$ edges contain a copy of any tree with k edges (see [33, p. 2]), Kalai (personal communication to the authors of [55]) conjectured that for any tight k-tree $\mathcal{T}$ on v vertices (i.e., with $|\cup\mathcal{T}| = v$) one has

$$ex(n, \mathcal{T}) \leqslant \frac{v-k}{k}\binom{n}{k-1}. \tag{10}$$

[15] We may say that the case $k = 4, d = 2$ corresponds to the harder case of Theorem 31 when $k = 2\ell + 2$, which we haven't discussed. In this case, there is no uniqueness in Lemma 32, and one requires extra arguments. The key idea is that any family not of the type as in Lemma 32 contributes not one, but two own subsets, which allows to bound the overall contribution of such families.

Note that $v - k = |\mathcal{T}| - 1$. Rödl's packing theorem [112] allows to show that, if true, this bound is tight up to $(1 + o(1))$ term. Indeed, consider a packing of $(v - 1)$-element sets such that any $(k - 1)$-element set is contained in at most one of them. There are $(1 + o(1))\binom{n}{k-1}/\binom{v-1}{k-1}$ sets in such a packing. Then, consider the shadow on level k of this packing (cf. Sects. 4.1, 5.1). It will have size $(1 + o(1))\binom{n}{k-1}\binom{v-1}{k}/\binom{v-1}{k-1} = (1 + o(1))\frac{v-k}{k}\binom{n}{k-1}$. Note that this example will have the size equal to the RHS of (10) if it is a perfect packing.

Using a weighting argument, Frankl and Füredi proved the conjecture (10) for *star-shaped trees.*

Theorem 49 (Frankl and Füredi [55]) *The conjecture* (10) *holds for all k-trees $\mathcal{T}$ for which there is an edge E such that any other edge F satisfies* $|E \cap F| = k - 1$.

Moreover, they managed to show that equality holds only if the family is the k-th shadow of a perfect packing of the type described above!

6 Stability, Structure, and Supersaturation Via Delta-Systems

6.1 Stability

A powerful and very popular method of proving results in extremal combinatorics is via stability results. In a large sense, it can be any 'bootstrapping' strategy. We start by discerning some structure in all large/nearly-optimal examples for a particular problem. This structure, sometimes together with more restrictive assumptions on how close is the size of the structure to extremal, allows us to discern even more structure. We may repeat this process several times until eventually getting the exact extremal object. For results in extremal set theory, typically, there are at most 3 steps in such proof strategy: (i) a coarse structural result, which in some way allows to compactly describe the structure of most of the object; (ii) a 99%-stability result, which states that almost all nearly-extremal families fall inside the expected extremal example; (iii) proof of the exact result.

Stability approach has rich history, going back to works of Erdős and Simonovits (see [115]). Their method is discussed in the survey of Keevash [82]. For the problems discussed in this survey, stability was pushed forward, extensively used and popularized in the works of Mubayi and coauthors, see [108]. In these works, stability results are made into a separate statement, which since then became a common practice. Stability results for extremal questions in a different parameter regime were also developed in the Boolean Analysis perspective in the works of Keller, Lifshitz and coauthors (see [32, 86]) and by Zakharov and the author [98]. (See Sect. 8 for a brief discussion of the last two methods.) In this survey, I wanted to point out that stability results are a natural part of the Delta-system method and, as such, stability method was used from the early days of the Delta-system method.

The earlier variant of the Delta-system method decomposition (such as, say, in Theorem 10) by its nature gives a low-uniformity family that contains most of the family. I.e., if the (n, k, L)-family in Theorem 10 in the Delta-system decomposition does not have a set of uniformity ℓ_1, then it has size $O_k(n^{|L|-1})$. Next, all but $O_k(n^{|L|-2})$ sets of the family are covered by sets of sizes ℓ_1, ℓ_2 etc. This opens up possibilities for a series of coarse structural results. We also refer the reader to Sect. 7, where we touch upon decompositions as in the early variant of the Delta-system method, as well as to the paper of Kostochka and Mubayi [89] for more examples.

I wanted to focus on post-Theorem 21 variant of the method, since it seems that inherent stability method, contained in the proof of Frankl and Füredi [54] is somewhat overlooked. As an example, Keevash et al. [85] prove a 99% stability result for the forbidden intersection 1, in the following form:

Theorem 50 *For any $\epsilon > 0$ there exists $\delta > 0$ such that if $|\mathcal{F}| \geqslant (1-\delta)\binom{n-2}{k-2}$ and $\mathcal{F}$ avoids intersection* 1*, then there is X, $|X| = 2$, such that $|\mathcal{F} \setminus \mathcal{F}[X]| \leqslant \epsilon n^{k-2}$.*

Let us explain, how the proof of Frankl and Füredi from Sect. 4.3 almost immediately implies the following stronger result.

Theorem 51 *Assume that $n \geqslant n_0(k)$. There exists $C = C(k)$ such that the following holds for any $\epsilon = \epsilon(n)$. Assume that $k \geqslant 2\ell + 3$ and $\mathcal{F} \subset \binom{[n]}{k}$ avoids intersection ℓ and $|\mathcal{F}| \geqslant (1-\epsilon)\binom{n-\ell-1}{k-\ell-1}$. Then there is X, $|X| = \ell + 1$, such that*

$$|\mathcal{F} \setminus \mathcal{F}[X]| \leqslant \Big(\frac{\ell\epsilon}{(1-\epsilon)(k-2\ell-1)}\Big)^{\frac{k-\ell-1}{\ell}} \binom{n-\ell-1}{k-\ell-1} + C_k n^{k-\ell-2}. \tag{11}$$

We remark that the same could be done for $k = 2\ell + 2$, but since we haven't treated this case in Sect. 4.3, we omit it here as well. The proof we present is essentially the same as in [54], just with some explicit calculations filled in.

Note the dependence on ϵ in the error term in (11). For example, in the Keevash–Mubayi–Wilson setting, we get $\epsilon^{k-2}\binom{n-2}{k-2} + O(n^{k-3})$ error term. We exploit the same phenomenon as for the proof of the fact that every monotone property has a threshold by Bollobás and Thomason [11]. A very similar argument was used by Keller and Lifshitz in [86].

Proof In the proof, we use O-notation as if k is fixed and $n \to \infty$. Going through Sect. 4.3, we first can limit ourselves to $\mathcal{F}^* := \mathcal{F}_1^* \cup \ldots \cup \mathcal{F}_{m-1}^*$ by removing at most $C_k(n^{k-\ell-2})$ sets (cf. (6)). Then we group the sets in $\mathcal{F}^*$ by their centers C_i, and form families $\mathcal{G}_i$. We see that their shadows are disjoint, and get the following two inequalities. First, as in (8),

$$\sum_{i\in[h]} |\partial^{(\ell)}(\mathcal{G}_i(C_i))| \leqslant \binom{n}{\ell}. \tag{12}$$

At the same time, the inequality $|\mathcal{F}^*| \geqslant (1-\epsilon)\binom{n-\ell-1}{k-\ell-1} - C_k n^{k-\ell-2}$ translates into

$$\sum_{i\in[h]} |\mathcal{G}_i(C_i)| \geqslant (1-\epsilon)\binom{n-\ell-1}{k-\ell-1} - O(n^{k-\ell-2})$$
$$= (1-\epsilon)\binom{n}{k-\ell-1} - O(n^{k-\ell-2}). \quad (13)$$

In Sect. 4.3 we said that

> We then apply the Kruskal–Katona theorem to $\mathcal{G}_i(C_i)$ and conclude that the last two displayed inequalities are compatible only if one of the families, say $\mathcal{G}_1$, contains a bulk of the sets: if $|\mathcal{F} \setminus \mathcal{G}_1| = O(n^{k-\ell-2})$.

Here, we will give the precise argument (in our slightly more general setting). Assume that $|\mathcal{G}_i(C_i)| = \binom{x_i}{k-\ell-1}$ with $x_1 \geqslant x_2 \geqslant \ldots \geqslant x_h$. Then, by the Kruskal–Katona theorem we get that

$$|\partial^{(\ell)}(\mathcal{G}_i(C_i))| \geqslant \binom{x_i}{\ell} = \prod_{j=\ell}^{k-\ell-2} \frac{k-1-j}{x_i-j}\binom{x_i}{k-\ell-1} = \prod_{j=\ell}^{k-\ell-2} \frac{k-1-j}{x_i-j}|\mathcal{G}_i(C_i)|.$$

In order to conveniently substitute it into (12) and (13), we multiply (12) by $\binom{n}{k-\ell-1}/\binom{n}{\ell} = \prod_{j=\ell}^{k-\ell-2} \frac{n-j}{k-j-1}$:

$$\begin{aligned}
\binom{n}{k-\ell-1} &\geqslant \sum_{i\in[h]} \frac{\binom{n}{k-\ell-1}}{\binom{n}{\ell}} |\partial^{(\ell)}(\mathcal{G}_i(C_i))| \\
&\geqslant \sum_{i\in[h]} \prod_{j=\ell}^{k-\ell-2} \frac{n-j}{k-j-1} \cdot \frac{k-j-1}{x_i-j} |\mathcal{G}_i(C_i)| \\
&\geqslant \prod_{j=\ell}^{k-\ell-2} \frac{n-j}{x_1-j} \sum_{i\in[h]} |\mathcal{G}_i(C_i)| \\
&\geqslant \left(\frac{n}{x_1}\right)^{k-2\ell-1} \sum_{i\in[h]} |\mathcal{G}_i(C_i)| \\
&\geqslant \left(\frac{n}{x_1}\right)^{k-2\ell-1} (1-\epsilon-O(1/n))\binom{n}{k-\ell-1}. \qquad (14)
\end{aligned}$$

The last inequality is (13). These simple calculations imply that $\left(\frac{n}{x_1}\right)^{k-2\ell-1}(1-\epsilon-O(1/n)) \leqslant 1$, which implies that $x_1 \geqslant (1-\frac{\epsilon}{(1-\epsilon)(k-2\ell-1)})n - O(1)$. From here we can derive that $|\mathcal{G}_1| \geqslant \left(1-\frac{(k-\ell-1)\epsilon}{(1-\epsilon)(k-2\ell-1)}\right)\binom{n}{k-\ell-1} - O(n^{k-\ell-2})$. Combined with (7), we infer a stability with the number of sets not contained in $\mathcal{G}_1$ bounded by $\frac{(k-\ell-1)\epsilon}{(1-\epsilon)(k-2\ell-1)}\binom{n}{k-\ell-1} + O(n^{k-\ell-2})$. Instead, let us use the Kruskal–Katona theorem once again and get a much better bound.

We have

$$|\partial^{(\ell)}\mathcal{G}_1(C_1)| \geqslant \binom{x_1}{\ell} \geqslant \binom{(1-\frac{\epsilon}{(1-\epsilon)(k-2\ell-1)}n - O(1)}{\ell}$$
$$\geqslant \binom{n}{\ell} - \frac{n\epsilon + O(1)}{(1-\epsilon)(k-2\ell-1)}\binom{n}{\ell-1}.$$

Thus,

$$\sum_{i=2}^{h}|\partial^{(\ell)}(\mathcal{G}_i(C_i))| \leqslant \frac{n\epsilon+O(1)}{(1-\epsilon)(k-2\ell-1)}\binom{n}{\ell-1} = \frac{\ell(\epsilon+O(1/n))}{(1-\epsilon)(k-2\ell-1)}\binom{n}{\ell} = \binom{y}{\ell},$$

where $y \leqslant n\Big(\frac{\ell\epsilon+O(1/n)}{(1-\epsilon)(k-2\ell-1)})\Big)^{1/\ell}$. Using the Kruskal–Katona theorem, for each $i \geqslant 2$ we have $x_i \leqslant y$. We substitute the inequality above into the same chain of inequalities as (14):

$$\frac{\ell(\epsilon+O(1/n))}{(1-\epsilon)(k-2\ell-1)}\binom{n}{k-\ell-1} \geqslant \sum_{i=2}^{h}\frac{\binom{n}{k-\ell-1}}{\binom{n}{\ell}}|\partial^{(\ell)}(\mathcal{G}_i(C_i))|$$
$$\geqslant \sum_{i=2}^{h}\prod_{j=\ell}^{k-\ell-2}\frac{n-j}{k-j-1}\cdot\frac{k-j-1}{x_i-j}|\mathcal{G}_i(C_i)|$$
$$\geqslant \prod_{j=\ell}^{k-\ell-2}\frac{n-j}{y-j}\sum_{i=2}^{h}|\mathcal{G}_i(C_i)|$$
$$\geqslant \Big(\frac{n}{y}\Big)^{k-2\ell-1}\sum_{i=2}^{h}|\mathcal{G}_i(C_i)|$$

Rearranging the resulting inequality and substituting the bound on y, we get

$$\sum_{i=2}^{h}|\mathcal{G}_i(C_i)| \leqslant \Big(\frac{\ell(\epsilon+O(1/n))}{(1-\epsilon)(k-2\ell-1)}\Big)^{\frac{k-\ell-1}{\ell}}\binom{n}{k-\ell-1}.$$

Adding the remainder $|\mathcal{F}\setminus\mathcal{F}^*|$, we get the statement. □

Can we say something similar for more complicated hypergraphs? Both coarse structure results and 99% stability for a large class of graphs that are embeddable into 2-contractible k-trees (see Definition 64) was obtained in a work of Füredi and Jiang [69, Theorem 4.2]. For this problem, the extremal function is of the order $\sigma_1(\mathcal{H})\binom{n-1}{k-1}$, where $\sigma_1(\mathcal{H})$ is the size of cross-cut minus 1, as defined in Sect. 5.2. In this 'quasi-dense' regime the intersection structure $\mathcal{M}$ is easier to control.

Can this be extended further? Let us take an example of Theorem 37. The problem is that the argument at the end of the proof is a certain double-counting argument, which arrives at a contradiction in case some of the $(k-\ell-1)$-sets that are own in some $\mathcal{F}_i^*$ can be extended by too many centers (i.e., the same subset is appearing as an own subset in too many $\mathcal{F}_i^*$). In an upcoming note with Noskov, we are going to show, that this obstacle is possible to bypass. Let us sketch the argument, which shows that, provided the family $\mathcal{F} \subset \binom{[n]}{k}$ as in Theorem 37 satisfies

$$|\mathcal{F}| \geqslant (\phi(\ell+1,s)-\epsilon)\binom{n-\ell-1}{k-\ell-1}, \tag{15}$$

then there is a family $C \subset \binom{[n]}{\ell+1}$ with no $\Delta(s+1)$-system, and such that $|\mathcal{F} \setminus \mathcal{F}[C]| < \delta\binom{n-\ell-1}{k-\ell-1}$.

Consider the Johnson graph $J(n, k-\ell-1)$ on vertices $X \in \binom{[n]}{k-\ell-1}$, that is, connect two sets if their symmetric difference has size 2. Take two sets X_1, X_2 that form an edge in J. Then, looking at the argument in Sect. 5.1, we can infer that not only $\mathcal{G}_{X_1}, \mathcal{G}_{X_2}$ cannot contain a $\Delta(s+1)$-system individually, but also does their union $\mathcal{G}_{X_1} \cup \mathcal{G}_{X_2}$. Provided that the family $\mathcal{F}$ satisfies (15), for all but an ϵ-fraction of sets $X \in \binom{[n]}{k-\ell-1}$ we must have $|\mathcal{G}_X| = \phi(\ell+1,s)$. But then any such two sets X_1, X_2 that form an edge must also have $\mathcal{G}_{X_1} = \mathcal{G}_{X_2}$. We can then use expansion properties of $J(n, k-\ell-1)$ that there should be a 'giant component' in J of size $(1-\delta)\binom{n}{k-\ell-1}$ on the sets X such that $|\mathcal{G}_X| = \phi(\ell+1,s)$. All of them will have the same family $\mathcal{G}_X$. This family $\mathcal{G}_X$ is the desired family C.

6.2 A More Quantitative View, Structure, and Supersaturation

In a series of papers [66, 69, 77], Delta-system method was adapted to give coarse structural results, in the sense of the previous section. Let us discuss the paper of Jiang and Longbrake [77], in which they took a new point of view on the Delta-system method, which allows to get more quantitative results. First, they managed to prove a structural result in the spirit of Lemma 32, but gave a decomposition of the family without directly using any forbidden structure. They also made Füredi's Theorem 21 more quantitative. These two developments together allowed them to prove *supersaturation* results. A supersaturation result for the function $ex(n, \mathcal{H})$ is a result that states that, whenever we have a family $\mathcal{F}$ of size just above $ex(n, \mathcal{H})$, not only we have one copy of $\mathcal{H}$, but we actually have many. Precise definitions of 'just above' and 'many' depend on the situation. These questions were popular in combinatorics in the recent years, and in the context of extremal set theory, it was particularly well-studied for Kneser/disjointness graphs. See, e.g., [10, 23].

In this section, we describe the contributions of [77]. First, the authors provide the following quantitative variant of Füredi's theorem.

Theorem 52 *Let $s, k \geqslant 2$ be integers with $s \geqslant 2k$. Then there exists a positive constant $c(s,k)$ (akin to the constant in Theorem 21), such that the following holds. Let $\mathcal{F} \subset \binom{[n]}{k}$. Then there is a k-partite subfamily $\mathcal{F}^* \subset \mathcal{F}$, such that the following holds:*

- $|\mathcal{F}^*| \geqslant c(k,s)|\mathcal{F}|$;
- $\pi(\mathcal{F}^*|_A) = \pi(\mathcal{F}^*|_B)$ *for any* $A, B \in \mathcal{F}$;
- *For every* $J = A \cap B$, *such that* $A, B \in \mathcal{F}^*$, *and every* $x \in [n] \setminus J$ *we have* $|\mathcal{F}^*(J \cup \{x\})| \leqslant \frac{1}{s}|\mathcal{F}^*(J)|$;
- *For each* J *as above we have* $|\mathcal{F}^*(J)| \geqslant \max\left\{s, \frac{1}{2k}\frac{|\mathcal{F}^*|}{n^{|J|}}\right\}$.

The proof is the same as that of Theorem 21, with the additional use of the fact that the sets which violate $|\mathcal{F}^*(J)| \geqslant \frac{1}{2k}\frac{|\mathcal{F}^*|}{n^{|J|}}$ for J all together cover no more than half of the sets from the family, and so we can get rid of them.

Second, the authors prove the following intersection structure lemma.

Lemma 53 *Let $k \geqslant 3$ be an integer. Let $\mathcal{J} \subset 2^{[k]}$ be a family of proper subsets of $[k]$ that is closed under intersection and of rank at least $k-1$. Then one of the following must hold.*

1. *There exists $B \subset [k]$ of size $k-2$, such that $2^B \subset \mathcal{J}$.*
2. *there exists a unique i, such that for any F such that $i \in F$, we have $F \in \mathcal{J}$. Moreover, for every $D \subset [k] \setminus \{i\}$ with $|D| \geqslant k-2$ we have $D \notin \mathcal{J}$.*

Previously, very similar statements were proved in [55, 69], with the small difference that in the second case nothing was stated concerning sets of size $k-2$. (We note that this was mentioned in passing in its analogue for the forbidden one intersection [54, Lemma 5.5a].) This is a key point for [77]. Using it, the authors obtain a variant of Propositions 34 and 35, and then give a variant of the disjoint shadows argument that exploits the Kruskal–Katona theorem as in that same Sect. 4.3. As a result, they obtain the following structural result.

Theorem 54 *Let $k, a, \epsilon > 0$ be constants. Consider a family $\mathcal{F} \subset \binom{[n]}{k}$ of size $a\binom{n}{k-1}$. Then one of the following should hold.*

1. *There is a subfamily $\mathcal{F}^* \subset \mathcal{F}$ as in Theorem 52 and with intersection structure as in part 1 of Lemma 53.*
2. *There is a subfamily $\mathcal{F}' \subset \mathcal{F}$ and $W \subset [n]$ such that (i) $|W| = O_k(1)$; (ii) $|\mathcal{F}'| \geqslant (1-\epsilon)|\mathcal{F}|$; (iii) for every $F \in \mathcal{F}'$ we have $|W \cap F| = 1$.*

As the authors point out, previously similar results were obtained in [66, 69], but under the assumption that $\mathcal{F}$ avoids a certain subfamily, which we cannot assume for the supersaturation problem. Now assume that we are trying to prove that in a family $\mathcal{F}$ of size $(1+\epsilon)ex(n,\mathcal{H})$ for some $\mathcal{H}$ there are many copies of $\mathcal{H}$. If $\mathcal{H}$ is chosen right, then in the first case we should be able to find many copies of $\mathcal{H}$ because of the form of intersection structure (having the 2^B part comes handy). In the second case, we get a good understanding of the coarse structure of the family

$\mathcal{F}$, and then exploit it to find many copies of $\mathcal{H}$. Note that for any $x \in W$ the ratio $|\mathcal{F}(x)|/\binom{n-1}{k-1}$ is constant, i.e., we are in the 'dense regime'. In the dense regime, it is often easy to provide embeddings or cross-embeddings of most structures using counting arguments.

6.3 A Weaker Version of Füredi's Structural Theorem with Better Dependencies

In this subsection, we discuss some of the contributions of the recent paper by Janzer et al. [76]. We focus on two major developments for the delta-system method.

First, the authors showed that $c(k, s)$ in Theorem 21 can be at most $2^{-2^{(1+o(1))k}}$ in some situations. The reason for this is the property that all sets in $\mathcal{F}^*$ from Füredi's theorem should have the same intersection structure. The authors of [76] basically combine families on disjoint ground sets with different possible intersection structures (specifically, of the form $\bigcup_{i=0}^{\lceil k/2\rceil-1}\binom{[k]}{i} \cup \mathcal{I}$ for an arbitrary $\mathcal{I} \subset \binom{[k]}{\lceil k/2\rceil}$) and show that essentially all sets in $\mathcal{F}^*$ must come from one of such families. In order to show the 'essentially' part, they require that, for any given set $A \in \mathcal{I}$, the families can be decomposed into large delta-systems with centers that are projected into A. The construction of such families is, however, non-trivial. The latter 'delta-system decomposition' property suggests that families defined by subspaces should be handy. Indeed, they construct their families using subspaces in high-dimensional spaces $\mathbb{F}_p^d$. They encode which sets from $\binom{[k]}{\lceil k/2\rceil}$ to include in $\mathcal{A}$ via linear dependencies.

Second, they prove the following theorem.

Theorem 55 ([76]) *Suppose $\mathcal{F} \subset \binom{[n]}{k}$ and let $s \geqslant 2$. Then, writing $\beta_{k,s} = (25 \cdot 2^k ks)^{-k}$, there exists a subfamily $\mathcal{F}' \subset \mathcal{F}$ with $|\mathcal{F}| \geqslant \beta_{k,s}|\mathcal{F}|$ such that*

- *For every distinct $F_1, F_2 \in \mathcal{F}'$ there is a $\Delta(s)$-system in $\mathcal{F}$ with kernel $F_1 \cap F_2$;*
- *in fact, for every $F \in \mathcal{F}'$, there exists an intersection-closed set system $\mathcal{I}_F$ such that $\mathcal{I}(F, \mathcal{F}') := \{F \cap F' : F' \in \mathcal{F}'\} \subset \mathcal{I}_F \subset 2^F \setminus \{F\}$, and every element of $\mathcal{I}_F$ is the kernel of a $\Delta(s)$-system in $\mathcal{F}$.*

It should be clear that the second property implies the first one. The proof of the theorem is actually simple, and we sketch it here, with a slightly worse constant. Somewhat simplifying, we produce a random partition into much more than k parts and choose only sets between some k of these parts, showing that they satisfy the properties.

Proof First, color the ground set randomly into $N = 25 \cdot 2^k ks$ colors and note that, for a set $F \in \mathcal{F}$ with probability at least $1/2$ we have the following:

- the set F is rainbow, i.e., all vertices get different colors;
- Assume that $A \subset F$ is such that $\mathcal{F}(A)$ has no matching of size s, and let $X = X(A)$ be a hitting set for $\mathcal{F}(A)$, $|X| \leqslant k(s-1)$. Then, for all $A \subset F$ and for all corresponding $X(A)$ the colors of $X(A) \setminus F$ are disjoint from colors used for F.

Next, take a subfamily $\mathcal{F}'' \subset \mathcal{F}$ consisting only of sets that satisfy the two properties above, and note that $2|\mathcal{F}''| \geqslant |\mathcal{F}|$. Among possible $\binom{N}{k}$ palettes of k colors, choose the palette with the largest number of sets in $\mathcal{F}''$. Let $\mathcal{F}' \subset \mathcal{F}''$ be a subfamily of these sets with a fixed palette, and note that $|\mathcal{F}'| \geqslant |\mathcal{F}|/2\binom{N}{k}$. We claim that $\mathcal{F}'$ satisfies the claimed properties. Of course, the condition on the size is satisfied.

Now, put

$$\mathcal{I}_F := \{F \cap F_1 \cap \ldots \cap F_t : F_1, \ldots, F_t \in \mathcal{F}' \setminus \{F\}\}.$$

Clearly, $\mathcal{I}(F, \mathcal{F}') \subset \mathcal{I}_F$ and the latter is intersection-closed. Next, take $A \in \mathcal{I}_F$ and assume that A is not a kernel of a $\Delta(s)$-system. Note that, for any $F' \in \mathcal{F}'$ such that $F' \supset A$, the set $X(A) \setminus F'$ must have colors that are disjoint from colors of F'. But F' is rainbow and we have k colors overall. Therefore, the part $X'(A)$ of $X(A)$ that is colored with the chosen k colors must lie inside F'. We have that if $A = F \cap F_1 \cap \ldots \cap F_t$ for some $F_1, \ldots, F_t \in \mathcal{F}'$, then $X'(A) \subset F \cap F_1 \cap \ldots \cap F_t$, a contradiction since $X'(A) \neq \emptyset$ and $X'(A) \cap A = \emptyset$. We conclude that any set in $\mathcal{I}_F$ is a kernel of a $\Delta(s)$-system. □

The family $\mathcal{F}'$ from Theorem 55 has weaker properties than the family $\mathcal{F}^*$ from Theorem 21, but for many applications of the Δ-system method it is sufficient. In particular, it is sufficient for the forbidden one intersection Theorem 31 and for the forbidden Δ-system with a fixed core Theorem 37.

7 Approaches to Constructing Bases

We have given a description of the 'original' sunflower base construction in Lemma 14. In this section, we discuss the proof of that lemma and give several other possible approaches. Let us abstract the setup of Lemma 14. The goal of the lemma was to give a construction of a base $\mathcal{B}$ such that

- the upper closure of $\mathcal{B}$ contains the family $\mathcal{F}$;
- the family $\mathcal{B}$ has the same properties as $\mathcal{F}$ (i.e., L-intersecting);
- the number of sets of different sizes in $\mathcal{B}$ is controlled.

In general, these are the properties that we would like to see when constructing a base.

The bounds on $|\mathcal{B}^{(k)}|$ in Lemma 14 were rather bad: we argued that it did not have a $\Delta(k^r)$-system. The proof is what we could call a 'bottom to top construction, top to bottom argument': when constructing the base family, we start from finding smaller kernels and then go to larger kernels. When seeking for contradiction, to the contrary, we start with a larger Δ-system with bigger kernels and get a smaller Δ-system with a smaller kernel.

A more efficient approach is to run a 'top to bottom construction, bottom to top argument'. Let us describe a possible procedure. Start by including in $\mathcal{B}$ the sets from the original family. Then, we simplify $\mathcal{B}$ as follows: if some sets in the current family form a $\Delta(k+1)$-system, then replace them by the kernel of the Δ-system. We perform this exchange as long as it is possible. Clearly, the resulting family has no $\Delta(k+1)$-systems (a much better restriction than no $\Delta(k^r)$-systems).

Claim 56 *If the original family $\mathcal{F} \subset \binom{[n]}{\leqslant k}$ had all pairwise intersections in L, then the resulting family $\mathcal{B}$ has all intersections in L.*

Proof The proof is by induction on the step of the procedure. Assume that the current simplified family is $\mathcal{B}'$, and $A, B \in \mathcal{B}'$ satisfy $|A \cap B| \notin L$. It cannot be that both sets A, B belong to $\mathcal{F}$, and thus one of them, say A, is a kernel of a $\Delta(k+1)$-system $A_1, \ldots, A_{k+1}$ that was replaced by A at some earlier step of the procedure. Moreover, we may assume that B appeared in the simplified family before A. But then for some i we have $B \cap A_i = B \cap A$, and thus $|B \cap A_i| \notin L$, a contradiction with the inductive assumption. □

In recent papers, Zakharov and the author [98] and then the author [95] developed the *peeling-simplification procedure*, probably the most efficient approach for the construction of base-like structures for the family and for subsequent analysis of the structure of the family. First, we need to recall the notion of r-spread families. A family $\mathcal{F}$ is r-spread for some $r > 1$ if $|\mathcal{F}(X)| < r^{-|X|}|\mathcal{F}|$ for any set X. We will see that r-spread families in many ways behave like sunflowers with r petals, albeit they are much easier to find. Let us first give an analogue of the 'top to bottom construction' above.

Start by including in $\mathcal{B}$ the sets from the original family $\mathcal{F} \subset \binom{[n]}{\leqslant k}$. Then, we simplify $\mathcal{B}$ as follows: if there is a set X and a family of sets $\mathcal{G} \subset \mathcal{B}$, such that, first, all sets in $\mathcal{G}$ contain X and, second, $\mathcal{G}(X)$ is k-spread, then replace $\mathcal{G}$ by X in the current family. We perform this exchange as long as it is possible.[16] Clearly, the resulting family has no k-spread subfamilies.

Claim 57 *If the original family $\mathcal{F} \subset \binom{[n]}{\leqslant k}$ had all pairwise intersections in L, then the resulting family $\mathcal{B}$ has all intersections in L.*

Proof The argument is very similar to the previous one. The proof is by induction on the step of the procedure. Assume that the current simplified family is $\mathcal{B}'$, and $A, B \in \mathcal{B}'$ satisfy $|A \cap B| \notin L$. Find the family $\mathcal{G}$ that was replaced by A in the procedure. Then

[16] Note that, technically, a family $\mathcal{B}(X)$ with $X \in \mathcal{B}$ being inclusion-maximal satisfies the condition since it has size 1, but any restriction would have size 0. But replacing $\mathcal{G}(X) = \{X\}$ with X does not change the family, so we stop the replacement if these are the only replacements left.

$$|\mathcal{G}| - \sum_{x \in B \setminus A} |\mathcal{G}(x)| > \left(1 - \frac{|B \setminus A|}{k}\right)|\mathcal{G}| \geqslant \left(1 - \frac{k}{k}\right)|\mathcal{G}| = 0.$$

Thus, we can find a set $A' \in \mathcal{G}$ such that $A' \cap B = A \cap B$. Consequently, $|B \cap A'| \notin L$, a contradiction with the inductive assumption. □

Let us now bound the sizes of layers of $\mathcal{B}$. It is not difficult to see that the following observation is valid.

Observation 58 *If $\mathcal{G} \subset \binom{[n]}{\ell}$ is such that there is no X such that $\mathcal{G}(X)$ is r-spread, then $|\mathcal{G}| \leqslant r^{\ell}$.*

Proof Take inclusion-maximal X that violates r-spreadness: a set X such that $|\mathcal{G}(X)| \geqslant r^{-|X|}|\mathcal{G}|$. Then $\mathcal{G}(X)$ is r-spread by the maximality of X. Since $\mathcal{G}$ has no such (non-trivial) subfamily, it means that $|X| = \ell$, and so $1 = |\mathcal{G}(X)| \geqslant r^{-\ell}|\mathcal{G}|$, or $|\mathcal{G}| \leqslant r^{\ell}$. □

This bound is already better than the bound coming from not containing a sunflower. Indeed, the observation shows that, as a result of the described procedure, we get a family $\mathcal{B}$ such that $|\mathcal{B}^{(\ell)}| \leqslant k^{\ell}$. This bound, however, may be inefficient in some cases. In what follows, let us limit ourselves to the case of t-intersecting families for simplicity. Then $|\mathcal{B}^{(t)}|$ should be either 0 or 1, but the bound that is guaranteed by the observation is only k^t. It is possible to somewhat improve it, but such improvement is insufficient for many purposes. In order to deal with this issue, we added another ingredient, called *peeling*, which we describe below. We present the construction for t-intersecting families, but it can be extended to many other properties. A very brief description is that we iteratively simplify the family and 'peel off' the top layer of the family. At the next step, we simplify again, but with a better parameter r.

First, we record the following easy observation.

Observation 59 *For any t-intersecting family $\mathcal{S}$ there exists a t-intersecting family $\mathcal{T}$ such that $\mathcal{T}$ is obtained from $\mathcal{S}$ by replacing some of the sets in $\mathcal{S}$ by their subsets, and such that $\mathcal{T}$ is maximal w.r.t. the property of being t-intersecting: for any proper subset $X \subsetneq T$ of a set $T \in \mathcal{T}$ there exists $T' \in \mathcal{T}$ such that $|X \cap T'| < t$.*

Next, we describe the peeling procedure. Assume that we are given a t-intersecting family $\mathcal{F} \subset \binom{[n]}{\leqslant k}$.

1. Let $\mathcal{T}_k$ be a family given by Observation 59 when applied to $\mathcal{F}$.
2. For $i = k, k-1 \dots, t$ we put $\mathcal{W}_i = \mathcal{T}_i \cap \binom{[n]}{i}$ and let $\mathcal{T}_{i-1}$ be the family given by Observation 59 when applied to the family $\mathcal{T}_i \setminus \mathcal{W}_i$ playing the role of $\mathcal{S}$.

Remark that $\mathcal{T}_i$ is t-intersecting for each $i = k, \dots, t$ by definition. We summarize the properties of these series of families in the following lemma.

Lemma 60 ([95]) *The following properties hold for each $i = k, k-1\dots, t$ and any family $\mathcal{A} \subset 2^{[n]}$.*

(i) All sets in $\mathcal{T}_i$ have size at most i.
(ii) We have $\mathcal{A}[\mathcal{T}_i] \subset \mathcal{A}[\mathcal{T}_{i-1}] \cup \mathcal{A}[\mathcal{W}_i]$.
(iii) The family $\mathcal{T}_i$ does not have a subfamily $\mathcal{G}$ and a set X such that $\mathcal{G}(X)$ satisfies $|\mathcal{G}(X)| > 1$ and is $(i-t+1)$-spread.
(iv) We have $|\mathcal{W}_i| \leqslant (6i)^{i-t}$.
(v) If $\mathcal{T}_i$ consists of a single t-element set X and this is not the case for $\mathcal{T}_{i+1}$ then $|\mathcal{A}[\mathcal{T}_{i+1} \setminus \mathcal{W}_{i+1}]| \leqslant \frac{k}{r}|\mathcal{A}[T]|$.

Let us comment on the lemma. Part (i) assures that the uniformity decreases as we progress. Part (ii), in particular, guarantees the first property from the beginning of this section: that the upper closure of the resulting family $\mathcal{B}$ contains $\mathcal{F}$. Note that $\mathcal{B}$ here is the union of some $\mathcal{T}_i$ and $\cup_{j>i}\mathcal{W}_j$. It is up to us to decide, at which step to stop the procedure. Part (iii) guarantees that $\mathcal{T}_i$, and consequently $\mathcal{W}_i$, does not contain spread subfamilies. As a matter of fact, we could have replaced the use of Observation 59 with the following subprocedure: at step i, find any subfamily $\mathcal{G} \subset \mathcal{T}_i \setminus \mathcal{W}_i$ and a set X such that, first, X is contained in all sets of $\mathcal{G}$ and, second, $\mathcal{G}(X)$ is $(i-t+1)$-spread. Remove $\mathcal{G}(X)$ from $\mathcal{T}_i \setminus \mathcal{W}_i$ and add to $\mathcal{T}_{i-1}$. Repeat until no such $\mathcal{G}$ can be found. From here, we could derive that $\mathcal{T}_i$ stays t-intersecting.

Part (iii) ultimately allows to prove the bound in (iv). It is much better than the simpler bound $|\mathcal{W}_i| \leqslant k^i$ that we could obtain using our previous 'top to bottom argument'. In particular, for $i = t$ (i.e., for sets from $\mathcal{W}_t$ that have size t) it gives the upper bound 1 on the number of such sets, which is the correct number, and it gives constant bounds for sizes of $\mathcal{W}_{t+const}$. Parts (iv) and (v) allow to efficiently bound the size of $\mathcal{A}(\mathcal{S})$ depending on the structure of $\mathcal{S}$. In order to get (iv) from (iii) and Observation 58, we first fix a subset X of size t in any of the sets $W \in \mathcal{W}_i$ that is contained in at least $\binom{i}{t}^{-1}$ ratio of sets in $\mathcal{W}_i$ (such X exists by the t-intersecting property of $\mathcal{W}_i$), and then apply part (iii) to $\mathcal{W}_i(X)$.

One important difference between other base constructions and peeling-simplification is that in the former we do not actually create a remainder during the process, only potentially after the end (when 'truncating' the base). In the latter, we gradually send sets into the remainder and take advantage of this during the process.

Next, let us discuss the bases (*nuclei*) used by Füredi for intersecting families in [62]. They are based on the following argument, essentially due to Erdős and Lovász [39]. Take an intersecting family $\mathcal{F}$ that is maximal w.r.t. that property (in the sense of Observation 59). Let us take a set $F = \{x_1, \dots, x_a\} \in \mathcal{F}$. Any set in $\mathcal{F}$ intersects it. Take x_i and any set $F_i \in \mathcal{F}$ such that $x_i \notin F_i$. All sets from $\mathcal{F}$ that contain x_i must additionally intersect F_i. Thus, any set in $\mathcal{F}$ contains one of the pairs $\{(x_i, y) : y \in F_i\}$. We may go on and construct 3-element sets etc. In particular, this allows to efficiently bound the sets in $\mathcal{F}$ of different sizes. Seen from another perspective, the produced sets (once we cannot find a set F' that is disjoint with a generated sequence of elements) give exactly the family of minimal covers for the sets in $\mathcal{F}$. These are the bases of the type used by Frankl in [44]. In a recent paper of Frankl [52], the family of minimal covers is efficiently analyzed in order to bound the maximal diversity of an intersecting family. We may also say that this is essentially a simpler variant of the peeling-simplification procedure, with just one round of simplification and no peeling.

This notion of a base is close to the notion of a *generating set* from the celebrated paper of Ahlswede and Khachatrian [2]. Given a family $\mathcal{F} \subset \binom{[n]}{k}$, a *generating set* $g(\mathcal{F})$ is a family in $\binom{[n]}{\leqslant k}$ such that the upper closure $\mathcal{U}$ of $g(\mathcal{F})$ satisfies $\mathcal{U} \cap \binom{[n]}{k} = \mathcal{F}$. The difference with the notion of Frankl and Füredi is that if we take Frankl's base or Füredi's nucleus, and all k-element sets that contain at least one of the sets in the base, then we get a family $\mathcal{F}'$ that contains, but is not necessarily equal to, $\mathcal{F}$. However, if the family is inclusion-maximal (which is typically a natural assumption), then the construction of Frankl or Füredi gives a particular generating set. In the proof of the Complete t-Intersecting Theorem [2], the authors use generating sets somewhat differently, as compared to the Delta-system method. Since we can use shifting, generating sets are much more structured. The authors can then perform a very careful analysis of the contribution of different sets from the base to the size of the family, depending on the form of a particular set from the base, and then gradually modify the family, bringing it closer to one of the extremal examples.

8 Other Regimes, Other Decompositions

Before, we have mostly treated the case when $n > n_0(k, t)$, where t encodes, say, the size of the center of the forbidden configuration (i.e., the number of vertices of degree $\geqslant 2$). There are other regimes in which general results could be proved. Below, we assume that k, t have this meaning.

In [86], Keller and Lifshitz developed a junta approximation approach based on Boolean Analysis tools that allows to approximate the structure of the family by a smaller family with the same properties, but that lives on $f(t)$ coordinates. This allowed them to extend some of the results from Sects. 3 and 5 to the regime $k > f(t), n > f(t)k$. Essentially, they solved the cases when the forbidden structure is a k-expansion, and the extremal configuration is either all sets containing a given t-element set (the Complete t-Intersection Theorem setting), or intersecting a given t-element set (the Erdős Matching Conjecture setting). In [37], Ellis, Keller and Lifshitz managed to solve the forbidden one intersection problem (no intersection $t-1$) in the regime $k > f(t, \epsilon), \epsilon n < k < (1/2 - \epsilon)n$. In this regime the extremal configuration is no longer the family of all sets containing a given t-element set, but a configuration of the Ahlswede-Khachatrian type: $\{F \in \binom{[n]}{k} : |F \cap [t + 2i]| \geqslant t + i\}$ for a constant i.

In [83], Keevash, Lifshitz, Long and Minzer developed an alternative, more efficient approach that uses hypercontractivity. They managed to prove the following result.

Theorem 61 (Keevash et al. [83]) *For any $r, \Delta \geqslant 2$ there is $C > 0$ such that the following holds for any r-graph $\mathcal{G}$ with s edges, maximum degree $\Delta(\mathcal{G}) \leqslant \Delta$ and $\sigma_1(\mathcal{G}) \geqslant 1$. For any $C \leqslant k \leqslant n/Cs$ we have $ex(n, \mathcal{H}) = (1 \pm \epsilon)\Big(\binom{n}{k} - \binom{n-\sigma_1(\mathcal{G})}{k}\Big)$, where $\mathcal{H}$ is the k-expansion of $\mathcal{G}$.*

That is, the main difference with the previous result is that now the unspecified dependence is on the maximum degree and the uniformity of the underlying structure, and no longer on the number of edges (this dependence is linear). That is, it applies for the Erdős Matching Conjecture in the regime $n > Csk$ with some absolute constant C. However, it still does not allow to resolve, say, forbidden intersection ℓ problem for ℓ that is not constant, or forbidden sunflowers with a fixed and non-empty core and non-constant number of leaves.

We should say that both Boolean Analysis approaches mentioned above can be seen as exploiting the sharp threshold phenomenon. As we have seen, Δ-system method does also exploits this, albeit to a much smaller extent. Zakharov and the author recently introduced a *spread approximation* method [98], which is also using this philosophy.

Two other important ingredients is the simplification/peeling procedure, described in Sect. 7, and the notion of *τ-homogeneous families.*[17] The latter definition allowed to extend the method from t-intersection type problems to the forbidden t-intersection type problems, and also ultimately allowed to work with any sufficiently quasirandom families, instead of just, say, $\binom{[n]}{k}$. We say that a family $\mathcal{F} \subset \mathcal{A}$ is *τ-homogeneous with respect to $\mathcal{A}$*, if for any set X we have $\frac{|\mathcal{F}(X)|}{|\mathcal{F}|} \leqslant \tau^{|X|}\frac{|\mathcal{A}(X)|}{|\mathcal{A}|}$. If we use the counting measure notation, where $\mu(\mathcal{F}) = |\mathcal{F}|/|\mathcal{A}|$ and $\mu(\mathcal{F}(X)) = |\mathcal{F}(X)|/|\mathcal{A}(X)|$, then it transforms into $\mu(\mathcal{F}(X)) \leqslant \tau^{|X|}\mu(\mathcal{F})$. We use this notion with τ very close to 1, say, with $\tau = (1+\epsilon)^{1/t}$. Ultimately, this allows to control the behaviour of the restrictions on $\mathcal{F}(X)$ at level $\leqslant t$.

Using this method, one can solve general problems for $n > poly(s,t)k\log k$.

9 Survey of Future Works

In this section, we summarize the developments in the problems that were treated earlier in this survey. These include forbidden one intersection problem, (n, k, L)-systems, families with different forbidden configurations: sunflowers with fixed kernel, different simplices, expansions and contractions of families of small uniformity, hypertrees. We end with references to other surveys on closely related topics that we haven't covered.

---

[17] It is an unfortunate clash of notation that τ-homogeneous is used both here and in Füredi's theorem. We resolved it here by calling the latter a *homogeneous structure*, rather than a family.

9.1 Forbidden One Intersection Problem

The following question was asked by Erdős [36], generalizing the question of Erdős and Sós for $\ell = 1$: what is the largest family $\mathcal{F} \subset \binom{[n]}{k}$ such that no two sets $A, B \in \mathcal{F}$ satisfy $|A \cap B| = \ell$? These families are called *ℓ-avoiding*. We treated this question in Sects. 2.4 and 4. Some of the key results here are, of course, Theorems 27 and 31. In [85], Keevash, Mubayi and Wilson found the exact size of the largest 1-avoiding family for $k = 4$ and all n.

In the aforementioned results, the extremal families are families of sets containing a given $(\ell + 1)$-element set. For $k \leqslant 2\ell + 1$, as we have seen in Sect. 4.1, extremal examples are expected to be of Steiner system-type. In [49], using linear-algebraic approach, Frankl showed that for $k \leqslant 2t\ell + 1$ and $k - t$ prime, the largest ℓ-avoiding family has size as suggested by Conjecture 1 (without the $(1 + o(1))$), moreover, equality is only possible when $\mathcal{F}$ is a Steiner system.

This problem is very interesting in other regimes, when k and ℓ are comparable to n. In [61], Frankl and Wilson gave a very strong upper bound $\binom{n}{k-\ell-1}$ on the size of ℓ-avoiding families that works for essentially all interesting regimes of n, ℓ, k with $k > 2\ell$, but with a restriction that $k - \ell$ is a prime power. Their bound with k, ℓ being linear in n was used to solve several discrete-geometric questions. Frankl and Rödl [59], motivated by further geometric applications, gave a worse exponential upper bound, but without number-theoretic restrictions. Recently, Kupavskii et al. [97] found an efficient way on how to deduce Frankl and Rödl's result from the result of Frankl and Wilson.

Recently, exact results were proved in other regimes of parameters. Unless mentioned otherwise, the extremal family is the family of all sets containing a given $(\ell + 1)$-element set. In [86], Keller and Lifshitz solved the problem for $k > k(\ell)$ and $n > C(\ell)k$. In [37], Ellis, Keller and Lifshitz managed to solve the forbidden one intersection problem in the regime $k > k(\ell, \epsilon)$, $\epsilon n < k < (1/2 - \epsilon)n$. In this regime, the extremal configuration is no longer the family of all sets containing a given $(\ell + 1)$-element set, but *Frankl configurations*. These are extremal configurations in the famous result of Ahlswede and Khachatrian [2] on ℓ-intersecting families: $\{F \in \binom{[n]}{k} : |F \cap [\ell + 2i]| \geqslant \ell + i\}$. In the paper [98], Zakharov and the author solved the problem in the regime when all parameters are polynomially dependent: in particular, for $n = k^{\alpha}$ and $t = k^{\beta}$ with $0 < \beta < 1/2$ and $\alpha > 1 + 2\beta$, provided $k > k_0$. In a recent paper [16], Cherkashin found an exact answer for the $\ell = 1$ case when $n = k^2 - k + 1$ and all k.

We should also note that the size of the largest ℓ-avoiding system $\mathcal{F} \subset \binom{[n]}{k}$ is the independence number of the generalized Johnson graph $J(n, k, \ell)$ with vertex set $\binom{[n]}{k}$ and sets connected by an edge if they have intersection exactly ℓ. Zakharov [120] obtained results concerning the chromatic number of Johnson graphs. Many other properties, including supersaturation for edges and cliques, transference etc. were studied in recent papers by Raigorodskii and his students, see, e.g., [29].

9.2 (n, k, L)-Systems

Let $m(n, k, L)$ stand for the size of the largest family $\mathcal{F} \subset \binom{[n]}{k}$ such that $|F_1 \cap F_2| \in L$ for any distinct $F_1, F_2 \in \mathcal{F}$. The problem of determining $m(n, k, L)$ for general L was proposed by Deza et al. [26] and independently by Larman [99] in 1978. It was addressed for $|L| = 2$ by Deza et al. [27] and then by Deza et al. [26] for general L. We discussed their main result, Theorem 10, and its precursors in Sect. 2. In 1975, Ray-Chaudhuri and Wilson [15] in the context of studying designs obtained a generalization of Ficher's inequality, which can be used to get the following general bound: $m(n, k, L) \leqslant \binom{n}{s}$ for all $n \geqslant k$ and $|L| = s$. The proof uses linear algebra (and could be done either using higher order incidence matrices or polynomials).

In the paper [51], Frankl showed that the problem of determining the order of magnitude of the largest (n, k, L)-system is very rich. Namely, he showed that any rational number α can appear as the exponent: for any α there exists k and a set L such that $m(n, k, L) = \Theta(n^\alpha)$. He used Delta-systems to show the upper bound. Continuing the work of Frankl [47], Frankl et al. [58] determined exponents of largest (n, k, L)-systems for all $k \leqslant 12$ except two cases. For upper bounds, they used Delta-systems together with different reductions, such as Theorem 26 and reductions akin to the ones in Sect. 2.4. Lower bounds use different algebraic and geometric constructions. This work was completed by Tokushige [119], who resolved the remaining two cases by providing the lower bound constructions.

We have discussed several reductions between $m(n, k, L)$ with different parameters throughout this survey. As we have seen in the proof of Theorem 10, Deza et al. [26] showed that, for fixed k,

$$m(n, k, \{\ell_1, \ldots, \ell_r\}) \sim m(n, k - \ell_1, \{\ell_2 - \ell_1, \ldots, \ell_r - \ell_1\}).$$

Another reduction is given in Theorem 26. Also, Theorem 25 characterizes the case when $m(n, k, L) = O(n)$. Frankl [51] showed that $m(n, k, L) \leqslant m(f(n), f(k), f(L))$, where f is a polynomial $f(x) = \sum_{i=1}^{d} a_i \binom{x}{i}$ with non-negative integer coefficients, and that in some cases the exponent for the two functions is the same.

Frankl and Wilson [61] proved the following modular analogue of the result of Ray-Chaudhuri and Wilson:

Theorem 62 *Fix $L \subset [0, k-1]$. If there is an integer-valued polynomial f of degree r and a prime number p such that $p | f(\ell)$ for every $\ell \in L$, and $p \nmid f(k)$, then $m(n, k, L) \leqslant \binom{n}{r}$.*

The proof uses linear algebra method. Füredi in [65] gives a very nice argument that allows us to compare the Δ-system method with the linear-algebraic method. We recite it here.

Theorem 63 *Fix $L \subset [0, k-1]$. If $\mathcal{F}$ is an (n, k, L)-system, and there is an integer-valued polynomial f of degree r and a positive integer q such that $q \mid f(\ell)$ for every $\ell \in L$, and $q \nmid f(k)$, then $|\mathcal{F}| \leqslant C_k\binom{n}{r}$, where C_k depends on k only.*

Note that the bound is weaker (working for large n only), but we drop the assumption that q is prime.

Proof Pass from $\mathcal{F}$ to a homogeneous structure $\mathcal{F}^*$. Let us show that the rank of $\mathcal{F}^*$ is at most r. This will imply $|\mathcal{F}^*| \leqslant \binom{n}{r}$, and thus the bound in the theorem.

Arguing indirectly, assume that the intersection structure $\mathcal{I}(\mathcal{F})$ covers all r-element subsets. Apply inclusion-exclusion to each individual i-element set in $[k]$ and sum it up over all sets, getting

$$\binom{k}{i} = \sum_{I \in \mathcal{I}} \binom{|I|}{i} - \sum_{I \in \mathcal{I}} \sum_{I' \in \mathcal{I}, I' \neq I} \binom{|I \cap I'|}{i} + \dots.$$

As f is an integer-valued polynomial, we can write it in the form $f = \sum_{i=0}^{r} a_i \binom{x}{i}$ with integer coefficients a_i. Multiply the display by a_i and sum it up over i. On the left hand side, we get $f(k)$, which is non-zero modulo q. On the right hand side, we get $\sum f(|I|) - \sum\sum f(|I \cap I'|) + \dots$, which is 0 modulo q, a contradiction. □

Füredi [65] conjectured that if there is a family $\mathcal{I} \subset 2^{[k]}$ of sets of size $\ell \in L$ that is closed under intersection and such that its rank is at least $d+1$, then for some positive $\epsilon = \epsilon(k)$ we have $m(n, k, L) = \Omega(n^{d+\epsilon})$. It is true for $d = 1$ via a construction of Frankl [50]. Rödl and Tengan [113] verified this conjecture for $L = \{0, 1, \dots, d-1, q\}$ and a large class of families $\mathcal{I}$, that, in particular, including complete designs (Steiner systems): each q-element set of $[k]$ is contained in exactly one p-element set. In this setup, $\epsilon > 0$ from the conjecture is essential: if we ask for $\Omega(n^d)$ only, then we may simply ignore intersection q and consider a (partial) Steiner system $S(n, k, d)$ in which each d-element set is contained in (at most) one k-set. However, I do not know any non-trivial lower bounds for the conjecture of Füredi for general L.

9.3 Sunflowers, Unavoidable Hypergraphs

Let us return to the question, discussed in Sect. 5.1. In 1977, Duke and Erdős [30] asked the following variant of the question of Erdős and Rado: what is the largest size $f(n, k, \ell, s)$ of the family $\mathcal{F} \subset \binom{[n]}{k}$ that does not contain a $\Delta(s)$-system with kernel of size ℓ?[18] For $s = 2$ this is the forbidden one intersection problem, discussed

[18] The authors of different papers used different permutations for the 4 parameters of the function. The size n of the ground set is always the first, while the other indices got at least three different permutations in [30, 45, 55]. Ours is the one used in [55].

in Sect. 9.1, and for $\ell = 0$ this is the famous Erdős Matching Conjecture [34] (see [57] for a survey). In Sect. 5.1 we mentioned that $f(n, k, \ell, s) = O_{k,s}(f(n, k, \ell, 2))$ follows from Theorem 21 of Füredi [63], and that Frankl and Füredi determined the asymptotic of $ex(n, \mathcal{S}_\ell^{(k)}(s))$ for fixed s, k and $k \geqslant 2\ell + 3$, see Theorem 37. The case of $k \leqslant 2\ell + 2$ resisted progress so far.

What happens if k, s are not fixed? Duke and Erdős [30] in their paper showed that $f(n, k, 1, s) \leqslant O_k(s^2 n^{k-2})$. It is the right dependence on s and n since we can use the same type of construction as in Sect. 5.1, that is an 'expansion' of a construction of a 2-graph on $(s-1)^2$ vertices with no sunflower (a complete bipartite graph with parts of size $(s-1)$). Using weight functions, Chung and Frankl [19] determined the value of $f(n, 3, 1, s)$ exactly for $n > Ck^3$. It improved upon an earlier work of Frankl [45], and Chung [17], who determined the answer up to lower-order terms.

In cite [30], Duke and Erdős noted that $f(n, 3, 2, s) \sim \frac{1}{6} sn^2$ for constant s (and which can be easily extended to $s = o(n)$). Note that we are asking for the largest hypergraph such that no pair of elements is contained in s triples. The upper bound is a simple double-counting, and the lower bound follows by using a connection to Steiner triple systems (e.g., we may find a partial triple system using Rödl's result, and then combine s randomly permuted copies of those). In particular, we get $f(n, 3, 2, s) = \Theta(sn^2)$ for all s. Much more recently, Bucić et al. [12] showed that $f(n, 4, 2, s) = \Theta(s^2 n^2)$ and $f(n, 4, 3, s) = \Theta(sn^3)$. Their result was generalized by Bradač et al. [13], who showed that $f(n, k, \ell, s) = \Theta_k(s^{\ell+1} n^{k-\ell-1})$ for $k \geqslant 2\ell + 2$ and $f(n, k, \ell, s) = \Theta_k(s^{k-\ell} n^\ell)$ for $k \leqslant 2\ell + 1$. They mostly followed the methods of Frankl and Füredi [54]. One difficulty that they managed to overcome is that Theorem 21 cannot be applied directly here, since s is assumed to be large. They overcome it by constructing an intersection structure-like family, as it would look after the application of Theorem 21, without actually going through the same 'filtering' process.

In a recent paper [96], Noskov and the author managed to find the correct asymptotic of $f(n, k, \ell, s)$ for k linear in n: $n > poly(s, \ell)k$, $n \geqslant n_0(s, t)$ and $k \geqslant 2\ell + 3$. (We note here that $n \geqslant n_0(s, \ell)$ with at least exponential n is unavoidable here because of our poor understanding of the original Erdős–Rado sunflower problem, cf. examples from Sect. 5.1.)

One motivation for studying $f(n, k, \ell, s)$ for s that grows with n is *unavoidable hypergraphs*. A k-uniform hypergraph $\mathcal{H}$ is (n, e)-*unavoidable* if every k-uniform hypergraph with e edges contains a copy of $\mathcal{H}$. Let us denote $un_k(n, e)$ the maximum number of edges in an (n, e)-unavoidable hypergraph. Chung and Erdős [18] determined $un_3(n, e)$ up to a multiplicative constant for the full range of e. The case $k = 4$ was resolved (again, up to a multiplicative constant) by Bucić et al. [12].

9.4 Simplices, Clusters

As we have discussed in Sect. 5.3, Erdős [35] conjectured that the largest family in $\binom{[n]}{k}$ without a triangle has size $\binom{n-1}{k-1}$ and is a star, and the question was generalized to d-dimensional simplices by Chvátal [20] (see Definition 41 and Conjecture 3): if $\mathcal{F}$ contains no d-simplex, $k > d$, and $n \geqslant (d+1)k/d$, then $|\mathcal{F}| \leqslant \binom{n-1}{k-1}$. Frankl proved it for $n \leqslant dk/(d-1)$ [41], as well as for $d = 2, k \geqslant 5$ and $k \geqslant 3d+1$, provided $n \geqslant n_0(k)$ [48]. Then Frankl and Füredi proved it for any $k > d$, provided $n \geqslant n_0(k)$. The case $d = 2$ (i.e., the original question of Erdős) was settled for all n, k by Mubayi and Verstraëte [108]. Keevash and Mubayi [84] proved it for any $\zeta > 0$ and $k_0(d) \leqslant k \leqslant (\frac{1}{2} - \zeta)n$, provided $n > n_0(d, \zeta)$. Keller and Lifshitz [86] proved it for all $k > k_0(d)$. This, combined with the result of Frankl and Füredi, covered all cases $n > n_0(d)$. Shortly after, Currier [22] made impressive progress on the problem and proved it for almost all values of parameters: for $d \geqslant 3$ and $n \geqslant 2k - d + 2$. Together with the previous results, the following cases of the conjecture are left: $k \geqslant 3, dk/(d-1) < n < 2k - d + 2$, for d that grow with k.

Actually, there are two very different types of simplices that the aforementioned works dealt with. In Sect. 5.3, we have discussed the very structured d-simplices that are called *special simplices* (see Definition 43). Theorem 44 by Frankl and Füredi [55] shows that for $k \geqslant d+3$ and $n \geqslant n_0(k)$ we have $ssim(n, k, d) \leqslant \binom{n-1}{k-1}$, i.e., it is enough to forbid only special simplices in order to get the same answer. They conjectured that the same should hold for $k = d+1, d+2$, and proved this for $d = 2$. They also conjectured that for $n \geqslant 2k$ $ssim(n, k, k-1) \leqslant \binom{n-1}{k-1}$. Some cases of it are resolved by Bermond and Frankl in [8]. Keller and Lifshitz [86] showed the validity of the conjecture for $k \geqslant k_0(d)$ and $n \geqslant C_0(d)k$. Using spread approximation method, Noskov and the author (in preparation) verified the conjecture for $n \geqslant poly(d, k)$.

The second type of simplices that was considered are much less structured and are called *simplex-clusters*. Let us first define clusters. A *d-cluster* is a collection of d distinct sets in $\binom{[n]}{k}$ with empty common intersection and with union of size at most $2k$. A 2-cluster is just two disjoint sets, and thus 2-cluster-free families are the same as intersecting families. As another possible generalization of the Erdős–Ko–Rado theorem, Katona conjectured that the largest 3-cluster-free family in $\binom{[n]}{k}$ has size $\binom{n-1}{k-1}$ (see [53]). Frankl and Füredi [53] proved Katona's conjecture for $n \geqslant k^2 + 3k$. Mubayi [104] proved the conjecture for 3-cluster-free families in $\binom{[n]}{k}$ has size at most $\binom{n-1}{k-1}$ for all $n, k \geqslant 3$, and conjectured that the same should hold for any $d \geqslant 4$. In a follow-up paper, he proved the asymptotic upper bound $(1 + o(1))\binom{n-1}{k-1}$, as well as resolved the case $k = 4$ and n sufficiently large. The aforementioned result of Keevash and Mubayi [84] applies to $(d+1)$-cluster-free sets as well. Then Mubayi and Ramadurai [106], and independently [73] showed that the conjecture holds for $k > 4$ and n large enough. The case $n < 2k$ (where the union condition in the definition of a cluster is automatically fulfilled) was resolved by Frankl [41]. Finally, Currier [21] made impressive progress and proved Mubayi's conjecture for the whole range of parameters: $d \leqslant k$ and $n \geqslant dk/(d-1)$.[19] The latter condition is needed so that there are d sets in $[n]$ with empty common intersection.

[19] His work on Chvátal's conjecture builds upon this work.

For $d = k + 1$, Füredi and Özkahya [73] gave an elegant argument showing that a $(k + 1)$-cluster-free family $\mathcal{F} \subset \binom{[n]}{k}$ must have size at most $\binom{n-1}{k-1}$. First, note that for any $F \in \mathcal{F}$ there must be a subset $X(F)$ of size $k - 1$ that is not contained in any other set $F' \in \mathcal{F}$. Otherwise, if $F = \{a_1, \ldots, a_k\}$ and $F_i \cap F = F \setminus \{a_i\}$, then $F \cap F_1 \cap \ldots \cap F_k = \emptyset$ and $|F \cup F_1 \cup \ldots \cup F_k| \leqslant 2k$. Thus, these sets form a $(k + 1)$-cluster. Thus the pairs $(A_F, B_F) := (X(F), [n] \setminus F)$, $F \in \mathcal{F}$, satisfy the property $A_F \cap B_{F'} = \emptyset$ iff $F = F'$. Applying the Bollobás set-pair inequality [9] to these pairs, we conclude that $|\mathcal{F}| \leqslant \binom{|A_F|+|B_F|}{A_F} = \binom{n-1}{k-1}$. We should also note that if in the definition of a d-cluster we change $2k$ to $2k - 1$, then a complete k-partite hypergraph has no such substructure. Frankl and Füredi [53] conjectured that for $d = 3$ the extremum is given by a complete k-partite hypergraph and proved it for $k = 3$, $n > n_0$. A related notion for a 3-uniform hypergraph is being *cancellative*: not having three sets A, B, C such that $A \Delta B \subset C$.

Merging the two definitions, that of a d-simplex and of a $(d + 1)$-cluster, we get a d-simplex-cluster. Keevash and Mubayi made, quoting, 'ambitious conjecture' that for $k \geqslant d + 1 > 2$, $n > k(d + 1)/d$ and a family in $\binom{[n]}{k}$ avoiding a d-simplex-cluster, its size is at most $\binom{n-1}{k-1}$, with equality only possible for stars. Lifshitz [100] proved it for $n \geqslant n_0(d)$. Shortly after, Currier proved it in [22] for $n \geqslant 2k - d + 2$ (this is actually the main result of [22], which immediately implies the result for Chvátal's conjecture).

Let us now discuss the variants of these questions that arose in the literature while studying simplices and clusters. A triangle, and, more generally, a d-simplex are the simplest examples of non-trivial intersecting families, that is, intersecting families with covering number 2. In [108], Mubayi and Verstraëte actually proved that if $\mathcal{F} \subset \binom{[n]}{k}$ does not contain a non-trivial intersecting family on $d + 1$ sets, $n \geqslant (d + 1)k/d$ and $k \geqslant d + 1$, then $|\mathcal{F}| \leqslant \binom{n-1}{k-1}$, with equality for stars only. They conjectured that the same should hold without the assumption $k \geqslant d + 1$, provided n is large enough. This was proved by Liu [101], who actually showed that is enough to forbid a certain configuration, which is called a $(\mathbf{a}, p)$-Δ-system. If the dimension of $\mathbf{a}$ is d, then $(\mathbf{a}, p)$-Δ-*system* is any family of the following type: start with a partition of a host set F_0 into d parts $A_1, \ldots, A_d$. Let $B_i := F_0 \setminus A_i$. Now extend each B_i into $b_i > 0$ pairwise disjoint k-sets, so that, first, all extensions of all B_i's are pairwise disjoint and, second, $\sum b_i = p$. This construction actually generalized the construction of Füredi and Özkahya [73], who used it to prove the d-cluster conjecture of Mubayi for sufficiently large n. They used $b_1 = \ldots = b_d = 1$, in which case each B_i has a unique extension B_i'. It is not difficult to see in this case that the union of all sets has size exactly $2k$, as well as they do not have a common intersection, so they indeed form a d-cluster. The proof of Füredi and Özkahya, as well as that of Liu, involved Delta-systems.

Keevash and Mubayi [84] actually proved their result for families avoiding a d-cluster and for families avoiding a *strong d-simplex*. It is a collection of $d + 2$ sets $A, A_1, \ldots, A_{d+1}$, such that $A_1, \ldots, A_{d+1}$ is a d-simplex, and A intersects $\cap_{j \neq i} A_i$ for each $i \in [d + 1]$. E.g., a strong 1-simplex is a path of length 3, i.e., three sets A_1, A, A_2, such that $A_1 \cap A_2 = \emptyset$ and $A_1 \cap A$, $A \cap A_2 \neq \emptyset$.

9.5 Hypertrees, Expanded Hypergraphs

In Sect. 5.2 we discussed a general result of Frankl and Füredi on expanded hypergraphs, as well as introduced an important notion of a cross-cut. Let us give a variant of the definition of an expanded hypergraph.

Definition 64 (*t-contractible hypergraphs*) For a positive integer $t \leqslant k-1$, a k-uniform family $\mathcal{H}$ is *t-contractible* if each set of $\mathcal{H}$ contains t vertices of degree 1. A t-contraction of $\mathcal{H}$ is the $(k-t)$-uniform multi-family obtained by deleting t degree 1 vertices from each edge of $\mathcal{H}$.

The notion of t-contractible hypergraphs is slightly is more general than that of the k-expanded hypergraphs, see Definition 38. The (only) difference is that the former allows edges of multiplicity more than 1 in the contraction.

The following general result was proved by Füredi and Jiang [69] using Δ-systems.

Theorem 65 (Jiang and Füredi, [69]) *Let $k \geqslant 4$ be an integer. For any k-graph $\mathcal{H}$ that is a subgraph of a 2-contractible k-tree, we have*

$$ex(n, \mathcal{H}) = (\sigma_1(\mathcal{H}) + o(1))\binom{n}{k-1}.$$

(We remind the reader that $\sigma_1(\mathcal{H})$ is the size of the 1-crosscut minus 1.) It was shown in [69] that the conclusion of Theorem 65 is no longer valid for all 1-contractible trees.

This theorem extended several previous results on the subject, which we discuss below. First, we should note that for k-expansions of graph trees with at most $k-1$ edges Theorem 40 gives the same bound. Then, Füredi [66] in 2014 proved this for k-expansions of graph trees, also under the condition of $k \geqslant 4$. (These are $(k-2)$-contractible k-trees, if we ignore the multi-edge situation.)

Definition 66 (*Linear paths and cycles*) A k-uniform family is a *linear path* $\mathcal{P}_\ell^{(k)}$ of length ℓ (*linear cycle* $\mathcal{C}_\ell^{(k)}$ of length ℓ) if it is a k-expansion of a graph path of vertex length ℓ (cycle of length ℓ).

That is, Füredi, in particular, determined the asympototic for $ex(n, \mathcal{T})$ for linear paths. The exact value for linear paths of length ℓ for $k \geqslant 4$ and $n \geqslant n_0(k, \ell)$ was determined by Füredi et al. [71]. Both papers used the Delta-system method. The case $k = 3$ was settled by Kostochka et al. [90] using the 'random sampling from the shadow' method, together with some Ramsey-type arguments. In the follow-up paper [91], they extended the result of Füredi [66] concerning expansions of graph trees to the case of $k = 3$. The case of forests consisting of linear paths was resolved for large n in [14] using induction and the aforementioned results.

Another type of expansions that are covered by Theorem 65 are linear cycles. Let us show that for $k \geqslant 5$ we can construct a 2-contractible k-tree that contains a linear cycle of length ℓ. Take the graph cycle $\{v_1, \ldots, v_\ell\}$ of length ℓ. Now we add 3-edges that will turn this graph cycle a tight 3-tree: add $\{v_1, v_2, v_3\}$, $\{v_1, v_3, v_4\}, \ldots,$ $\{v_1, v_{\ell-1}, v_\ell\}$ in this order. Then add extra vertices $u_1, \ldots, u_\ell$ so that $\{v_i, u_i, v_{i+1}\}_{i\in[\ell]}$ is the set of hyperedges of $C_\ell^{(3)}$. (We can add them in any order.) It is not difficult to check that the resulting 3-uniform hypergraph is a tight 3-tree. Now we may add $k-3$ distinct vertices to each hyperedge and get a $(k-3)$-contractible k-tree that contains $C_\ell^{(k)}$.

Thus, Theorem 65 implies the asymptotic behavior of $ex(n, C_\ell^{(k)})$ for $k \geqslant 5$. The exact value of $ex(n, C_\ell^{(k)})$ for large n was actually determined by Füredi and Jiang for $k \geqslant 5$ in an earlier paper [68]. They used Delta-systems. Then the remaining cases $k = 3, 4$ were resolved by Kostochka et al. [90] using the 'random sampling from the shadow' method.

We should note that Füredi and Jiang [69] actually proved sharper bounds. For 2-reducible k-trees the error term in Theorem 65 is $O(n^{k-2})$. For $(k-2)$-reducible k-trees $\mathcal{T}$ (i.e., essentially k-expansions of graph trees), the upper bound on $ex(n, \mathcal{T})$ for large n depends on the value of $f(n, k, 1, s)$ (extremal function for forbidden sunflower with core of size 1), and is sharp in some cases (cf. [69, Theorem 4.5]).

3-expansions of different 2-graphs were treated by Kostochka, Mubayi and Verstraëte in [91, 92]. In particular, in [91] the authors showed that for large n $ex(n, \mathcal{H}) \leqslant \binom{n}{2}$ for any 3-uniform hypergraph that is an expansion of a 2-graph with cross-cut 2, and in [92] they showed that $ex(n, \mathcal{H}) = O(n^2)$ for graphs with cross-cut 3. They also showed that $ex(n, \mathcal{H}) = \Theta(n^2)$ for the 3-expansion of the graph of the 3-dimensional cube. In [102] the authors sharpened some of the results of [91]. We refer to [108] for a survey on the topic of extremal problems for graph expansions.

In the paper [70] the authors determined the asymptotic of $ex(n, C)$, where C is a $(a+b)$-uniform path, where consecutive edges alternately share a and b vertices. They developed an averaging technique which they called an asymmetric version of Katona's circle method.

9.6 Other Surveys

In this short review, we covered only a small portion of the field. We refer to several other excellent surveys in the field. A survey by Füredi [65] has a significant overlap with this survey; Mubayi and Verstraëte [108] treat Turán problems for expansions; intersection theorems are treated in a survey by Ellis [31]; methods for solving (mostly dense) hypergraph Turan problems are surveyed by Keevash in [82]. In this survey, we did not treat Berge hypergraphs. See, e.g., a recent paper [72] for a brief overview and an interesting result. Applications of sunflowers and the spread lemma are discussed in an expository paper by Rao [110]. There are also two books on extremal set theory: by Frankl and Tokushige [60] and by Gerbner and Patkós [74].

Acknowledgements I am indebted to Liza Iarovikova, Fedor Noskov, Nikolay Terekhov, Georgii Sokolov and Yakov Shubin for proofreading earlier versions of the manuscript and providing numerous suggestions that improved the presentation, as well as to Peter Frankl, Tao Jiang and the anonymous referee for giving multiple helpful comments on the final version of the text. The research is supported by the Ministry of Science and Higher Education of the Russian Federation, project No. FSMG-2024-0011.

References

1. Abbott, H.L., Hanson, D., Sauer, N.: Intersection theorems for systems of sets. J. Comb. Theory **12**, 381–389 (1972)
2. Ahlswede, R., Khachatrian, L.H.: The complete intersection theorem for systems of finite sets. Eur. J. Comb. **18**, 125–136 (1997)
3. Alon, N., Frankl, P., Lovász, L.: The chromatic number of Kneser hypergraphs. Trans. Am. Math. Soc. **298**, 359–370 (1986)
4. Alon, N., Boppana, R.B.: The monotone circuit complexity of boolean functions. Combinatorica **7**, 1–22 (1987)
5. Alweiss, R., Lovett, S., Wu, K., Zhang, J.: Improved bounds for the sunflower lemma (2019). arXiv:1908.08483
6. Babai, L., Frankl, P.: Note on set intersections. J. Combin. Theory A **28**, 103–105 (1980)
7. Bell, T., Chueluecha, S., Warnke, L.: Note on sunflowers. Discrete Math. **344**, N7 (2021)
8. Bermond, J.C., Frankl, P.: On a conjecture of Chvátal on m -intersecting hypergraphs. Bull. Lond. Math. Sot **9**, 310–312 (1977)
9. Bollobás, B.: On generalized graphs. Acta Math. Acad. Sci. Hungar. **16**, 447–452 (1965)
10. Bollobás, B., Leader, I.: Set systems with few disjoint pairs. Combinatorica **23**, 559–570 (2003)
11. Bollobás, B., Thomason, A.G.: Threshold functions. Combinatorica **7**, 35–38 (1987)
12. Bucić, M., Draganić, N., Sudakov, B., Tran, T.: Unavoidable hypergraphs. J. Comb. Theory, Ser. B **151**, 307–338 (2021)
13. Bradač, D., Bucić, M., Sudakov, B.: Turán numbers of sunflowers (2021). https://doi.org/10.48550/arXiv.2110.11319
14. Bushaw, N., Kettle, N.: Turán numbers for forests of paths in hypergraphs. SIAM J. Discrete Math. **28**, 711–721 (2014)
15. Ray-Chaudhuri, D.K., Wilson, R.M.: On t-designs. Osaka J. Math. **12**, 737–744 (1975)
16. Cherkashin, D.: On set systems without singleton intersections (2024). arXiv:2408.00484
17. Chung, F.R.K.: Unavoidable stars in 3-graphs. J. Comb. Theory A **35**, 252–262 (1983)
18. Chung, F.R.K., Erdős, P.: On unavoidable hypergraphs. J. Graph Theory **11**, 251–263 (1987)
19. Chung, F.R.K., Frankl, P.: The maximum number of edges in a 3 -graph not containing a given star. Graphs Comb. **3**, 111–126 (1987)
20. Chvátal, V.: An extremal set-intersection theorem. J. Lond. Math. Soc. **9**, 355–359 (1974)
21. Currier, G.: On the d-cluster generalization of Erdős-Ko-Rado. J. Comb. Theory Ser. A **182**, 105464 (2021)
22. Currier, G.: New results on simplex-clusters in set systems. Combinatorica **41**, 495–506 (2021)
23. Das, S., Gan, W., Sudakov, B.: The minimum number of disjoint pairs in set systems and related problems. Combinatorica **36**, 623–660 (2016)
24. Deza, M.: Solution d'un problème de Erdős-Lovász. J. Comb. Theory Ser. B **16**, 166–67 (1974)
25. Deza, M.: Une propriété extrémale des plans projectifs dans une classe de codes équidistants. Discrete Math. **6**, 343–352 (1973)

26. Deza, M., Erdős, P., Frankl, P.: Intersection properties of systems of finite sets. Proc. Lond. Math. Soc. **36**, N3, 369–384 (1978). https://doi.org/10.1112/plms/s3-36.2.369
27. Deza, M., Erdős, P., Singhi, N.M.: Combinatorial problems on subsets and their intersections, studies in foundations and combinatorics. Adv. Math. Suppl. Stud. **1**, 259–265 (1978)
28. Deza, M., Frankl, P.: Every large set of equidistant $(0, +1, -1)$ -vectors forms a sunflower. Combinatorica **1**, 225–231 (1981)
29. Dubinin, N.A., Neustroeva, E.A., Raigorodskii, A.M., Shubin, Y.K.: Lower and upper bounds for the minimum number of edges in some subgraphs of the Johnson graph. Sb. Math. **215**, 634–657 (2024)
30. Duke, R.A., Erdős, P.: Systems of finite sets having a common intersection. In: Proceedings 8th S-E Conference Combinatorics, Graph Theory Computing, pp. 247–252 (1977)
31. Ellis, D.: Intersection Problems in Extremal Combinatorics: Theorems, Techniques and Questions Old and New. Surveys in combinatorics, pp. 115–173 (2022)
32. Erdős, P., Kleitman, D.J.: On coloring graphs to maximize the proportion of multicolored k-edges. J. Comb. Theory **5**, 164–169 (1968)
33. Erdős, P.: Extremal Problems in Graph Theory, Theory of Graphs and its Applications (Proc. Sympos. Smolenice, 1963) , pp. 29–36. Publ. House Czech. Acad. Sci., Prague (1964)
34. Erdős, P.: A problem on independent r-tuples. Ann. Univ. Sci. Budapest. **8**, 93–95 (1965)
35. Erdős, P.: Topics in combinatorial analysis. In: Mullin, R.C., et al. (eds.) Proceedings of Second Louisiana Conference on Combinatorics, Graph Theory and Computing, pp. 2–20. Louisiana State Univ., Baton Rouge (1971)
36. Erdős, P.: Problems and results in graph theory and combinatorial analysis. In: Proceedings Fifth British Combinatorial Conference, pp. 169–172 (1975) (University of Aberdeen, Aberdeen: Congressus Numerantium, 15. Winnipeg, Utilitas Mathematica, 1976)
37. Erdős, P., Kleitman, D.J.: On coloring graphs to maximize the proportion of multicolored k-edges. J. Comb. Theory **5**, 164–169 (1968)
38. Erdős, P., Ko, C., Rado, R.: Intersection theorems for systems of finite sets. The Q. J. Math. **12**(N1), 313–320 (1961)
39. Erdős, P., Lovász, L.: Problems and results on 3-chromatic hypergraphs and some related questions, infinite and finite sets (Colloq., Keszthely, 1973; dedicated to P. Erdős on his 60th birthday), Vol. II; Colloq. Math. Soc. János Bolyai, vol. 10, 609–627, North-Holland, Amsterdam (1975)
40. Erdős, P., Rado, R.: Intersection theorems for systems of sets. J. Lond. Math. Soc. **35**(N1), 85–90 (1960)
41. Frankl, P.: Sperner systems satisfying an additional condition. J. Comb. Theory Ser. A **20**, 1–11 (1976)
42. Frankl, P.: On families of finite sets no two of which intersect in a singleton. Bull. Austral. Math. Soc. **17**(N1), 125–134 (1977)
43. Frankl, P.: A constructive lower bound for some Ramsey numbers. Ars Comb. **3**, 297–302 (1977)
44. Frankl, P.: On intersecting families of finite sets. J. Comb. Theory Ser. A **24**, 146–161 (1978)
45. Frankl, P.: An extremal problem for 3-graphs. Acta Mathematica Academiae Scientiarum Hungaricae **32**, 157–160 (1978)
46. Frankl, P.: Extremal problems and coverings of the space. Eur. J. Comb. **1**, 101–106 (1980)
47. Frankl, P.: Families of finite sets with prescribed cardinalities for pairwise intersections. Acta Mathematica Academiae Scientiarum Hungaricae **35**, 351–360 (1980)
48. Frankl, P.: On a problem of Chvátal and Erdős on hypergraphs containing no generalized simplex. J. Comb. Theory Ser. A **30**, 169–182 (1981)
49. Frankl, P.: An extremal set theoretical characterization of some Steiner systems. Combinatorica **3**(N2), 193–199 (1983)
50. Frankl, P.: Families of finite sets with three intersections. Combinatorica **4**, 141–148 (1984)
51. Frankl, P.: All rationals occur as exponents. J. Comb. Theory Ser. A **42**, 200–206 (1986)
52. Frankl, P.: Antichains of fixed diameter. Moscow J. Combin. Number Theory **7**(N3) (2017)

53. Frankl, P., Füredi, Z.: A new generalization of the Erdős-Ko-Rado theorem. Combinatorica **3**, 341–349 (1983)
54. Frankl, P., Füredi, Z.: Forbidding just one intersection. J. Comb. Theory Ser. A **39**, 160–176 (1985)
55. Frankl, P., Füredi, Z.: Exact solution of some Turán-type problems. J. Comb. Theory Ser. A **45**, 226–262 (1987)
56. Frankl, P., Katona, G.O.H.: If the intersection of any r sets has a size $\neq r - 1$. Studia Sci. Math. Hungar. **14**, 47–49 (1979)
57. Frankl, P., Kupavskii, A.: The Erdős matching conjecture and concentration inequalities. J. Comb. Theory Ser B. **157**, 366–400 (2022)
58. Frankl, P., Ota, K., Tokushige, N.: Exponents of uniform L-systems. J. Comb. Theory Ser. A **75**, 23–43 (1996)
59. Frankl, P., Rödl, V.: Forbidden intersections. Trans. Am. Math. Soc. **300**, 259–286 (1987)
60. Frankl, P., Tokushige, N.: Extremal Problems for Finite Sets. American Mathematical Society, Providence, Rhode Island (2018)
61. Frankl, P., Wilson, R.: Intersection theorems with geometric consequences. Combinatorica **1**, 357–368 (1981)
62. Füredi, Z.: Erdős–Ko–Rado Type Theorems with Upper Bounds on the Maximum Degree, pp. 177–207. Colloquia Math. Soc. J. Bolyai 25, Szeged (1978)
63. Füredi, Z.: On finite set-systems whose every intersection is a kernel of a star. Disceret Math. **47**, 129–132 (1983)
64. Füredi, Z.: Set systems with three intersections. Combinatorica **5**(N1), 27–31 (1985)
65. Füredi, Z.: Turán Type Problems. In: Keedwell, A.D. (ed.) Surveys in Combinatorics, 1st edn., pp. 253–300 Cambridge University Press (1991)
66. Füredi, Z.: Linear trees in uniform hypergraphs. Eur. J. Comb. **35**, 264–272 (2014)
67. Füredi, Z., Gerbner, D.: Hypergraphs without exponents. J. Comb. Theory Ser. A **184** (2021)
68. Füredi, Z., Jiang, T.: Hypergraph Turán numbers of linear cycles. J. Comb. Theory Ser. A **123**(N1), 252–270 (2014)
69. Füredi, Z., Jiang, T.: Turán numbers of hypergraph trees (2015). doi:https://doi.org/10.48550/arXiv.1505.03210
70. Füredi, Z., Jiang, T., Kostochka, A., Mubayi, D., Verstraëte, J.: Extremal problems for hypergraph blowups of trees. SIAM J. Discret. Math. **37**(N4), 2397–2416 (2023)
71. Füredi, Z., Jiang, T., Seiver, R.: Exact solution of the hypergraph Turán problem for k-uniform linear paths. Combinatorica **34**, 299–322 (2014)
72. Füredi, Z., Luo, R.: Induced Turán problems and traces of hypergraphs. Eur. J. Comb. **111** (2023)
73. Füredi, Z., Özkahya, L.: Unavoidable subhypergraphs: a-clusters. J. Comb. Theory Ser. A **118**, 2246–2256 (2011)
74. Gerbner, D., Patkós, B.: Extremal Finite Set Theory. Chapman and Hall/CRC (2018)
75. Hu, L.: Entropy Estimation via Two Chains: Streamlining the Proof of the Sunflower Lemma (2021). https://web.archive.org/web/20220502190641, https://theorydish.blog/2021/05/19/entropy-estimation-via-two-chains-streamlining-the-proof-of-the-sunflower-lemma/
76. Janzer, B., Jin, Z., Sudakov, B., Wu, K.: Sunflowers and Ramsey Problems for Restricted Intersections (2025). arXiv:2504.15264
77. Jiang, T., Longbrake, S.: On the Number of H-Free Hypergraphs (2024). https://doi.org/10.48550/arXiv.2409.06810
78. Jiang, T., Pikhurko, O., Yilma, Z.: Set systems without a strong simplex. SIAM J. Discrete Math. **24**, 1038–1045 (2010)
79. Jukna, S.: Extremal Combinatorics. Springer, Berlin, Heidelberg (2011)
80. Katona, G.O.H.: A theorem of finite sets, theory of graphs. Proc. Coll. Tihany,: Akad, Kiado, Budapest 1968. Classic Papers in Combinatorics **1987**, 381–401 (1966)
81. Keevash, P.: The existence of designs. arXiv:1401.3665
82. Keevash, P.: Hypergraph Turán problems. In: Chapman, R. (ed.) Surveys in Combinatorics 2011, pp. 83–140. Cambridge University Press, Cambridge (2011)

83. Keevash, P., Lifshitz, N., Long, E., Minzer, D.: Global Hypercontractivity and Its Applications (2021). arXiv:2103.04604
84. Keevash, P., Mubayi, D.: Set systems without a simplex or a cluster. Combinatorica **30**, 175–200 (2010)
85. Keevash, P., Mubayi, D., Wilson, R.M.: Set systems with no singleton intersection. SIAM J. Discrete Math. **20**, 1031–1041 (2006)
86. Keller, N., Lifshitz, N.: The Junta method for hypergraphs and Chvátal's simplex conjecture. Adv. Math. **392** (2021)
87. Kostochka, A.V.: An intersection theorem for systems of sets. Random Struct. Algorithms **9**, 213–221 (1996)
88. Kostochka, A.: Extremal problems on Δ-systems. In: Numbers, Information and Complexity, pp. 143–150. Springer, US, Boston, MA (2000)
89. Kostochka, A., Mubayi, D.: The structure of large intersecting families. Proc. Am. Math. Soc. **145**, 2311–2321 (2016)
90. Kostochka, A., Mubayi, D., Verstraëte, J.: Turán problems and shadows I: paths and cycles. J. Comb. Theory Ser. A **129**, 57–79 (2015)
91. Kostochka, A., Mubayi, D., Verstraëte, J.: Turán problems and shadows II: trees. J. Comb. Theory Ser. B **122**, 457–478 (2017)
92. Kostochka, A., Mubayi, D., Verstraete, J.: Turán Problems and Shadows III: Expansions of Graphs (2014). https://doi.org/10.48550/arXiv.1407.2588
93. Kostochka, A.V., Rödl, V., Talysheva, L.A.: On systems of small sets with no large Δ-subsystems. Comb. Probab. Comput. **8**, 265–268 (1999)
94. Kruskal, J.B.: The number of simplices in a complex. Math Optim Tech **251**, 251–278 (1963)
95. Kupavskii, A.: Erdős–Ko–Rado type results for partitions via spread approximations (2023). arXiv:2309.00097
96. Kupavskii, A., Noskov, F.: Linear dependencies, polynomial factors in the Duke–Erdős forbidden sunflower problem (2025). arXiv:2410.06156
97. Kupavskii, A., Sagdeev, A., Zakharov, D.: Cutting Corners (2022). arXiv:2211.17150
98. Kupavskii, A., Zakharov, D.: Spread approximations for forbidden intersections problems, to appear in Advances in Mathematics. arxiv:2203.13379
99. Larman, D.C.: A note on the realization of distances within sets in Euclidean space. Comment. Math. Helv. **53**, 529–535 (1978)
100. Lifshitz, N.: On set systems without a simplex-cluster and the junta method. J. Comb. Theory Ser. A. **170** (2020)
101. Liu, X.: A note on hypergraphs without non-trivial intersecting subgraphs (2020). arXiv:2007.11055
102. Liu, X., Song, J., Yuan, L.T.: Exact results for some extremal problems on expansions I (2024). arXiv:2310.01736
103. Mossel, E., Niles-Weed, J., Sun, N., Zadik, I.: A second moment proof of the spread lemma (2022). arXiv:2209.11347
104. Mubayi, D.: Erdős—Ko—Rado for three sets. J. Comb. Theory Ser. A **113**, 547–550 (2006)
105. Mubayi, D.: An intersection theorem for four sets. Adv. Math. **215**(N2), 601–615 (2007)
106. Mubayi, D., Ramadurai, R.: Set systems with union and intersection constraints. J. Comb. Theory, Ser. B **99**, 639–642 (2009)
107. Mubayi, D., Verstraëte, J.: Proof of a conjecture of Erdős on triangles in set-systems. Combinatorica **25**, 599–614 (2005)
108. Mubayi, D., Verstraëte, J.: A survey of Turán problems for expansions. In: Recent Trends in Combinatorics, pp. 117–143. Springer International Publishing, Cham (2016)
109. Rao, A.: Coding for sunflowers. Discrete Anal. 11877 (2020)
110. Rao, A.: Sunflowers: from soit to oil. Bull. Am. Math. Soc. **60**, 29–38 (2023)
111. Razborov, A.A.: Lower bounds for the monotone complexity of some Boolean functions, Dokl. Ak. Nauk. SSSR, 281,: 798–801 (in Russian). English translation. In: Sov. Math. Dokl. **31**(1985), 354–357 (1985)
112. Rödl, V.: On a packing and covering problem European. J. Comb. **6**(N1), 69–78 (1985)

113. Rödl, V., Tengan, E.: A note on a conjecture by Füredi. J. Comb Theory, Ser A **113**, 1214–1218 (2006)
114. Ruzsa, I.Z., Szemerédi, E.: Triple systems with no six points carrying three triangles. In: Combinatorics, Colloq. Marh. Sot. J. Bolyai 18, Vol. II, pp. 939–945, North-Holland, Amsterdam, 1978
115. Simonovits, M.: A method for solving extremal problems in graph theory, stability problems. In: Theory of Graphs (Proc. Colloq. Tihany,: Academic Press, New York, and Akad. Kiadó, Budapest **1968**, 279–319 (1966)
116. Spencer, J.: Intersection theorems for systems of sets. Can. Math. Bull. **20**(N2), 249–254 (1977)
117. Stoeckl, M.: Lecture notes on recent improvements for the sunflower lemma. https://mstoeckl.com/notes/research/sunflower_notes.html
118. Tao, T.: The sunflower lemma via shannon entropy. https://terrytao.wordpress.com/2020/07/20/the-sunflower-lemma-via-shannon-entropy/
119. Tokushige, N.: An L-system on the small Witt design. J. Comb. Theory, Ser. A **113**, 420–434 (2006)
120. Zakharov, D.: Chromatic numbers of Kneser-type graphs (2019). arXiv:1811.10567

Recent Advances in Arrow Relations and Traces of Sets

Mingze Li, Jie Ma, and Mingyuan Rong

Abstract The arrow relation, a central concept in extremal set theory, captures quantitative relationships between families of sets and their traces. Formally, the arrow relation $(n, m) \to (a, b)$ signifies that for any family $\mathcal{F} \subseteq 2^{[n]}$ with $|\mathcal{F}| \geqslant m$, there exists an a-element subset $T \subseteq [n]$ such that the trace $\mathcal{F}_{|T} = \{F \cap T : F \in \mathcal{F}\}$ contains at least b distinct sets. This survey highlights recent progress on a variety of problems and results connected to arrow relations. We explore diverse topics, broadly categorized by different extremal perspectives on these relations, offering a cohesive overview of the field.

1 Introduction

For a positive integer n, we let $[n] = \{1, 2, \ldots, n\}$. We denote the power set of $[n]$ as $2^{[n]} = \{T : T \subseteq [n]\}$. The study of finite sets and their extremal properties is a fundamental area in combinatorics. A key concept in this field is the *trace* of a family of sets. For a family $\mathcal{F} \subseteq 2^{[n]}$, the trace of $\mathcal{F}$ on a subset $T \subseteq [n]$ is defined as $\mathcal{F}_{|T} = \{F \cap T : F \in \mathcal{F}\}$. We say T is *shattered* by $\mathcal{F}$ if $\mathcal{F}_{|T} = 2^T$. A natural problem arises: How large must a family $\mathcal{F} \subseteq 2^{[n]}$ be to guarantee that it shatters some k-element subset?

M. Li · J. Ma (✉) · M. Rong
School of Mathematical Sciences, University of Science and Technology of China, Hefei, Anhui, China
e-mail: jiema@ustc.edu.cn

M. Li
e-mail: lmz10@mail.ustc.edu.cn

M. Rong
e-mail: rong_ming_yuan@mail.ustc.edu.cn

J. Ma
Yau Mathematical Sciences Center, Tsinghua University, Beijing, China

G. O. H. Katona et al. (eds.), *Sum(m)it280*, Bolyai Society Mathematical Studies 32,
https://doi.org/10.1007/978-3-032-18810-6_14

This question was resolved in the early 1970s through independent work by Sauer [65], Perles and Shelah [58], and Vapnik and Chervonenkis [67, 68], confirming a conjecture of Erdős [33]. Their result, now known as the Sauer-Shelah Lemma, gives a tight bound on the size of $\mathcal{F}$.

Theorem 1 (Sauer-Shelah Lemma) *Let $n \geqslant k \geqslant 1$. For any family $\mathcal{F} \subseteq 2^{[n]}$ satisfying*

$$|\mathcal{F}| \geqslant \sum_{i=0}^{k-1} \binom{n}{i} + 1,$$

there exists a k-element set $T \subseteq [n]$ with $\mathcal{F}|_T = 2^T$. Moreover, this bound is tight.

In other words, this establishes that families of size at least $1 + \sum_{i=0}^{k-1} \binom{n}{i}$ must contain a trace of size 2^k on some k-element set. To study the general problem on traces of any size, Hajnal [18] introduced the arrow relation through the following definition:

Definition 1 For positive integers n, m, a, b, the *arrow relation* $(n, m) \to (a, b)$ denotes that for every family $\mathcal{F} \subseteq 2^{[n]}$ with $|\mathcal{F}| \geqslant m$, there exists some a-element set $T \subseteq [n]$ such that $|\mathcal{F}_{|T}| \geqslant b$.

Note that the Sauer-Shelah Lemma can be expressed via the following arrow relation

$$\left(n,\ 1 + \sum_{i=0}^{k-1} \binom{n}{i}\right) \to (k,\ 2^k).$$

For convenience, for a family $\mathcal{F} \subseteq 2^{[n]}$, we also write $\mathcal{F} \to (a, b)$ if there exists an a-element subset $T \subseteq [n]$ with $|\mathcal{F}_{|T}| \geqslant b$. Conversely, $\mathcal{F} \nrightarrow (a, b)$ signifies that for every a-element subset T, we have $|\mathcal{F}_{|T}| < b$. With these notions, $(n, m) \to (a, b)$ is equivalent to that $\mathcal{F} \to (a, b)$ holds for every family $\mathcal{F} \subseteq 2^{[n]}$ with $|\mathcal{F}| \geqslant m$.

Since the introduction of arrow relations, this framework has spurred a wide array of problems and results in extremal combinatorics. In recent years, research has primarily concentrated on exploring various extremal aspects of arrow relations. A natural direction is to determine the minimum size of families to ensure a trace of given size, that is to determine the minimum m such that $(n, m) \to (a, b)$ holds when n, a and b are given. Since the case $b = 2^a$ is solved by the Sauer-Shelah Lemma, it remains to consider $b < 2^a$. The first step was made by Frankl [27], who determined this bound for the 3-vertex trace of size 7. Recently, Frankl and Wang [31] investigated the cases of 4-vertex traces. Researchers are also interested in determining the maximum b such that $(n, m) \to (a, b)$ holds when n, m and a are given. This line of inquiry led to the introduction of the so-called *trace function*, which was proposed and studied by Bollobás and Radcliffe [17]. Further advancements were achieved by Alon et al. [4]. Apart from these, another important direction is to consider the maximum m such that $(n, m) \to (n-1, m-s)$ holds when n and s are given. Bondy [18] and Bollobás [50] first studied this problem, and later Frankl [27] presented a general method, resolving the cases where s is just below powers of 2. Very recently, several new results in this area were presented in [49, 59, 63]. There are several other topics closely related to arrow relations, which will be introduced in Sects. 6, 7, and 8.

This survey aims to provide an accessible entry point for researchers new to the field, serving as a useful resource for both orientation and reference. While we have made every effort to include representative works, the rapidly evolving nature of this area may have led to the omission of some valuable contributions.

We structure the discussion by grouping results based on their shared research themes. The paper is organized into the following sections: Sect. 1. Introduction, Sect. 2. A fundamental lemma of Frankl, Sect. 3. Defect Sauer results (a term introduced in [17]), Sect. 4. Single-element removal, Sect. 5. Trace functions, Sect. 6. Extremal families, Sect. 7. Forbidden configurations in matrices, Sect. 8. Turán numbers for traces, and Sect. 9. Open problems.

Notation

Throughout this survey, we use the following notations:

- Let $\mathbb{N} = \{0, 1, 2, \dots\}$ denote the set of all non-negative integers, $\mathbb{N}^+ = \{1, 2, 3, \dots\}$ denote the set of all positive integers and $\mathbb{R}$ denote the set of all real numbers.
- For $n \in \mathbb{N}^+$, let $[n] = \{1, 2, \dots, n\}$ denote the set of the first n positive integers. For $a, b \in \mathbb{N}^+$ with $a \leqslant b$, let $[a : b] = \{a, a+1, \dots, b\}$ denote the set of consecutive integers from a to b.
- For $s \in \mathbb{N}^+$ and a set T, let 2^T denote the collection of all subsets of T. Moreover, $\binom{T}{s}$ denotes the collection of all subsets of T with exactly s elements, and $\binom{T}{\leqslant s}$ denotes the collection of all subsets of T with at most s elements.
- For two sets A and B, we write $A \triangle B = (A \setminus B) \cup (B \setminus A)$ as the *symmetric difference* between A and B.
- For a family $\mathcal{F} \subseteq 2^{[n]}$ and $v \in [n]$, let $d_{\mathcal{F}}(v) = |\{F \in \mathcal{F} : v \in F\}|$ denote the *degree* of v in $\mathcal{F}$ and $\delta(\mathcal{F}) = \min_{w \in [n]} d_{\mathcal{F}}(w)$ denote the *minimum degree* of $\mathcal{F}$.
- For a graph G, $\mathrm{ex}(n, G)$ denotes the *Turán number* of G, which is the maximum number of edges in an n-vertex graph that does not contain G as a subgraph. For integers $n \geqslant k \geqslant 2$, the *Turán graph* $T(n, k)$ denotes the unique balanced complete k-partite graph on n vertices.
- Similarly, for a k-uniform hypergraph $\mathcal{F}$, $\mathrm{ex}_k(n, \mathcal{F})$ denotes the maximum number of edges in an n-vertex k-uniform hypergraph which does not contain $\mathcal{F}$ as a subhypergraph.
- The *Turán density* of a k-uniform hypergraph $\mathcal{F}$ is the limit $\pi(\mathcal{F}) = \lim_{n\to\infty} \binom{n}{k}^{-1} \mathrm{ex}_k(n, \mathcal{F})$.
- For integers n, m, a and b, the notion $(n, m) \nrightarrow (a, b)$ denotes the negation of $(n, m) \rightarrow (a, b)$.

- Let $\mathbf{0}_q$ denote the $q \times 1$ matrix (i.e., a column vector of length q) whose entries are all 0, $\mathbf{1}_p$ denote the $p \times 1$ matrix with all entries equal to 1, and $\mathbf{I}_k$ denote the $k \times k$ identity matrix.

2 A Fundamental Lemma of Frankl

We begin by introducing a basic result of Frankl [27] on the study of arrow relation through the lens of a special but fundamental class of families: *hereditary* families, which are closed under taking subsets. Formally, a family $\mathcal{F}$ is called *hereditary* if for every $F' \subseteq F \in \mathcal{F}$, it holds that $F' \in \mathcal{F}$. Frankl [27] showed that to establish an arrow relation, it suffices to verify it for hereditary families, which are often more convenient to handle. This is formalized in the following lemma.

Lemma 1 (Frankl [27]) *The following statements are equivalent for any integers $n, m, a, b \in \mathbb{N}^+$:*

- $(n, m) \to (a, b)$.
- *For every hereditary family $\mathcal{F} \subseteq 2^{[n]}$ with $|\mathcal{F}| = m$, $\mathcal{F} \to (a, b)$ holds.*

In the rest of this section, we will present a proof of this lemma due to Frankl [27]. We first need to introduce an operation on families called *squashing*. Frankl [27] and Alon [2] used squashing techniques to generalize Theorem 1. It is worth mentioning that this operation is also known as *shifting* in contexts such as [6, 7]. Formally, given a family $\mathcal{F} \subseteq 2^{[n]}$ and a vertex $v \in [n]$, the squash operation at v transforms $\mathcal{F}$ into the following new family:

$$S_v(\mathcal{F}) = \{F_v : F \in \mathcal{F}\},$$

where each modified set F_v is defined as:

$$F_v = \begin{cases} F \setminus \{v\} & \text{if } v \in F \in \mathcal{F} \text{ and } F \setminus \{v\} \notin \mathcal{F}, \\ F & \text{otherwise.} \end{cases}$$

By definition, it is easy to see that $|S_v(\mathcal{F})| = |\mathcal{F}|$, and $\sum_{F \in S_v(\mathcal{F})} |F| \leqslant \sum_{F \in \mathcal{F}} |F|$, where this inequality becomes strict if a change is made in this operation. Therefore, we observe that it takes a finite number of squash operations, transforming $\mathcal{F}$ into a family $\tilde{\mathcal{F}}$, which satisfies $S_v(\tilde{\mathcal{F}}) = \tilde{\mathcal{F}}$ for any $v \in [n]$. We call such a family $\tilde{\mathcal{F}}$ a *squashed family* obtained from $\mathcal{F}$. The following proposition follows directly from the definition.

Proposition 1 *A squashed family is hereditary.*

In fact, in order to get a squashed family from $\mathcal{F}$, it is enough to apply the squash S_v once for every $v \in [n]$. More precisely, let $\mathcal{F} \subseteq 2^{[n]}$ and σ be a permutation of $[n]$, then the transformed family

$$S_\sigma(\mathcal{F}) := S_{\sigma(n)}(S_{\sigma(n-1)}(\dots(S_{\sigma(1)}(\mathcal{F}))\dots))$$

is a squashed family, thus it is a hereditary family. It is worth noting that $S_\sigma(\mathcal{F})$ may vary depending on the choice of the permutation σ.

The following propositon reveals a close relation between squash operations and trace of sets.

Proposition 2 (Frankl [27]) *Let $\mathcal{F} \subseteq 2^{[n]}$ and $v \in [n]$. For every $T \subseteq [n]$, we have*

$$|S_v(\mathcal{F})_{|T}| \leqslant |\mathcal{F}_{|T}|.$$

Proof If $v \notin T$, we obviously have $S_v(\mathcal{F})_{|T} = \mathcal{F}_{|T}$. We now assume that $v \in T$. For any subset $A \subseteq T \setminus \{v\}$, we claim that

$$|S_v(\mathcal{F})_{|T} \cap \{A, A \cup \{v\}\}| \leqslant |\mathcal{F}_{|T} \cap \{A, A \cup \{v\}\}|.$$

If $A \notin S_v(\mathcal{F})_{|T}$ and $A \cup \{v\} \notin S_v(\mathcal{F})_{|T}$, this inequality is trivial. If $A \cup \{v\} \in S_v(\mathcal{F})_{|T}$, then both A and $A \cup \{v\}$ are members of $S_v(\mathcal{F})_{|T} \cap \mathcal{F}_{|T}$. If $A \in S_v(\mathcal{F})_{|T}$ and $A \cup \{v\} \notin S_v(\mathcal{F})_{|T}$, we have either $A \in \mathcal{F}_{|T}$ or $A \cup \{v\} \in \mathcal{F}_{|T}$. Thus we prove our claim. Then we have

$$|S_v(\mathcal{F})_{|T}| = \sum_{A \subseteq T \setminus \{v\}} |S_v(\mathcal{F})_{|T} \cap \{A, A \cup \{v\}\}| \leqslant \sum_{A \subseteq T \setminus \{v\}} |\mathcal{F}_{|T} \cap \{A, A \cup \{v\}\}| = |\mathcal{F}_{|T}|,$$

which proves the proposition. □

The following proof of Lemma 1, building on Proposition 2, is due to Frankl [27].

Proof of Lemma 1 By definition, the first statement obviously implies the second. Therefore, it is enough to show that if the second statement holds, then for any family $\mathcal{F} \subseteq 2^{[n]}$ with $|\mathcal{F}| \geqslant m$, we have $\mathcal{F} \to (a, b)$. Suppose the contrary that there exists such a family $\mathcal{F} \subseteq 2^{[n]}$ with $\mathcal{F} \not\to (a, b)$, and subject to this, we may assume $\mathcal{F}$ minimizes $\sum_{F \in \mathcal{F}} |F|$. By assumption, $\mathcal{F}$ is not hereditary. Therefore, there exists $v \in [n]$ such that $\sum_{F \in S_v(\mathcal{F})} |F| < \sum_{F \in \mathcal{F}} |F|$. The minimality of $\sum_{F \in \mathcal{F}} |F|$ implies that $S_v(\mathcal{F}) \to (a, b)$. Then there exists $T \subseteq [n]$ with $|T| = a$ such that $|S_v(\mathcal{F})_{|T}| \geqslant b$. By Proposition 2, we know that

$$|\mathcal{F}_{|T}| \geqslant |S_v(\mathcal{F})_{|T}| \geqslant b,$$

which is a contradiction with $\mathcal{F} \not\to (a, b)$. So we can conclude that $(n, m) \to (a, b)$. □

3 Defect Sauer Results

The Sauer-Shelah Lemma motivates the study of "defect Sauer results" (a term introduced in [17]), which study the minimum size required for a family $\mathcal{F}$ to ensure a trace of size at least b on some a-element set. Expressed through arrow relation terminology, this problem reduces to determining the minimum m_0 such that the relation $(n, m) \to (a, b)$ holds for all $m \geqslant m_0$ given fixed parameters a and b. This is formalized by the following notation.

Definition 2 Let $m(n, a, b)$ denote the minimum integer m such that $(n, m) \to (a, b)$.

This notation allows us to express the Sauer-Shelah Lemma (Theorem 1) concisely as

Theorem 2 (Sauer-Shelah Lemma) *For* $k \geqslant 1$,

$$m(n, k, 2^k) = \sum_{i=0}^{k-1} \binom{n}{i} + 1.$$

In 1983, Frankl [27] pioneered the study of defect Sauer results by resolving a conjecture of Bondy and Lovász, proving the following theorem.

Theorem 3 (Frankl [27])

$$m(n, 3, 7) = \left\lfloor \frac{n^2}{4} \right\rfloor + n + 2.$$

The proof of this theorem is as follows, which combines Mantel's theorem on $\mathrm{ex}(n, K_3)$ with the hereditary family reduction of the arrow relation via Lemma 1.

Proof of Theorem 3 We establish the upper bound by contradiction. Suppose $m(n, 3, 7) > \lfloor \frac{n^2}{4} \rfloor + n + 2$. By Lemma 1, we may assume there exists a hereditary family $\mathcal{F} \subseteq 2^{[n]}$ such that $|\mathcal{F}| = \lfloor \frac{n^2}{4} \rfloor + n + 2$ and $\mathcal{F} \nrightarrow (3, 7)$. As $\mathcal{F}$ is a hereditary family, if there exists set $T \in \mathcal{F}$ and $|T| = 3$, then $|\mathcal{F}_{|T}| = |2^T| = 8$, a contradiction. Thus we may assume that every set F in the hereditary family $\mathcal{F}$ has $|F| \leqslant 2$. Moreover, if $\{\{a, b\}, \{a, c\}, \{b, c\}\} \subseteq \mathcal{F}$ for some $a, b, c \in [n]$, then $|\mathcal{F}_{|\{a,b,c\}}| \geqslant 7$, contradicting the assumption $\mathcal{F} \nrightarrow (3, 7)$. Thus the collection of 2-element sets in $\mathcal{F}$ constitutes a K_3-free graph. Applying Mantel's theorem yields that

$$\begin{aligned} 2 + n + \left\lfloor \frac{n^2}{4} \right\rfloor = |\mathcal{F}| &= \left|\mathcal{F} \cap \binom{[n]}{0}\right| + \left|\mathcal{F} \cap \binom{[n]}{1}\right| + \left|\mathcal{F} \cap \binom{[n]}{2}\right| \\ &\leqslant 1 + n + \mathrm{ex}(n, K_3) = 1 + n + \left\lfloor \frac{n^2}{4} \right\rfloor, \end{aligned}$$

a contradiction. This establishes the upper bound $m(n, 3, 7) \leqslant \lfloor \frac{n^2}{4} \rfloor + n + 2$.

For the lower bound, let $\mathcal{F} = \binom{[n]}{0} \cup \binom{[n]}{1} \cup \mathcal{G}$, where $\mathcal{G} \subseteq \binom{[n]}{2}$ denotes the edge set of the Turán graph $T(n, 2)$. Then $|\mathcal{F}| = \lfloor \frac{n^2}{4} \rfloor + n + 1$ and $\mathcal{F} \nrightarrow (3, 7)$, which yields $m(n, 3, 7) \geqslant \lfloor \frac{n^2}{4} \rfloor + n + 2$. □

Employing similar arguments with a more detailed analysis, one can determine

$$m(n, 4, 11) = \mathrm{ex}(n, K_4) + n + 2 = \left\lfloor \frac{n^2}{3} \right\rfloor + n + 2.$$

Following these developments, Bollobás and Radcliffe [17] further determined the exact value for $m(n, 4, 12)$ in 1995.

Theorem 4 (Bollobás and Radcliffe [17]) *For the case* $n = 6$, $m(6, 4, 12) = 24$. *And for* $n \neq 6$,

$$m(n, 4, 12) = \binom{n}{2} + n + 2.$$

In 2023, Frankl and Wang [31] established the exact value of $m(n, 4, 13)$ in the following theorem.

Theorem 5 (Frankl and Wang [31]) *For* $n \geqslant 25$,

$$m(n, 4, 13) = 1 + \left\lfloor \frac{n+5}{3} \right\rfloor \left\lfloor \frac{n+4}{3} \right\rfloor \left\lfloor \frac{n+3}{3} \right\rfloor.$$

Besides, Frankl and Wang [31] also investigated the value of $m(n, 4, b)$ for $6 \leqslant b \leqslant 15$. Combining with other known Turán type problems on graphs and hypergraphs, the following table summarizes the values on $m(n, a, b)$ for $a \in \{2, 3, 4\}$ and $1 \leqslant b \leqslant 2^a$. Here, we let C_3^+ denote the graph formed by attaching a pendant edge to a triangle, $K_4^{(3)}$ denote the complete 3-uniform hypergraph on 4 vertices, and $K_4^{(3)-}$ denote the 3-uniform hypergraph obtained from $K_4^{(3)}$ by deleting an edge (Table 1).

Note that all values in this table are precisely established except for $m(n, 4, 9)$, $m(n, 4, 14)$ and $m(n, 4, 15)$. As for $m(n, 4, 9)$, the exact value of $\mathrm{ex}(n, \{C_3^+, C_4\})$ remains undetermined. The *Zarankiewicz number* $z(n, C_4)$, extensively studied in extremal graph theory [21, 35, 40, 52, 56], is defined as the maximum number of edges in an n-vertex C_4-free bipartite graph. It is known that $z(n, C_4) = \left(\frac{n}{2}\right)^{3/2} + o(n^{3/2})$, which implies $\mathrm{ex}(n, \{C_3^+, C_4\}) \geqslant z(n, C_4) = \left(\frac{n}{2}\right)^{3/2} + o(n^{3/2})$. Similarly, the classical result $\mathrm{ex}(n, C_4) = \frac{1}{2}n^{3/2} + O(n)$ (see [45]) yields $\mathrm{ex}(n, \{C_3^+, C_4\}) \leqslant \mathrm{ex}\,(n, C_4) = \frac{1}{2}n^{3/2} + O(n)$. Consequently, we obtain the bounds:

$$\left(\frac{n}{2}\right)^{3/2} + o(n^{3/2}) \leqslant m(n, 4, 9) \leqslant \frac{1}{2}n^{3/2} + O(n).$$

Table 1 Values of $m(n, a, b)$ for $a \in \{2, 3, 4\}$ and $1 \leqslant b \leqslant 2^a$

	a = 2	a = 3	a = 4
$b = 1$	1	1	1
$b = 2$	2	2	2
$b = 3$	3	3	3
$b = 4$	$n + 2$	4	4
$b = 5$	–	$n + 2$	5
$b = 6$	–	$\lfloor \frac{3}{2}n \rfloor + 2$	$n + 2$
$b = 7$	–	$\lfloor \frac{n^2}{4} \rfloor + n + 2$	$n + 3$
$b = 8$	–	$\binom{n}{2} + n + 2$	$\lfloor \frac{5}{3}n \rfloor + 2$
$b = 9$	–	–	$\mathrm{ex}(n, \{C_3^+, C_4\}) + n + 2$
$b = 10$	–	–	$\lfloor \frac{n^2}{4} \rfloor + n + 2$
$b = 11$	–	–	$\lfloor \frac{n^2}{3} \rfloor + n + 2$
$b = 12, n = 6$	–	–	24
$b = 12, n \neq 6$	–	–	$\binom{n}{2} + n + 2$
$b = 13, n \geqslant 25$	–	–	$\lfloor \frac{n+5}{3} \rfloor \lfloor \frac{n+4}{3} \rfloor \lfloor \frac{n+3}{3} \rfloor + 1$
$b = 14$	–	–	$\mathrm{ex}_3(n, K_4^{(3)-}) + \binom{n}{2} + n + 2$
$b = 15$	–	–	$\mathrm{ex}_3(n, K_4^{(3)}) + \binom{n}{2} + n + 2$
$b = 16$	–	–	$\binom{n}{3} + \binom{n}{2} + n + 2$

Regarding $m(n, 4, 14)$ and $m(n, 4, 15)$, their exact values depend on the hypergraph Turán problems $\mathrm{ex}_3(n, K_4^{(3)-})$ and $\mathrm{ex}_3(n, K_4^{(3)})$, which are notorious open problems in extremal graph theory. For complete 3-uniform hypergraph, Turán [66] proposed the following conjecture.

Conjecture 1 (*Turán* [66]) For any integer $k \geqslant 4$, we have

$$\pi(K_k^{(3)}) = 1 - \left(\frac{2}{k-1}\right)^2.$$

In the particular case $k = 4$, Conjecture 1 asserts that $\pi(K_4^{(3)}) = \frac{5}{9}$. For its lower bound, there have been several constructions (see, e.g., [19, 26, 44]) showing $\pi(K_4^{(3)}) \geqslant \frac{5}{9}$. As for the upper bound, progress has been made in [20, 22]. The best known upper bound $\pi(K_4^{(3)}) \leqslant 0.561666$ was obtained by Razborov [62] using flag algebra. For $K_4^{(3)-}$, the current best bounds are $\frac{2}{7} \leqslant \pi(K_4^{(3)-}) \leqslant 0.286889$, where the lower bound is due to Frankl and Füredi [29], and the upper bound was obtained by Falgas-Ravry and Vaughan [25]. Consequently, we have the following bounds:

$$\frac{2}{7} \times \frac{n^3}{6} + o(n^3) \leqslant m(n, 4, 14) \leqslant 0.286889 \times \frac{n^3}{6} + o(n^3),$$

$$\frac{5}{9} \times \frac{n^3}{6} + o(n^3) \leqslant m(n, 4, 15) \leqslant 0.561666 \times \frac{n^3}{6} + o(n^3).$$

4 Single-Element Removal

By definition, the arrow relation is inherently transitive: if $(n, m) \to (a, b)$ and $(a, b) \to (c, d)$, then $(n, m) \to (c, d)$. Therefore, it is essential to investigate the special arrow relation $(n, m) \to (n - 1, m')$. Since single-element removal typically causes bounded trace reduction, the relation should be parameterized by a controlled loss rate s rather than an arbitrary m'. This leads to the following refined formulation presented in the book by Frankl and Tokushige [30].

Problem 1 (*Frankl and Tokushige* [30],*Problem 3.8*) For any non-negative integers n and s, determine or estimate the maximum value $m = m(n, s)$ such that $(n, m) \to (n - 1, m - s)$.

In view of Lemma 1, Problem 1 is equivalent to the following problem:

Problem 2 Find the maximum $m = m(n, s)$ such that every hereditary family $\mathcal{F}$ on n vertices with m edges satisfies $\delta(\mathcal{F}) \leqslant s$.

The investigation of Problem 1 initially focused on determining $m(n, s)$ for small values of s. Bondy [18] and Bollobás [50] solved the cases of $s = 0$ and $s = 1$ respectively.

Theorem 6 (Bondy [18]) *Let n be a positive integer. Then*

$$m(n, 0) = n.$$

Theorem 7 (Bollobás [50]) *Let n be a positive integer. Then*

$$m(n, 1) = \left\lceil \frac{3}{2} n \right\rceil.$$

In 1983, Frankl [27] proved the first result for an infinite family of cases, specifically for $s = 2^{d-1} - 1$, where d is any positive integer.

Theorem 8 (Frankl [27]) *Let n and d be positive integers.*

$$m(n, 2^{d-1} - 1) \geqslant \left\lceil \frac{2^d - 1}{d} n \right\rceil.$$

The proof of Theorem 8 relies on the well-known Kruskal-Katona Theorem, named after Kruskal [47] and Katona [42]. For two finite sets $A, B \subseteq \mathbb{N}^+$, we say that A precedes B in the *colexicographic order*, which is denoted by $A \prec_{col} B$, if $\max(A \triangle B) \in B$. For $m \in \mathbb{N}$, we define $\mathcal{R}(m)$ to be the family containing the first m finite subsets of $\mathbb{N}^+$ according to the colexicographic order. For $2^k \leqslant m < 2^{k+1}$, we have

$$\mathcal{R}(m) = 2^{[k]} \cup \{R \cup \{k+1\} : R \in \mathcal{R}(m - 2^k)\}.$$

In particular, $\mathcal{R}(0) = \emptyset$ and $\mathcal{R}(2^k) = 2^{[k]}$. The following theorem due to Katona [43] is a generalization of the Kruskal-Katona theorem.

Theorem 9 (Katona [43]) *Let $f : \mathbb{N} \to \mathbb{R}$ be a monotone non-increasing function and let $\mathcal{F}$ be a hereditary family with $|\mathcal{F}| = m$. Then*

$$\sum_{F \in \mathcal{F}} f(|F|) \geq \sum_{R \in \mathcal{R}(m)} f(|R|).$$

We refer readers to Chap. 8.2 of [38] for a proof of Theorem 9. We now present the proof of Theorem 8, originally established by Frankl [27].

Proof of Theorem 8 By Lemma 1 and the definition of arrow relation, it suffices to show that a hereditary family $\mathcal{F} \subseteq 2^{[n]}$ with $\delta(\mathcal{F}) \geqslant 2^{d-1}$ must satisfy $|\mathcal{F}| > \left\lceil \frac{2^d-1}{d} n \right\rceil$. For any $x \in [n]$, let $\mathcal{F}(x) = \{F \setminus \{x\} : x \in F \in \mathcal{F}\}$ denote the link of x. Then $|\mathcal{F}(x)| \geqslant 2^{d-1}$ for any $x \in [n]$. It is easy to see that the fact that $\mathcal{F}$ is hereditary implies that $\mathcal{F}(x)$ is also hereditary. Applying Theorem 9 for $\mathcal{F}(x)$ and $f(k) = \frac{1}{k+1}$, we can obtain that for any $x \in [n]$,

$$\sum_{H \in \mathcal{F}(x)} \frac{1}{|H|+1} \geqslant \sum_{R \in \mathcal{R}(|\mathcal{F}(x)|)} \frac{1}{|R|+1} \geqslant \sum_{R \in 2^{[d-1]}} \frac{1}{|R|+1} = \sum_{i=0}^{d-1} \frac{\binom{d-1}{i}}{i+1} = \frac{2^d-1}{d}.$$

Thus by the method of double-counting, we obtain

$$|\mathcal{F} \setminus \{\emptyset\}| = \sum_{\emptyset \neq F \in \mathcal{F}} \sum_{x \in F} \frac{1}{|F|} = \sum_{x \in [n]} \sum_{x \in F \in \mathcal{F}} \frac{1}{|F|} = \sum_{x \in [n]} \sum_{H \in \mathcal{F}(x)} \frac{1}{|H|+1} \geqslant \frac{2^d-1}{d} n.$$

This implies that $|\mathcal{F}| \geqslant \frac{2^d-1}{d} n + 1 > \left\lceil \frac{2^d-1}{d} n \right\rceil$, as desired. □

The upper bound of $m(n, 2^{d-1} - 1)$ when $d \mid n$ is given by the following general construction:

Construction 10 Let d and c be positive integers with $1 \leqslant c \leqslant d-1$. Assume that $n = dk$ for some positive integer k. Let $U_1, \ldots, U_k$ form a partition of $[n]$ into sets of size d. For every $i \in [k]$, arbitrarily pick a family $\mathcal{G}_i \subseteq 2^{U_i} \setminus \{\emptyset\}$ with $|\mathcal{G}_i| = c-1$. Define

$$\mathcal{F}(n,d,c) := \left\{F \subseteq [n] : F \in 2^{U_i} \setminus \mathcal{G}_i \text{ for some } i \in [k]\right\}.$$

For $\mathcal{F} = \mathcal{F}(n,d,c)$ in this construction, one can verify that $|\mathcal{F}| = \frac{2^d-c}{d}n + 1$ and for any $T \in \binom{[n]}{n-1}$, we have $|\mathcal{F}_{|T}| < |\mathcal{F}| - 2^{d-1} + c$. This implies that $m(n, 2^{d-1} - c) \leqslant \frac{2^d-c}{d}n$ whenever $d \mid n$. In particular, by letting $c = 1$, we now can obtain

$$m(n, 2^{d-1}-1) = \frac{2^d-1}{d}n \text{ for } d \mid n.$$

Since tight constructions matching the lower bound exist only when $d \mid n$, determining $m(n,s)$ for arbitrary values of n becomes less tractable. This led researchers to shift focus toward the asymptotic behavior of $m(n,s)$ instead. In 1994, Watanabe and Frankl [71] established that for any $s \geqslant 0$, the limit $\lim_{n\to\infty} \frac{m(n,s)}{n}$ exists, and they introduced the notation $m(s)$ for this limit.

Theorem 11 (Watanabe and Frankl [71]) *For any $s \geqslant 0$, the limit $\lim_{n\to\infty} \frac{m(n,s)}{n}$ exists.*

Definition 3 (Watanabe and Frankl [71]) Let $m(s) = \lim_{n\to\infty} \frac{m(n,s)}{n}$ denote the limit.

Under this notation, the above results of Bondy [18], Bollobás [50], and Frankl [27] can be restated as $m(0) = 1$, $m(1) = \frac{3}{2}$, and $m(2^{d-1}-1) = \frac{2^d-1}{d}$ for $d \geqslant 1$, respectively. Watanabe and Frankl [71] further determined the values of $m(2^{d-1}-2)$ and $m(2^{d-1})$ for $d \geqslant 1$ as follows.

Theorem 12 (Watanabe and Frankl [71]) *Let n and d be positive integers. Then*

$$m(2^{d-1}-2) = \frac{2^d-2}{d} \quad \textit{and} \quad m(2^{d-1}) = \frac{2^d-1}{d} + \frac{1}{2}.$$

The lower bounds in both cases are established using arguments similar to those in Theorem 8. For the upper bounds, $m(2^{d-1}-2)$ is derived from $\mathcal{F}(n,d,2)$ in Construction 10, while $m(2^{d-1})$ is obtained from the following construction:

Construction 13 Let d and k be positive integers and let $n = 2dk$. Define

$$\mathcal{F} = \{\{j, j+dk\} : j \in [dk]\} \cup \bigcup_{i=0}^{2k-1} 2^{[id+1:id+d]}.$$

One can verify that $|\mathcal{F}| = \left(\frac{2^d-1}{d} + \frac{1}{2}\right) n + 1$ and for any $T \in \binom{[n]}{n-1}$, we have $|\mathcal{F}_{|T}| < |\mathcal{F}| - 2^{d-1}$. This demonstrates that $m(2^{d-1}) \leqslant \frac{2^d-1}{d} + \frac{1}{2}$ for $d \geqslant 1$.

Beyond these precise formulas for $m(s)$ with $s \in \{2^{d-1}, 2^{d-1}-1, 2^{d-1}-2\}$, researchers have also determined the exact values for several specific cases. In 1991, Watanabe [69] established that $m(5) = \frac{13}{4}$. Further advancing this line of inquiry, Watanabe and Frankl [71] proved $m(9) = \frac{65}{14}$, while Watanabe [70] later determined that $m(10) = 5$ and $m(13) = \frac{29}{5}$.

A recent result of Piga and Schülke [59] determined $m(12)$ and $m(2^{d-1} - c)$ for a range of c.

Theorem 14 (Piga and Schülke [59])

$$m(12) = \frac{28}{5}.$$

Theorem 15 (Piga and Schülke [59]) *Let* $1 \leqslant c \leqslant \frac{d}{4}$. *Then,*

$$m(2^{d-1} - c) = \frac{2^d - c}{d}.$$

Piga and Schülke [59] also provided the following construction.

Construction 16 Let $d \geqslant 3$ and k be positive integers and let $n = 2dk$. Define

$$\mathcal{F} = \{[jd-d+1 : jd-1] : j \in [2k]\} \cup \{\{2dt-d, 2dt\} : t \in [k]\} \cup \bigcup_{i=0}^{2k-1} \binom{[id+1 : (i+1)d]}{\leqslant d-2}.$$

One can verify that $|\mathcal{F}| = \frac{2^d-d-\frac{1}{2}}{d} n + 1$ and for any $T \in \binom{[n]}{n-1}$, we have $|\mathcal{F}_{|T}| < |\mathcal{F}| - 2^{d-1} + d$. This demonstrates that $m(2^{d-1} - d) \leqslant \frac{2^d-d-\frac{1}{2}}{d}$ for $d \geqslant 3$.

In 2024, Li et al. [49] investigated this problem, by determining the precise values of $m(11)$ as well as $m(2^{d-1} - c)$ whenever $1 \leqslant c \leqslant d$ and $d \geqslant 50$. The result $m(11) = \frac{53}{10}$ confirmed a conjecture of Watanabe and Frankl [71] from 1994, while the latter one solved a problem posted by Piga and Schülke [59].

Theorem 17 (Li et al. [49])

$$m(11) = \frac{53}{10}.$$

Theorem 18 (Li et al. [49]) *Let* $d \geqslant 50$ *and* $1 \leqslant c \leqslant d - 1$. *Then*

$$m(2^{d-1} - c) = \frac{2^d - c}{d} \quad \text{and} \quad m(2^{d-1} - d) = \frac{2^d - d - \frac{1}{2}}{d}.$$

In their proof of Theorem 18, they partitioned the ground set $[n]$ into several "piles", which are neighborhoods of vertices with small weight. The proof focused on estimating the average weight in "intersecting piles" and "non-intersecting piles". The result $m(11) = \frac{53}{10}$ was proved by performing a perturbation on the weight function. With the determination of $m(11)$, all the values of $m(s)$ for $s \leqslant 16$ are now known, and for convenience we list them all in the following table.

$s =$	0	1	2	3	4	5	6	7	8	9	10	11	12	13	14	15	16
$m(s) =$	1	$\frac{3}{2}$	2	$\frac{7}{3}$	$\frac{17}{6}$	$\frac{13}{4}$	$\frac{7}{2}$	$\frac{15}{4}$	$\frac{17}{4}$	$\frac{65}{14}$	5	$\frac{53}{10}$	$\frac{28}{5}$	$\frac{29}{5}$	6	$\frac{31}{5}$	$\frac{67}{10}$

Very recently, Reiher and Schülke [63] independently obtained similar results. In their proof, they grouped vertices together into "conglomerates", which are d-element sets and contain most of hyperedges at some vertices. Then they managed to show that the average weight in each conglomerate is sufficiently large and any two conglomerates intersect in at most one vertex. Using their approach, they proved that $m(11) = \frac{53}{10}$ and established the following result on $m(2^{d-1} - c)$ for $1 \leqslant c \leqslant d - 1$, which closes the gap for $6 \leqslant d \leqslant 49$ in Theorem 18.

Theorem 19 (Reiher and Schülke [63]) *Let d and c be positive integers with $1 \leqslant c \leqslant d - 1$. Then*

$$m(2^{d-1} - c) = \frac{2^d - c}{d}.$$

These aforementioned theorems determine the value of $m(s)$ when s is slightly smaller than a power of 2. The problem of determining $m(s)$ remains open when s lies further away from powers of two, or when s slightly exceeds powers of two.

5 Trace Functions

The framework of arrow relation $(n, m) \to (a, b)$ naturally extends to a variety of extremal problems in combinatorial set theory. In Sect. 3, we examined defect Sauer results, which focus on determining the minimum m attainable under fixed parameters n, a, and b. This section focuses on a complementary extremal problem of equal significance: for prescribed values of n, m, and a, what is the maximum b such that the relation $(n, m) \to (a, b)$ holds?

In order to formalize the optimization of b in $(n, m) \to (a, b)$, let us introduce the following definition.

Definition 4 For a family $\mathcal{F} \subseteq 2^{[n]}$ and $1 \leqslant a \leqslant n$, the *trace function* is defined as $\mathrm{Tr}(\mathcal{F}, a) = \max_{T \in \binom{[n]}{a}} |\mathcal{F}_{|T}|$. Furthermore, let $\mathrm{Tr}(n, m, a)$ denote the maximum size of traces on a-element sets guaranteed in any families of size m, that is

$$\mathrm{Tr}(n, m, a) = \min_{\substack{\mathcal{F} \subseteq 2^{[n]} \\ |\mathcal{F}| = m}} \mathrm{Tr}(\mathcal{F}, a) = \min_{\substack{\mathcal{F} \subseteq 2^{[n]} \\ |\mathcal{F}| = m}} \max_{T \in \binom{[n]}{a}} |\mathcal{F}_{|T}|.$$

Combining with the arrow relation, we have the following proposition.

Proposition 3 *For positive integers* n, m *and* a, $\mathrm{Tr}(n, m, a)$ *is precisely the maximum* b *for which the arrow relation* $(n, m) \to (a, b)$ *holds.*

A fundamental contribution to this direction was established by Bollobás and Radcliffe [17], who determined a general lower bound for $\mathrm{Tr}(n, m, a)$:

Theorem 20 (Bollobás and Radcliffe [17]) *For any positive integers* $a \leqslant n$ *and* $1 \leqslant m \leqslant 2^n$, *we have*

$$\mathrm{Tr}(n, m, a) \geqslant m^{a/n}.$$

Although Theorem 20 is concise and useful, it does not necessarily give the best possible bound for b. Specifically, let $m = n^r$ and $a = \alpha n$, Theorem 20 shows that $(n, n^r) \to (\alpha n, n^{\alpha r})$. The following result of Bollobás and Radcliffe [17] improves this bound by using probabilistic method. Let $H(x) = -x \log_2 x - (1-x)\log_2(1-x)$ denote the binary entropy function.

Theorem 21 **(Bollobás and Radcliffe** *[17]) For any integer* $r \geqslant 2$ *and real number* $0 < \alpha < 1$, *we have*

$$\mathrm{Tr}(n, n^r, \alpha n) \geqslant (1 - o(1))n^{\lambda r},$$

where we set $\lambda_0 = \log_2(1 + \alpha)$, *and* λ *is defined as*

$$\lambda = \begin{cases} \frac{\lambda_0}{H(\lambda_0)} & \text{if } \alpha \in (0, \sqrt{2} - 1); \\ \lambda_0 & \text{if } \alpha \in [\sqrt{2} - 1, 1). \end{cases}$$

Theorem 21 states that $\mathrm{Tr}(n, n^r, \alpha n) = \Omega(n^{\lambda r})$. Additionally, there are several influential results on estimating trace functions. One notable example is the following classical result by Kahn, Kalai and Linial [41].

Theorem 22 **(Kahn, Kalai and Linial** *[41]) For any integer* $n \geqslant 1$ *and any* $0 < \alpha, \beta < 1$, *we have*

$$\mathrm{Tr}(n, \beta 2^n, \alpha n) \geqslant (1 - n^{-c})2^{\alpha n},$$

where $c = c(\alpha, \beta)$ *depends only on* α *and* β.

To pursue the upper bound, it can be argued that among families $\mathcal{F} \subseteq 2^{[n]}$ of size m, the extremal family $\mathcal{F}$ satisfying $\mathrm{Tr}(\mathcal{F}, k) = \mathrm{Tr}(n, m, k)$ should exhibit some form of symmetry. Thus, it is natural to conjecture that the extremal family includes all sets up to a certain size. However, Bollobás and Radcliffe [17] proved the following theorem, demonstrating that this is not the case.

Theorem 23 (Bollobás and Radcliffe [17]) *For any integer* $r \geqslant 2$ *and real number* $0 < \varepsilon < \frac{1}{2}$, *there exists an* $n_0 = n_0(r, \varepsilon)$, *such that for any* $n \geqslant n_0$, *there exists a hereditary family* $\mathcal{H} \subseteq 2^{[n]}$ *with* $|\mathcal{H}| \geqslant \sum_{i=0}^{r} \binom{n}{i}$ *and*

$$\mathrm{Tr}(\mathcal{H}, n/2) \leqslant \sum_{i=0}^{r} \binom{n/2}{i} - (1-\varepsilon)2^{-r}\binom{n}{r}.$$

As a corollary of Theorem 23, we can derive $\mathrm{Tr}(n, \sum_{i=0}^{r}\binom{n}{i}, \frac{n}{2}) = o(n^r)$ for any integer $r \geqslant 2$. Moreover, if we set $r = 2$ and $\alpha = 1/2$, then Theorems 21 and 23 bound $\mathrm{Tr}(n, n^2, n/2)$ as follows:

$$\Omega(n^{1.1699\ldots}) = \Omega(n^{2\log_2 \frac{3}{2}}) \leqslant \mathrm{Tr}(n, n^2, n/2) \leqslant o(n^2).$$

Throughout this example, the gap between the lower and upper bounds is evident. Recently, Alon et al. [4] determined the value of $\mathrm{Tr}(n, n^r, \alpha n)$ up to logarithmic factors whenever r and α are constants. They introduced a new version of the Kruskal-Katona Theorem, referred to as the Sparse Kruskal-Katona Theorem, and derived the following theorem from it.

Theorem 24 (Alon et al. [4]) *For any integer $r \geqslant 1$ and real number $\alpha \in (0, 1]$, if $r, \alpha^{-1} \leqslant n^{o(1)}$ then* $\mathrm{Tr}(n, n^r, \alpha n) = n^{\mu(1-o(1))}$ *where*

$$\mu = \mu(r, \alpha) = \frac{r + 1 - \log(1+\alpha)}{2 - \log(1+\alpha)}.$$

Moreover, if $r = O(1)$ and $\alpha^{-1} \leqslant (\log(n))^{O(1)}$, then $\mathrm{Tr}(n, n^r, \alpha n) = \tilde{\Theta}(n^{\mu})$.[1]

In particular, by setting $r = 2$ and $\alpha = 1/2$, Theorem 24 implies that

$$\mathrm{Tr}(n, n^2, n/2) = \tilde{\Theta}(n^{1+\frac{1}{3-\log_2 3}}) = \tilde{\Theta}(n^{1.706695\ldots}).$$

6 Extremal Families

Characterizing combinatorial structures that optimize the given parameters is a central problem in extremal combinatorics. This section focuses on structural results on the largest families with trace constraints. Before proceeding, we introduce some basic definitions.

Definition 5 A family $\mathcal{F} \subseteq 2^{[n]}$ is said to *trace* (or *shatter*) a set $T \subseteq [n]$ if $\mathcal{F}_{|T} = 2^T$. We denote the collection of traced sets by $tr(\mathcal{F}) = \{T : \mathcal{F} \text{ traces } T\}$.

Definition 6 The *Vapnik-Chervonenkis dimension* (or *VC-dimension*) of a family $\mathcal{F} \subseteq 2^{[n]}$, denoted by $\dim_{\mathrm{VC}}(\mathcal{F})$, is defined as the maximum cardinality among sets in $tr(\mathcal{F})$, that is

$$\dim_{\mathrm{VC}}(\mathcal{F}) = \max_{T \in tr(\mathcal{F})} |T|.$$

[1] Here, all logarithms are in base 2, and we use the following standard notation: for two functions $f(n)$ and $g(n)$, we write $f = \tilde{\Theta}(g)$ if $f = \Theta(g \log^c(g))$ for some absolute constant $c > 0$.

With these notions, the Sauer-Shelah lemma can be rewritten as following.

Theorem 25 (Sauer [65], Perles and Shelah [58], Vapnik and Chervonenkis [67, 68]) *For any family* $\mathcal{F} \subseteq 2^{[n]}$*, the inequality* $\dim_{\mathrm{VC}}(\mathcal{F}) \geqslant k+1$ *holds whenever* $|\mathcal{F}| \geqslant 1 + \sum_{i=0}^{k} \binom{n}{i}$.

The subsequent theorem from Pajor [57] strengthens this result.

Theorem 26 (Pajor [57]) *For any family* $\mathcal{F} \subseteq 2^{[n]}$*, the inequality* $|\mathcal{F}| \leqslant |tr(\mathcal{F})|$ *holds.*

In other words, a set system $\mathcal{F}$ must trace at least $|\mathcal{F}|$ distinct sets. It is evident to observe that Theorem 26 directly implies Theorem 25, as there are exactly $\sum_{i=0}^{k} \binom{n}{i}$ subsets of $[n]$ with cardinality at most k. A significant direction of research focuses on the extremal set systems addressed in Theorems 25 and 26. Specifically, the former concerns the characterization of maximum families that trace no $(k+1)$-sets, while the latter deals with the characterization of families $\mathcal{F}$ satisfying $|\mathrm{tr}(\mathcal{F})| = |\mathcal{F}|$. For the former, Frankl [27] and Dudley [23] characterized families $\mathcal{F} \subseteq 2^{[n]}$ of size $n+1$ with $\dim_{\mathrm{VC}}(\mathcal{F}) = 1$. In comparison, the latter has been the subject of more extensive study in the literature.

Similar to Definition 5, Bollobás and Radcliffe [17] proposed the following definition.

Definition 7 We say that a family $\mathcal{F} \subseteq 2^{[n]}$ *strongly traces* a set $T \subseteq [n]$, if there exists a set $S \subseteq [n]$ disjoint from T such that $\mathcal{F} \supseteq \{A \cup S : A \in 2^T\}$. The set S is called a *support* of T. The family of all sets strongly traced by $\mathcal{F}$ is denoted by $str(\mathcal{F})$.

By definition, it is easy to see that $str(\mathcal{F}) \subseteq tr(\mathcal{F})$ for any family $\mathcal{F} \subseteq 2^{[n]}$. Bollobás et al. [16] proved the following theorem, which turns out to be equivalent to Theorem 26.

Theorem 27 (Bollobás et al. [16]) *Any family* $\mathcal{F} \subseteq 2^{[n]}$ *satisfies* $|\mathcal{F}| \geqslant |str(\mathcal{F})|$.

This theorem states that a family $\mathcal{F}$ can strongly trace at most $|\mathcal{F}|$ sets. For this reason, it is often referred to as the *reverse Sauer inequality*. In fact, there is a certain symmetry between $\mathrm{tr}(\mathcal{F})$ and $\mathrm{str}(\mathcal{F})$. Specifically, the sets that $\mathcal{F}$ fails to trace are precisely the complements of the sets that $2^{[n]} \setminus \mathcal{F}$ strongly traces. To formalize this observation, we state it as a theorem.

Theorem 28 (see [17]) *For any family* $\mathcal{F} \subseteq 2^{[n]}$*, we have*

$$2^{[n]} \setminus tr(\mathcal{F}) = \{T^c : T \in str(2^{[n]} \setminus \mathcal{F})\}.$$

The advantage of Theorem 28 is that it allows us to apply the various criteria from [16] for the condition $|\mathrm{tr}(\mathcal{F})| = |\mathcal{F}|$ to hold, as will be discussed later. Although Theorem 28 shows that $|tr(\mathcal{F})| \geqslant |\mathcal{F}|$ and $|str(\mathcal{F})| \leqslant |\mathcal{F}|$ are equivalent, the equivalence between $|tr(\mathcal{F})| = |\mathcal{F}|$ and $|str(\mathcal{F})| = |\mathcal{F}|$ is not trivial. Recall that by applying the squash operation once for every $i \in [n]$, we can get a hereditary family $S_\sigma(\mathcal{F})$, where σ is a permutation of $[n]$ and $S_\sigma(\mathcal{F}) = S_{\sigma(n)}(S_{\sigma(n-1)}(\ldots(S_{\sigma(1)}(\mathcal{F}))\ldots))$. In general, $S_\sigma(\mathcal{F})$ is not identical for every permutation σ. The following result shows that the conditions for equality in Sauer's inequality and the reverse Sauer inequality are equivalent.

Theorem 29 (Bollobás and Radcliffe [17]) *For any family $\mathcal{F} \subseteq 2^{[n]}$, the following four properties are equivalent:*

(1) $|tr(\mathcal{F})| = |\mathcal{F}|$.
(2) $|str(2^{[n]} \setminus \mathcal{F})| = 2^n - |\mathcal{F}|$.
(3) There is a unique hereditary family $\tilde{\mathcal{F}}$ that can be achieved from $\mathcal{F}$ by squash operations. In other words, $S_\sigma(\mathcal{F}) = S_\pi(\mathcal{F})$ for any permutations σ, π of $[n]$.
(4) $|str(\mathcal{F})| = |\mathcal{F}|$.

Fix sets $J \subseteq I \subseteq [n]$. Following [16], we call a subfamily of the form $C = \{F \in \mathcal{F} : F \cap I = J\}$ a *chunk* of $\mathcal{F}$. A chunk C is *self-complementary* if $\{C \triangle I^c : C \in C\} = C$. In other words, having fixed the behaviour inside I, we check and see whether on I^c the chunk looks like a self-complementary family. A chunk C is called *trivial* if it is either empty or of maximal size, that is $C = \{J \cup F : F \subseteq I^c\}$. For any $A, B \subseteq [n]$, we call $\{F \in \mathcal{F} : A \cap B \subseteq F \subseteq A \cup B\}$ the chunk of $\mathcal{F}$ they span. In particular, this is a chunk of $\mathcal{F}$, since

$$\{F \in \mathcal{F} : A \cap B \subseteq F \subseteq A \cup B\} = \{F \in \mathcal{F} : F \cap (A \triangle B)^c = A \cap B\}.$$

Let $\mathcal{F} \subseteq 2^{[n]}$ and $I \subseteq [n]$. We write $\mathcal{F}(I) = \{F \in \mathcal{F} : F \subseteq I^c \text{ and } F \cup J \in \mathcal{F} \text{ for any } J \subseteq I\}$. Let $G_\mathcal{F}$ denote the inclusion graph of $\mathcal{F} \subseteq 2^{[n]}$, that is the graph with vertex set $\mathcal{F}$ and edge set $\{(F, F') \in \mathcal{F} \times \mathcal{F} : |F \triangle F'| = 1\}$. We say that $\mathcal{F}$ is connected if $G_\mathcal{F}$ is connected. The following are several criteria given in [16] for the family $\mathcal{F}$ with $|tr(\mathcal{F})| = |\mathcal{F}|$.

Theorem 30 (Bollobás et al. [16]) *Suppose that $\mathcal{F} \subseteq 2^{[n]}$.*

(1) If $|tr(\mathcal{F})| = |\mathcal{F}|$, then for any chunk C of $\mathcal{F}$, $|tr(C)| = |C|$.
(2) $|tr(\mathcal{F})| = |\mathcal{F}|$ if and only if $\mathcal{F}(I)$ is connected for every $I \subseteq [n]$.
(3) $|tr(\mathcal{F})| = |\mathcal{F}|$ if and only if every chunk of every $\mathcal{F}(I)$ is connected. In particular, if $I \subseteq [n]$ and $|tr(\mathcal{F})| = |\mathcal{F}|$, then any two elements of $\mathcal{F}(I)$ can be connected in the chunk of $\mathcal{F}(I)$ they span.

Furthermore, Bollobás and Radcliffe [17] proved the following theorem.

Theorem 31 (Bollobás and Radcliffe [17]) *$|tr(\mathcal{F})| = |\mathcal{F}|$ if and only if $\mathcal{F}$ contains no non-trivial self-complementary chunks.*

We can also find different types of characterization of families with $|tr(\mathcal{F})| = |\mathcal{F}|$ in [46, 48, 53, 54]. For instance, in [53], Mészáros and Rónyai investigated families of VC dimension at most 1 with $|tr(\mathcal{F})| = |\mathcal{F}|$ from the view of the inclusion graphs of $\mathcal{F}$. In [54], Mészáros and Rónyai characterized families of VC dimension 2 with $|tr(\mathcal{F})| = |\mathcal{F}|$ in terms of their inclusion graphs. In [48], Kusch and Mészáros discussed an approach to study the extremal families using Sperner families.

In general, we have $|str(\mathcal{F})| \leqslant |\mathcal{F}| \leqslant |tr(\mathcal{F})|$ for any $\mathcal{F} \subseteq 2^{[n]}$. There is a version of shattering that always results in equality, which is called *order-shattering* and is introduced by Anstee et al. [12]. They define this concept in an inductive way on the size of T. Note that this is just the set version of $C(s)$ given in [1]. For $T = \emptyset$, all we need is a single set from $\mathcal{F}$. For $T = \{t_1, t_2, \ldots, t_{|T|}\}$ with $|T| \geqslant 1$ and $t_1 < t_2 < \cdots < t_{|T|}$, we say that T is order-shattered by $\mathcal{F}$ if there are $2^{|T|}$ sets from $\mathcal{F}$ divided into two families $\tilde{\mathcal{F}}_0$ and $\tilde{\mathcal{F}}_1$ so that if we define $T' = \{t_{|T|}+1, t_{|T|}+2, \ldots, n\}$ (if $t_{|T|} = n$, then $T' = \emptyset$) we have that

$$\begin{cases} T' \cap C = T' \cap D, \\ t_{|T|} \notin C, \\ t_{|T|} \in D \end{cases}$$

for all $C \in \tilde{\mathcal{F}}_0$, $D \in \tilde{\mathcal{F}}_1$ and both $\tilde{\mathcal{F}}_0$ and $\tilde{\mathcal{F}}_1$ individually order-shatter $T \setminus \{t_{|T|}\}$. We define

$$\operatorname{osh}(\mathcal{F}) = \{T \subseteq [n] : \mathcal{F} \text{ order-shatters } T\}.$$

Note that $\operatorname{osh}(\mathcal{F})$ is a hereditary family and $\operatorname{osh}(\mathcal{F}) \subseteq tr(\mathcal{F})$. Anstee et al. [12] proved that $\mathcal{F}$ order-shatters exactly $|\mathcal{F}|$ sets.

Theorem 32 (Anstee et al. [12]) *Let $\mathcal{F} \subseteq 2^{[n]}$. Then we have*

$$|\operatorname{osh}(\mathcal{F})| = |\mathcal{F}|.$$

Another intriguing problem is to determine the number of extremal families. Let $f(n,k)$ denote the number of families $\mathcal{F} \subseteq 2^{[n]}$ of VC dimension k with $|\mathcal{F}| = \sum_{i=0}^{k} \binom{n}{i}$. Frankl [28] raised the problem of estimating $f(n,k)$ and showed that $2^{\binom{n-1}{k}} \leqslant f(n,k) \leqslant 2^{n\binom{n-1}{k}}$. Better upper and lower bounds were obtained by Alon et al. [3]. They showed that for any integer $k \geqslant 2$, as $n \to \infty$, we have

$$n^{(1+o(1))\frac{1}{k+1}\binom{n}{k}} \leqslant f(n,k) \leqslant n^{(1+o(1))\binom{n}{k}}.$$

Recently, Balogh et al. [14] closed the gap and showed that the upper bound is asymptotically tight, even if we allow k to grow as $k = n^{o(1)}$.

Theorem 33 (Balogh et al. [14]) *Let $k = n^{o(1)}$. Then $f(n,k) = n^{(1+o(1))\binom{n}{k}}$ as $n \to \infty$.*

Similarly, let $\operatorname{ExVC}(n,k)$ be the number of families $\mathcal{F} \subseteq 2^{[n]}$ of VC dimension k with $|tr(\mathcal{F})| = |\mathcal{F}|$. Theorem 25 shows that for a family $\mathcal{F} \subseteq 2^{[n]}$ of VC dimension k, $|\mathcal{F}| = \sum_{i=0}^{k} \binom{n}{i}$ implies $|tr(\mathcal{F})| = |\mathcal{F}|$. Thus we have $f(n,k) \leqslant \operatorname{ExVC}(n,k)$. In fact, Balogh, Mészáros and Wagner [14] proved a further result. An induced matching is a matching such that no endpoints of two edges of the matching are joined by an edge of the graph. Let $\operatorname{IndMat}(n,k)$ denote the number of induced matchings in Q_n between the layers $\binom{[n]}{k}$ and $\binom{[n]}{k+1}$, where Q_n denotes the hypercube of dimension n.

Balogh, Mészáros and Wagner [14] showed that the asymptotics of the logarithm of these two quantities and of $f(n, k)$ have the same value, as follows.

Theorem 34 (Balogh et al. [14]) *Let $k = n^{o(1)}$. Then we have*

$$n^{(1+o(1))\binom{n}{k}} \leqslant \mathrm{IndMat}(n, k) \leqslant f(n, k) \leqslant \mathrm{ExVC}(n, k) \leqslant n^{(1+o(1))\binom{n}{k}}.$$

7 Forbidden Configurations in Matrices

The main topic of this section is to reformulate the Sauer-Shelah Lemma through the language of forbidden configurations in *simple matrices*. Before proceeding, we require some basic definitions.

A *simple matrix* is a $(0, 1)$-matrix with no repeated columns. Such matrices naturally correspond to set families in set theory. For a family $\mathcal{F} \subseteq 2^{[n]}$, it corresponds to a unique $n \times |\mathcal{F}|$ simple matrix $\mathbf{A} = (a_{ij})$, where rows index elements of $[n]$, columns index sets in $\mathcal{F}$, and $a_{ij} = 1$ if and only if element $i \in [n]$ belongs to the j-th set of $\mathcal{F}$. This matrix $\mathbf{A}$ is called the *incidence matrix* of $\mathcal{F}$. Conversely, a simple matrix also corresponds to a unique set family.

Definition 8 Given a fixed k-row $(0, 1)$-matrix $\mathbf{F}$, we say a $(0, 1)$-matrix $\mathbf{A}$ *contains configuration* $\mathbf{F}$ if there is a submatrix of $\mathbf{A}$ that is a row and column permutation of $\mathbf{F}$. We refer to $\mathbf{F}$ as a *configuration* of $\mathbf{A}$ and write $\mathbf{F} \prec \mathbf{A}$.

It can be checked that $\prec$ is a partial order on the set of all $(0, 1)$-matrices. The following proposition shows that configurations align precisely with traces in set theory:

Proposition 4 *Let* $\mathbf{F}$ *and* $\mathbf{G}$ *be the incidence matrices of families $\mathcal{F} \subseteq 2^{[k]}$ and $\mathcal{G} \subseteq 2^{[n]}$ respectively, where $k \leqslant n$ are two positive integers. Then* $\mathbf{G}$ *avoids* $\mathbf{F}$ *as a configuration if and only if $\mathcal{G}$ forbids any isomorphic copy of $\mathcal{F}$ as a trace on some k-element subset of $[n]$.*

Definition 9 For a $(0, 1)$-matrix $\mathbf{F}$, we denote $\mathrm{forb}(m, \mathbf{F})$ as the largest value such that there exists a simple $m \times \mathrm{forb}(m, \mathbf{F})$ matrix $\mathbf{A}$ such that $\mathbf{A}$ has no configuration $\mathbf{F}$.

Note that $\mathbf{F}$ is not required to be simple when considering $\mathrm{forb}(m, \mathbf{F})$. The following are two straightforward yet useful propositions for $\mathrm{forb}(m, \mathbf{F})$.

Proposition 5 *Assume that* $\mathbf{F}$ *is a* $(0, 1)$*-matrix.*

(1) Let $\mathbf{F}^c$ denote the $(0, 1)$-complement of $\mathbf{F}$. Then we have $\mathrm{forb}(m, \mathbf{F}) = \mathrm{forb}(m, \mathbf{F}^c)$.
(2) If $\mathbf{F}$ has $\mathbf{F}'$ as a configuration, then we have $\mathrm{forb}(m, \mathbf{F}') \leqslant \mathrm{forb}(m, \mathbf{F})$.

Let $\mathbf{K}_k$ denote the complete configuration which contains all 2^k binary columns on k rows. Then Sauer-Shelah Lemma can be restated as follows:

Theorem 35 (Sauer-Shelah Lemma) *For the complete configuration* $\mathbf{K}_k$, *we have*

$$\mathrm{forb}(m, \mathbf{K}_k) = \sum_{i=0}^{k-1} \binom{m}{i} = \Theta(m^{k-1}) \tag{1}$$

Let $\mathbf{K}_k^s$ denote the $k \times \binom{k}{s}$ simple matrix consisting of all possible columns with a column sum of s. Since $\mathbf{K}_k$ has $\mathbf{K}_k^s$ as a configuration, it is easy to see that $\mathrm{forb}(m, \mathbf{K}_k^s) \leqslant \mathrm{forb}(m, \mathbf{K}_k)$. Füredi and Quinn [34] proved that these two functions are eventually equal.

Theorem 36 (Füredi and Quinn [34]) *Let k and s be positive integers with* $0 \leqslant s \leqslant k$. *Then*

$$\mathrm{forb}(m, \mathbf{K}_k^s) = \binom{m}{k-1} + \binom{m}{k-2} + \cdots + \binom{m}{0}.$$

Let $[\mathbf{A} \mid \mathbf{B}]$ denote the matrix obtained from concatenating the two matrices $\mathbf{A}$ and $\mathbf{B}$. For any simple matrix $\mathbf{A}$ and a positive integer k, let $k \cdot \mathbf{A}$ to denote the matrix $[\mathbf{A}|\mathbf{A}| \cdots \mid \mathbf{A}]$ consisting of k copies of $\mathbf{A}$ concatenated together. The following theorem of Gronau [39] gives the exact value of $\mathrm{forb}(m, 2 \cdot \mathbf{K}_k)$, which is equal to $\mathrm{forb}(m, \mathbf{K}_{k+1})$.

Theorem 37 (Gronau [39]) *We have*

$$\mathrm{forb}(m, 2 \cdot \mathbf{K}_k) = \binom{m}{k} + \binom{m}{k-1} + \cdots + \binom{m}{0}.$$

The following theorem by Anstee and Füredi [8] determines an upper bound of $\mathrm{forb}(m, t \cdot \mathbf{K}_k)$ for general t.

Theorem 38 (Anstee and Füredi [8]) *We have*

$$\mathrm{forb}(m, t \cdot \mathbf{K}_k) = \mathrm{forb}(m, t \cdot \mathbf{K}_k^k) \leqslant \frac{t-2}{k+1}\binom{m}{k}(1 - o(1)) + \binom{m}{k} + \binom{m}{k-1} + \cdots + \binom{m}{0}.$$

For $t = 1$, Theorem 35 shows that $\mathrm{forb}(m, \mathbf{K}_k) = \Theta(m^{k-1})$. For $t \geqslant 2$, since $\mathrm{forb}(m, t \cdot \mathbf{K}_k) \geqslant \mathrm{forb}(m, 2 \cdot \mathbf{K}_k) = \mathrm{forb}(m, \mathbf{K}_{k+1})$, Theorem 38 gives that $\mathrm{forb}(m, t \cdot \mathbf{K}_k) = \Theta(m^k)$. For further results on the exact bounds of $\mathrm{forb}(m, \mathbf{F})$ for specific configurations $\mathbf{F}$, we refer interested readers to the survey by Anstee [7].

Recall that Theorem 36 shows that $\mathrm{forb}(m, \mathbf{K}_k) = \mathrm{forb}(m, \mathbf{K}_k^s)$. This inspires us to think of what is the minimal $\mathbf{F}'$ under the partial order $\prec$ such that $\mathrm{forb}(m, \mathbf{F}') = \mathrm{forb}(m, \mathbf{F})$. To be more precise, Anstee and Karp [9] introduced the following definition.

Definition 10 A *critical substructure* of a configuration $\mathbf{F}$ as a minimal configuration $\mathbf{F}'$ under the partial order $\prec$ such that $\mathbf{F}$ contains configuration $\mathbf{F}'$ and $\mathrm{forb}(m, \mathbf{F}') = \mathrm{forb}(m, \mathbf{F})$.

Raggi [61] verified that the only k-rowed critical substructures of $\mathbf{K}_k$ are $\mathbf{K}_k^s$ for $s = 0, 1, \ldots k$. Besides, Raggi [61] also determined the complete list of critical substructures for $\mathbf{K}_4$.

Theorem 39 (Raggi [61]) *The critical substructures of* $\mathbf{K}_4$ *are* $\mathbf{0}_4$, $\mathbf{I}_4$, $\mathbf{K}_4^2$, $\mathbf{I}_4^c$, $\mathbf{1}_4$, $2 \cdot \mathbf{0}_3$ *and* $2 \cdot \mathbf{1}_3$.

We mention that the critical substructures for $\mathbf{K}_5$ have not been fully determined.

On the other hand, it is also worth considering the opposite scenario, namely, configurations $\mathbf{F}'$ that contain $\mathbf{F}$ and satisfy $\text{forb}(m, \mathbf{F}') = \text{forb}(m, \mathbf{F})$. Let $\mathbf{1}_p\mathbf{0}_q$ denote the column consisting of p 1s at the top and q 0s at the bottom. Anstee and Meehan [10] showed that a column $\mathbf{1}_p\mathbf{0}_q$ can be added to $\mathbf{K}_k$ without changing the bound.

Theorem 40 (Anstee and Meehan [10]) *Assume that* $p + q = k$ *and* $p, q \geqslant 2$. *Then*

$$\text{forb}(m, [\mathbf{K}_k | \mathbf{1}_p\mathbf{0}_q]) = \text{forb}(m, \mathbf{K}_k).$$

Let $\mathbf{A}$ be an $m_1 \times n_1$ matrix and $\mathbf{B}$ be an $m_2 \times n_2$ matrix. Let $\mathbf{A} \times \mathbf{B}$ denote the $(m_1 + m_2) \times (n_1 n_2)$ matrix with all possible columns formed from one column from $\mathbf{A}$ placed on top of one column from $\mathbf{B}$. We let $\mathbf{K}_2^T$ be the following matrix:

$$\mathbf{K}_2^T = \begin{pmatrix} 1 & 1 \\ 1 & 0 \\ 0 & 1 \\ 0 & 0 \end{pmatrix}.$$

Anstee and Nikov [11] also proved the following theorem.

Theorem 41 (Anstee and Nikov [11]) *Assume* $k \geqslant 4$ *and* $t \geqslant 1$. *There exists an* $m_{k,t}$ *such that for* $m \geqslant m_{k,t}$ *we have*

$$\text{forb}(m, [\mathbf{K}_k | t \cdot (\mathbf{K}_2^T \times \mathbf{K}_{k-4})]) = \text{forb}(m, \mathbf{K}_k).$$

8 Turán Numbers for Traces

Turán's theorem [66], a cornerstone of extremal graph theory, determines the Turán numbers for all cliques. Building on this, Turán-type problems seek to determine $\text{ex}(n, F)$ for arbitrary graphs F, a framework that naturally extends to hypergraphs. Specifically, for a family $\mathfrak{G}$ of r-uniform hypergraphs, $\text{ex}_r(n, \mathfrak{G})$ represents the maximum number of hyperedges in an n-vertex r-uniform hypergraph that avoids all members of $\mathfrak{G}$. However, for general r, this problem remains open and is widely regarded as highly challenging. Therefore, researchers have focused on more tractable variants of the problem. A notable example is the *Berge copy*, a concept due to Berge [15]. The following formal definition can be found in Gerbner and Palmer [37].

Definition 11 Let $\mathcal{F}$ and $\mathcal{T}$ be uniform hypergraphs (possibly with different uniformities) with $V(\mathcal{F}) \subseteq V(\mathcal{T})$. We say that $\mathcal{T}$ is a *Berge copy* of $\mathcal{F}$ if there exists a bijection $\phi : E(\mathcal{F}) \to E(\mathcal{T})$ such that for every edge $e \in E(\mathcal{F})$, we have $e \subseteq \phi(e)$. Moreover, we say that $\mathcal{T}$ is an *induced Berge copy* of $\mathcal{F}$ (or a trace of $\mathcal{F}$) if there exists a bijection $\phi : E(\mathcal{F}) \to E(\mathcal{T})$ such that for every edge $e \in E(\mathcal{F})$, we have $\phi(e) \cap V(\mathcal{F}) = e$.

The following establishes a connection between induced Berge copies and traces of families.

Proposition 6 *If $\mathcal{T}$ is an induced Berge copy of $\mathcal{F}$, then we have $\mathcal{T}_{|V(\mathcal{F})} = \mathcal{F}$.*

To proceed, we require the following definition provided by Luo and Spiro [51].

Definition 12 Given a hypergraph $\mathcal{F}$, let $\mathrm{B}_r(\mathcal{F})$ denote the set of all r-uniform hypergraphs that are Berge copies of $\mathcal{F}$ up to isomorphism. Similarly, let $\mathrm{Tr}_r(\mathcal{F})$ denote the set of all r-uniform induced Berge copies of $\mathcal{F}$ up to isomorphism.

It is evident that $\mathrm{Tr}_r(\mathcal{F}) \subseteq \mathrm{B}_r(\mathcal{F})$. Taking an example, if $\mathcal{F} = K_3$ and $V(\mathcal{F}) = [3]$, then for $r = 3$ we have

$$\mathrm{Tr}_3(K_3) = \{\{124, 134, 234\}, \{124, 134, 235\}, \{124, 135, 236\}\}, \text{ and}$$
$$\mathrm{B}_3(K_3) = \{\{124, 134, 234\}, \{124, 134, 235\}, \{124, 135, 236\}, \{123, 134, 235\}\}.$$

We also define the Turán number for these graph-based hypergraphs.

Definition 13 Assume that F is a graph, let $\mathrm{ex}(n, \mathrm{B}_r(F))$ (resp. $\mathrm{ex}(n, \mathrm{Tr}_r(F))$) be the maximum number of edges of an n-vertex r-uniform hypergraph that does not contain a subhypergraph isomorphic to a Berge copy (resp. an induced Berge copy) of F.

The Turán problem for graph-based hypergraphs is closely related to the so-called *generalized Turán problems*, which were first pioneered by Erdős [24]. The generalized Turán problems were systematically studied by Alon and Shikhelman [5].

Definition 14 Let F be a graph. Given two graphs H and G, let $\mathcal{N}(H, G)$ be the number of copies of H contained in G. We let $\mathrm{ex}(n, H, F) = \max\{\mathcal{N}(H, G) :$ G is an n-vertex F-free graph$\}$.

For a graph G, we define $\mathcal{N}_r(G)$ to be the r-uniform hypergraph that consists of all r-element vertex sets which span a copy of K_r in G. Obviously, we have $\mathrm{ex}(n, K_r, F) = \max\{|\mathcal{N}_r(G)| : G \text{ is } F\text{-free}\}$. If G is F-free, then $\mathcal{N}_r(G)$ is $\mathrm{B}_r(F)$-free, implying $|\mathcal{N}_r(G)| \leqslant \mathrm{ex}(n, \mathrm{B}_r(F))$. Therefore, we can get the following relationship between these three functions:

$$\mathrm{ex}(n, K_r, F) \leqslant \mathrm{ex}(n, \mathrm{B}_r(F)) \leqslant \mathrm{ex}(n, \mathrm{Tr}_r(F)).$$

Sali and Spiro [64] determined the order of magnitude of $\mathrm{ex}(n, \mathrm{Tr}_r(K_{s,t}))$ when $t \geqslant (s-1)! + 1$ and $s \leqslant 2r - 4$. Later, Füredi and Luo [32] determined the order of magnitude of $\mathrm{ex}(n, \mathrm{Tr}_r(F))$ for all graphs F in terms of the generalized Turán numbers.

Theorem 42 (Füredi and Luo [32]) *Suppose that $r \geqslant 2$, F is a graph with $E(F) \neq \emptyset$. Then, as $n \to \infty$, we have*

$$\mathrm{ex}(n, \mathrm{Tr}_r(F)) = \Theta(\max_{2 \leqslant s \leqslant r} \{\mathrm{ex}(n, K_s, F)\}).$$

The first to study forbidden induced Berge copies of graphs were Mubayi and Zhao [55]. They determined the asymptotic value of $\mathrm{ex}(n, \mathrm{Tr}_r(K_s))$ for $s \in \{3, 4\}$ or $r \in \{s, s+1\}$, as follows.

Theorem 43 (Mubayi and Zhao [55]) *Suppose that $s \in \{3, 4\}$ or $r \in \{s, s+1\}$. Then*

$$\mathrm{ex}(n, \mathrm{Tr}_r(K_s)) = \left(\frac{n}{s-1}\right)^{s-1} + o(n^{s-1}).$$

Mubayi and Zhao [55] also conjectured that $\mathrm{ex}(n, \mathrm{Tr}_r(K_s)) \sim (\frac{n}{s-1})^{s-1}$ holds for $s \geqslant 5$. In fact, they conjectured an extremal construction for $\mathrm{ex}(n, \mathrm{Tr}_r(K_s))$ for sufficiently large n when $r \geqslant s$ and they proved this for the case $r = 3$. Gerbner [36] confirmed this conjecture for the case $s = 3$. He [36] also extended the asymptotic bound to book graphs $K_{1,1,t}$ in the 3-uniform case, which gave that $\mathrm{ex}(n, \mathrm{Tr}_r(K_{1,1,t})) = (1 + o(1))n^2/4$. The research of $\mathrm{ex}(n, \mathrm{Tr}_r(F))$ for specific graphs F, such as stars and $K_{2,t}$, can be found in [32, 36, 51, 60].

9 Open Problems

9.1 Generalizing Theorem 5

Recall that $m(n, a, b) = \min\{m : (n, m) \to (a, b)\}$. The following construction given by Frankl and Wang [31] shows that $m(n, \ell+1, 2^{\ell+1} - 2^{\ell-1} + 1) > \prod_{0 \leqslant i < \ell} \left\lfloor \frac{n+\ell+i}{\ell} \right\rfloor$ for any positive integer ℓ.

Construction 44 Let ℓ be a positive integer. Let $X_0, \ldots, X_{\ell-1}$ form a partition of $[n]$ with $|X_i| = \left\lfloor \frac{n+i}{\ell} \right\rfloor$ for $0 \leqslant i < \ell$. Define

$$\mathcal{F}(n, \ell) = \{F \subseteq [n] : |F \cap X_i| \leqslant 1, 0 \leqslant i < \ell\}.$$

It is easy to see that $|\mathcal{F}(n, \ell)| = \prod_{0 \leqslant i < \ell} \left\lfloor \frac{n+\ell+i}{\ell} \right\rfloor$. For any $T \in \binom{[n]}{\ell+1}$, by Pigeonhole Principle, there exist two vertices $x, y \in [n]$ such that $x, y \in X_{i_0}$ for some $0 \leqslant i_0 < \ell$. Since $\mathcal{F}(n, \ell)_{|T} \subseteq 2^T \setminus \{F \subseteq T : F \supseteq \{x, y\}\}$, we can conclude that

$|\mathcal{F}(n,\ell)_{|T}| \leqslant 2^{\ell+1} - 2^{\ell-1}$. Therefore, for any positive integer ℓ, this construction shows that

$$\left(n, \prod_{0 \leqslant i < \ell} \left\lfloor \frac{n+\ell+i}{\ell} \right\rfloor \right) \nrightarrow (\ell+1, 2^{\ell+1} - 2^{\ell-1} + 1),$$

providing the desired lower bound for $m(n, \ell+1, 2^{\ell+1} - 2^{\ell-1} + 1)$. Frankl and Wang [31] made the following conjecture, which asks whether the above lower bound is best possible.

Conjecture 2 (Frankl and Wang [31]) For any positive integers $n > \ell > 0$, we have

$$\left(n, 1 + \prod_{0 \leqslant i < \ell} \left\lfloor \frac{n+\ell+i}{\ell} \right\rfloor \right) \rightarrow (\ell+1, 2^{\ell+1} - 2^{\ell-1} + 1).$$

Note that Conjecture 2 holds true for $\ell \leqslant 3$, as follows: For $\ell = 1$, $(n, n+2) \rightarrow (2, 4)$ is just from Sauer-Shelah Lemma; for $\ell = 2$, $\left(n, \left\lfloor \frac{n^2}{4} \right\rfloor + n + 2\right) \rightarrow (3, 7)$ is proved by Frankl [27]; finally, for $\ell = 3$, Frankl and Wang [31] show that $(n, 1 + \left\lfloor \frac{n+5}{3} \right\rfloor \left\lfloor \frac{n+4}{3} \right\rfloor \left\lfloor \frac{n+3}{3} \right\rfloor) \rightarrow (4, 13)$ for $n \geqslant 25$.

9.2 Shattering in Antichains

Recall that a family $\mathcal{F}$ is an *antichain* if for any $F, F' \in \mathcal{F}$, $F \nsubseteq F'$ holds. The following longstanding conjecture was proposed by Frankl [28] thirty years ago.

Conjecture 3 (Frankl [28]) Let k be a non-negative integer and $n \geqslant 2k$. Suppose that $\mathcal{F} \subseteq 2^{[n]}$ is an antichain with $\mathcal{F} \nrightarrow (k+1, 2^{k+1})$. Then $|\mathcal{F}| \leqslant \binom{n}{k}$.

Conjecture 3 was proved by Frankl [28] for $k \leqslant 2$ and by Anstee and Sali [13] for $k = 3$.

9.3 The Structure of Extremal Families

A family $\mathcal{F} \subseteq 2^{[n]}$ is *shattering-extremal*, or *s-extremal* for short, if $|tr(\mathcal{F})| = |\mathcal{F}|$. The main problem concerning s-extremal families is to find good characterizations of them. A potential approach to this problem was proposed by Mészáros and Rónyai [53] in the following conjecture.

Conjecture 4 (Mészáros and Rónyai [53]) For every nonempty s-extremal family $\mathcal{F} \subseteq 2^{[n]}$, there exists $F_0 \in \mathcal{F}$ such that $\mathcal{F} \setminus \{F_0\}$ is still s-extremal.

By Theorem 29, we know that $\mathcal{F}$ is s-extremal if and only if $2^{[n]} \setminus \mathcal{F}$ is also s-extremal. Therefore, it is equivalent to consider whether there exists $F_0 \notin \mathcal{F}$ such that $\mathcal{F} \cup \{F_0\}$ is s-extremal. It is obvious that hereditary families trivially satisfy Conjecture 4. In [53, 54], Mészáros and Rónyai confirmed this conjecture for all families with VC dimension at most 2.

9.4 Forbidden Traces of Complete Graphs

Let $K_s^{(t)}$ be the complete s-vertex t-uniform hypergraph. Recall that for a hypergraph F, we let $\mathrm{Tr}_r(F)$ denote the set of all r-uniform induced Berge copies of F up to isomorphism. Consider the family $\mathrm{Tr}_r(K_s^{(t)})$, let $H_{t,s}^r$ be the member of $\mathrm{Tr}_r(K_s^{(t)})$ with the maximum number of vertices. In other words, $H_{t,s}^r$ is the r-uniform hypergraph obtained from $K_s^{(t)}$ by adding $r-t$ new vertices to each hyperedge such that each new vertex is added to only one hyperedge. In particular, $H_{r,s}^r = K_s^{(r)}$.

By definition, we have $\mathrm{ex}(n, \mathrm{Tr}_r(K_s^{(t)})) \leqslant \mathrm{ex}(n, H_{t,s}^r)$. In order to determine the value of $\mathrm{ex}(n, \mathrm{Tr}_r(K_s^{(t)}))$, Mubayi and Zhao [55] presented the following conjecture.

Conjecture 5 (Mubayi and Zhao [55]) Given positive integers n, r, s and t with $2 \leqslant t < \min\{r, s\}$. There exists $n_0 > 0$ such that for $n > n_0$, we have

$$\mathrm{ex}(n, \mathrm{Tr}_r(K_s^{(t)})) = \begin{cases} \mathrm{ex}(n, H_{t,s}^r) & \text{if } r < s, \\ \mathrm{ex}(n-r+s-1, \mathrm{Tr}_{s-1}(K_s^{(t)})) = \mathrm{ex}(n-r+s-1, H_{t,s}^{s-1}) & \text{if } r \geqslant s. \end{cases}$$

Mubayi and Zhao [55] observed that for $r < s$ and sufficiently large n, T_{s-1}^r is an extremal family for both $\mathrm{ex}(n, \mathrm{Tr}_r(K_s))$ and $\mathrm{ex}(n, H_{2,s}^r)$, where T_ℓ^r is the complete ℓ-partite r-uniform hypergraph. In this case, it holds that $\mathrm{ex}(n, \mathrm{Tr}_r(K_s)) = \mathrm{ex}(n, H_{2,s}^r)$. So Conjecture 5 holds for the cases when $t = 2$ and $r < s$. As for the cases when $r \geqslant s$, Mubayi and Zhao [55] showed that Conjecture 5 holds for $t = 2$ and $r = s = 3$, that is, $\mathrm{ex}(n, \mathrm{Tr}_3(K_3)) = \mathrm{ex}(n-1, K_3) = \left\lfloor \frac{(n-1)^2}{4} \right\rfloor$. Besides, they also showed that $\mathrm{ex}(n, \mathrm{Tr}_r(K_3)) = n^2/2 + o(n^2)$, which confirmed Conjecture 5 asymptotically if $s = 3$ and $t = 2$. Later, Gerbner [36] solved the case when $t = 2$ and $s = 3$, by showing for sufficiently large n, $\mathrm{ex}(n, \mathrm{Tr}_r(K_3)) = \left\lfloor \frac{(n-r+2)^2}{4} \right\rfloor$.

References

1. Aldred, R.E., Anstee, R.P.: On the density of sets of divisors. Discret. Math. **137**(1–3), 345–349 (1995)
2. Alon, N.: On the density of sets of vectors. Discret. Math. **46**(2), 199–202 (1983)
3. Alon, N., Moran, S., Yehudayoff, A.: Sign rank versus Vapnik-Chervonenkis dimension. Sbornik: Mathematics **208**(12), 1724–1757 (2017)

4. Alon, N., Moshkovitz, G., Solomon, N.: Traces of hypergraphs. J. Lond. Math. Soc. **100**(2), 498–517 (2019)
5. Alon, N., Shikhelman, C.: Many T copies in H-free graphs. J. Comb. Theory Ser. B **121**, 146–172 (2016)
6. Anstee, R.P.: A forbidden configuration theorem of Alon. J. Comb. Theory Ser. A **47**(1), 16–27 (1988)
7. Anstee, R.P.: A survey of forbidden configuration results. Electron. J. Comb. **1000** (2013)
8. Anstee, R.P., Füredi, Z.: Forbidden submatrices. Discret. Math. **62**(3), 225–243 (1986)
9. Anstee, R.P., Karp, S.N.: Forbidden configurations: exact bounds determined by critical substructures. Electron. J. Comb. **17**(1) (2010)
10. Anstee, R.P., Meehan, C.: Forbidden configurations and repeated induction. Discret. Math. **311**(20), 2187–2197 (2011)
11. Anstee, R.P., Nikov, N.A.: Shattering and more: extending the complete object. Electron. J. Comb. **29**(2) (2022)
12. Anstee, R.P., Rónyai, L., Sali, A.: Shattering news. Graphs Comb. **18**(1), 59–73 (2002)
13. Anstee, R.P., Sali, A.: Sperner families of bounded VC-dimension. Discret. Math. **175**(1–3), 13–21 (1997)
14. Balogh, J., Mészáros, T., Wagner, A.Z.: Two results about the hypercube. Discret. Appl. Math. **247**, 322–326 (2018)
15. Berge, C.: Hypergraphes: Combinatoire Des Ensembles Finis. Gauthier-Villars, Paris (1987)
16. Bollobás, B., Leader, I., Radcliffe, A.J.: Reverse kleitman inequalities. Proc. Lond. Math. Soc **s3-58**(1), 153–168 (1989)
17. Bollobás, B., Radcliffe, A.: Defect Sauer results. J. Comb. Theory Ser. A **72**(2), 189–208 (1995)
18. Bondy, J.: Induced subsets. J. Comb. Theory Ser. B **12**(2), 201–202 (1972)
19. Brown, W.G.: On an open problem of Paul Turán concerning 3-graphs. In: Studies in Pure Mathematics, pp. 91–93. Birkhäuser Basel, Basel (1983)
20. Chung, F., Lu, L.: An upper bound for the Turán number $t_3(n, 4)$. J. Comb. Theory Ser. A **87**(2), 381–389 (1999)
21. Damásdi, G., Héger, T., Szőnyi, T.: The Zarankiewicz problem, cages and geometries. Annales Universitatis Scientiarum Budapestinensis de Rolando Eötvös Nominatae. Sectio mathematica. Eötvös Loránd Tudományegyetem **LVI**, 3–37 (2013)
22. de Caen, D.: The current status of Turán's problem on hypergraphs. Extremal Problems Finite Sets **3**, 187–197 (1991)
23. Dudley, R.M.: The structure of some Vapnik-Chervonenkis classes. In: Proceedings of the Berkeley Conference in Honor of Jerzy Neyman, vol. 2, pp. 495–507 (1985)
24. Erdős, P.: On the number of complete subgraphs contained in certain graphs. Magyar Tudományos Akadémia Matematikai Kutató Intézetének Közleményei **7**, 459–464 (1962)
25. Falgas-Ravry, V., Vaughan, E.R.: Applications of the semi-definite method to the turán density problem for 3-graphs. Comb. Probab. Comput. **22**(1), 21–54 (2013)
26. Fon-Der-Flaas, D.G.: Method for construction of (3, 4)-graphs. Matematicheskie Zametki **44**(4), 546–550 (1988)
27. Frankl, P.: On the trace of finite sets. J. Comb. Theory Ser. A **34**(1), 41–45 (1983)
28. Frankl, P.: Traces of antichains. Graphs Comb. **5**(1), 295–299 (1989)
29. Frankl, P., Füredi, Z.: An exact result for 3-graphs. Discret. Math. **50**, 323–328 (1984)
30. Frankl, P., Tokushige, N.: Extremal Problems for Finite Sets, vol. 86. American Mathematical Soc. (2018)
31. Frankl, P., Wang, J.: Four-vertex traces of finite sets. Graphs Comb. **40**(1) (2024)
32. Füredi, Z., Luo, R.: Induced Turán problems and traces of hypergraphs. Eur. J. Comb. **111**, 103692 (2023)
33. Füredi, Z., Pach, J.: Traces of finite sets: extremal problems and geometric applications. Extremal Problems Finite Sets **3**, 255–282 (1991)
34. Füredi, Z., Quinn, F.: Traces of finite sets. Ars Comb. **18**, 195–200 (1984)
35. Füredi, Z., Simonovits, M.: The history of degenerate (Bipartite) extremal graph problems. In: Bolyai Society Mathematical Studies, pp. 169–264. Springer, Berlin (2013)

36. Gerbner, D.: On forbidding graphs as traces of hypergraphs (2023). arXiv:2310.05601
37. Gerbner, D., Palmer, C.: Extremal results for berge hypergraphs. SIAM J. Discret. Math. **31**(4), 2314–2327 (2017)
38. Gerbner, D., Patkós, B.: Extremal Finite Set Theory, 1 edn. Chapman and Hall/CRC (2018)
39. Gronau, H.O.F.: An extremal set problem. Studia Sci. Math. Hungar **15**, 29–30 (1980)
40. He, J., Ma, J., Yang, T.: Some exact results on 4-cycles: stability and supersaturation. CSIAM Trans. Appl. Math. **4**(1), 74–128 (2023)
41. Kahn, J., Kalai, G., Linial, N.: The influence of variables on Boolean functions. In: [Proceedings 1988] 29th Annual Symposium on Foundations of Computer Science, pp. 68–80. IEEE, White Plains, NY, USA (1988)
42. Katona, G.O.H.: A theorem on finite sets. In: Theory of Graphs, Proceedings of the Colloquium Held at Tihany, Hungary, pp. 187–207. Academic (1968)
43. Katona, G.O.H.: Optimization for order ideals under a weight assignment. In: Problèmes Combinatoires et Théorie Des Graphes, pp. 257–258 (1978)
44. Kostochka, A.V.: A class of constructions for Turán's (3, 4)-problem. Combinatorica **2**(2), 187–192 (1982)
45. Kővári, T., Sós, V., Turán, P.: On a problem of K. Zarankiewicz. Colloquium Mathematicum **3**(1), 50–57 (1954)
46. Kozma, L., Moran, S.: Shattering, graph orientations, and connectivity. Electron. J. Comb. **20**(3) (2013)
47. Kruskal, J.B.: The number of simplices in a complex. Math. Optim. Tech. **10**, 251–278 (1963)
48. Kusch, C., Mészáros, T.: Shattering-extremal set systems from Sperner families. Discret. Appl. Math. **276**, 92–101 (2020)
49. Li, M., Ma, J., Rong, M.: Exact results on traces of sets (2024). arXiv:2406.18870
50. Lovász, L.: Combinatorial Problems and Exercises, vol. 361. American Mathematical Soc. (2007)
51. Luo, R., Spiro, S.: Forbidding $K_{2,t}$ traces in triple systems. Electron. J. Comb. **28**(2), P2.4 (2021)
52. Ma, J., Yang, T.: Upper bounds on the extremal number of the 4-cycle. Bull. Lond. Math. Soc. **55**(4), 1655–1667 (2023)
53. Mészáros, T., Rónyai, L.: Shattering-extremal set systems of small VC-dimension. ISRN Comb. **2013**, 1–8 (2013)
54. Mészáros, T., Rónyai, L.: Shattering-extremal set systems of VC dimension at most 2. Electron. J. Comb. **21**(4) (2014)
55. Mubayi, D., Zhao, Y.: Forbidding complete hypergraphs as traces. Graphs Comb. **23**(6), 667–679 (2007)
56. Naor, A., Verstraëte, J.: A note on bipartite graphs without 2k-cycles. Comb. Probab. Comput. **14**(5–6), 845–849 (2005)
57. Pajor, A.: Sous Espaces Ln1 Des Espaces de Banach. Editions Hermann (1985)
58. Perles, M.A., Shelah, S.: A combinatorial problem; stability and order for models and theories in infinitary languages. Pac. J. Math. **41**(1), 247–261 (1972)
59. Piga, S., Schülke, B.: On extremal problems concerning the traces of sets. J. Comb. Theory Ser. A **182**, 105447 (2021)
60. Qian, B., Ge, G.: A note on the Turán number for the traces of hypergraphs (2022). arXiv:2206.05884
61. Raggi, M.: Forbidden configurations. Ph.D. thesis, University of British Columbia, Vancouver, BC, Canada (2011)
62. Razborov, A.A.: On 3-Hypergraphs with forbidden 4-vertex configurations. SIAM J. Discret. Math. **24**(3), 946–963 (2010)
63. Reiher, C., Schülke, B.: Minimum degree in simplicial complexes (2025). arXiv:2501.01294
64. Sali, A., Spiro, S.: Forbidden families of minimal quadratic and cubic configurations. Electron. J. Comb. **24**(2) (2017)
65. Sauer, N.: On the density of families of sets. J. Comb. Theory Ser. A **13**(1), 145–147 (1972)
66. Turán, P.: On an external problem in graph theory. Mat. Fiz. Lapok **48**, 436–452 (1941)

67. Vapnik, V.N.: On the uniform convergence of relative frequencies of events to their probabilities. In: Doklady Akademii Nauk USSR, vol. 181, pp. 781–787 (1968)
68. Vapnik, V.N., Chervonenkis, A.Ya.: On the uniform convergence of relative frequencies of events to their probabilities. In: Measures of Complexity, pp. 11–30. Springer International Publishing, Cham (2015)
69. Watanabe, M.: Arrow relations on families of finite sets. Discret. Math. **94**(1), 53–64 (1991)
70. Watanabe, M.: Some best possible bounds concerning the traces of finite sets II. Graphs Comb. **11**(3), 293–303 (1995)
71. Watanabe, M., Frankl, P.: Some best possible bounds concerning the traces of finite sets. Graphs Comb. **10**(2–4), 283–292 (1994)

Transversal Structures in Graph Systems: A Survey

Wanting Sun, Guanghui Wang, and Lan Wei

Abstract Given a system $\mathcal{G} = \{G_1, G_2, \ldots, G_m\}$ of graphs/digraphs/hypergraphs on the common vertex set V of size n, an m-edge graph/digraph/hypergraph H with vertices in V is *transversal* in $\mathcal{G}$ if there exists a bijection $\phi : E(H) \to [m]$ such that $e \in E(G_{\phi(e)})$ for all $e \in E(H)$. In this survey, we consider extremal problems for transversal structures in graph systems. More precisely, we summarize some sufficient conditions that ensure the existence of transversal structures in graph/digraph/hypergraph systems, which generalize several classical theorems in extremal graph theory to transversal versions. We also include a number of conjectures and open problems.

1 Introduction

How do global parameters of a graph, such as its degree or chromatic number, influence its local substructures? Is there some sufficiently large average degree or chromatic number that forces a given substructure to appear? Questions of this type are one of the most natural problems in graph theory, and there are many profound and interesting results. These are collectively referred to as *extremal graph theory*. It is a branch of combinatorial mathematics that studies the maximal or minimal properties of graphs under certain conditions.

Many important problems in extremal graph theory can be framed as subgraph containment problems, which ask for conditions on a graph G that ensure it contains a

W. Sun
Data Science Institute, Shandong University, Jinan, China
e-mail: wtsun@sdu.edu.cn

G. Wang (✉)
School of Mathematics, Shandong University, Jinan, China
e-mail: ghwang@sdu.edu.cn

L. Wei
Mathematical Research Center, Shandong University, Jinan, China
e-mail: lanwei@mail.sdu.edu.cn

G. O. H. Katona et al. (eds.), *Sum(m)it280*, Bolyai Society Mathematical Studies 32,
https://doi.org/10.1007/978-3-032-18810-6_15

copy of a general graph H. For example, Mantel's theorem states that every n-vertex graph with more than $\lfloor n^2/4 \rfloor$ edges contains a triangle; Turán [140] generalized it by determining the maximum number of edges in an n-vertex graph that forbids K_r, where K_r denotes a complete graph on r vertices; Dirac [40] proved that every n-vertex ($n \geq 3$) graph with minimum degree at least $n/2$ contains a Hamilton cycle. For more results, the reader is referred to the book of Bollobás [15].

In this paper, we survey some recent developments on transversal problems in graph systems. The notion of transversal appears in diverse branches of mathematics. A *transversal* X over a system $\mathcal{F} = \{F_1, \ldots, F_m\}$ of objects refers to an object that intersects each F_i exactly once. Here, the object can be any mathematical object, such as sets, spaces, set systems, and matroids, etc. Several classical theorems, which guarantee the existence of certain objects, have been reiterated in the context of transversals, demonstrating the possibility and impossibility of obtaining such an object as a transversal over a certain system. To name a few, transversals are extensively studied for Carathéodory's theorem [14], Helly's theorem [81], Erdős–Ko–Rado theorem [7], and Rota's basis conjecture [77, 125], etc.

Building on this trend, there have been numerous studies dedicated to generalizing classical results in extremal graph theory to transversal versions. In addition to being interesting in their own right, they often provide a strengthening of the original results, which also reflects the robustness of graph properties. A transversal in a graph system is also referred to as a "rainbow subgraph" in much literature, and it was formally introduced in the recent work of Joos and Kim [80]. For a positive integer n, we write $[n] := \{1, 2, \ldots, n\}$.

Definition 1 For a given system $\mathcal{G} = \{G_1, \ldots, G_m\}$ of graphs/digraphs/hypergraphs with the same vertex set V, an m-edge graph/digraph/hypergraph H with vertices in V is a *$\mathcal{G}$-transversal* (or a transversal inside $\mathcal{G}$) if there exists a bijection $\phi : E(H) \to [m]$ such that $e \in E(G_{\phi(e)})$ for all $e \in E(H)$.

By interpreting each G_i as the set of edges colored with the color i, the function ϕ in the above definition is referred to as a *rainbow coloring* and H is a *rainbow subgraph* inside $\mathcal{G}$. We shall emphasize that in what follows, when we say that $\mathcal{G}$ is a graph system with order n, then all graphs in $\mathcal{G}$ are not necessarily distinct and they have the same vertex set V of size n. Many classical results in extremal graph theory have been extended to transversal settings, exhibiting interesting phenomena.

Perhaps the most famous transversals are those related to Latin squares, which were studied by Euler. In 1782, Euler [49] considered a Latin square of order n, which is an $n \times n$ array filled with symbols $1, \ldots, n$, where each symbol appears exactly once in each row and column, see Fig. (a). A *partial transversal* of a Latin square is a set of cells such that every pair of cells shares no row, column or symbol, in particular a *transversal* is a partial transversal with n cells. The famous Ryser–Brualdi–Stein conjecture [22, 131, 133] is one of the most important problems on transversals in Latin squares.

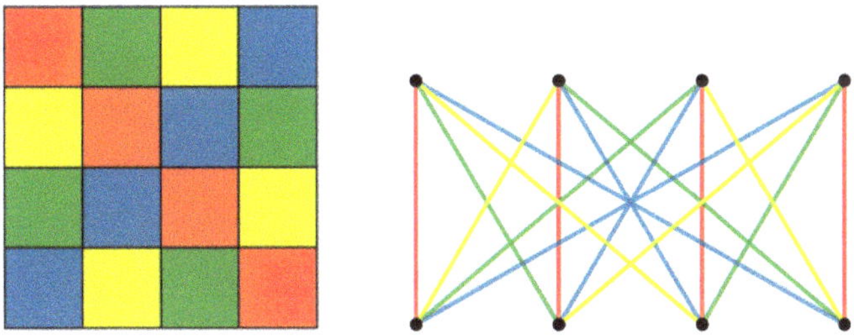

(a) Latin square of order 4. (b) Edge-colored bipartite graph.

Conjecture 1 (Ryser–Brualdi–Stein conjecture, [22, 131, 133]) *Every Latin square of order n has a partial transversal with $n-1$ cells, and a transversal with n cells if n is odd.*

In 2023, Montgomery [120] introduced the first techniques to identify and exploit the possible algebraic properties behind the entries in a Latin square, and proved the following.

Theorem 1 ([120]) *There is some $n_0 \in \mathbb{N}$ such that every Latin square of order $n \geq n_0$ contains a partial transversal with $n-1$ cells.*

For more relevant results of transversals in Latin squares, see the recent survey by Montgomery [121], which collected results from the past decade.

Considering the rows and columns of the Latin square as a complete bipartite graph $K_{n,n}$, where each symbol represents a color and each cell represents a colored edge, the Latin square naturally corresponds to a properly edge-colored $K_{n,n}$, see Fig. (b). The edge-colored $K_{n,n}$ can be viewed as a set of graphs $\mathcal{G} = \{G_1, \ldots, G_n\}$, where each G_i consists of edges with color i, then a rainbow matching of $\mathcal{G}$ corresponds to a transversal of the Latin square. We refer the reader to see the nice survey [141]. These studies have inspired research on the existence of transversal structures in a system of graphs. The central question in the study of transversals can be generalized and depicted as follows.

Problem 1 ([80]) Let H be a graph/digraph/hypergraph with m edges, and let $\mathcal{G} = \{G_1, G_2, \ldots, G_m\}$ be a system of graphs/digraphs/hypergraphs. Which property imposed on $\mathcal{G}$ will ensure a transversal copy of H?

Since we do not require the objects in the family $\mathcal{G}$ to be distinct, a necessary condition for a positive answer to the above problem is that every n-vertex graph/digraph/hypergraph satisfying the property contains a copy of H. In this survey, it will be shown by some examples that this necessary condition is not sufficient in some cases.

In this paper, we aim to survey known results for transversal problems in graph/random graph/digraph/hypergraph systems. Throughout the survey, we include a number of conjectures and open problems. It is of course impossible to cover all the known results for graph systems in a single survey; thus, the choice of results we present is inevitably subjective. Nevertheless, we hope to describe enough results from this fascinating area to engage many researchers in extremal and probabilistic combinatorics, and motivate further study of this subject.

2 Turán-Type Problem in Graph Systems

One of the central topics in extremal combinatorics is to determine the maximum number of edges of a graph on n vertices that does not contain a copy of a given graph H as a subgraph, such a maximum number is denoted by $\mathrm{ex}(n, H)$. The most notable result is Turán's theorem [140], which states that $\mathrm{ex}(n, K_{r+1}) = t_r(n)$, where $t_r(n)$ is the number of edges of a balanced complete r-partite graph on n vertices (denoted by $T_r(n)$). Erdős, Stone and Simonovits [47, 48] found a connection between $\mathrm{ex}(n, H)$ and the chromatic number $\chi(H)$ of H.

In addition to the Turán problem itself, there are also many variants involving colorings. For example, Diwan and Mubayi [41] considered a generalization of Turán's theorem in a graph system containing two graphs. It is of course possible to consider similar questions with more than two graphs. In this section, we survey results on the Turán-type problems in graph systems. In other words, we are to determine the "largest" n-vertex graph systems that do not contain a copy of H as a rainbow subgraph.

Define the *k-rainbow extremal number* $\mathrm{ex}_k^*(n, H)$ of a graph H to be the maximum $\min_{i\in[k]} |E(G_i)|$ among all rainbow H-free graph systems $\mathcal{G} = \{G_1, \ldots, G_k\}$ with order n. The simplest case here is a rainbow triangle. Diwan and Mubayi [41] proposed a problem: determine the value of $\mathrm{ex}_3^*(n, K_3)$. Magnant [118] answered this problem under the assumption that the multigraphs (i.e., graphs formed by taking the union of the edge sets $E(G_i)$) in question contain at least one edge between every pair of vertices. In 2020, Aharoni et al. [3] solved the above problem completely.

Theorem 2 ([3]) $\mathrm{ex}_3^*(n, K_3) = \lfloor \frac{26-2\sqrt{7}}{81} \rfloor n^2$.

Interestingly, the constant $\frac{26-2\sqrt{7}}{81}$ is an irrational number and larger than $\frac{1}{4}$. Therefore, compared to Mantel's theorem, the transversal triangle problem requires a stronger size condition. This signifies the difference between the transversal setting and the classical setting, and renders the rainbow extremal number more intriguing. Aharoni et al. [3] proposed a potential interesting direction to proceed.

Problem 2 ([3]) For what real numbers $\alpha_1, \alpha_2, \alpha_3 > 0$, is it true that every graph system $\{G_1, G_2, G_3\}$ satisfying $|E(G_i)| > \alpha_i n^2$ must have a transversal triangle?

Falgas-Ravry, Markström and Räty [51, Theorem 1.13] solved this problem when n is sufficiently large. Recall that Turán's theorem [140] generalizes Mantel's theorem by determining the maximum number of edges in n-vertex K_r-free ($r \geq 3$) graphs. Analogously, one may consider the following.

Problem 3 ([3]) Determine the value of $\mathrm{ex}^*_{\binom{r}{2}}(n, K_r)$ for $r \geq 4$.

In general cases, the problem becomes more complicated and there are only few known results related to the k-rainbow extremal number $\mathrm{ex}^*_k(n, H)$. As a corollary of Keevash et al. [89], we have $\mathrm{ex}^*_k(n, K_3) \leq n^2/4 + o(n^2)$ when $k \geq 4$. Frankl [57] gave a new proof of this result. Babiński and Grzesik [12] determined the value of $\mathrm{ex}^*_k(n, P_4)$ up to an additive $o(n^2)$ error term for every $k \geq 3$, where P_4 denotes the path on 4 vertices.

The *k-rainbow Turán density* of a graph H is defined as follows:

$$\pi^*_k(H) = \lim_{n\to\infty} \frac{\mathrm{ex}^*_k(n, H)}{\binom{n}{2}}.$$

Im et al. [78] showed that the limit exists for every graph H. They [78] conjectured that the rainbow Turán density is always an algebraic number, and they proved the conjecture for all trees. Furthermore, they [78] gave an answer to an analogue of Problem 2 for trees.

Theorem 3 ([78]) *Let $\alpha_1, \ldots, \alpha_k$ be nonnegative reals such that $\sum_{i=1}^k (1 - \sqrt{\alpha_i}) < 1$. For every $\epsilon > 0$, there exists an integer n_0 such that for every $n \geq n_0$ and every k-edge tree T, every graph system $\mathcal{G} = \{G_1, \ldots, G_k\}$ of order n with $|E(G_i)| \geq (\alpha_i + \epsilon)\binom{n}{2}$ contains a transversal copy of T.*

Except for the rainbow Turán number of a graph, depending on the contexts, different measures for the "size" of graph systems have been introduced, for example, the sum $\sum_i |E(G_i)|$ [25, 89], the product $\Pi_i |E(G_i)|$ [57, 74] and other measures.

The *k-color Turán number*, denoted by $\mathrm{ex}_k(n, H)$, is the maximum value of $\sum_{i\in[k]} |E(G_i)|$ among all graph systems $\mathcal{G} = \{G_1, \ldots, G_k\}$ of order n and without rainbow copy of H. We call such a graph system $\mathcal{G}$ satisfying $\sum_{i\in[k]} |E(G_i)| = \mathrm{ex}_k(n, H)$ the *k-color extremal graph system of H*. To determine $\mathrm{ex}_k(n, H)$, there are two natural lower bounds that arise from the following constructions. One is that each G_i is the n-vertex H-free graph with $\mathrm{ex}(n, H)$ edges; the other one is that there are $|E(H)| - 1$ copies of the complete graph K_n and all the other graphs are empty.

A graph is called *r-color-critical* if it has chromatic number r, and it has an edge whose removal reduces the chromatic number to $r - 1$. In 2004, Keevash et al. [89] proposed the following conjecture.

Conjecture 2 ([89]) *Suppose $r \geq 3$ and $k \geq h$. Let H be an r-color-critical graph with h edges. Then, there exists an $n_0 = n_0(H) > 0$ such that for all $n \geq n_0$, the k-color extremal graph system of H either consists of exactly $h - 1$ nonempty graphs where each of them is a copy of K_n or consists of k graphs where all of them are identical copies of $T_{r-1}(n)$. In particular,*

$$\mathrm{ex}_k(n, H) = \begin{cases} (h-1)\binom{n}{2}, & \text{if } h \leq k < \frac{r-1}{r-2}(h-1), \\ k \cdot t_{r-1}(n), & \text{if } k \geq \frac{r-1}{r-2}(h-1). \end{cases}$$

They provided evidence for this conjecture by proving it holds when H is a complete graph or $r = 3$. Recently, Chakraborti et al. [25] showed that the conjecture holds for every 4-color-critical graph H and almost all r-color-critical graphs H with $r \geq 5$. Furthermore, they [25] proved that if H is not r-color-critical for some $r \geq 3$ and n is sufficiently large, then neither of the constructions in Conjecture 2 is extremal. Li et al. [110] proved Conjecture 2 holds when $k \geq 2(r-1)(h-1)/r$, significantly improving earlier results. Furthermore, their proof combined the stability argument of Chakraborti et al. [25] with a novel graph packing technique for embedding multigraphs.

Theorem 4 ([110]) *Let $r \geq 5$ and H be an r-color-critical graph with h edges. If n is sufficiently large and $k \geq \frac{2(r-1)(h-1)}{r}$, then $\mathrm{ex}_k(n, H) = k \cdot t_{r-1}(n)$, and the unique n-vertex k-color extremal graph system of H consists of k colors all of which are identical copies of $T_{r-1}(n)$.*

There is also some research focusing on maximizing the product $\Pi_i |E(G_i)|$ of the number of edges of a graph system without a rainbow copy of H. Frankl [57] conjectured that a graph system $\{G_1, G_2, G_3\}$ on a common vertex set of order n with $\Pi_{i\in[3]}|E(G_i)| > \lfloor n^2/4 \rfloor^3$ contains a transversal triangle. This conjecture was disproved by Frankl et al. [74], and they further provided the following conjecture.

Conjecture 3 ([74]) *Let $\mathcal{G} = \{G_1, G_2, G_3\}$ be a graph system of order n. If $\mathcal{G}$ contains no transversal triangle, then*

$$\Pi_{i\in[3]}|E(G_i)| \leq \gamma n^6(1 + o(1)),$$

where γ is the maximum of $\frac{x^2}{2}\left(\frac{x^2}{2} + \frac{(1-x)^2}{2}\right)\left(x(1-x) + \frac{(1-x)^2}{2}\right)$ on $[0, 1]$.

Recall that Falgas-Ravry et al. [51] solved Problem 2 for sufficiently large n. As a direct consequence, Conjecture 3 holds when n is sufficiently large.

3 Dirac-Type Problem in Graph Systems

Let $G = (V, E)$ be a graph. For a vertex $v \in V$, denote by $N_G(v)$ the neighborhood of v and let $d_G(v) = |N_G(v)|$ be the degree of v. The subscripts are omitted when G is clear from the context. Let $\delta(G)$ and $\Delta(G)$ be the minimum degree and maximum degree of G respectively. Let $\mathcal{G} = \{G_1, G_2, \ldots, G_m\}$ be a graph system. Denote by $\delta(\mathcal{G}) = \min\{\delta(G_i) : i \in [m]\}$ the minimum degree of $\mathcal{G}$.

In this section, we investigate the minimum degree condition that ensures the existence of transversal spanning structures in graph systems. When one is interested in embedding a spanning graph, it makes sense to consider minimum degree conditions rather than size conditions, as, for example, a graph could be very dense but contain an isolated vertex, and therefore not contain a copy of any given spanning graph with minimum degree at least 1.

Given the difficulty of obtaining exact results in (rainbow) graph embedding problems, many results have focused on determining the minimum d such that for all $\epsilon > 0$ and sufficiently large integer n, each graph (system) of order n with minimum degree at least $(d + \epsilon)n$ contains a (rainbow) copy of H. Note that when we consider this problem in graph systems, d is at least the value required to guarantee a copy of H in a single graph, since all graphs in the system could be identical.

Definition 2 Let $\mathbf{F}$ be an infinite family of graphs. Denote by $\delta_{\mathbf{F}}$ (if it exists) the smallest real number δ such that for all $\alpha > 0$ and for all but finitely many $F \in \mathbf{F}$ the following holds. Let $n = |V(F)|$ and H be any n-vertex graph with $\delta(H) \geq (\delta + \alpha)n$. Then H contains a copy of F.

Denote by $\delta_{\mathbf{F}}^{rb}$ (if it exists) the smallest real number δ such that for all $\alpha > 0$ and for all but finitely many $F \in \mathbf{F}$ the following holds. Let $n = |V(F)|$ and $m = |E(F)|$, and let $\mathcal{H} = \{H_1, \ldots, H_m\}$ be any graph system of order n with $\delta(\mathcal{H}) \geq (\delta + \alpha)n$. Then $\mathcal{H}$ contains a transversal copy of F.

If the above two values exist, then it is clear that $\delta_{\mathbf{F}}^{\mathrm{rb}} \geq \delta_{\mathbf{F}}$. Moreover, Montgomery et al. [122] observed that $\delta_{\mathbf{F}}^{rb}$ can be much larger than $\delta_{\mathbf{F}}$. Set $\mathbf{F} = \{k \times (K_{2,3} \cup C_4) : k \in \mathbb{N}\}$, where $k \times G$ denotes the graph obtained by taking k vertex-disjoint copies of G. Kühn and Osthus [101] showed that $\delta_{\mathbf{F}} = 4/9$. Consider the graph system $\mathcal{H} = \{H_1, \ldots, H_m\}$ on V obtained in the following way. Partition V into two almost equal vertex subsets, say A and B, and for each $i \in [m-1]$, assume that H_i is the disjoint union of a clique on A and a clique on B. Suppose that H_m is a complete bipartite graph between A and B. Observe that if $\mathcal{H}$ contains a transversal copy of some $F \in \mathbf{F}$, the edge of $K_{2,3}$ or C_4 that belongs to H_m would be a bridge of F. However, neither $K_{2,3}$ nor C_4 contains a bridge. Hence, $\delta_{\mathbf{F}}^{rb} \geq \delta(\mathcal{H}) \geq 1/2$.

On the other hand, there are many natural instances with $\delta_{\mathbf{F}}^{\mathrm{rb}} = \delta_{\mathbf{F}}$, for example, the family of n-cycles (see [80]). When this equality holds, we say that the corresponding $\mathbf{F}$ is *color-blind*.

3.1 Hamilton Path and Hamilton Cycle

A *Hamilton path* (*cycle*) in a graph is a path (cycle) that visits every vertex exactly once. Hamiltonicity is one of the most well-studied properties in graph theory. Dirac [40] showed that every graph on $n \geq 3$ vertices with minimum degree at least $n/2$ contains a Hamilton cycle. An n-vertex graph is a *Dirac graph* if its minimum degree is at least $n/2$. Later on, Ore [124] relaxed this condition and proved that every n-vertex graph G with $\sigma_2(G) \geq n$ contains a Hamilton cycle, where $n \geq 3$ and $\sigma_2(G) = \min\{d(x) + d(y) : xy \notin E(G)\}$. Moon and Moser [123] established the bipartite analogue of Dirac's theorem.

In 2020, Aharoni et al. [3] conjectured that Dirac's theorem can be extended to a transversal version. Cheng et al. [34] proved an asymptotic version of this conjecture; Joos and Kim [80] confirmed the conjecture completely.

Theorem 5 ([80]) *Let $n \in \mathbb{N}$ and $n \geq 3$. Suppose that $\mathcal{G} = \{G_1, \ldots, G_n\}$ is a graph system of order n and $\delta(\mathcal{G}) \geq \frac{n}{2}$. Then $\mathcal{G}$ contains a transversal Hamilton cycle.*

Bradshaw [20] gave a degree condition to guarantee the existence of transversal Hamilton cycles in bipartite graph systems, which generalizes the result of Moon and Moser [123].

The *distance* of two distinct vertices in a graph is the length of a shortest path connecting them. The *k-th power* G^k of a graph G is the graph obtained from G by joining every pair of vertices with distance at most k in G. Pósa (for $r = 3$, [44]) and Seymour (for $r > 3$, [132]) conjectured that for all $r \geq 3$ and all $n \geq r$, every n-vertex graph G with $\delta(G) \geq (r-1)n/r$ must contain the $(r-1)$-th power of a Hamilton cycle. This conjecture has been confirmed by Komlós et al. [96, 98] for sufficiently large n. Recently, Gupta et al. [65] determined the asymptotically tight bound on $\delta(\mathcal{G})$ for the existence of transversal powers of Hamilton cycles, and they proposed a problem for the exact version of transversal Pósa–Seymour conjecture.

Problem 4 ([65]) Let $\mathcal{G} = \{G_1, \ldots, G_{(r-1)n}\}$ be a graph system of order n. If $\delta(\mathcal{G}) \geq (1 - \frac{1}{r})n$, is this already sufficient for a transversal copy of the $(r-1)$-th power of a Hamilton cycle?

Bowtell et al. [19] proposed a further generalization of the transversal problem, where one seeks a copy of H with a given coloring.

Problem 5 ([19]) Given a graph system $\mathcal{G} = \{G_1, \ldots, G_m\}$ and a graph H, what conditions imposed on $\mathcal{G}$ will guarantee that for every coloring $\chi : E(H) \to \{1, \ldots, m\}$, there is a copy of H such that the image of e is an edge of $G_{\chi(e)}$ for all $e \in E(H)$?

When H is a Hamilton cycle, Bowtell et al. [19] gave an asymptotically optimal minimum degree condition for Problem 5.

Theorem 6 ([19]) *For every $\alpha > 0$ there exists n_0 such that for every $n \geq n_0$ the following holds. Let $m \in \mathbb{N}$ and let $\mathcal{G} = \{G_1, \ldots, G_m\}$ be a graph system of order n. If $\delta(\mathcal{G}) \geq (\frac{1}{2} + \alpha)n$, then for every $\chi : [n] \to [m]$, there is a Hamilton cycle C such that the image of e is an edge of $G_{\chi(e)}$ for all $e \in E(C)$.*

Determining the exact minimum degree condition to guarantee all Hamilton cycle patterns remains an intriguing open problem. Bowtell et al. [19] provided a construction demonstrating that $n/2 + 1$ is a necessary lower bound and conjectured this is an exact threshold. In light of this, it seems more pertinent to either obtain results that improve this $o(n)$ term to an explicit and relatively small constant, or find counterexamples to this being possible.

It is known that one can measure the robustness of Dirac graphs with respect to Hamiltonicity by counting the number of distinct Hamilton cycles that a Dirac graph must contain. For more details about the robustness, we refer the reader to see the nice survey of Sudakov [134]. This method can be extended to measure the robustness of Dirac graph systems.

Bradshaw et al. [21] showed that given a system of n Dirac graphs with order n, there are at least $(cn/e)^{cn}$ different transversal Hamilton cycles for some constant $c \geq 1/68$. Theorem 6 recovers a stronger lower bound, that there exist at least $n! \sim \sqrt{2\pi n}(n/e)^n$ different transversal Hamilton cycles in a graph system $\mathcal{G}$ consisting of n asymptotically Dirac graphs (i.e., $\delta(\mathcal{G}) \geq (1/2+\alpha)n$ for $\alpha > 0$). Anastos and Chakraborti [11] improved this by showing that there exists a constant $C > 0$ such that there are $(Cn)^{2n}$ different transversal Hamilton cycles in a graph system consisting of n Dirac graphs, which is tight up to the choice of C. They [11] also showed that every graph system consisting of n Dirac graphs with order n spans $n/2$ pairwise edge-disjoint transversal Hamilton cycles, and this is best possible.

Cheng and Staden [32] established a stability result for transversal Hamilton cycles. Given integers $a, b \geq 0$, let $\mathcal{H}_a^b$ be a graph system obtained by taking a copies of $K_{\frac{n}{2}} \cup K_{\frac{n}{2}}$ and b copies of $K_{\frac{n}{2},\frac{n}{2}}$ where they are defined on the same equitable partition $A \cup B$. For two disjoint vertex subsets $X, Y \subseteq V$, let $G[X]$ denote the subgraph of G induced by X, let $G[X, Y]$ be the subgraph of G with vertex set $X \cup Y$ and edge set $\{xy : x \in X,\ y \in Y \text{ and } xy \in E(G)\}$. We say that a graph system $\mathcal{J}$ on the vertex set V of order n is *half-split* if there is $A \subseteq V$ with $|A| = \lfloor n/2 \rfloor + 1$ such that $J[A] = \emptyset$ and $J[A, V \setminus A]$ is complete for all $J \in \mathcal{J}$. Given $\kappa > 0$, we say that two graph systems $\mathcal{G}$ and $\mathcal{G}'$ on a common vertex set of order n are *κ-close* if $\mathcal{G}'$ can be obtained from $\mathcal{G}$ by adding or deleting at most κn^3 edges.

Theorem 7 ([32]) *For all $\kappa > 0$, there exist $\alpha > 0$ and n_0 such that the following holds for all integers $n \geq n_0$. Let $\mathcal{G} = \{G_1, \ldots, G_n\}$ be a graph system of order n and $\delta(\mathcal{G}) \geq \frac{1-\alpha}{2}n$. If $\mathcal{G}$ contains no transversal Hamilton cycles, then either $\mathcal{G}$ is κ-close to $\mathcal{H}_{n-t}^t$ for some odd integer $t \in [n]$, or $\mathcal{G}$ is κ-close to a half-split graph system.*

The proof of Theorem 7 was a combination of their newly developed transversal regularity-blow-up method [31] (see Theorem 15 in the next section) and the absorption method. As a continuation of the transversal version of Dirac's theorem (Theorem 5) and the stability result for transversal Hamilton cycles (Theorem 7), Cheng et al. [33] characterized all graph systems with order n and minimum degree at least $n/2 - 1$, each of which does not contain transversal Hamilton cycles.

Li et al. [108] studied the existence of rainbow Hamilton paths in a graph system $\mathcal{G} = \{G_i : i \in [n]\}$ under Ore-type conditions. However, the edges of a rainbow Hamilton path only come from $n-1$ graphs of $\mathcal{G}$. Thus, it is worth studying whether $n-1$ graphs can guarantee the existence of a rainbow Hamilton path in a graph system $\mathcal{G} = \{G_i : i \in [n-1]\}$. It is well-known that Dirac's theorem implies that every n-vertex graph with minimum degree at least $(n-1)/2$ contains a Hamilton path. Cheng et al. [33] proved a transversal version of this result.

Theorem 8 ([33]) *Let $n \in \mathbb{N}$ and $n \geq 3$. Suppose $\mathcal{G} = \{G_1, \ldots, G_{n-1}\}$ is a graph system of order n. If $\delta(\mathcal{G}) \geq \frac{n-1}{2}$, then $\mathcal{G}$ contains a transversal Hamilton path.*

Similar to the case of Hamilton cycles, Cheng and Staden [32] established a stability result for transversal Hamilton paths; Cheng et al. [33] characterized all graph systems with order n and minimum degree at least $(n-3)/2$, each of which does not contain transversal Hamilton paths.

In 1971, Bondy [16] posed a meta-conjecture which states that almost all nontrivial sufficient conditions on a graph which ensure that the graph is Hamiltonian also imply that the graph is pancyclic, or there might be a simple family of exceptional graphs. Some concepts similar to pancyclic have also attracted extensive attention, such as, vertex-pancyclic and panconnected. Here, we only illustrate those concepts in graph systems. Let $\mathcal{G} = \{G_1, \ldots, G_n\}$ be a system of graphs on a common vertex set V of order n. We say

- $\mathcal{G}$ is *transversal pancyclic* if there is a rainbow cycle of length ℓ in $\mathcal{G}$ for each integer $\ell \in [3, n]$;
- $\mathcal{G}$ is *transversal vertex-pancyclic* if each vertex of V is contained in a rainbow cycle of length ℓ for all $\ell \in [3, n]$;
- $\mathcal{G}$ is *rainbow panconnected* if for any two vertices $u, v \in V$, $\mathcal{G}$ contains a rainbow path of length ℓ joining u and v for all $\ell \in [d_{\mathcal{G}}(u, v) + 1, n]$, where $d_{\mathcal{G}}(x, y)$ is the length of the shortest rainbow path (if it exists) connecting u and v in $\mathcal{G}$.

Li et al. [108, 109] obtained Dirac-type conditions for transversal pancyclicity, transversal vertex-pancyclicity and rainbow panconnectedness respectively. The analogue results in bipartite graph systems were obtained by Bradshaw [20] and Hu et al. [75].

In addition to the Dirac-type condition, there are many other sufficient conditions to guarantee the Hamiltonicity of a graph, such as Ore's condition [124] and Pósa's condition [127]. Bradshaw [20] and Li et al. [108] proposed that it is also interesting to consider those conditions in graph systems. We list one of them here.

Problem 6 ([20, 108]) Let $\mathcal{G} = \{G_1, \ldots, G_n\}$ be a graph system of order n. If $\sigma_2(G_i) \geq n$ for all $i \in [n]$, does $\mathcal{G}$ contain a transversal Hamilton cycle?

3.2 *F-factors*

Given two graphs F and G, an *F-tiling* is a set of vertex-disjoint copies of F in G. A *perfect F-tiling* (or an *F-factor*) in G is an F-tiling covering all vertices of G. Characterizing sufficient conditions for the existence of an F-factor is one of the central areas of extremal graph theory. The most celebrated one is the Hajnal–Szemerédi theorem.

Theorem 9 (Hajnal–Szemerédi [66], Corrádi–Hajnal [35] for $k = 3$) *Every n-vertex graph G with $n \in k\mathbb{N}$ and $\delta(G) \geq (1 - \frac{1}{k})n$ has a K_k-factor. Moreover, the minimum degree condition is sharp.*

For any r-vertex graph F, Kühn and Osthus [101] determined the smallest value of δ such that there is an $n_0 \in \mathbb{N}$ and a constant $c = c(F)$ such that any graph with

$n \geq n_0$ vertices and minimum degree at least $\delta n + c$ contains an F-factor if $r|n$. For more results, please see the survey of Kühn and Osthus [100].

Now, we survey some results on minimum degree conditions for the existence of a transversal F-factor in graph systems. Joos and Kim [80] established an exact minimum degree condition to guarantee the existence of a transversal perfect matching (i.e., K_2-factor) in graph systems.

Theorem 10 ([80]) *Let $n \in \mathbb{N}$ and $n \geq 2$ even. Suppose $\mathcal{G} = \{G_1, \ldots, G_{\frac{n}{2}}\}$ is a graph system of order n and $\delta(\mathcal{G}) \geq \frac{n}{2}$. Then $\mathcal{G}$ contains a transversal perfect matching.*

Bradshaw [20] proved the analogous result in bipartite graph systems, which is equivalent to a special case of a matching theorem for r-partite r-uniform hypergraphs obtained by Aharoni et al. [4], for the case $r = 3$.

Cheng et al. [28] gave an asymptotic version of the transversal Hajnal–Szemerédi theorem. Furthermore, they proposed a conjecture for the exact version of transversal Hajnal–Szemerédi theorem.

Conjecture 4 ([28]) *Let $\mathcal{G} = \{G_1, \ldots, G_{\frac{n}{k}\binom{k}{2}}\}$ be a graph system of order n. If $\delta(\mathcal{G}) \geq (1 - \frac{1}{k})n$, then $\mathcal{G}$ contains a transversal K_k-factor.*

Montgomery et al. [122] showed that, in most cases, the same minimum degree bound is asymptotically sufficient for the existence of transversal F-factors, which generalized the Kühn–Osthus theorem [101] on factors. For each graph F, let δ_F^f be the smallest real number $\delta \geq 0$ such that, for each $\epsilon > 0$ there is some n_0 such that, for every $n \geq n_0$ with $|V(F)|$ dividing n, if an n-vertex graph H has minimum degree at least $(\delta + \epsilon)n$, then H contains an F-factor.

Theorem 11 ([122]) *Let $\varepsilon > 0$ and let F be a graph on r vertices with t edges. If $\delta_F^f \geq \frac{1}{2}$ or F has a bridge, then let $\delta_F^T = \delta_F^f$, and otherwise let $\delta_F^T = \frac{1}{2}$. Then, there is some n_0 such that the following holds for all $n \geq n_0$.*

Assume that $\mathcal{G} = \{G_1, \ldots, G_{tn}\}$ is a graph system of order rn and $\delta(\mathcal{G}) \geq (\delta_F^T + \varepsilon)rn$. Then, $\mathcal{G}$ contains a transversal F-factor.

Suppose $V_1, V_2, \ldots, V_k$ are disjoint vertex sets each of order n, and G is a k-partite graph on vertex classes $V_1, V_2, \ldots, V_k$. Denote the *partite minimum degree* of G by

$$\delta'(G) := \min_{i \in [k]} \min_{v \in V_i} \min_{j \in [k] \setminus \{i\}} |N(v) \cap V_j|.$$

Fischer [54] conjectured that the k-partite graph G contains a K_k-factor if $\delta'(G) \geq (1 - 1/k)n + 1$ (actually his original conjecture did not include the term $+1$, but this stronger conjecture is known to be false). An approximate version of this conjecture was demonstrated independently by Keevash and Mycroft [87], and by Lo and Markström [111]. Later, Keevash and Mycroft [88] showed that the conjecture holds for any sufficiently large graph (in fact, they obtained a more general result for r-partite graphs with $r \geq k$). Cheng et al. [28] extended the approximate result of Fischer's conjecture to the transversal version.

Theorem 12 ([28]) *For every $\epsilon > 0$ and integer k, there exists $n_0 \in \mathbb{N}$ such that the following holds for all integers $n \geq n_0$. If $\mathcal{G} = \{G_1, \ldots, G_{n\binom{k}{2}}\}$ is a system of k-partite graphs with a common vertex partition $V_1, \ldots, V_k$ each of size n such that $\delta'(G_i) \geq (1 - \frac{1}{k} + \varepsilon)n$, then $\mathcal{G}$ contains a transversal K_k-factor.*

3.3 Graphs with Bounded Bandwidth

Montgomery et al. [122] determined an asymptotically tight bound on $\delta(\mathcal{G})$ for the existence of a transversal spanning tree with maximum degree at most $cn/\log n$, generalizing the tree-embedding theorem of Komlós et al. [99].

It is noteworthy that many graphs, including F-factors, powers of Hamilton cycles, and trees, exhibit a somewhat bounded maximum degree and low connectivity. This property of low connectivity can be captured by the following notion of bandwidth. A graph H has a *bandwidth* at most b if there exists an ordering $x_1, \ldots, x_n$ of $V(H)$ such that all edges $x_i x_j \in E(H)$ satisfy $|i - j| \leq b$. Böttcher et al. [18] proved the celebrated bandwidth theorem, which determines the asymptotically optimal minimum degree condition for a graph to contain a spanning subgraph H with bounded maximum degree and low bandwidth.

Since the bandwidth theorem provides a common generalization of many classical results, it is natural to give such a generalization for the above transversal version results. Chakraborti et al. [24] proved that such a generalization holds by showing the following bandwidth theorem for graph transversals.

Theorem 13 ([24]) *For every $\epsilon > 0$ and positive integers Δ, k, there exist $\alpha > 0$ and $h_0 > 0$ satisfying the following for every $h \geq h_0$. Let H be an n-vertex graph with h edges and bandwidth at most αn such that $\Delta(H) \leq \Delta$ and $\chi(H) \leq k$. If $\mathcal{G} = \{G_1, \ldots, G_h\}$ is a graph system of order n and $\delta(\mathcal{G}) \geq \left(1 - \frac{1}{k} + \varepsilon\right) n$, then $\mathcal{G}$ contains a transversal copy of H.*

This covers several of the above results, and is asymptotically best possible for many graphs H. Note that by taking each G_i in Theorem 13 to be the same graph, the bandwidth theorem of Böttcher et al. [18] can be deduced immediately. Therefore, Theorem 13 generalizes and strengthens the original bandwidth theorem.

4 Transversal via Regularity

The classical "regularity-blow-up method" has been used to prove many results concerning the embedding of a given spanning subgraph H in a large graph G. Such proofs are typically developed using two important tools: Szemerédi's regularity lemma [136] and blow-up lemma of Komlós et al. [97]. Cheng and Staden [31] provided a tool needed to obtain transversal embeddings by applying the regularity-blow-up method. More precisely, they proved a transversal blow-up lemma that embeds separable graphs into graph systems.

Definition 3 (*Regularity and superregularity*) Suppose that $\mathcal{G} = \{G_c : c \in \mathscr{C}\}$ is a system of bipartite graphs on a common vertex partition $V_1 \cup V_2$ with color set $\mathscr{C}$. We say that

- $\mathcal{G}$ is (ε, d)-*regular* if whenever $V_i' \subseteq V_i$ with $|V_i'| \geq \varepsilon|V_i|$ for $i = 1, 2$ and $\mathscr{C}' \subseteq \mathscr{C}$ with $|\mathscr{C}'| \geq \varepsilon|\mathscr{C}|$, we have

$$\left| \frac{\sum_{c\in\mathscr{C}'} |E(G_c[V_1', V_2'])|}{|\mathscr{C}'||V_1'||V_2'|} - \frac{\sum_{c\in\mathscr{C}} |E(G_c)|}{|\mathscr{C}||V_1||V_2|} \right| < \varepsilon$$

  and $\sum_{c\in\mathscr{C}} |E(G_c)| \geq d|\mathscr{C}||V_1||V_2|$.
- $\mathcal{G}$ is (ε, d)-*superregular* if it is (ε, d)-regular and $\sum_{c\in\mathscr{C}} d_{G_c}(x) \geq d|\mathscr{C}||V_{3-i}|$ for all $x \in V_i$ with $i \in [2]$, and $|E(G_c)| \geq d|V_1||V_2|$ for all $c \in \mathscr{C}$.

Note that if every G_c with $c \in \mathscr{C}$ is the same, then $\mathcal{G}$ is (ε, d)-regular if and only if G_c is (ε, d)-regular; and $\mathcal{G}$ is (ε, d)-superregular if and only if G_c is (ε, d)-superregular.

Theorem 14 (Regularity lemma for graph systems, [31]) *For all integers $L_0 \geq 1$ and every $\varepsilon, \delta > 0$, there is an $n_0 = n_0(\varepsilon, \delta, L_0)$ such that for every $d \in [0, 1)$ and every graph system $\mathcal{G} = \{G_c : c \in \mathscr{C}\}$ on vertex set V of size $n \geq n_0$ with $\delta n \leq |\mathscr{C}| \leq \frac{n}{\delta}$, there exists a partition of V into $V_0, V_1, \ldots, V_L$, of $\mathscr{C}$ into $\mathscr{C}_0, \mathscr{C}_1, \ldots, \mathscr{C}_M$ and a spanning subgraph G_c' of G_c for each $c \in \mathscr{C}$ such that the following properties hold:*

(i) *$L_0 \leq L, M \leq n_0$ and $|V_0| + |\mathscr{C}_0| \leq \varepsilon n$;*
(ii) *$|V_1| = \cdots = |V_L| = |\mathscr{C}_1| = \cdots = |\mathscr{C}_M|$;*
(iii) *$\sum_{c\in\mathscr{C}} d_{G_c'}(v) > \sum_{c\in\mathscr{C}} d_{G_c}(v) - (\frac{3d}{\delta^2} + \varepsilon)n^2$ for all $v \in V$ and $|E(G_c')| > |E(G_c)| - (\frac{3d}{\delta^2} + \varepsilon)n^2$ for all $c \in \mathscr{C}$;*
(iv) *if, for $c \in \mathscr{C}$, the graph G_c' has an edge with both vertices in a single vertex cluster V_i for some $i \in [L]$, then $c \in \mathscr{C}_0$;*
(v) *for all triples $(\{h, i\}, j) \in \binom{[L]}{2} \times [M]$, we have that either $G_c'[V_h, V_i] = \emptyset$ for all $c \in \mathscr{C}_j$, or $\mathcal{G}_{hi,j}' := \{G_c'[V_h, V_i] : c \in \mathscr{C}_j\}$ is (ε, d)-regular.*

The sets V_i are called vertex clusters and the sets $\mathscr{C}_j$ are called color clusters, while V_0 and $\mathscr{C}_0$ are the exceptional vertices and colors sets respectively.

Definition 4 (*Reduced graph system*) Given a graph system $\mathcal{G} = \{G_c : c \in \mathscr{C}\}$ on V and parameters $\varepsilon > 0$, $d \in [0, 1)$ and $L_0 \geq 1$, the *reduced graph system* $\mathcal{R} = \mathcal{R}(\varepsilon, d, L_0)$ is defined as follows. Apply Theorem 14 to $\mathcal{G}$ with parameters ε, d, L_0 to obtain a graph system $\mathcal{G}' = \{G_c' : c \in \mathscr{C}\}$, a partition $V_0, \ldots, V_L$ of V and $\mathscr{C}_0, \ldots, \mathscr{C}_M$ of $\mathscr{C}$, where $V_0, \mathscr{C}_0$ are the exceptional sets, $V_1, \ldots, V_L$ are the vertex clusters and $\mathscr{C}_1, \ldots, \mathscr{C}_M$ are the color clusters. Then $\mathcal{R} = \{R_1, \ldots, R_M\}$ is a graph system consisting of M graphs each on the same vertex set $[L]$, where $hi \in R_j$ if and only if $\mathcal{G}_{hi,j}'$ is (ε, d)-regular for each $(\{h, i\}, j) \in \binom{[L]}{2} \times [M]$.

An n-vertex graph H is *μ-separable* if there is $X \subseteq V(H)$ of size at most μn such that $H - X$ consists of components of size at most μn. With suitable small μ, separable graphs include many natural graphs, for example, F-factors for a fixed F, 2-regular graphs, trees, powers of Hamilton cycles and graphs of small bandwidth.

Theorem 15 (Transversal blow-up lemma, [31]) *For all $\nu, d, \Delta, r > 0$, where $r \geq 2$ is an integer, there exist $\epsilon, \mu, \alpha > 0$ and $m_0 \in \mathbb{N}$ such that the following holds for all integers $m \geq m_0$. Assume $\mathcal{G} = \{G_i : i \in \mathscr{C}\}$ is a graph system with the following properties.*

- *There is a graph R with vertex set $[r]$ and a partition $\mathscr{C} = \cup_{e \in E(R)} \mathscr{C}_e$ where $|\mathscr{C}_e| \geq \delta m$ for all $e \in E(R)$;*
- *for all $ij \in E(R)$ and $c \in \mathscr{C}_{ij}$, G_c is bipartite with parts V_i, V_j, and V is a vertex set of size n with partition $V = V_1 \cup V_2 \cup \ldots \cup V_r$, where $m \leq |V_i| \leq \frac{m}{\delta}$ for each $i \in [r]$;*
- *for all $ij \in E(R)$,*
  - *for all $V_h' \subseteq V_h$ $(h = i, j)$ and $\mathscr{C}_{ij}' \subseteq \mathscr{C}_{ij}$ with $|V_h'| \geq \epsilon |V_h|$ $(h = i, j)$ and $|\mathscr{C}_{ij}'| \geq \epsilon |\mathscr{C}_{ij}|$, we have that*

$$\sum_{c \in \mathscr{C}_{ij}'} |E(G_c[V_i', V_j'])| \geq d |\mathscr{C}_{ij}'| |V_i'| |V_j'|;$$

  - *for every $v \in V_h$ with $h = i, j$ we have $\sum_{c \in \mathscr{C}_{ij}'} d_{G_c}(v) \geq d|\mathscr{C}_{ij}'| m$, and for every $c \in \mathscr{C}_{ij}$, we have $|E(G_c)| \geq dm^2$.*

Suppose that H is a graph with the following properties.

- *$\Delta(H) \leq \Delta$;*
- *H is μ-separable;*
- *H has vertex partition $A_1 \cup A_2 \cup \ldots \cup A_r$ such that $|A_i| = |V_i|$ for all $i \in [r]$, and for every $xy \in E(H)$ there is $ij \in E(R)$ such that $x \in A_i$ and $y \in A_j$;*
- *$|E(H[A_i, A_j])| = |\mathscr{C}_{ij}|$ for all $ij \in E(R)$;*
- *for $i \in [r]$, there is a set $U_i \subseteq A_i$ with $|U_i| \leq \alpha |A_i|$ and for each $x \in U_i$, there exists a target set $T_x \subseteq V_i$ with $|T_x| \geq \nu m$.*

Then $\mathcal{G}$ contains a transversal copy of H such that every $x \in U_i$ is embedded inside T_x for $i \in [r]$.

By utilizing the transversal blow-up lemma, Cheng and Staden [31] proved a stability result for transversal Hamilton cycles, and characterized a large class of spanning 3-uniform linear hypergraphs H such that any sufficiently large uniformly dense n-vertex 3-uniform hypergraph with minimum vertex degree $\Omega(n^2)$ contains H as a subhypergraph.

5 Random Graph Systems

In this section, we survey some results on the existence of transversal structures in a system of random graphs in various settings. Let $0 \leq p \leq 1$ and $G(n, p)$ be the binomial *random graph*. For graphs $G = (V, E)$ and $G' = (V', E')$, we set $G \cap G' = (V \cap V', E \cap E')$. Given an n-vertex graph G, we call $G(p) := G \cap G(n, p)$ the *random subgraph* of G. In particular, $G(p) = G(n, p)$ when G is the complete graph K_n. We say that $G(p)$ has a graph property $\mathcal{P}$ with high probability, or whp for brevity, if the probability that $G(p)$ has $\mathcal{P}$ approaches 1 as n tends to infinity. Sudakov [134] pointed out that $G(p)$ can be used to measure the robustness of graph properties.

A natural question concerns the existence of transversal spanning structures in random graph systems. A graph property is called *monotone increasing* if it is closed under the addition of edges. For a monotone increasing graph property, the *resilience* quantifies the robustness in terms of the number of edges one must delete from a graph, locally or globally, in order to destroy the property. Ferber et al. [52] proved a resilient result for a system of n graphs with order n where each graph is an independent copy of $G(n, p)$ provided that $np = \omega(\log n)$, and they also noted that the bound for p is probably not tight.

Theorem 16 ([52]) *For every $\epsilon > 0$, there exists an absolute constant $c = c(\epsilon)$ such that for any sufficiently large n and $p \geq \frac{c \log n}{n}$ the following holds. Let $\mathcal{G} = \{G_1, \ldots, G_n\}$ be a system of n independent copies of $G(n, p)$ on the same vertex set V of size n. For all $i \in [n]$, let $H_i \subseteq G_i$ be such that $d_{H_i}(v) \geq (\frac{1}{2} + \epsilon)d_{G_i}(v)$ for all $v \in V$. Then, whp $\{H_1, \ldots, H_n\}$ contains a transversal Hamilton cycle.*

A similar result for transversal perfect matchings was also obtained in [52]. Note that the above theorem implies that if $\mathcal{G} = \{G_1, \ldots, G_n\}$ is a graph system of order n with $\delta(\mathcal{G}) \geq (1/2 + \epsilon)n$, and $\mathcal{F} = \{F_1, \ldots, F_n\}$ where $F_i = G_i(p)$ for some $p \geq c \log n/n$, then whp $\mathcal{F}$ contains a transversal Hamilton cycle. Anastos and Chakraborti [11] improved this by removing the error term in the minimum degree condition with spread techniques.

Theorem 17 ([11]) *There exists an absolute c such that the following holds. Suppose that $\mathcal{G} = \{G_1, \ldots, G_n\}$ is a graph system of order n with $\delta(\mathcal{G}) \geq \frac{n}{2}$, and $p \geq \frac{c \log n}{n}$. Let $\mathcal{F} = \{F_1, \ldots, F_n\}$ where $F_i = G_i(p)$. Then, whp $\mathcal{F}$ contains a transversal Hamilton cycle.*

In the above theorem, each graph in the given system intersects with the same random graph $G(n, p)$. They [11] also considered a system that consists of n independent random subgraphs of a Dirac graph.

Theorem 18 ([11]) *There exists an absolute constant c such that the following holds. Suppose G is a graph on n vertices with $\delta(G) \geq \frac{n}{2}$, and $p \geq \frac{c \log n}{n^2}$. Let $\mathcal{G} = \{G_1, \ldots, G_n\}$ be a family of graphs where each G_i is independently distributed as $G(p)$, for $i \in [n]$. Then, whp $\mathcal{G}$ contains a transversal Hamilton cycle.*

In Theorems 17 and 18, the lower bound on p is optimal up to a constant factor c. Naturally, they [11] hoped for a combined generalization of the above two theorems.

Problem 7 ([11]) Does there exist a constant c such that the following holds? Let $\mathcal{G} = \{G_1, \dots, G_n\}$ be a graph system of order n with $\delta(\mathcal{G}) \geq \frac{n}{2}$, and $p \geq \frac{c \log n}{n^2}$. For each $i \in [n]$, let $F_i = G_i(p)$ and let $\mathcal{F} = \{F_1, \dots, F_n\}$. Then, whp $\mathcal{F}$ contains a transversal Hamilton cycle.

In the same paper, they [11] showed that Problem 7 has a positive answer when n is odd, whereas it turns out to be false when n is even.

Han et al. [68] obtained a transversal robust version of the Hajnal–Szemerédi theorem [66], where the bound for p is not tight.

Theorem 19 ([68]) *Let $r \geq 3$ be an integer and let $\gamma > 0$. There exists a constant $c = c(r, \gamma)$ such that for any sufficiently large $n \in r\mathbb{N}$ and $p \geq cn^{-\frac{2}{r}}(\log n)^{1/\binom{r}{2}}$ the following holds. Suppose that $\mathcal{G} = \{G_1, \dots, G_{\frac{n}{r}\binom{r}{2}}\}$ is a graph system of order n with $\delta(\mathcal{G}) \geq (1 - \frac{1}{r} + \gamma)n$. Let $\mathcal{F} = \{F_1, \dots, F_{\frac{n}{r}\binom{r}{2}}\}$ where $F_i = G_i(p)$. Then whp $\mathcal{F}$ contains a transversal K_r-factor.*

The transversal robust version of the multipartite Hajnal–Szemerédi theorem [66] was also obtained in [68].

6 Digraph Systems

In this section, we consider transversal problems in digraph systems. Let D be a *digraph* with no loops and at most one edge in each direction between every pair of vertices. An *oriented graph* is a digraph with no cycle of length two. Let $\delta^+(D)$ and $\delta^-(D)$ be the *minimum out-degree* and *minimum in-degree*, respectively. Let $\delta^0(D) := \min\{\delta^+(D), \delta^-(D)\}$ be the *minimum semi-degree* of D. Let $\mathcal{D} = \{D_1, D_2, \dots, D_m\}$ be a system of not necessarily distinct digraphs with common vertex set V. Define $\delta^+(\mathcal{D}) := \min\{\delta^+(D_i) : i \in [m]\}$ and $\delta^0(\mathcal{D}) := \min\{\delta^0(D_i) : i \in [m]\}$.

A *tournament* is an oriented graph where every pair of distinct vertices are adjacent. We call a tournament T *strongly connected* if for every pair of vertices x, y in T, there is a directed path from x to y. We say that a digraph D is *transitive* if xy and yz are arcs in D with $x \neq z$, then the arc xz is also in D.

Let T_k be the transitive tournament on k vertices and $\mathcal{T}_k$ be the family of all tournaments on k vertices. Czygrinow et al. [36] proved that every n-vertex digraph D with $\delta^+(D) \geq (1 - 1/k)n$ contains a T_k-factor for integers n, k with $k|n$. Treglown [139] proved that for $T \in \mathcal{T}_k$ with $k \geq 3$ and sufficiently large integer n with $k|n$, every digraph D on n vertices with $\delta^0(D) \geq (1 - 1/k)n$ contains a T-factor.

In light of Theorem 12, it is natural to seek an analogue of the transversal Hajnal–Szemerédi theorem [66] in digraph systems. Cheng et al. [28] established the minimum out-degree and semi-degree condition for the existence of transversal tournament factors in digraph systems.

Theorem 20 ([28]) *For every integer $k \geq 3$, $T \in \mathcal{T}_k$ and real $\varepsilon > 0$, there exists $n_0 \in \mathbb{N}$ such that the following holds for all integers $n \geq n_0$ and $n \in k\mathbb{N}$. If $\mathcal{D} = \{D_1, \dots, D_m\}$, $m = \frac{n}{k}\binom{k}{2}$, is a system of n-vertex digraphs on the same vertex set such that $\delta^+(\mathcal{D}) \geq (1 - \frac{1}{k} + \varepsilon)n$ (resp. $\delta^0(\mathcal{D}) \geq (1 - \frac{1}{k} + \varepsilon)n$), then $\mathcal{D}$ contains a transversal T_k-factor (resp. T-factor).*

It is well-known that every tournament contains a Hamilton path, and every strongly connected tournament contains a Hamilton cycle. Chakraborti et al. [26] proved the following result concerning the existence of transversal directed Hamilton paths and cycles in a system of tournaments.

Theorem 21 ([26]) *For every sufficiently large n,*

(i) *every system of $(n-1)$ tournaments with order n contains a transversal directed Hamilton path;*
(ii) *every system $\mathcal{T}$ of n tournaments with order n satisfies the following. If all tournaments in $\mathcal{T}$ possibly except one are strongly connected, then $\mathcal{T}$ contains a transversal directed Hamilton cycle.*

Thomason [138] proved that every sufficiently large tournament contains Hamilton paths and cycles with all possible orientations, except possibly the consistently oriented Hamilton cycle. Chakraborti et al. [27] established the transversal generalization of this result.

Theorem 22 ([27]) *For every sufficiently large n,*

(i) *every system of $(n-1)$ tournaments with order n contains a transversal of every orientation of the Hamilton path;*
(ii) *every system of n tournaments with order n contains a transversal of every orientation of the Hamilton cycle except possibly the directed one.*

Babiński et al. [13] studied the transversal version of Mantel's theorem in digraph systems and oriented graph systems. More precisely, they determined the minimum number of edges in each graph that ensures the existence of transversal transitive triangles or transversal directed triangles in digraph systems and oriented graph systems. Gerbner et al. [60] determined the minimum number of edges required in each digraph of a digraph system that guarantees the existence of any rainbow directed star.

Ghouila-Houri [61] showed that any n-vertex digraph with minimum semi-degree at least $n/2$ contains a directed Hamilton cycle. Cheng et al. [30] provided a transversal generalization of Ghouila-Houri's theorem, which solved a problem proposed by Chakraborti et al. [27].

Theorem 23 ([30]) *Let $\mathcal{D} = \{D_1, \dots, D_n\}$ be a system of digraphs on a common vertex set V of sufficiently large order n. If $\delta^0(\mathcal{D}) \geq \frac{n}{2}$, then $\mathcal{D}$ contains a transversal directed Hamilton cycle.*

DeBiasio and Molla [39] proved the anti-directed Hamilton cycle (an orientation in which consecutive edges alternate direction) is guaranteed to appear if $\delta^0(D) \geq n/2 + 1$. DeBiasio et al. [38] showed that for sufficiently large n, every n-vertex digraph D with $\delta^0(D) \geq n/2$ contains every orientation of a Hamilton cycle except, possibly, the anti-directed one. It would be interesting to generalize the above two results to the transversal setting.

Keevash et al. [83] proved the oriented version of Dirac's theorem, which states that for sufficiently large n, every n-vertex digraph with minimum semi-degree at least $(3n - 4)/8$ contains a directed Hamilton cycle. Considering a transversal version of this theorem would be interesting.

Problem 8 ([30]) Let $\mathcal{D} = \{D_1, \ldots, D_n\}$ be a system of oriented graphs with common vertex set V of size n. For sufficiently large n, if $\delta^0(\mathcal{D}) \geq \lceil \frac{3n-4}{8} \rceil$, does $\mathcal{D}$ contain a transversal directed Hamilton cycle?

7 Hypergraph Systems

In this section, we will survey results in k-uniform hypergraph systems. For a set V, we denote by $\binom{V}{k}$ the set of subsets of size k of V. A *k-uniform hypergraph* (or *k-graph*) $H = (V, E)$ consists of a vertex set V and an edge set E which is a family of k-element subsets of V, i.e., $E \subseteq \binom{V}{k}$. For any $S \subseteq V$, the degree of S in H, denoted by $\deg_H(S)$, is the number of edges containing S. For any integer $s \geq 0$, define the *minimum s-degree* $\delta_s(H)$ to be $\min\{\deg_H(S) : S \in \binom{V(H)}{s}\}$. Let $\mathcal{H} = \{H_1, H_2, \ldots, H_m\}$ be a system of not necessarily distinct k-graphs with common vertex set V. Define $\delta_s(\mathcal{H}) = \min\{\delta_s(H_i) : i \in [m]\}$.

The problem of characterizing k-graphs which contain a Hamilton cycle or a perfect matching is perhaps the most prominent one in hypergraph theory. In the remainder of this section, we focus on those two most well studied structures.

7.1 Hamilton Cycle

We say that a k-graph C is an *ℓ-cycle* if there exists a cyclic ordering of the vertices of C such that every edge of C consists of k consecutive vertices and every two consecutive edges (in the natural ordering of the edges) intersects in precisely ℓ vertices. We say that C is a *Hamilton ℓ-cycle* of a k-graph H if C is a spanning sub-k-graph of H. We call C a *tight Hamilton cycle* if $\ell = k - 1$, a *loose Hamilton cycle* if $\ell = 1$. A Hamilton cycle of a k-graph H is a cycle which is a spanning subhypergraph of H. Note that if a k-graph H on n vertices contains a Hamilton ℓ-cycle then $(k - \ell)|n$, since every edge of the cycle contains exactly $k - \ell$ vertices which were not contained in the previous edge.

In 1999, Katona and Kierstead [82] proved a Dirac-type theorem in hypergraphs: every n-vertex k-graph H with $\delta_{k-1}(H) > (1 - \frac{1}{2k})n + 4 - k - \frac{5}{2k}$ contains a tight Hamilton cycle. They anticipated that a much stronger result is true.

Conjecture 5 ([82]) *Let H be a k-graph on $n \geq k + 1 \geq 4$ vertices. If $\delta_{k-1}(H) \geq \lfloor \frac{n-k+3}{2} \rfloor$, then H has a tight Hamilton cycle.*

Rödl et al. [128, 129] confirmed this asymptotically, and they [130] also gave the exact version for $k = 3$.

Theorem 24 ([128–130]) *Let $k \geq 3$, $\gamma > 0$ and H be an n-vertex k-graph, where n is sufficiently large. If $\delta_{k-1}(H) \geq (\frac{1}{2} + \gamma)n$, then H contains a tight Hamilton cycle. Furthermore, when $k = 3$ it is enough to have $\delta_2(H) \geq \lfloor \frac{n}{2} \rfloor$.*

However, Kühn and Osthus [102] and Zhao [144] noted that it is much more difficult to determine the minimum d-degree condition for tight Hamilton cycle for $d \in [k-2]$. The best general bound for such a problem was given by Lang and Sanhueza-Matamala [106], and Polcyn et al. [126] independently.

Theorem 25 ([106, 126]) *Let $k \geq 3$, $\gamma > 0$ and H be an n-vertex k-graph, where n is sufficiently large. If $\delta_{k-2}(H) \geq (\frac{5}{9} + \gamma)\binom{n}{2}$, then H contains a tight Hamilton cycle.*

Let $1 \leq d, \ell \leq k-1$. For $n \in (k-\ell)\mathbb{N}$, define $h_d^{\ell}(k, n)$ to be the smallest integer h such that every n-vertex k-graph H satisfying $\delta_d(H) \geq h$ contains a Hamilton ℓ-cycle. Han and Zhao [69] proved that

$$h_d^{k-1}(k, n) \geq \left(1 - \binom{t}{\lfloor \frac{t}{2} \rfloor} \frac{\lceil \frac{t}{2} \rceil^{\lceil \frac{t}{2} \rceil} (\lfloor \frac{t}{2} \rfloor + 1)^{\lfloor \frac{t}{2} \rfloor}}{(t+1)^t + o(1)}\right) \binom{n}{t}, \tag{1}$$

where $d \in [k-1]$ and $t = k - d$. In particular, $h_{k-3}^{k-1}(k, n) \geq (5/8 + o(1))\binom{n}{3}$. Lang and Sanhueza-Matamala [106] proved $h_d^{k-1}(k, n) \leq 2^{-1/(k-d)}$ and they conjectured that the minimum d-degree threshold for k-uniform tight Hamilton cycles coincides with the lower bounds in (1).

Recently, Cheng et al. [29] and Tang et al. [137] extended Theorems 24 and 25 to transversal versions, respectively.

Theorem 26 ([29, 137]) *For every $k \geq 3$, $\gamma > 0$, there exists n_0 such that the following holds for $n \geq n_0$. Let $\mathcal{H} = \{H_1, \ldots, H_n\}$ be a k-graph system of order n.*

(i) *If $\delta_{k-1}(\mathcal{H}) \geq (\frac{1}{2} + \gamma)n$, then $\mathcal{H}$ admits a transversal tight Hamilton cycle.*
(ii) *If $\delta_{k-2}(\mathcal{H}) \geq (\frac{5}{9} + \gamma)\binom{n}{2}$, then $\mathcal{H}$ admits a transversal tight Hamilton cycle.*

Cheng et al. [29] suspected that the threshold for transversal tight Hamilton cycles in a k-graph system is the same as the threshold conjectured for tight Hamilton cycles in a single k-graph (i.e., Conjecture 5) and provided the following conjecture.

Conjecture 6 ([29]) *Suppose* $\mathcal{H} = \{H_1, \ldots, H_n\}$ *is a* k*-graph system of order* $n\,(n \geq k+1 \geq 4)$ *and* $\delta_{k-1}(\mathcal{H}) \geq \lfloor \frac{n-k+3}{2} \rfloor$, *then* $\mathcal{H}$ *contains a transversal tight Hamilton cycle.*

Motivated by (1), Tang et al. [137] proposed the following conjecture, which concerns the minimum d-degree condition for transversal tight Hamilton cycles.

Conjecture 7 ([137]) *For every* $k \geq d+1$, $\mu > 0$, *there exists an integer* n_0 *such that the following holds for* $n \geq n_0$. *Let* $\mathcal{H} = \{H_1, \ldots, H_n\}$ *be a* k*-graph system of order* n.

(i) *If* $k \geq 4$ *and* $\delta_{k-3}(\mathcal{H}) \geq (\frac{5}{8} + \mu)\binom{n}{3}$, *then* $\mathcal{H}$ *admits a transversal tight Hamilton cycle.*
(ii) *If* $\delta_d(\mathcal{H}) \geq h_d^{k-1}(k, n) + \mu\binom{n}{d}$, *then* $\mathcal{H}$ *admits a transversal tight Hamilton cycle.*

Furthermore, the problem of giving sufficient conditions for transversal Hamilton ℓ-cycles with $\ell \in [k-2]$ is still open. It would be very nice to settle this problem.

Theorem 26 implies that if we consider $(k-1)$-degree and $(k-2)$-degree conditions, then tight Hamilton cycles are color-blind. Gupta et al. [65] provided a unified approach to find color-blind graphs, which is general enough to recover many previous results and obtain novel transversal variants of several classical Dirac-type results. The precise statement of their main result is quite technical, we refer the reader to see [65, Theorem 2.6]. The following theorem lists some applications in their paper and they believed that their setting can capture even more families of hypergraphs.

Theorem 27 ([65]) *Under* d*-degree condition, the following families of hypergraphs are all color-blind.*

(i) *The family of the* r*-th powers of Hamilton cycles for* $d = 1$ *and fixed* $r \geqslant 2$.
(ii) *The family of* k*-uniform Hamilton* ℓ*-cycles for the following ranges of* k, ℓ *and* d.

- $1 < \ell < \frac{k}{2}$ *and* $d = k-2$;
- $1 \leq \ell < \frac{k}{2}$ *or* $\ell = k-1$, *and* $d = k-1$;
- $\ell = \frac{k}{2}$ *and* $\frac{k}{2} < d \leq k-1$ *with* k *even.*

7.2 Matching and Factor

A t*-matching* M in a hypergraph H is a subset of $E(H)$ consisting of t pairwise disjoint edges, which is *perfect* if $V(M) = V(H)$. We denote by $\binom{[n]}{r}$ the set of subsets (also called edges) of size r of $[n]$. For a set X, let 2^X be the set of all subsets of X. Let F be a subfamily of 2^X, denote by $\nu(F)$ its *matching number*, that is, the largest m such that F contains m pairwise disjoint sets. A classical problem in extremal set theory is to determine the maximum size of F with fixed $\nu(F)$. In 1965, Erdős [45] proposed the following conjecture, which is known as Erdős matching conjecture.

Conjecture 8 (Erdős matching conjecture, [45]) *If $n \geq (s+1)k$ and a family $F \subset \binom{[n]}{k}$ satisfies $\nu(F) \leq s$, then $|F| \leq \max\{\binom{n}{k} - \binom{n-s}{k}, \binom{(s+1)k-1}{k}\}$.*

Notice that the bound in Conjecture 8 is tight. In fact, one may consider the edge set of the complete k-graph on $(s+1)k-1$ vertices or the k-graph on n vertices in which every edge intersects a fixed set of s vertices. Setting $s = 1$ in Conjecture 8, then it corresponds to the well-known Erdős–Ko–Rado theorem [46]. There are numerous papers devoted to the Erdős matching conjecture. For recent development on this conjecture, see [10, 56, 67, 95].

Many scholars have attempted to generalize the Erdős matching conjecture to the transversal version in hypergraph systems. Matsumoto and Tokushige [119] generalized the Erdős–Ko–Rado theorem to two families.

Theorem 28 ([119]) *Let n, k, ℓ be integers such that $n \geq \max\{2k, 2\ell\}$. Assume that $\mathcal{H}_1 \subseteq \binom{[n]}{k}$ and $\mathcal{H}_2 \subseteq \binom{[n]}{\ell}$. If $|\mathcal{H}_1||\mathcal{H}_2| > \binom{n-1}{k-1}\binom{n-1}{\ell-1}$, then there exist two disjoint subsets $X_1 \in \mathcal{H}_1$, $X_2 \in \mathcal{H}_2$.*

It is natural to extend the Erdős–Ko–Rado theorem to more than two hypergraphs. For this, Aharoni and Howard [6] and Huang et al. [76] proposed the following conjecture independently.

Conjecture 9 ([6, 76]) *Let $\mathcal{H} = \{H_1, \ldots, H_t\}$ be a system of k-graphs on a common vertex set of size n. If $|E(H_i)| > \max\left\{\binom{n}{k} - \binom{n-t+1}{k}, \binom{kt-1}{k}\right\}$ for all $1 \leq i \leq t$, then $\mathcal{H}$ has a transversal t-matching.*

When $k = 2$, this conjecture was confirmed by Meshulam (see [6]). Indeed, the case $k = 2$ of this conjecture follows directly from an earlier result established by Akiyama and Frankl [8], as restated in [55]. Huang et al. [76] proved that Conjecture 9 holds for $n > 3k^2t$. Keller and Lifshitz [90] proved that Conjecture 9 holds when $n \geq f(t)k$ for some large constant $f(t)$ which only depends on t, and this was further improved the bound on n to $12kt\log(e^2t)$ by Frankl and Kupavskii [58]. Lu et al. [115] improved this bound to $n > 2kt$ for sufficiently large t. By using the sharp threshold techniques developed in [85], Keevash et al. [84] proved a more general result with $n = \Omega(kt)$, which is the optimal order of magnitude. Kupavskii [104] gave the concrete dependencies on the parameters by showing the conjecture holds for $n > 3ekt$ with $t > 10^7$. Recently, Gao et al. [59] showed that Conjecture 9 is true when $k = 3$ and n is sufficiently large.

Aharoni and Howard [6] proposed the following conjecture in k-partite k-graph systems.

Conjecture 10 ([6]) *Let $\mathcal{H} = \{H_1, \ldots, H_t\}$ be a k-partite k-graph system with n vertices in each partition class. If $|E(H_i)| \geq (t-1)n^{k-1}$ for all $i \in [t]$, then $\mathcal{H}$ admits a transversal t-matching.*

Aharoni and Howard proved that Conjecture 10 holds when $t = 2$ [6] or $k \leq 3$ [7]. Lu and Yu [116] proved Conjecture 10 for $n > 3(k-1)(t-1)$. Kiselev and Kupavskii [94] showed that this conjecture is true for $t \geq t_0$ where t_0 is a constant.

The degree conditions for the existence of large matchings in uniform hypergraphs have been extensively studied. Kühn et al. [103] (also see Khan [91]) and Khan [92] determined the minimum 1-degree condition for an n-vertex 3-graph and 4-graph to contain a perfect matching, respectively. Lu et al. [117] and Lu et al. [114] proved the transversal version of the above results.

Theorem 29 ([114, 117]) *Let n be a sufficiently large integer and $k \in \{3, 4\}$ with $k|n$. If $\mathcal{H} = \{H_1, \ldots, H_{\frac{n}{k}}\}$ is a k-graph system of order n with $\delta_1(\mathcal{H}) > \binom{n-1}{k} - \binom{(k-1)n/k}{k}$, then $\mathcal{H}$ contains a transversal perfect matching.*

Recently, Lu and Wang [113] determined the minimum 1-degree condition that guarantees a 3-partite 3-graph system admits transversal perfect matchings. This generalized a result of Lo and Markström [112] on the 1-degree threshold for the existence of perfect matchings in 3-partite 3-graphs. Moreover, in their proof, they used fractional transversal matching theory obtained by Aharoni et al. [5].

It is well-known that perfect matchings are closely related to its fractional counterpart. Given a k-graph H, a *fractional matching* is a function $f : E(H) \to [0, 1]$, subject to the requirement that $\sum_{e:v\in e} f(e) \leq 1$, for every $v \in V(H)$. Furthermore, if equality holds for every $v \in V(H)$, then we call the fractional matching *perfect*. Denote the maximum size of a fractional matching of H by $\nu^*(H) = \max_f \Sigma_{e\in E(H)} f(e)$. Let $c_{k,d}$ be the minimum d-degree threshold for perfect fractional matchings in k-graphs, namely, for every $\varepsilon > 0$ and sufficiently large $n \in \mathbb{N}$, every n-vertex k-graph H with $\delta_d(H) \geq (c_{k,d} + \varepsilon)\binom{n-d}{k-d}$ contains a perfect fractional matching.

Aharoni et al. [5] studied the existence of rainbow fractional matchings in a k-graph system.

Theorem 30 ([5]) *Let $k \geq 2$ be an integer, and let n be a positive rational number. Assume that $\mathcal{H} = \{H_1, \ldots, H_{\lceil kn \rceil}\}$ is a k-graph system with $\nu^*(H_i) \geq n$ for all $i \in [\lceil kn \rceil]$. Then $\mathcal{H}$ has a rainbow fractional matching of size n.*

When the k-graph is k-partite and n is an integer, the number of graphs in $\mathcal{H}$ required in Theorem 30 goes down from kn to $kn - k + 1$. Theorem 30 provided a fractional version of the corresponding problem concerning rainbow matchings, which was solved by Drisko [43] and by Aharoni and Berger [1] for bipartite graphs (the rainbow matching problem remains open for general graphs as well as for k-partite hypergraphs when $k > 2$).

Alon et al. [9] established the relationship between perfect matching and perfect fractional matching by showing that every n-vertex k-graph H with $\delta_d(H) \geq$

$(\max\{c_{k,d}, 1/2\} + o(1))\binom{n-d}{k-d}$ has a perfect matching, and this condition is asymptotically best possible. Cheng et al. [28] proved that the d-degree condition that ensures the existence of a perfect matching in a k-graph also guarantees the existence of a transversal perfect matching in a k-graph system for $d \in [k-1]$.

Theorem 31 ([28]) *For every $\varepsilon > 0$ and integer $d \in [k-1]$, there exists $n_0 \in \mathbb{N}$, such that the following holds for all integers $n \geq n_0$ and $n \in k\mathbb{N}$. Every n-vertex k-graph system $\mathcal{H} = \{H_1, \ldots, H_{\frac{n}{k}}\}$ with $\delta_d(\mathcal{H}) \geq (\max\{c_{k,d}, \frac{1}{2}\} + \varepsilon)\binom{n-d}{k-d}$ contains a transversal perfect matching.*

Lang [105] gave a common generalization of the above theorem. Let G_1 and G_2 be k-graphs on the same vertex set V, and let F be a k-graph. A *color covering homomorphism* from F to (G_1, G_2) is a function $\phi : V(F) \to V$ with an edge $f \in E(F)$ such that $\phi(f) \in E(G_1)$ and ϕ is a homomorphism from F' to G_2, where F' is obtained from F by deleting f from its edge set. We define $\delta_d^{C_F}$ to be the smallest real number δ such that for all $\alpha > 0$ and n large enough any two k-graphs G_1, G_2 on the same n vertices with $\delta_d(G_1), \delta_d(G_2) \geq (\delta + \alpha)\binom{n-d}{k-d}$ admit a color covering homomorphism from F to (G_1, G_2). Denote by δ_d^F the smallest real number δ such that for all $\alpha > 0$ and sufficiently large n, every n-vertex k-graph G with $\delta_d(G) \geq (\delta + \alpha)\binom{n-d}{k-d}$ contains a copy of F-factor.

Theorem 32 ([105]) *Let F be a k-graph with r vertices and t edges. For every $\varepsilon > 0$ and integer $d \in [k-1]$, there exists $n_0 \in \mathbb{N}$ such that the following holds for all integers $n \geq n_0$. Every rn-vertex k-graph system $\mathcal{H} = \{H_1, \ldots, H_{tn}\}$ with $\delta_d(\mathcal{H}) \geq \left(\max\left\{\delta_d^{C_F}, \delta_d^F\right\} + \varepsilon\right)\binom{n-d}{k-d}$ contains a transversal F-factor.*

This is essentially a hypergraph version of the work of Montgomery, Müyesser and Pehova (see Theorem 11). As an application of Theorem 32, Lang [105] deduced the minimum degree condition that guarantees the existence of transversal K_4^3-factors in a 3-graph or transversal power of a Hamilton cycles in a k-graph, where K_4^3 is the complete 3-graph on 4 vertices.

8 Further Results

There are some additional results on transversal problems in graph systems whose details we will not discuss in this survey. For convenience of the reader we give a concise list of such results together with relevant references.

- The rainbow Turán problem can be viewed as a special case of Turán problems in graph systems. Let H be a graph. The *rainbow Turán number* of H, denoted by $\mathrm{ex}^*(n, H)$, is defined to be the maximum number of edges in a properly edge-colored n-vertex graph G with no rainbow copies of H. Notice that this properly edge-colored graph G can be viewed as a system $\mathcal{G} = \{G_1, \ldots, G_{\chi(G)}\}$ of graphs where each G_i is the graph consisting of edges with color i. Then finding a rainbow copy of H in G is equivalent to finding a transversal copy of H inside $\mathcal{G}$.

The rainbow Turán number of some classical families of graphs were studied by Keevash et al. [86]. See more results in [17, 37, 79, 93].

- In Sect. 3, we mainly survey results on the minimum degree condition for transversal spanning structures in graph systems. In fact, some studies have considered the problem of embedding transversal non-spanning structures in a graph system. Li et al. [108] studied the existence of transversal cliques in graph systems under Dirac-type conditions. Falgas-Ravry et al. [50] studied the minimum degree version of Problem 2, and they found that the extremal behavior in the minimum degree condition differs strikingly from that seen in the density condition, with discrete jumps as opposed to continuous transitions.
- The *adjacency matrix* of G is $A(G) = (a_{ij})$, where $a_{xy} = 1$ if $xy \in E(G)$ and 0 otherwise. The largest eigenvalue of $A(G)$ is called the *spectral radius* of G. The maximum spectral radius of graphs in a graph system that forbids transversal perfect matchings, Hamilton paths or Hamilton cycles was studied in [64, 73, 143].
- One may consider the problem of embedding a rainbow graph H in a graph system $\mathcal{G}$ provided that H and all graphs in $\mathcal{G}$ belong to the same class. For example, Aharoni et al. [2] proved that every family of $2\lceil n/2\rceil - 1$ odd cycles of order n contains a rainbow odd cycle. For more results, we refer the reader to [42, 63, 107].
- Another classical and closely related concept is the independent transversal. Given a graph G and a partition of its vertex set, an *independent transversal* is an independent set of G that contains one vertex from each partition class. It also concerns selecting one element from each of several sets while satisfying a global structural constraint. This is a very general notion, many combinatorial problems can be formulated by asking whether a certain graph with a certain vertex partition has an independent transversal. For more related results, see [23, 53, 62, 70–72, 135, 142].

9 Conclusion

Although we have made an effort to provide a systematic coverage of recent results in graph systems, it is inevitable that some topics were left out of this survey due to constraints of space and time. Nevertheless, we would like to believe that we have presented enough results demonstrating how one can generalize known results in extremal graph theory to transversal versions. Therefore we hope this survey will inspire further research in this area.

Acknowledgements We would like to express our gratitude to the anonymous reviewer for his/her valuable comments that greatly improved the presentation of this paper. We also thank Stijn Cambie, Yangyang Cheng, Jie Hu and Richard Lang for helpful comments and valuable suggestions.

Competing Interests Wanting Sun is supported by the China Postdoctoral Science Foundation (2023M742092) and the Natural Science Foundation of Shandong Province (ZR2024QA023). Guanghui Wang is supported by the National Key Research and Development Program (2023YFA1009603) and the Natural Science Foundation of China (12231018).
The authors have no conflicts of interest to declare that are relevant to the content of this chapter.

References

1. Aharoni, R., Berger, E.: Rainbow matchings in r-partite r-graphs. Electron. J. Combin. **16**(1), 9 (2009). Research Paper 119
2. Aharoni, R., Briggs, J., Holzman, R., Jiang, Z.: Rainbow odd cycles. SIAM J. Discrete Math. **35**(4), 2293–2303 (2021)
3. Aharoni, R., DeVos, M., de la Maza, S.G.H, Montejano, A., Šámal, R.: A rainbow version of Mantel's theorem. Adv. Comb. 12 (2020). Paper No. 2
4. Aharoni, R., Georgakopoulos, A., Sprüssel, P.: Perfect matchings in r-partite r-graphs. Eur. J. Comb. **30**(1), 39–42 (2009)
5. Aharoni, R., Holzman, R., Jiang, Z.: Rainbow fractional matchings. Combinatorica **39**(6), 1191–1202 (2019)
6. Aharoni, R., Howard, D.: Size conditions for the existence of rainbow matchings (2011). https://math.colgate.edu/~dmhoward/rsc.pdf
7. Aharoni, R., Howard, D.: A rainbow r-partite version of the Erdős-Ko-Rado theorem. Combin. Probab. Comput. **26**(3), 321–337 (2017)
8. Akiyama, J., Frankl, P.: On the size of graphs with complete factors. J. Graph Theory **9**(1), 197–201 (1985)
9. Alon, N., Frankl, P., Huang, H., Rödl, V., Ruciński, A., Sudakov, B.: Large matchings in uniform hypergraphs and the conjecture of Erdős and Samuels. J. Combin. Theory Ser. A **119**(6), 1200–1215 (2012)
10. Alon, N., Huang, H., Sudakov, B.: Nonnegative k-sums, fractional covers, and probability of small deviations. J. Combin. Theory Ser. B **102**(3), 784–796 (2012)
11. Anastos, M., Chakraborti, D.: Robust Hamiltonicity in families of Dirac graphs (2023). arXiv:2309.12607
12. Babiński, S., Grzesik, A.: Graphs without a rainbow path of length 3. SIAM J. Discret. Math. **38**(1), 629–644 (2024)
13. Babiński, S., Grzesik, A., Prorok, M.: Directed graphs without rainbow triangles. J. Graph Theory **109**(3), 269–281 (2025). https://doi.org/10.1002/jgt.23224
14. Bárány, I.: A generalization of Carathéodory's theorem. Discret. Math. **40**(2–3), 141–152 (1982)
15. Bollobás, B.: Extremal Graph Theory. London Mathematical Society Monographs, vol. 11. Academic Press Inc, London, New York (1978)
16. Bondy, J.A.: Pancyclic graphs. I. J. Combin. Theory Ser. B **11**, 80–84 (1971)
17. Bondy, J.A., Simonovits, M.: Cycles of even length in graphs. J. Combin. Theory Ser. B **16**, 97–105 (1974)
18. Böttcher, J., Schacht, M., Taraz, A.: Proof of the bandwidth conjecture of Bollobás and Komlós. Math. Ann. **343**(1), 175–205 (2009)
19. Bowtell, C., Morris, P., Pehova, Y., Staden, K.: Universality for transversal Hamilton cycles. Bull. Lond. Math. Soc. **57**(3), 711–729 (2025)

20. Bradshaw, P.: Transversals and bipancyclicity in bipartite graph families. Electron. J. Combin. **28**(4), 20 (2021). Paper No. 4.25
21. Bradshaw, P., Halasz, K., Stacho, L.: From one to many rainbow Hamiltonian cycles. Graphs Combin. **38**(6), 21 (2022). Paper No. 188
22. Brualdi, R.A., Ryser, H.J.: Combinatorial Matrix Theory. Encyclopedia of Mathematics and its Applications, vol. 39. Cambridge University Press, Cambridge (1991)
23. Cambie, S., Haxell, P., Kang, R.J., Wdowinski, R.: A precise condition for independent transversals in bipartite covers. SIAM J. Discret. Math. **38**(2), 1451–1461 (2024)
24. Chakraborti, D., Im, S., Kim, J., Liu, H.: A bandwidth theorem for graph transversals (2023). arXiv:2302.09637
25. Chakraborti, D., Kim, J., Lee, H., Liu, H., Seo, J.: On a rainbow extremal problem for color-critical graphs. Random Struct. Algorithms **64**(2), 460–489 (2024)
26. Chakraborti, D., Kim, J., Lee, H., Seo, J.: Hamilton transversals in tournaments. Combinatorica **44**(6), 1381–1400 (2024)
27. Chakraborti, D., Kim, J., Lee, H., Seo, J.: Transversal cycles and paths in tournaments (2024). arXiv:2407.14300
28. Cheng, Y., Han, J., Wang, B., Wang, G.: Rainbow spanning structures in graph and hypergraph systems. Forum Math. Sigma **11**, 20 (2023). Paper No. e95
29. Cheng, Y., Han, J., Wang, B., Wang, G., Yang, D.: Transversal Hamilton cycle in hypergraph systems. SIAM J. Discret. Math. **39**(1), 55–74 (2025)
30. Cheng, Y., Li, H., Sun, W., Wang, G.: Transversal Hamilton cycles in digraph collections. Combin. Probab. Comput. (2026). https://doi.org/10.1017/S0963548326100388
31. Cheng, Y., Staden, K.: Transversals via regularity. J. Combin. Theory Ser. B **176**, 498–550 (2026)
32. Cheng, Y., Staden, K.: Stability of transversal Hamilton cycles and paths. Electron. J. Combin. **32**(4), 35 (2025). Paper No. 4.36
33. Cheng, Y., Sun, W., Wang, G., Wei, L.: Transversal Hamilton paths and cycles (2024). arXiv:2406.13998
34. Cheng, Y., Wang, G., Zhao, Y.: Rainbow pancyclicity in graph systems. Electron. J. Combin. **28**(3), 9 (2021). Paper No. 3.24
35. Corrádi, K., Hajnal, A.: On the maximal number of independent circuits in a graph. Acta Math. Hungar. **14**(3–4), 423–439 (1963)
36. Czygrinow, A., DeBiasio, L., Kierstead, H.A., Molla, T.: An extension of the Hajnal-Szemerédi theorem to directed graphs. Combin. Probab. Comput. **24**(5), 754–773 (2015)
37. Das, S., Lee, C., Sudakov, B.: Rainbow Turán problem for even cycles. Eur. J. Comb. **34**(5), 905–915 (2013)
38. DeBiasio, L., Kühn, D., Molla, T., Osthus, D., Taylor, A.: Arbitrary orientations of Hamilton cycles in digraphs. SIAM J. Discret. Math. **29**(3), 1553–1584 (2015)
39. DeBiasio, L., Molla, T.: Semi-degree threshold for anti-directed Hamiltonian cycles. Electron. J. Combin. **22**(4), 23 (2015). Paper 4.34
40. Dirac, G.A.: Some theorems on abstract graphs. Proc. London Math. Soc. **3**(2), 69–81 (1952)
41. Diwan, A.A., Mubayi, D.: Turán's theorem with colors (2006). https://homepages.math.uic.edu/~mubayi/papers/coloredturan.pdf
42. Dong, Z., Xu, Z.: Rainbow even cycles. SIAM J. Discret. Math. **38**(2), 1269–1284 (2024)
43. Drisko, A.A.: Transversals in row-Latin rectangles. J. Combin. Theory Ser. A **84**(2), 181–195 (1998)
44. Erdős, P.: Problem 9. Theory of graphs and its applications. In: Proceedings of the Symposium held in Smolenice in June 1963, vol. 13, pp. 85–90 (1964)
45. Erdős, P.: A problem on independent r-tuples. Ann. Univ. Sci. Budapest. Eötvös Sect. Math. **8**, 93–95 (1965)
46. Erdős, P., Ko, C., Rado, R.: Intersection theorems for systems of finite sets. Quart. J. Math. Oxford Ser. **2**(12), 313–320 (1961)
47. Erdős, P., Simonovits, M.: A limit theorem in graph theory. Studia Sci. Math. Hungar. **1**, 51–57 (1966)

48. Erdős, P., Stone, A.H.: On the structure of linear graphs. Bull. Am. Math. Soc. **52**, 1087–1091 (1946)
49. Euler, L.: Recherches sur un nouvelle espéce de quarrés magiques. Verh. Zeeuwsch. Gennot. Weten. Vliss **9**, 85–239 (1782)
50. Falgas-Ravry, V., Markström, K., Räty, E.: Minimum-degree conditions for rainbow triangles. J. Graph Theory **107**(2), 298–329 (2024)
51. Falgas-Ravry, V., Markström, K., Räty, E.: Rainbow variations on a theme by Mantel: extremal problems for Gallai colouring templates. Combinatorica **44**(5), 977–1010 (2024)
52. Ferber, A., Han, J., Mao, D.: Dirac-type problem of rainbow matchings and Hamilton cycles in random graphs (2022). arXiv:2211.05477
53. Dai, T., Liu, W., Zhang, X.: Independent transversal blow-up of graphs (2025). arXiv:2502.19682
54. Fischer, E.: Variants of the Hajnal-Szemerédi theorem. J. Graph Theory **31**(4), 275–282 (1999)
55. Frankl, P.: The shifting technique in extremal set theory. In: Surveys in Combinatorics 1987 (New Cross, 1987). London Mathematical Society Lecture Note Series, vol. 123, pp. 81–110. Cambridge University Press, Cambridge (1987)
56. Frankl, P.: Improved bounds for Erdős' matching conjecture. J. Combin. Theory Ser. A **120**(5), 1068–1072 (2013)
57. Frankl, P.: Graphs without rainbow triangles (2022). arXiv:2203.07768
58. Frankl, P., Kupavskii, A.: Simple juntas for shifted families. Discrete Anal. 18 (2020). Paper No. 14
59. Gao, J., Lu, H., Ma, J., Yu, X.: On the rainbow matching conjecture for 3-uniform hypergraphs. Sci. China Math. **65**(11), 2423–2440 (2022)
60. Gerbner, D., Grzesik, A., Palmer, C., Prorok, M.: Directed graphs without rainbow stars. Electron. J. Combin. **31**(4), 15 (2024). Paper No. 4.70
61. Ghouila-Houri, A.: Une condition suffisante d'existence d'un circuit hamiltonien. C. R. Acad. Sci. Paris **251**, 495–497 (1960)
62. Glock, S., Sudakov, B.: An average degree condition for independent transversals. J. Combin. Theory Ser. B **154**, 370–391 (2022)
63. Goorevitch, I., Holzman, R.: Rainbow triangles in families of triangles (2022). arXiv:2209.15493
64. Guo, M., Lu, H., Ma, X., Ma, X.: Spectral radius and rainbow matchings of graphs. Linear Algebra Appl. **679**, 30–37 (2023)
65. Gupta, P., Hamann, F., Müyesser, A., Parczyk, O., Sgueglia, A.: A general approach to transversal versions of Dirac-type theorems. Bull. Lond. Math. Soc. **55**(6), 2817–2839 (2023)
66. Hajnal, A., Szemerédi, E.: Proof of a conjecture of P. Erdős. In: Combinatorial Theory and its Applications, I–III (Proceedings of the Colloquium, Balatonfüred, 1969), Colloquia Mathematica Societatis János Bolyai, vol. 4, pp. 601–623. Amsterdam, North-Holland, London (1970)
67. Han, J.: Perfect matchings in hypergraphs and the Erdős matching conjecture. SIAM J. Discret. Math. **30**(3), 1351–1357 (2016)
68. Han, J., Hu, J., Yang, D.: A robust version of the multipartite Hajnal–Szemerédi theorem (2023). arXiv:2311.00950
69. Han, J., Zhao, Y.: Forbidding Hamilton cycles in uniform hypergraphs. J. Combin. Theory Ser. A **143**, 107–115 (2016)
70. Haxell, P.: Independent transversals and hypergraph matchings—an elementary approach. In: Recent Trends in Combinatorics. IMA Volumes in Mathematics and its Applications, vol. 159, pp. 215–233. Springer, Cham (2016)
71. Haxell, P.: Topological connectedness and independent sets in graphs. In: Surveys in Combinatorics 2019. London Mathematical Society Lecture Note Series, vol. 456, pp. 89–113. Cambridge University Press, Cambridge (2019)
72. Haxell, P., Szabó, T., Tardos, G.: Bounded size components–partitions and transversals. J. Combin. Theory Ser. B **88**(2), 281–297 (2003)

73. He, X., Li, Y., Feng, L.: Spectral radius and rainbow Hamilton paths of a graph. Discret. Math. **347**(10), 11 (2024). Paper No. 114128
74. He, Z., Frankl, P., Győri, E., Lv, Z., Salia, N., Tompkins, C., Varga, K., Zhu, X.: Extremal results for graphs avoiding a rainbow subgraph. Electron. J. Combin. **31**(1), 11 (2024). Paper No. 1.28
75. Hu, J., Li, L., Li, X., Xu, N.: Vertex-bipancyclicity in a bipartite graph collection. Discret. Math. **347**(7), 10 (2024). Paper No. 113980
76. Huang, H., Loh, P.S., Sudakov, B.: The size of a hypergraph and its matching number. Combin. Probab. Comput. **21**(3), 442–450 (2012)
77. Huang, R., Rota, G.C.: On the relations of various conjectures on Latin squares and straightening coefficients. Discret. Math. **128**(1–3), 225–236 (1994)
78. Im, S., Kim, J., Lee, H., Seo, H.: On rainbow Turán densities of trees. Random Struct. Algorithms **66**(3), 20 (2025). Paper No. e70005
79. Johnston, D., Palmer, C., Sarkar, A.: Rainbow Turán problems for paths and forests of stars. Electron. J. Combin. **24**(1), 15 (2017). Paper No. 1.34
80. Joos, F., Kim, J.: On a rainbow version of Dirac's theorem. Bull. Lond. Math. Soc. **52**(3), 498–504 (2020)
81. Kalai, G., Meshulam, R.: A topological colorful Helly theorem. Adv. Math. **191**(2), 305–311 (2005)
82. Katona, G.Y., Kierstead, H.A.: Hamiltonian chains in hypergraphs. J. Graph Theory **30**(3), 205–212 (1999)
83. Keevash, P., Kühn, D., Osthus, D.: An exact minimum degree condition for Hamilton cycles in oriented graphs. J. Lond. Math. Soc. (2) **79**(1), 144–166 (2009)
84. Keevash, P., Lifshitz, N., Long, E., Minzer, D.: Global hypercontractivity and its applications (2021). arXiv:2103.04604
85. Keevash, P., Lifshitz, N., Long, E., Minzer, D.: Hypercontractivity for global functions and sharp thresholds. J. Am. Math. Soc. **37**(1), 245–279 (2024)
86. Keevash, P., Mubayi, D., Sudakov, B., Verstraëte, J.: Rainbow Turán problems. Combin. Probab. Comput. **16**(1), 109–126 (2007)
87. Keevash, P., Mycroft, R.: A geometric theory for hypergraph matching. Mem. Am. Math. Soc. **233**(1098), vi+95 (2015)
88. Keevash, P., Mycroft, R.: A multipartite Hajnal-Szemerédi theorem. J. Combin. Theory Ser. B **114**, 187–236 (2015)
89. Keevash, P., Saks, M., Sudakov, B., Verstraëte, J.: Multicolour Turán problems. Adv. Appl. Math. **33**(2), 238–262 (2004)
90. Keller, N., Lifshitz, N.: The junta method for hypergraphs and the Erdős-Chvátal simplex conjecture. Adv. Math. **392**, 95 (2021). Paper No. 107991
91. Khan, I.: Perfect matchings in 3-uniform hypergraphs with large vertex degree. SIAM J. Discret. Math. **27**(2), 1021–1039 (2013)
92. Khan, I.: Perfect matchings in 4-uniform hypergraphs. J. Combin. Theory Ser. B **116**, 333–366 (2016)
93. Kim, J., Lee, J., Liu, H., Tran, T.: Rainbow cycles in properly edge-colored graphs. Combinatorica **44**(4), 909–919 (2024)
94. Kiselev, S., Kupavskii, A.: Rainbow matchings in k-partite hypergraphs. Bull. Lond. Math. Soc. **53**(2), 360–369 (2021)
95. Kolupaev, D., Kupavskii, A.: Erdős matching conjecture for almost perfect matchings. Discret. Math. **346**(4), 9 (2023). Paper No. 113304
96. Komlós, J., Sárközy, G.N., Szemerédi, E.: On the square of a Hamiltonian cycle in dense graphs. In: Proceedings of the Seventh International Conference on Random Structures and Algorithms, Atlanta, GA, 1995, vol. 9, pp. 193–211 (1996)
97. Komlós, J., Sárközy, G.N., Szemerédi, E.: Blow-up lemma. Combinatorica **17**(1), 109–123 (1997)
98. Komlós, J., Sárközy, G.N., Szemerédi, E.: Proof of the Seymour conjecture for large graphs. Ann. Comb. **2**(1), 43–60 (1998)

99. Komlós, J., Sárközy, G.N., Szemerédi, E.: Spanning trees in dense graphs. Combin. Probab. Comput. **10**(5), 397–416 (2001)
100. Kühn, D., Osthus, D.: Embedding large subgraphs into dense graphs. In: Surveys in Combinatorics 2009. London Mathematical Society Lecture Note Series, vol. 365, pp. 137–167. Cambridge University Press, Cambridge (2009)
101. Kühn, D., Osthus, D.: The minimum degree threshold for perfect graph packings. Combinatorica **29**(1), 65–107 (2009)
102. Kühn, D., Osthus, D.: Hamilton cycles in graphs and hypergraphs: an extremal perspective. In: Proceedings of the International Congress of Mathematicians—Seoul 2014, vol. 4, pp. 381–406. Kyung Moon Sa, Seoul (2014)
103. Kühn, D., Osthus, D., Treglown, A.: Matchings in 3-uniform hypergraphs. J. Combin. Theory Ser. B **103**(2), 291–305 (2013)
104. Kupavskii, A.: Rainbow version of the Erdős matching conjecture via concentration. Comb. Theory **3**(1), 19 (2023). Paper No. 1
105. Lang, R.: Tiling dense hypergraphs (2023). arXiv:2308.12281
106. Lang, R., Sanhueza-Matamala, N.: Minimum degree conditions for tight Hamilton cycles. J. Lond. Math. Soc. (2) **105**(4), 2249–2323 (2022)
107. Li, E.Y.H., Li, L., Li, P.: Transversals in a collection of stars or generic trees. Discret. Math. **349**(3), 10 (2026). Paper No. 114836
108. Li, L., Li, P., Li, X.: Rainbow structures in a collection of graphs with degree conditions. J. Graph Theory **104**(2), 341–359 (2023)
109. Li, L., Li, P., Li, X.: Rainbow pancyclicity in a collection of graphs under the Dirac-type condition. Acta Math. Appl. Sin. Engl. Ser. **40**(2), 269–274 (2024)
110. Li, X., Ma, J., Zheng, Z.: On the multicolor Turán conjecture for color-critical graphs. Canadian Journal of Mathematics, pp. 1–31 (2025)
111. Lo, A., Markström, K.: A multipartite version of the Hajnal-Szemerédi theorem for graphs and hypergraphs. Combin. Probab. Comput. **22**(1), 97–111 (2013)
112. Lo, A., Markström, K.: Perfect matchings in 3-partite 3-uniform hypergraphs. J. Combin. Theory Ser. A **127**, 22–57 (2014)
113. Lu, H., Wang, Y.: Rainbow perfect matchings in 3-partite 3-uniform hypergraphs (2024). arXiv:2408.08523
114. Lu, H., Wang, Y., Yu, X.: Rainbow perfect matchings for 4-uniform hypergraphs. SIAM J. Discret. Math. **36**(3), 1645–1662 (2022)
115. Lu, H., Wang, Y., Yu, X.: A better bound on the size of rainbow matchings. J. Combin. Theory Ser. A **195**, 19 (2023). Paper No. 105700
116. Lu, H., Yu, X.: On rainbow matchings for hypergraphs. SIAM J. Discret. Math. **32**(1), 382–393 (2018)
117. Lu, H., Yu, X., Yuan, X.: Rainbow matchings for 3-uniform hypergraphs. J. Combin. Theory Ser. A **183**, 21 (2021). Paper No. 105489
118. Magnant, C.: Density of Gallai multigraphs. Electron. J. Combin. **22**(1), 6 (2015). Paper 1.28
119. Matsumoto, M., Tokushige, N.: The exact bound in the Erdős-Ko-Rado theorem for cross-intersecting families. J. Combin. Theory Ser. A **52**(1), 90–97 (1989)
120. Montgomery, R.: A proof of the Ryser-Brualdi-Stein conjecture for large even n (2023). arXiv:2310.19779
121. Montgomery, R.: Transversals in Latin squares. In: Surveys in Combinatorics 2024. London Mathematical Society Lecture Note Series, vol. 493, pp. 131–158. Cambridge University Press, Cambridge (2024)
122. Montgomery, R., Müyesser, A., Pehova, Y.: Transversal factors and spanning trees. Adv. Comb. 25 (2022). Paper No. 3
123. Moon, J., Moser, L.: On Hamiltonian bipartite graphs. Israel J. Math. **1**, 163–165 (1963)
124. Ore, O.: Note on Hamilton circuits. Am. Math. Monthly **67**, 55 (1960)
125. Pokrovskiy, A.: Rota's basis conjecture holds asymptotically. Proc. Amer. Math. Soc. **153**(10), 4101–4118 (2025)

126. Polcyn, J., Reiher, C., Rödl, V., Schülke, B.: On Hamiltonian cycles in hypergraphs with dense link graphs. J. Combin. Theory Ser. B **150**, 17–75 (2021)
127. Pósa, L.: A theorem concerning Hamilton lines. Magyar Tud. Akad. Mat. Kutató Int. Közl. **7**, 225–226 (1962)
128. Rödl, V., Ruciński, A., Szemerédi, E.: A Dirac-type theorem for 3-uniform hypergraphs. Combin. Probab. Comput. **15**(1–2), 229–251 (2006)
129. Rödl, V., Ruciński, A., Szemerédi, E.: An approximate Dirac-type theorem for k-uniform hypergraphs. Combinatorica **28**(2), 229–260 (2008)
130. Rödl, V., Ruciński, A., Szemerédi, E.: Dirac-type conditions for Hamiltonian paths and cycles in 3-uniform hypergraphs. Adv. Math. **227**(3), 1225–1299 (2011)
131. Ryser, H.J.: Neuere probleme der kombinatorik. Vorträge über Kombinatorik, Oberwolfach **69**(91), 35 (1967)
132. Seymour, P.D.: Problem section, problem 3, combinatorics. In: London Mathematical Society Lecture Note Series (Proceedings of the British Combinatorial Conference, University College of Wales, Aberystwyth, 1973), pp. 201–202 (1974)
133. Stein, S.K.: Transversals of Latin squares and their generalizations. Pacific J. Math. **59**(2), 567–575 (1975)
134. Sudakov, B.: Robustness of graph properties. In: Surveys in combinatorics 2017. London Mathematical Society Lecture Note Series, vol. 440, pp. 372–408. Cambridge University Press, Cambridge (2017)
135. Szabó, T., Tardos, G.: Extremal problems for transversals in graphs with bounded degree. Combinatorica **26**(3), 333–351 (2006)
136. Szemerédi, E.: Regular partitions of graphs. In: Problèmes combinatoires et théorie des graphes (Colloque International du CNRS, University. Orsay, Orsay, 1976), Colloque International CNRS, vol. 260, pp. 399–401. CNRS, Paris (1978)
137. Tang, Y., Wang, B., Wang, G., Yan, G.: Transversal Hamilton cycle in the hypergraph system. Sci. China Math. (2026). https://doi.org/10.1007/s11425-024-2469-2
138. Thomason, A.: Paths and cycles in tournaments. Trans. Am. Math. Soc. **296**(1), 167–180 (1986)
139. Treglown, A.: On directed versions of the Hajnal-Szemerédi theorem. Combin. Probab. Comput. **24**(6), 873–928 (2015)
140. Turán, P.: Eine Extremalaufgabe aus der Graphentheorie. Mat. Fiz. Lapok **48**, 436–452 (1941)
141. Wanless, I.M.: Transversals in Latin squares: a survey. In: Surveys in Combinatorics. London Mathematical Society Lecture Note Series, vol. 392, pp. 403–437. Cambridge University Press, Cambridge (2011)
142. Yuster, R.: Independent transversals in r-partite graphs. Discret. Math. **176**(1–3), 255–261 (1997)
143. Zhang, Y., van Dam, E.R.: Rainbow Hamiltonicity and the spectral radius. Discret. Math. **348**(11), (2025). Paper No. 114600
144. Zhao, Y.: Recent advances on Dirac-type problems for hypergraphs. In: Recent Trends in Combinatorics. IMA Volumes in Mathematics and its Applications, vol. 159, pp. 145–165. Springer (2016)

Generalizations of the Crossing Lemma

Géza Tóth

Abstract The crossing number of a graph is the minimum number of crossings over all of its drawings on the plane. The Crossing Lemma, proved more than 40 years ago, is a tight lower bound on the crossing number of a graph in terms of the number of vertices and edges. It is definitely the most important inequality on crossing numbers. We review some generalizations and applications of the Crossing Lemma.

1 Introduction

The *crossing number* $\text{CR}(G)$ of G is the minimum number of edge crossings over all drawings of G on the plane. It has a fascinating history and lots of applications in theory and practice. It started with Turán's Brick Factory problem in 1944, also known as Zarankiewicz's problem [32]. Determining or even estimating the crossing number is a notoriously difficult problem [30, 62], and it has a huge number of practical and theoretical applications [20, 66] in VLSI design, geographic information systems, graph drawing, computer graphics, and many other fields. Even the crossing numbers of complete graphs are not known, despite the huge number of results, attempts, improvements [3, 6, 15, 33]. The Crossing Lemma of Ajtai, Chvátal, Newborn, Szemerédi and independently Leighton, 40 years ago was a real breakthrough. It is a general lower bound on the crossing number, it states that any simple graph with n vertices and $e \geq 4n$ edges has crossing number at least ce^3/n^2 for some $c > 0$.

Székely [66] and then others observed that crossing numbers, in particular the Crossing Lemma, also has important applications in the theory of incidences [65],

Supported by National Research, Development and Innovation Office, NKFIH, K-131529 and ERC Advanced Grant "GeoScape" 882971.

G. Tóth (✉)
HUN-REN Alfréd Rényi Institute of Mathematics, Budapest, Hungary

Department of Computer Science and Information Theory, Budapest University of Technology and Economics, Budapest, Hungary
e-mail: geza@renyi.hu

G. O. H. Katona et al. (eds.), *Sum(m)it280*, Bolyai Society Mathematical Studies 32,
https://doi.org/10.1007/978-3-032-18810-6_16

additive number theory [23], in the analysis of geometric algorithms [19] and many other fields.

In this paper we survey some results related to the Crossing Lemma and its generalizations. It Sect. 2 we review the history and improvements of the Crossing Lemma. It is easy to see that it does not hold for multigraphs in general, not even in any weaker form. In Sect. 3 we investigate its generalizations for multigraphs under different conditions on the drawing. In Sect. 4 we try to figure out, how far can we go with such a generalization, that is, what could be the weakest condition which implies some Crossing Lemma type statement. Finally, in Sect. 5 we introduce some other versions of the crossing number, the pair-crossing number and the odd-crossing number, study their relationships, and generalizations of the Crossing Lemma for these crossing numbers.

2 The Crossing Lemma

In a *drawing* of a graph (or multigraph) G, vertices are represented by points and edges are represented by curves connecting the corresponding points. If it does not lead to confusion, the points (curves), representing the vertices (edges) are also called vertices (edges). We assume that in a drawing an edge cannot contain a vertex in its interior, if two edges have a common point, then it is a common endpoint or a proper crossing, and three edges do not cross at the same point. For any graph G, $n(G)$ and $e(G)$ denote its number of vertices and edges, respectively. A graph (resp. multigraph), together with its drawing, is called a *topological* graph (resp. multigraph).

The most important, fundamental result on crossing numbers is the Crossing Lemma, first proved by Ajtai, Chvátal, Newborn, Szemerédi, and Leighton, 40 years ago [5, 42].

Theorem 1 (Crossing Lemma) *For any simple graph G, if $e(G) \geq 4n(G)$ then*

$$\text{CR}(G) \geq \frac{1}{64}\frac{e(G)^3}{n(G)^2}.$$

The original proofs were quite complicated and gave a much worse constant in the bound. The following beautiful proof is attributed to Chazelle, Sharir, and Welzl and appeared in "Proofs from THE BOOK" [10]. For the proof we need the following easy but very important statement (see e.g. [58]).

Observation 1 *For any simple graph G, in a drawing with exactly* CR(G) *crossings, (i) no edge crosses itself, (ii) two edges have at most one common point.*

Proof (i) Suppose that an edge α crosses itself at point p. Reverse the orientation of the part of α between the two occurrences of p. Then p becomes a self-touching

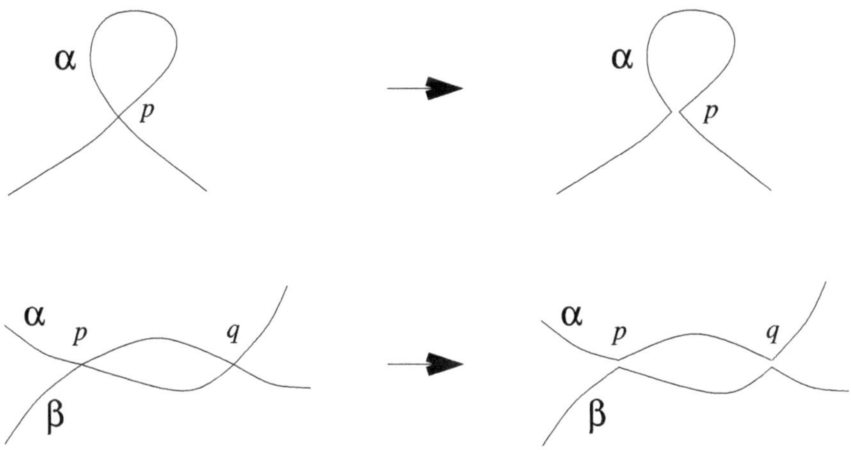

Fig. 1 Reducing the number of crossings

point. And now we can get rid of the touching by a slight perturbation at p. This way we get a drawing with fewer crossings, which is a contradiction. See Fig. 1.
(ii) Suppose that two edges α and β have two common points, p and q. One of them can be a crossing or a common endpoint, the other one is a crossing. Then swap the parts of the edges between p and q. Again, we get a drawing with fewer crossings, which is a contradiction. See Fig. 1. □

Proof of the Crossing Lemma The proof goes in two steps.
1. First we show that for any graph G

$$\text{CR}(G) \geq e(G) - 3n(G). \tag{1}$$

This can be proved by induction on $e(G)$. If $e(G) \leq 3n(G)$, then the statement holds trivially. Suppose that $e(G) > 3n(G)$ and we have already proved the statement for smaller values of $e(G)$. It is well-known consequence of Euler's formula that a planar graph of n vertices has at most $3n - 6$ edges. Therefore, $\text{CR}(G) > 0$. Remove an edge α from G which contains a crossing. Then $\text{CR}(G) \geq \text{CR}(G - \alpha) + 1$ and by the induction hypothesis, $\text{CR}(G - \alpha) \geq e(G) - 1 - 3n(G)$. Combining them, the statement follows.
2. Consider a drawing of G with $\text{CR}(G)$ crossings. Take a random induced subgraph G' of G by selecting each vertex independently with probability p (whose value is determined later). We obtain a random subset of vertices $U \subset V(G)$ and let G' be the subgraph induced by U. Draw the edges of G' as they were drawn in the original drawing. The number of vertices and edges of G', $n(G')$ and $e(G')$ are random variables, and $E[n(G')] = pn(G)$, $E[e(G')] = p^2e(G)$ since each edge of G survives with probability p^2. Let $X(G')$ be the number of crossings of G' in the obtained drawing. Then clearly $X(G') \geq \text{CR}(G')$. By Observation 1, only indepen-

dent edges can cross in the original optimal drawing of G, consequently, also in G'. A crossing is present in the drawing of G' if all four corresponding endpoints are selected and the probability of this is p^4. (A crossing of two edges with a common endpoint would be present with probability p^3.) Therefore, $E[X(G')] = p^4\text{CR}(G)$. By 1, $\text{CR}(G') \geq e(G') - 3n(G')$. Taking expectations, we get

$$p^4\text{CR}(G) = E[X(G')] \geq E[\text{CR}(G')] \geq E[e(G')] - 3E[n(G')] = p^2e(G) - 3pn(G)$$

so $\text{CR}(G) \geq e(G)/p^2 - 3n(G)/p^3$. Set $p = 4n(G)/e(G)$, by assumption $p \leq 1$. Substitute in the last formula and we get

$$\text{CR}(G) \geq \frac{1}{64}e(G)^3/n(G)^2.$$

□

This bound cannot be improved apart from the value of the constant. That is, there is a constant c with the following property. For any n, e, $e \leq \binom{n}{2}$, there is a graph G with $n(G) = n$, $e(G) = e$ and $\text{CR}(G) \leq ce^3/n^2$. The easiest example is—very roughly—the following. Take the disjoint union of $n^2/2e$ complete graphs, each of $2e/n$ vertices. This graph has approximately n vertices, e edges. Each component can be drawn such that any two edges cross at most once, and components can be separated from each other. Hence $\text{CR}(G) \leq (n^2/2e)((2e/n)^2/2)^2/2 = e^3/n^2$.

There is a better example by Pach and Tóth [56]. Suppose that $n \gg e \gg n^2$, that is, $\lim_{n\to\infty} e/n = \infty$ and $\lim_{n\to\infty} n^2/e = \infty$. We define a graph G of n vertices and e edges, together with its drawing. Let $V(G)$ be a set of n points arranged in a slightly perturbed unit square grid of size $\sqrt{n} \times \sqrt{n}$, so that the points are in general position. Let $r = \sqrt{2e/\pi n}$, so that $r^2\pi = 2e/n$. Connect two points by a straight-line segment if and only if their distance is at most r. Then, $n(G) = n$, the degrees of almost all vertices are close to $r^2\pi$, so $e(G) \approx nr^2\pi/2 = e$. One can approximate the number of crossings, which is an upper bound for the crossing number, by an integral. It turns out that

$$\text{CR}(G) \leq \left(8/9\pi^2 + o(1)\right) e^3/n^2 < (0.09006 + o(1))\, e^3/n^2. \tag{2}$$

In fact, the calculation in [56] includes a quadruple integral and there was an error in it, which was corrected in [48]. Recently, Czabarka et al. [18] found essentially the same bound with a much simpler and more general construction. Take n random points independently on the sphere. Connect a pair of points by the geodesic (the shortest connecting curve, which is an arc on a great circle) if it has length at most r. If we set the value of r properly, the expected number of edges is e and the expected number of crossings is $(8/9\pi^2 + o(1))e^3/n^2$. It is not hard to argue that there is an instance where both the number of edges and crossings are close to their expected values.

From the other direction there were several improvements. In general, for a better constant we have to pay with a stronger condition in place of $e \geq 4n$. First, in the proof above we can set $p = 4.5n/e$ but then we have to assume that $e \geq 4.5n$. We obtain the following statement:

$$\text{if } e(G) \geq 4.5n(G) \text{ then } \operatorname{CR}(G) \geq \frac{1}{60.75}\frac{e(G)^3}{n(G)^2}.$$

The next improvement was in [56]. The main idea is that equation (1) can be replaced by a stronger one. In Eq. 1 we only use that once a graph has more than $3n - 6$ edges, then there is an edge with a crossing on it. But if the number of edges is a little bit more, then there will be an edge with *two* or even more crossings.

For any $k \geq 0$, a graph is called *k-planar* if it has a drawing where every edge contains at most k crossings. Let $e_k(n)$ denote the maximum number of edges of a k-planar graph of n vertices. Clearly, 0-planar graphs are exactly the planar graphs, therefore, $m_0(n) = 3n - 6$. It was shown in [56] that $m_1(n) = 4n - 8$ for $n \geq 12$ and $m_2(n) \leq 5n - 10$ which is tight for infinitely many values of n, $m_3(n) \leq 6n - 12$ and $m_4(n) \leq 7n - 14$.

Therefore, for any graph G

$$\operatorname{CR}(G) \geq (e(G) - 3n(G)) + (e(G) - 4n(G)) + (e(G) - 5n(G))$$

$$+ (e(G) - 6n(G)) + (e(G) - 7n(G)) = 5e(G) - 25n(G).$$

By Observation 1, all crossings are between independent edges. Using this inequality in the probabilistic argument in the proof of the Crossing Lemma we get the following.

Theorem 2 ([56]) *For any simple graph G, if $e(G) \geq 7.5n(G)$ then* $\operatorname{CR}(G) \geq \frac{1}{33.75}\frac{e(G)^3}{n(G)^2}$.

Ten years later, it was shown by Pach, Radoičić, Tardos and Tóth [48] that $m_3(n) \leq 5.5n - 11$, which is tight apart from an additive constant. Using this bound, we already get an improvement, but they added some further structural observations and obtained that

$$\operatorname{CR}(G) \geq 7e(G)/3 - 25n(G)/3.$$

This implies the following result, by the usual probabilistic argument.

Theorem 3 ([48]) *For any simple graph G, if $e(G) \geq 17.16n(G)$ then* $\operatorname{CR}(G) \geq \frac{1}{31.1}\frac{e(G)^3}{n(G)^2}$.

The next improvement was by Ackerman [4]. He proved that $m_4(n) \leq 6n - 12$, which is also tight up to an additive constant. He obtained the following bound.

Theorem 4 ([4]) *For any simple graph G, if $e(G) \geq 6.95n(G)$ then* $\operatorname{CR}(G) \geq \frac{1}{29}\frac{e(G)^3}{n(G)^2}$.

Finally, the best known bound is a recent result of Büngener and Kaufmann [13]. They use the same approach and some additional structural results on near maximal 2-planar and 3-planar graphs. They obtained the following linear bound.

$$\text{CR}(G) \geq 5e(G) - 203n(G)/9$$

which implies.

Theorem 5 ([13]) *For any simple graph G, if $e(G) \geq 6.77n(G)$ then* $\text{CR}(G) \geq \frac{1}{27.48}\frac{e(G)^3}{n(G)^2}$.

Note that in each case we could apply Obseration 1 to conclude that all crossings are between independent edges. This is necessary for the probabilistic argument.

For larger k, $m_k(n) = \Theta(\sqrt{k}n)$, the upper bound is also a consequence of the Crossing Lemma [36].

So we do not know the optimal constant in the Crossing Lemma, but at least we know that it exists in the following sense. Let $\kappa(n, e)$ denote the minimum crossing number of a graph G with n vertices and at least e edges. That is,

$$\kappa(n, e) = \min_{\substack{n(G) = n \\ e(G) \geq e}} \text{CR}(G).$$

Verifying a conjecture of Erdős and Guy [22], Pach, Spencer and Tóth proved the following.

Theorem 6 ([55]) *If $n \ll e \ll n^2$, then*

$$\lim_{n \to \infty} \kappa(n, e)\frac{n^2}{e^3} = c > 0.$$

It is not known whether the condition $n \ll e \ll n^2$ can be replaced by the weaker condition $C_1 n \leq e \leq C_2 n^2$. If the answer is yes, then an easy calculation shows that $C_1 > 4$ and $C_2 < 1/2$. We call the constant $c > 0$ in Theorem 6 the *midrange crossing constant*. As we have seen, the value of c is still unknown but we know that $1/27.48 \leq c \leq 8/9\pi^2$. We have $1/27.48 \approx 0.036$ and $8/9\pi^2 \approx 0.09$, so the upper and lower bounds are within a factor of 2.5.

It was shown by Czabarka et al. [18] that there is a midrange crossing constant c_b, defined analogously, for *bipartite graphs* as well. The best known bounds are $0.055 \leq c_b \leq 0.18$ [2]. They also proved the existence of a midrange crossing constant for many other graph classes.

3 Multigraphs

The Crossing Lemma does not hold for multigraphs in general. To see this, just take two vertices and arbitrarily many parallel edges connecting them, with no crossings, or just take one vertex and many loops. So it is important to understand, under what conditions on the drawings does the statement of the Crossing Lemma, or a similar statement holds for multigraphs. For simplicity, we assume that there are no loops, but there could be parallel edges in the multigraphs considered in this section.

The first such attempt to generalize the Crossing Lemma for multigraphs was to limit the multiplicity of parallel edges [45, 56, 66]. In this case we have an easy, weaker, but tight bound. But this condition does not really give an insight.

Theorem 7 (Crossing Lemma for multigraphs with bounded edge-multiplicity) *For any multigraph G, if any two vertices are connected by at most m edges and $e(G) \geq 4mn(G)$ then*

$$\text{CR}(G) \geq \frac{1}{64}\frac{e(G)^3}{mn(G)^2}.$$

Proof Consider a drawing of G with $\text{CR}(G)$ crossings. Take a random induced subgraph G'' by the following random process. Let $p = 4mn/e$ which is at most 1, by the assumption. First select each vertex independently with probability p. We obtain a random subset of vertices $U \subset V(G)$ and let G' be the subgraph induced by U. Now for each pair of adjacent vertices, $u, v \in U$, de the following. Keep each of the edges between u and v with probability $1/m$, and delete the others, in such a way that these events are mutually exclusive. The resulting random graph is G''. The edges of G'' are drawn as they were drawn in the original drawing. The number of vertices and edges of G'', $n(G'')$ and $e(G'')$ are random variables, and $E[n(G'')] = pn(G)$, $E[e(G'')] = p^2e(G)/m$ since each edge of G survives with probability p^2/m. For the number of crossings $X(G'')$ of G'' in the obtained drawing we have $X(G'') \geq \text{CR}(G'')$. The probability that a crossing is present in the drawing of G'' is p^4/m^2, so $E[X(G'')] = p^4/m^2\text{CR}(G)$. Since G'' is a simple graph, by (1), $\text{CR}(G'') \geq e(G'') - 3n(G'')$, so

$$p^4/m^2\text{CR}(G) = E[X(G'')] \geq E[\text{CR}(G'')]$$

$$\geq E[e(G'')] - 3E[n(G'')] = p^2/me(G) - 3pn(G)$$

and we get that

$$\text{CR}(G) \geq \frac{1}{64}\frac{e(G)^3}{n(G)^2\, m}.$$

□

For $m = 1$, Theorem 7 gives the Crossing Lemma, but as m increases, the bound is getting weaker. Still, it is also tight up to a constant factor. Moreover, it was shown by Ágoston and Pálvölgyi [7] that Theorem 7 holds with the same constant as the the original Crossing Lemma, in the following sense. Let $\kappa(n, e, m)$ denote the minimum crossing number of a multigraph G with n vertices, at least e edges, and edge multiplicity at most m. In particular, $\kappa(n, e, 1) = \kappa(n, e)$.

Theorem 8 ([7]) *For any $m \geq 1$,*

$$m^2\kappa\left(n, \lfloor e/m \rfloor\right) \leq \kappa(n, e, m) \leq m^2\kappa\left(n, \lceil e/m \rceil\right).$$

Proof For the upper bound on $\kappa(n, e, m)$, take a simple graph with n vertices, $\lceil e/m \rceil$ edges drawn with $\kappa\left(n, \lceil e/m \rceil\right)$ crossings. Now replace each edge by at most m parallel edges, drawn very close to the original edge. This way we can get a multigraph of e edges and edge multiplicity at most m. Each original crossing corresponds to at most m^2 new crossings, so we have $\kappa(n, e, m) \leq m^2\kappa\left(n, \lceil e/m \rceil\right)$.

Now we prove the lower bound on $\kappa(n, e, m)$. Two parallel edges are called *homotopic* if they are homotopic in the vertex-punctured plane. That is, the closed curve formed by their union is contractible to a point. Take a multigraph G with n vertices, e edges, edge mutiplicity at most m, drawn with $\kappa(n, e, m)$ crossings. Let u, v be two vertices and take all uv edges. Let α be one with the minimum number of crossings. Redraw all other parallel edges very close to α. We did not increase the number of crossings, so we still have $\kappa(n, e, m)$ crossings. Do it with all pairs of vertices. Now we have a drawing with $\kappa(n, e, m)$ crossings such that parallel edges are homotopic.

An edge is called *full* if it has multiplicity m. Suppose that α and β are not parallel, α is not full, and it has at most as many crossings as β. Remove β and add an edge parallel to α, drawn very close to it. We obtain another multigraph with the same number of edges, edge multiplicity at most m, and we did not increase the number of crossings. Therefore, we still have $\kappa(n, e, m)$ crossings. By a repeated application of this step, we can obtain a multigraph G' with n vertices, e edges, edge mutiplicity at most m, drawn with $\kappa(n, e, m)$ crossings, such that the number of non-full edges is less than m, and they are all parallel.

Now delete all non-full edges and keep only one of each parallel m-tuples of the the remaining edges. Let G_1 be the obtained simple graph. It has n vertices, $\lfloor e/m \rfloor$ edges and in its drawing each crossing corresponds to m^2 original crossings. Therefore, $m^2\kappa\left(n, \lfloor e/m \rfloor\right) \leq \kappa(n, e, m)$. □

It was suggested by Bekos, Kaufmann, and Raftopoulou [39] that the dependence on the multiplicity might be eliminated if we restrict our attention to special classes of drawings. In particular, it is natural to assume that each "lens" enclosed by two parallel edges contains a vertex.

Definition 1 A *drawing* of a multigraph G in the plane is called
– **separated** if any pair of parallel edges form a simple closed curve and there is at least one vertex in its interior and exterior,
– **single-crossing** if any two independent edges cross at most once,
– **locally starlike** if adjacent edges do not cross.

It turned out that these conditions are sufficient to generalize the Crossing Lemma in some form. The first such result is by Pach and Tóth [59].

Theorem 9 ([59]) *Suppose that M is a separated, single-crossing, locally starlike topological multigraph and $e(M) \geq 4n(M)$. Then* $\mathrm{CR}(M) \geq c\frac{e(G)^3}{n(G)^2}$ *for some constant $c > 0$.*

Later it turned out that the “single-crossing” condition can be dropped.

Theorem 10 ([40]) *Suppose that M is a separated, locally starlike topological multigraph and $e(M) \geq 4n(M)$. Then*

$$\mathrm{CR}(M) \geq c\frac{e(G)^3}{n(G)^2}$$

for some constant $c > 0$.

Both of these results are asymptotically thight. If we have the “separated” and “single-crossing” conditions, then we have a slightly weaker result.

Theorem 11 ([27]) *Suppose that M is a separated, single-crossing topological multigraph and $e(M) \geq 4n(M)$. Then*

$$\mathrm{CR}(M) \geq c\frac{e(G)^3}{n(G)^2 \log(e/n)}$$

for some constant $c > 0$.

This bound is probably not tight, most likely the log factor can be removed. If we only have the “separated” condition, we have a weaker bound, surprisingly, it is also tight.

Theorem 12 ([40]) *Suppose that M is a separated topological multigraph and $e(M) \geq 4n(M)$. Then*

$$\mathrm{CR}(M) \geq c\frac{e(G)^{2.5}}{n(G)^{1.5}}$$

for some constant $c > 0$. This bound is asymptotically tight.

We show why this bound it asymptotically tight. For any m (divisible by 3), let H_m be the following topological multigraph. Let $V(H) = V_1 \cup V_2 \cup V_3$, each of size $m/3$. Place the points of V_i on the line $y = i$. Connect each point in V_1 to each point in V_3 with $m/3$ parallel edges, crossing the line $y = 2$ between different consecutice points of V_2. We can do it such that any two edges cross at most twice. This multigraph has m vertices, $m^3/27$ edges and less than m^6 crossings. Suppose now that $n > 0$ and $4n < e < cn^3$. Take roughly $n^{3/2}/e^{0.5}$ disjoint copies of H_m with $m \approx \sqrt{e/n}$. This topological multigraph M has n vertices, e edges and $\mathrm{CR}(M) = \Omega\left(\frac{e(G)^{2.5}}{n(G)^{1.5}}\right)$.

To prove these theorems, the most natural idea is to adapt the proof of the Crossing Lemma, Theorem 1 in Sect. 2 to bound the number of crossings in separated, (single-crossing, locally starlike) multigraphs.

It follows from Euler's formula, that a separated topological multigraph of n vertices and no edge crossings has at most $3n - 6$ edges. Observe, that if we delete an edge from a separated topological multigraph, it remains separated. Therefore, we can apply the induction in the first part of the proof of the Crossing Lemma, 1 and get that for any separated topological multigraph M, $\text{CR}(M) \geq e(M) - 3n(M)$.

But there is a problem with the second step. When we take a random induced subgraph, it might not be separated. Indeed, suppose that vertices u and v are connected by edges α and β. By the conditions, these edges together determine a simple closed curve, which contains some vertices $w_1, \ldots, w_k$. When we take a random set of vertices, it is possible that we keep u and v, but delete $w_1, \ldots w_k$. Despite many attempts, it is not known how to modify the random process to rule out this scenario. Now we sketch another proof of the Crossing Lemma, which can be used in these cases as well.

Theorem 13 (Weak Crossing Lemma) *For any simple graph G, if $e(G) \geq 4n(G)$ then*

$$\text{CR}(G) \geq 10^{-26} \frac{e(G)^3}{n(G)^2}.$$

For the (sketch of the) proof, we need some preparation. The *bisection width*, $b(G)$ of a graph G is the minimum number of edges whose removal splits the graph into two, roughly equal subgraphs. More precisely, $b(G)$ is the minimum number of edges of G between V_1 and V_2 over all partitions $V(G) = V_1 \cup V_2$, $|V_1|, |V_2| \geq |V(G)|/3$. Based on the Lipton-Tarjan separator theorem [44] for planar graphs, and its generalization by Leighton [42], Pach, Shahrokhi and Szegedy [51] proved the following relationship between the bisection width and the crossing number.

Theorem 14 ([51]) *Suppose that G is a graph of n vertices of degrees $d_1, \ldots, d_n$. Then*

$$b(G) \leq 10\sqrt{\text{CR}(G)} + 10\sqrt{\sum_{i=1}^{n} d_i^2}.$$

Sketch of the proof of the Weak Crossing Lemma Let G be a graph of n vertices and $e \geq 4n$ edges. If $e \leq 10^6 n$, then the statement follows from equation (1). Assume for a contradiction that $e > 10^6 n$ and $\text{CR}(G) < 10^{-26} \frac{e(G)^3}{n(G)^2}$.

By splitting large degree vertices of G we can get a graph with the same number of edges, at most twice as many vertices and with maximum degree at most twice the average. Here, for simplicity, we assume that G already has this property, its maximum degree is at most twice the average degree. This is important when we apply Theorem 14 and want to control the term $\sqrt{\sum_{i=1}^{n} d_i^2}$. By a repeated application of Theorem 14 we break H into smaller parts. We apply the following procedure.

DECOMPOSITION ALGORITHM

STEP 0. **Let** $G^0 = G$, $G_1^0 = G$, $M_0 = 1$, $m_0 = 1$. Suppose that we have already executed STEP i, and the resulting graph, G^i, consists of M_i parts, $G_1^i, G_2^i, \ldots, G_{M_i}^i$, each having at most $(2/3)^i n$ vertices. Assume without loss of generality that the first m_i parts of G^i have at least $(2/3)^{i+1}n$ vertices and the remaining $M_i - m_i$ have fewer.

STEP $i + 1$. **If**

$$(2/3)^i < \frac{e}{n^2}, \tag{3}$$

then STOP. (3) is called the *stopping rule*.

Else, for $j = 1, 2, \ldots, m_i$, apply Theorem 14 to G_j^i and delete $b(G_j^i)$ edges from it such that G_j^i falls into two parts, each of at most $(2/3)n(G_j^i)$ vertices. Let G^{i+1} denote the resulting graph on the original set of n vertices. Clearly, each part of G^{i+1} has at most $(2/3)^{i+1}n$ vertices.

Suppose that the DECOMPOSITION ALGORITHM terminates in STEP $k + 1$. Since we assumed that $\mathrm{CR}(G) < 10^{-26}\frac{e(G)^3}{n(G)^2}$, components G_j^i also have low crossing numbers. Therefore, when we apply Theorem 14, we do not remove too many edges. It can be verified by a rather technical but straightforward calculation [40, 55] that the total number of edges removed in the procedure is at most $e/2$.

So for the number of edges of the graph G^k obtained in the final step of the DECOMPOSITION ALGORITHM we have $e(G^k) \geq \frac{e}{2}$.

On the other hand, we know by the stopping rule that each component of G^k has at most e/n vertices. Therefore, even if each component is a complete graph, G^k has less than $e/2$ edges which is the desired contradiction. This concludes the proof. The details can be found in [55] and in [40]. □

This proof of the Crossing Lemma appeared first in the paper of Pach, Spencer and Tóth [55] in a more general context.

A graph property $\mathcal{P}$ is *monotone* if the following two conditions are satisfied: (i) if G satisfies $\mathcal{P}$, then every subgraph of G satisfies $\mathcal{P}$ as well, and (ii) if G_1 and G_2 satisfy $\mathcal{P}$, then their disjoint union also satisfies $\mathcal{P}$. Let $\mathrm{ex}(n, \mathcal{P})$ be the maximum number of edges of a simple graph of n vertices that satisfy property $\mathcal{P}$.

Theorem 15 ([55]) *Let $\mathcal{P}$ be a monotone graph property with $\mathrm{ex}(n, \mathcal{P}) = O(n^{1+\alpha})$. Then there are two constants, $c, c' > 0$ such that the crossing number of any simple graph G with property $\mathcal{P}$ with n vertices and $e > cn\log^2 n$ edges satisfies*

$$\mathrm{CR}(G) \geq \frac{e^{2+1/\alpha}}{n^{1+1/\alpha}}.$$

If $\mathrm{ex}(n, \mathcal{P}) = \Theta(n^{1+\alpha})$, then this bound is asymptotically sharp.

When $\mathcal{P}$ is just the property that the graph is simple, we get the Crossing Lemma. Very recently, Chen and Ma [17] proved a slightly stronger version of Theorem 15, they replaced the condition $e \geq cn\log^2 n$ by the weaker condition $e \geq cn$. The proof is a more general version of the proof of the Weak Crossing Lemma above, using the bisection width. We show now why is it asymptotically tight when $\text{ex}(n, \mathcal{P}) = \Theta(n^{1+\alpha})$. The construction is basically the same as in the Crossing Lemma or in Theorem 12. We take a disjoint union of extremal graphs of proper sizes. Let H_m be a graph of m vertices, $Cm^{1+\alpha}$ edges, satisfying Property $\mathcal{P}$. Clearly, $\text{CR}(H_m) < C^2m^{2+2\alpha}$. Let $n > 0$, $e > cn$ and $e = O(n^{1+\alpha})$. Let $m = (e/Cn)^{1/\alpha}$ and let G be n/m disjoint copies of H_m. This graph has n vertices, roughly e edges and $\text{CR}(G) = O\left(\frac{e^{2+1/\alpha}}{n^{1+1/\alpha}}\right)$.

Let $\mathcal{P}$ be the property that the graph does not contain some fixed graphs as subgraphs. This is a monotone property, so Theorem 15 can be applied and one can get very interesting results [55]. Here is a prominent example, when the forbidden subgraph is a C_4. It is known that $\text{ex}(n, C_4) \approx n^{3/2}/2$ [24].

Corollary 1 *Suppose that G does not contain C_4 as a subgraph and $e(G) \geq 4n(G)$. Then*

$$\text{CR}(G) \geq 10^{-12}\frac{e^4}{n^3}$$

and this bound is tight apart from the value of the constant.

The bisection method was used again by Kaufmann, Pach, Tóth and Ueckerdt [40] to prove a common generalization of Theorems 9, 10, 11, 12, 15.

A *drawing style* $\mathcal{D}$ is a property of topological multigraphs or drawings. A topological multigraph is in drawing style $\mathcal{D}$ if it satisfies property $\mathcal{D}$.

Let $\mathcal{D}$ be a drawing style and let M be a topological multigraph in drawing style $\mathcal{D}$. The $\mathcal{D}$-bisection width $b_{\mathcal{D}}(M)$ of M is the minimum number of edges whose removal splits M into two topological multigraphs, M_1 and M_2 in drawing style $\mathcal{D}$, and $|V(M_1)|, |V(M_2)| \geq |V(M)|/5$.

A *vertex split* in a topological multigraph is the following operation. Replace a vertex v by two vertices, v_1 and v_2, very close to it, and modify edges adjacent to v in a small neighborhood of v so that they are connected to either v_1 or v_2, without creating additional crossings. A drawing style $\mathcal{D}$ is *monotone*, if removing edges keeps the drawing in $\mathcal{D}$. It is *split-compatible*, is vertex splits keep the drawing in $\mathcal{D}$. The maximum degree in a topological multigraph M is denoted by $\Delta(M)$.

Theorem 16 ([40]) **Generalized Crossing Lemma** *Suppose that $\mathcal{D}$ is a monotone and split-compatible drawing style and there are constants $k_1, k_2, k_3, a > 0$ such that the following properties hold for any topological multigraph M in drawing style $\mathcal{D}$.*
(i) if $\text{CR}(M) = 0$, then $e(M) \leq k_1n(M)$,
(ii) $b_{\mathcal{D}} \leq k_2\sqrt{\text{CR}(M) + \Delta(M)e(M) + n(M)}$,
(iii) $e(M) \leq k_3n(M)^{1+a}$.

Then there is a constant $\alpha > 0$ such that for any topological multigraph M in drawing style $\mathcal{D}$, with $e(M) \geq (k_3 + 1)n(M)$ we have

$$\text{CR}(G) \geq \alpha\frac{e(M)^{2+1/a}}{n(M)^{1+1/a}}.$$

4 Non-homotopic Drawings

What is the "weakest possible" condition that guarantees many crossings in terms of the number of edges? We certainly have to rule out the trivial example from the previous section, that is, arbitrarily many loops or parallel edges, without crossings. A natural candidate is that we do not allow that an edge can be continuously deformed into another edge without passing through vertices. More precisely, consider the drawing on a vertex-punctured plane (but it could be the sphere as well) that is, vertices are removed, and we assume that no two parallel edges are homotopic.

In this case there is again a positive answer, shown by Pach, Tardos and Tóth [60]. A multigraph G, drawn in the plane, punctured at the vertices, with no homotopic edges is called a *non-homotopic topological multigraph*. If it is clear from the context, we do not make a notational distinction between a non-homotopic topological multigraph and its underlying abstract multigraph. The *crossing number* $\text{CR}(G)$ of a non-homotopic topological multigraph G is simply the number of edge crossings in its drawing. Define the function $\text{cr}(n, m)$ as the minimum crossing number of a non-homotopic multigraph with n vertices and m edges.

Theorem 17 ([60]) *The crossing number of a non-homotopic topological multigraph G with $n > 1$ vertices and $m > 5n$ edges satisfies $\text{cr}(G) \geq \frac{1}{40}\frac{m^2}{n}$. That is, $\text{cr}(n, m) \geq \frac{1}{40}\frac{m^2}{n}$.*

This bound is tight up to a polylogarithmic factor.

Theorem 18 ([60]) *For any $n \geq 2$, $m > 4n$, there exists a non-homotopic topological multigraph G with n vertices and m edges such that its crossing number satisfies $\text{cr}(G) \leq 100\frac{m^2}{n}\log_2^2\frac{m}{n}$. In other words, $\text{cr}(n, m) \leq 100\frac{m^2}{n}\log_2^2\frac{m}{n}$.*

Theorem 17 essentially says that if the number of non-homotopic edges is large, then an average pair of edges intersects $1/12n$ times. But it is natural to conjecture that the more edges we have the more intersections we need on an average pair of edges. That is, Theorem 17 can be improved. The following result is the first step in this direction.

Theorem 19 ([60]) *The minimum crossing number of a non-homotopic multigraph with $n \geq 2$ vertices and m edges is super-quadratic in m. That is, for any fixed $n \geq 2$, we have*

$$\lim_{m\to\infty}\frac{\text{cr}(n, m)}{m^2} = \infty.$$

Explicitly, we have

$$\frac{\text{cr}(n, m)}{m^2} = \begin{cases} \Omega\left(\frac{(\log m/\log\log m)^{1/6}}{n^8}\right) & \text{for } m > 4n, \\ \Omega\left(\log^{2/3} m\right) & \text{for fixed } n. \end{cases} \tag{4}$$

Recently, Hubard and Parlier [35] improved Theorem 19 and showed that the bound in Theorem 18 is asymptotically tight. Their proof uses the geometry of hyperbolic surfaces.

Theorem 20 ([35]) *The minimum crossing number of a non-homotopic multigraph with $n \geq 2$ vertices and m edges satisfies*

$$\frac{\mathrm{cr}(n,m)}{m^2} = \Omega\left(\log^2(m/n) - 256n\right) \tag{5}$$

A topological multigraph is called *k-crossing* if any two (one) edges (edge) can intersect (itself) at most k times. Estimating the minimum number of crossings in a non-homotopic multigraph turned out to be closely related to the following well studied topological problem.

Problem 1 What is the maximum number $f(n,k)$ of edges of a non-homotopic k-crossing multigraph of n vertices?

This problem, together with many related problems, has a long and interesing history, [1, 9, 21, 29, 31, 34, 38, 46, 47, 49, 50].

Juvan et al. [38] proved that $f(n,k)$ exists for any $n, k > 0$. They showed that

$$f(n,k) \leq (nk)^{O(nk^2)} \tag{6}$$

It was improved very recently by Girão, Illingworth, Scott, and Wood [29] to

$$f(n,k) \leq 6^{13n(k+1)}. \tag{7}$$

Aougab and Souto [9] showed that for any fixed n,

$$f(n,k) \leq 2^{O(\sqrt{k})}. \tag{8}$$

Juvan et al. [38] and Aougab, Souto [9] studied the problem in a more general context, not only on the plane but also on other surfaces.

Pach, Tardos and Tóth [60] noticed that any upper bound on $f(k,n)$, for every k, n, implies that

$$\lim_{m\to\infty} \frac{\mathrm{cr}(n,m)}{m^2} = \infty.$$

They proved a very weak upper bound on $f(n,k)$, namely, $f(n,k) \leq 2^{(2k)^{2n}}$ which appeared in the preliminary version of their paper [60]. This implies that

$$\frac{\mathrm{cr}(n,m)}{m^2} = \Omega\left(\frac{(\log m)^{1/6n}}{n^7}\right).$$

Then they learned about the above mentioned, much stronger results. The bound (6) of Juvan et al. [38] implies that

$$\frac{\mathrm{cr}(n,m)}{m^2} = \Omega\left(\frac{(\log m/\log\log m))^{1/6}}{n^8}\right).$$

The result (7) of Girão, Illingworth, Scott, and Wood ([29] implies that

$$\frac{\mathrm{cr}(n,m)}{m^2} = \Omega\left(\frac{(\log m)^{1/3}}{n^{10/3}}\right).$$

The proofs of these three upper bounds are similar, first they construct a "frame" from some of the edges and investigate how the other edges can cross this "frame".

Finally, for any fixed n, it follows from the bound of Aougab and Souto (8) that

$$\frac{\mathrm{cr}(n,m)}{m^2} = \Omega\left(\log^{2/3} m\right).$$

The best known general lower bound on $f(n,k)$ is the following.

Theorem 21 ([60]) *For any $k \geq n \geq 2$ integers we have*

$$f(n,k) \geq 2^{\sqrt{nk}/3}.$$

This result asymptotically matches the upper bound of Aougab and Souto (8) for any fixed n. It also implies Theorem 18.

Now we sketch the proofs Theorems 17, 18 and 21. The full proofs can be found in [60]. We start with some preparation.

We say that a topological multigraph is *loose* if no pair of *distinct* edges cross each other. An edge can cross itself. We start by finding the maximum number of edges in a loose non-homotopic multigraph in the plane. We will see that despite allowing parallel edges, loops, and self-intersections, loose non-homotopic multigraphs with $n > 2$ vertices have at most $4n-3$ edges, just slightly more than planar graphs. The proof is easier if we work on the sphere instead of the plane first. A non-homotopic multigraph on the sphere is defined analogously to those in the plane. We show that the maximum number of edges of a loose non-homotopic multigraph on the sphere with $n > 2$ vertices is $4n-6$.

A loop is trivial if it is contractible to a point. Clearly, a non-homotopic multigraph (in the plane or sphere) contains at most n trivial loops, one at each vertex. Note that here we use a slightly different terminology than in [60]. In [60] trivial loops were not allowed. Therefore, the bounds and calculations are slightly different.

Lemma 1 *On the sphere, any loose non-homotopic multigraph with $n > 2$ vertices has $m \leq 4n-6$ edges. For $n = 2$, the maximum number of edges is* 3.

Proof Clearly, in a maximal loose non-homotopic multigraph there are exactly n trivial loops, one at each vertex, therefore, the following statement is equivalent to the Lemma.

Proposition 1 *On the sphere, any loose non-homotopic multigraph with $n > 2$ vertices and no trivial loops has $m \leq 3n - 6$ edges. For $n = 2$, the maximum number of edges is* 1.

Now we prove Proposition 1. We omit the case $n = 2$ here, so let $n > 2$.

Assume for a contradiction that there is a loose non-homotopic multigraph H on the sphere, which is a counterexample to the proposition. Let H to be a counterexample with *minimum* number of crossings, and among those, the maximum number of edges. Let n and m be the number of vertices and edges of H.

Suppose first that there is no edge crossing in H, so we have a *plane* drawing. Then we can also assume that H is connected, otherwise two components could be joined by an extra edge without creating a crossing. If each face has at least three edges on its boundary (counted with multiplicity) then a standard application of Euler's formula shows that $m \leq 3n - 6$, a contradiction. Therefore, H must have a face bounded by one or two edges. Consequently, this face is bounded by a trivial loop or two parallel edges, or one edge on both sides. The first two possibilities contradict the assumptions. In the last case we have $n = 2$ which is again a contradiction.

For the rest of the proof we can assume that H has at least one self-intersecting edge e. Find a minimal interval γ of e between two occurrences of the same intersection point p. This simple closed curve partitions the sphere S^2 into two connected components. We call them (arbitrarily) the *left* and *right sides* of γ. Obviously, e is the only edge that may run between these sides. Let H_1 and H_2 be the subgraphs of $H - e$ induced by the vertices in the left and right sides of γ, respectively. Both of them are loose topological multigraphs, but they may contain homotopic edges, and a trivial loop. Add p to both of them as an isolated vertex, now both of them are non-homotopic with no trivial loops.

If an endpoint u of e lies in the left part, then by adding to H_1 a non-self-intersecting edge connecting p and u along e, we create no new intersection and do not violate the non-homotopic condition either. The resulting topological multigraph H_1' is a loose non-homotopic multigraph on the sphere with n_1 vertices and m_1 edges. Analogously, we can construct the loose non-homotopic multigraph H_2' from H_2. Denote its number of vertices and edges by n_2 and m_2, respectively. We have $n_1 + n_2 = n + 2$ and $m_1 + m_2 \geq m$. We eliminated a self-crossing (of e) and did not add any new crossings, so the number of crossings in both H_1' and H_2' are smaller than the number of crossings in H.

If $n_1, n_2 > 2$, then we have $m_1 \leq 3n_1 - 6$ and $m_2 \leq 3n_2 - 6$, by the minimality of H. Summing up these inequalities, we get $m \leq 3n - 6$, contradicting our assumption that H was a counterexample.

If $n_1 = 1$ or $n_2 = 1$, then all vertices of H lie on the same side of γ. In this case, by deleting γ from e, the homotopy class of e remains the same. Hence, the resulting topological multigraph is still a loose non-homotopic multigraph with n vertices and m edges, but its crossing number is smaller than that of H, contradicting the minimality of H.

Finally, consider the case $n_1 = 2$ or $n_2 = 2$. By symmetry, we can assume that $n_1 = 2$, $n_2 = n$, so we have a single vertex u of H on the left side of γ and $n - 1$

vertices on the right side. No edge of $H - e$ can lie in the left side since it would be a trivial loop. If e has at least one endpoint in the right part, then we have $m_2 = m$. This implies that H_2' is another counterexample to the lemma with fewer crossings, contradicting the minimality of H.

Therefore, e must be a loop at u. Since it is not a trivial loop, it separates a pair of vertices, $v, w \in V(H) - u$ from each other. But then we can draw another loop e' along some parts of e, with no self-intersection, so that it also separates v and w. Therefore, e' is not trivial either. See Fig. 2.

Fig. 2 The replacement of e by e' in the proof of Proposition 1

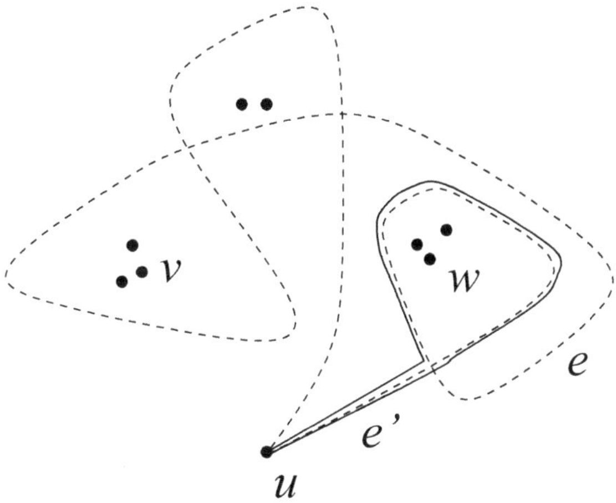

Let H' be the topological multigraph obtained from H by replacing e by e'. There is no other loop in H' homotopic to e', because there is no other loop at u. Hence, H' is a loose non-homotopic multigraph with no trivial loops, the same number of edges but fewer crossings than H, contradicting the minimality of H. For the lower bound, for $n \geq 3$, just take a triangulation, which has $3n - 6$ edges. This concludes the proof of the Proposition 1 and Lemma 1. □

Lemma 2 *In the plane, the maximum number of edges of a loose non-homotopic multigraph with $n > 1$ vertices is $4n - 3$.*

Proof Let H be a loose non-homotopic multigraph in the plane with $n \geq 1$ vertices and m edges. Consider the plane as the sphere S^2 with a point p^* removed. Add p^* to H as an additional vertex, and the trivial loop at p^* to obtain a loose non-homotopic multigraph H' with $n + 1$ vertices and $m + 1$ edges. If $n > 1$, applying Lemma 1 to H', we obtain that $m \leq 4n - 3$, as required. To see that the bound is tight, take a triangulation with n vertices and $3n - 6$ edges. Let uvw be the boundary of the unbounded face. Add another non-self-intersecting edge connecting u and v in the unbounded face, which is not homotopic with the arc uv of uvw. Add two further loops at u. First, a simple loop l that has all other edges and vertices (except u) in its interior, and then another loop l' outside of l, which goes twice around l. (Of course, l' must be self-intersecting.) And finally, add a trivial loop at each vertex. □

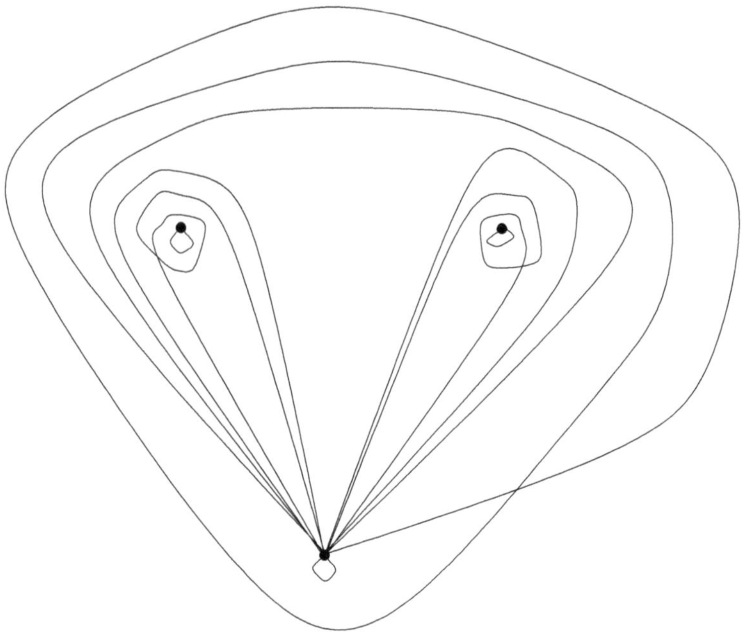

Fig. 3 A non-homotopic loose multigraph in the plane with 3 vertices and 9 edges, all of which are loops

There are many other, different constructions that show that the bounds in Lemmas 1 and 2 are tight. See also Fig. 3.

Proof of Theorem 17 Let G be a non-homotopic topological multigraph in the plane with $n > 1$ vertices and $m > 5n$ edges.

Let D denote the *non-crossing graph* of the edges of G, that is, let $V(D) = E(G)$ and connect two vertices of D by an edge if and only if the corresponding edges of G do not share an interior point. Any clique in D corresponds to a loose non-homotopic sub-multigraph of G. Therefore, by Lemma 2, D has no clique of size $4n - 2$. Thus, by Turán's theorem [68],

$$|E(D)| \leq \frac{|V(D)|^2}{2}\left(1 - \frac{1}{4n-3}\right) = \frac{m^2}{2}\left(1 - \frac{1}{4n-3}\right).$$

The crossing number $\mathrm{cr}(G)$ is at least the number of crossing pairs of edges in G, which is equal to the number of non-edges of D. Since $m > 5n$, we have

$$\mathrm{cr}(G) \geq \binom{m}{2} - \frac{m^2}{2}\left(1 - \frac{1}{4n-3}\right) < \frac{m^2}{8}\left(\frac{1}{n-1} - \frac{4}{m}\right) > \frac{m^2}{8}\left(\frac{1}{n} - \frac{4}{5n}\right) = \frac{1}{40}\frac{m^2}{n},$$

as claimed. □

The proof above gives a lower bound on the number of crossing *pairs of edges* in G, and in this respect it is tight up to a constant factor. Suppose that n is even and m is divisible by n. Let G_0 be a non-homotopic topological multigraph with two vertices and $\frac{2m}{n}$ non-homotopic loops on one of its vertices. Taking $\frac{n}{2}$ disjoint copies of G_0, we obtain a non-homotopic topological multigraph with n vertices, m edges, and fewer than $\frac{m^2}{n}$ crossing pairs of edges.

Now we construct many, pairwise homotopic loops in the punctured plane such that any two intersect at most k times.

Proof of Theorem 21 Let $S = \mathbb{R}^2 \setminus \{a_1, \dots, a_n\}$, where $a_1, \dots, a_n$ are distinct points in the plane, and let $x \in S$ be also fixed. Assume without loss of generality that $a_i = (i, 0)$, $1 \le i \le n$, and $x = (0, -1)$. An *x-loop* is a (possibly self-crossing) oriented path in S from x to x, i.e., a continuous function $f\colon [0, 1] \to S$ with $f(0) = f(1) = x$.

The homotopy group of S is the free group F_n generated by $g_1, \dots, g_n$, where g_i can be represented by a triangular x-loop around a_i, for example the one going from x to $(2i-1, 1)$, from here to $(2i+1, 1)$, and then back to x along three straight-line segments; see [43].

We define an *elementary loop* to be a polygonal x-loop with intermediate vertices

$$(1, \varepsilon_1),\ (2, \varepsilon_2),\ \dots,\ (n, \varepsilon_n),\ (n+1, -1),$$

in this order, where each ε_i $(1 \le i \le n)$ is equal either to $1/2$ or to $-1/2$. There are 2^n distinct elementary loops, depending on the choice of the ε_i. Each of them represents a distinct homotopy class of the form $g_{i_1} \cdots g_{i_t}$, where the indices form a strictly increasing sequence. By making infinitesimal perturbations on the interior vertices of the elementary loops, we can make sure that every pair of them intersect in at most $n-1$ points. Thus, we have $f(n, n) \ge 2^n$. See Fig. 4.

If $n \le k \le 9n$, we have $f(n, k) \ge f(n, n) \ge 2^n \ge 2^{\sqrt{nk}/3}$, and we are done.

In the case $k > 9n$, we consider all x-loops which can be obtained as the product (concatenation) of $j = \left\lfloor \sqrt{\frac{k-1}{n}} \right\rfloor \ge 3$ elementary loops. Unfortunately, some of these concatenated x-loops will be homotopic. For example, if the elementary loops l_1, l_2, l_3, and l_4 represent the homotopy classes g_1, g_2g_3, g_1g_2, and g_3, respectively, then l_1l_2 and l_3l_4 are homotopic. To avoid this complication, we only use the 2^{n-1} elementary loops that represent homotopy classes involving g_1 (that is, the ones with $(1, 1/2)$ as their first intermediate vertex). Concatenating j such elementary loops, we obtain $2^{j(n-1)}$ different x-loops, no pair of which are homotopic. By infinitesimal perturbation of the interior vertices of these x-loops (including the $j-1$ interior vertices at x), we can ensure that they do not pass through x, and no two polygonal paths corresponding to a single elementary loop intersect more than n times. Therefore, any pair of perturbed concatenated loops cross at most $j^2 n < k$ times, and the same bound holds for the number of self-intersections of any concatenated loop. This yields that $f(n, k) \ge 2^{j(n-1)} \ge 2^{\sqrt{nk}/3}$, completing the proof of the theorem. □

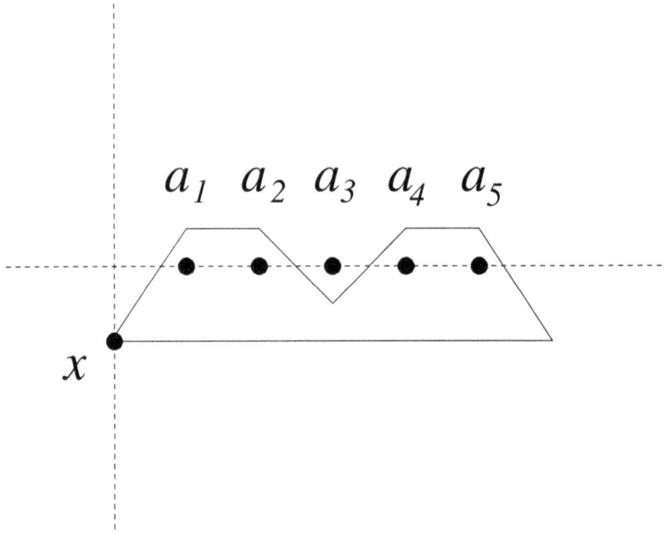

Fig. 4 An elementary x-loop ($n = 5$)

5 Other Crossing Numbers

The crossing number of a graph G is usually defined as "the minimum number of edge crossings in any drawing of G in the plane" [14]. But this definition can be interpreted in several different ways. Sometimes it is assumed that in a drawing two edges have at most one common point, which is a crossing or a common endpoint ([12, 70]). Many authors do not make this assumption ([30, 64, 69]). If two edges are allowed to cross several times, we may count their intersections with multiplicity or without [8]. We can impose some further restrictions on the drawings (e.g., the edges must be straight-line segments [37], or polygonal paths of length at most k [11]). No matter what definition we use, the determination of the crossing number of a graph appears to be an extremely difficult task ([12, 30]). In fact, we do not even know the value of *any* of the above quantities for the complete graph K_n with n vertices and for the complete bipartite graph $K_{n,n}$ with $2n$ vertices, for large n.

In his famous book about crossing numbers, Schaefer [62] lists more than 100 different versions of the crossing number. The systematic study of different versions of the crossing number and their relationships started with the work of Pach and Tóth [57, 58]. See also the dynamic survey of Schaefer [61].

The most commonly used version of the crossing number is simply called the *crossing number* of a graph G, $\text{CR}(G)$, is the minimum number of crossings (crossing points) over all drawings of G. (Here we assume that no three edges cross at the same point, or these crossings are counted with multiplicity, and not edge goes through a vertex.) The *pair-crossing number*, $\text{PCR}(G)$, is the minimum number of pairs of crossing edges over all drawings of G. As we have seen in Observation 1, in an

Table 1 Nine versions of the crossing number

Rule +	$\mathrm{OCR}_*(G) \le \mathrm{OCR}_+(G)$	$\mathrm{PCR}_+(G)$	$\mathrm{CR}(G)$
Rule 0	$\mathrm{OCR}(G)$	$\mathrm{PCR}(G)$	
Rule –	$\mathrm{OCR}_-(G)$	$\mathrm{PCR}_-(G)$	$\mathrm{CR}_-(G)$

optimal drawing for $\mathrm{CR}(G)$, any two edges cross at most once. Therefore, it is not easy to see the difference between these two definitions. Indeed, there was some confusion in the literature between these two notions, until the systematic study of their relationship. Clearly, $\mathrm{PCR}(G) \le \mathrm{CR}(G)$ for every G, and in fact, we cannot rule out the possibility, that $\mathrm{CR}(G) = \mathrm{PCR}(G)$ for every graph G. Undoubtedly, it is the most interesting open problem in this area. Any progress would be a great step in understanding the crossing number. From the other direction, after several steps, the best known bound is $\mathrm{CR}(G) = O(\mathrm{PCR}(G)^{3/2} \log \mathrm{PCR}(G))$ [62].

The *odd-crossing number*, $\mathrm{OCR}(G)$, is the minimum number of pairs of edges that cross an *odd* number of times, over all drawings of G. Clearly, for every graph G,

$$\mathrm{OCR}(G) \le \mathrm{PCR}(G) \le \mathrm{CR}(G).$$

This definition was partly inspired by the weak Hanani-Tutte theorem [16, 52], which states that if $\mathrm{OCR}(G) = 0$, then G is planar, that is, $\mathrm{OCR}(G) = \mathrm{PCR}(G) = \mathrm{CR}(G) = 0$. It was shown in [52] that for $k = 1, 2, 3$, if $\mathrm{OCR}(G) = k$, then $\mathrm{OCR}(G) = \mathrm{PCR}(G) = \mathrm{CR}(G) = k$. There are examples where OCR is different from PCR and CR, there is an infinite family of graphs with $\mathrm{OCR}(G) < 0.855 \cdot \mathrm{PCR}(G)$ [67], and there is also a slightly weaker ($\mathrm{OCR}(G) < 0.866 \cdot \mathrm{PCR}(G)$), completely different construction [53]. From the other direction we only have $\mathrm{PCR}(G) < 2\mathrm{OCR}(G)^2$ [57].

There are further distinctions of these crossing numbers according to the roles of crossings between adjacent edges. For each of CR, PCR, OCR, there are three natural counting rules:

Rule $+$: Only those drawings are considered, where adjacent edges cannot cross.
Rule 0: Adjacent edges are allowed to cross and their crossings are counted as well.
Rule $-$: Adjacent edges are allowed to cross and their crossings are not counted.

Combining these rules with the three crossing numbers, we get nine possibilities. But it is easy to see that $\mathrm{CR}_+ = \mathrm{CR}$ [57]. On the other hand, regarding Rule $+$ for the odd-crossing number, it seems more natural to assume that adjacent edges cross an *even number of times* than to assume that they do not cross at all. So, let $\mathrm{OCR}_*(G)$ be the minimum number of odd-crossing pairs of edges over all drawings of G where adjacent edges cross an even number of times. Therefore, we have nine versions, see Table 1. In this table, values do not decrease if we move to the right or up, and it was shown in [53] that $\mathrm{CR}(G) < 2\mathrm{OCR}_-(G)^2$. On the other hand, for the greatest surprise of our community there are graphs G, where $\mathrm{OCR}_-(G) < \mathrm{OCR}(G)$ [28].

For each of these crossing numbers, we can get rid of self-crossings just like in the proof of Observation 1. So we can assume without loss of generality, that *self-crossings are not allowed.*

We show that the basic version of the Crossing Lemma (Theorem 1) holds for all nine versions of the crossing number.

Theorem 22 (Crossing Lemma for other crossing numbers) *Let $c(G)$ be any of the parameters* $\mathrm{CR}(G)$, $\mathrm{CR}_-(G)$ $\mathrm{PCR}_+(G)$, $\mathrm{PCR}(G)$, $\mathrm{PCR}_-(G)$, $\mathrm{OCR}_*(G)$, $\mathrm{OCR}_+(G)$, $\mathrm{OCR}(G)$, $\mathrm{OCR}_-(G)$. *Then for any simple graph G, if $e(G) \geq 4n(G)$ then*

$$c(G) \geq \frac{1}{64}\frac{e(G)^3}{n(G)^2}.$$

For the proof we need the Strong Hanani-Tutte theorem [16, 63, 69].

Theorem 23 (Strong Hanani-Tutte theorem, [16]) *If a graph G can be drawn in the plane such that every pair of* independent *edges cross an even number of times, then it is planar.*

In other words, if $\mathrm{OCR}_-(G) = 0$, then it is planar so all other crossing numbers are also 0.

Proof of the Crossing Lemma for other crossing numbers. The proof is very similar to the proof of the original Crossing Lemma (Theorem 1). It goes again in two steps. It is enough to prove the statement for the smallest one, $\mathrm{OCR}_-(G)$.

1. Just like in the original proof, we can prove by induction on $e(G)$ that

$$\mathrm{OCR}_-(G) \geq e(G) - 3n(G). \tag{9}$$

If $e(G) \leq 3n(G)$, then the statement holds trivially. Suppose that $e(G) > 3n(G)$ and we already proved the statement for smaller values of $e(G)$. By the Strong Hanani-Tutte theorem, we have $\mathrm{OCR}_-(G) > 0$. Remove an edge that crosses an independent edge an odd number of times and use the induction hypothesis.

2. Consider a drawing of G with $\mathrm{OCR}_-(G)$ independent odd-crossing pairs of edges. Take a random induced subgraph G' of G by selecting each vertex independently with probability p. We obtain a random subset $U \subset V(G)$ and let $G' = G(U)$, consider it in the drawing inherited from the original drawing of G. We have $E[n(G')] = pn(G)$ and $E[e(G')] = p^2e(G)$. Let $ocr_-(G')$ be the number of pairs of odd-crossing independent edges of G'. Then clearly $ocr_-(G') \geq \mathrm{OCR}_-(G')$. The probability that an odd-crossing independent pair is present in the drawing of G' is p^4, so $E[ocr_-(G')] = p^4\mathrm{OCR}_-(G)$.

By (9), $\mathrm{OCR}_-(G') \geq e(G') - 3n(G')$. Just like in the proof of the Crossing Lemma, we take expectations of both sides, set $p = 4n(G)/e(G)$ and obtain that

$$\mathrm{OCR}_-(G) \geq \frac{1}{64}e(G)^3/n(G)^2.$$

□

Recall that the improvements of the coefficient in the original Crossing Lemma for $\text{CR}(G)$ are based on the following observation: if G has substantially more edges than $3n - 6$, then there is an edge with two, three, or even more crossings. See Sect. 2 for details. Using this idea, the linear inequality (1), $\text{CR}(G) \geq e(G) - 3n(G)$, has been replaced by stronger ones. Then the probabilistic argument is applied, analogous to Step 2 in the proof of Theorem 1. Observe that in this argument we can count only crossings between *independent* edges. Such a crossing appears in the random subgraph G' with probability p^4. A crossing between two *adjacent* edges appears with probability p^3, which is not enough (or in fact, too much) for the calculation. Fortunately, we have Observation 1 which guarantees that in a drawing with the minimum number of crossings, only independent edges cross.

For the pair-crossing number and the odd-crossing number, we have Theorem 22, and for the proof we needed the Strong Hanani-Tutte theorem instead of Observation 1 to guarantee crossings between independent edges. However, the proofs for the improvements do not work for the pair-crossing number and the odd-crossing number. As we will see, we can guarantee more crossings (crossing pairs, or odd-crossing pairs), but they might not be independent. In Sect. 2 we defined $m_k(n)$, the maximum number of edges of a simple graph of n vertices that has a drawing where every edge contains at most k crossings (these are the k-planar graphs). Ackerman and Schaefer [8] defined the following analogue: let $m_k''(n)$ be the maximum number of edges of a simple graph of n vertices that has a drawing where every edge is crossed by at most k other edges (no matter, how many times). By definition, $m_k''(n) \geq m_k(n)$. It is trivial that $m_0''(n) = 3n - 6$, and easy to see that $m_1''(n) = 4n - 8$ for $n \geq 12$. Ackerman and Schaefer proved that $m_2''(n) \leq 5n - 10$, which is tight for infinitely many values of n, and $m_3''(n) \leq 6n - 12$, which is probably not tight. (For comparision, we have $m_3(n) \leq 5.5n - 11$, which is tight apart from the additive constant.) The bounds for $m_k''(n)$ imply that

$$\text{PCR}(G) \geq 4e(G) - 18n(G) \tag{10}$$

Consequently, we also have

$$\text{PCR}_+(G) \geq 4e(G) - 18n(G) \tag{11}$$

In other words, if we do not allow non-independent crossings, then we have $4e - 18n$ independent crossings. And now we can apply the probabilistic argument to get the following result.

Theorem 24 (Improved Crossing Lemma for the pair-crossing number, [8]) *For any simple graph G, if $e(G) \geq 6.75n(G)$ then $\text{PCR}_+(G) \geq \frac{1}{34.2}\frac{e(G)^3}{n(G)^2}$.*

For the odd-crossing number, the situation is very similar. Let $m_k^{\text{odd}}(n)$ be the maximum number of edges of a simple graph of n vertices that has a drawing where every edge is crossed by at most k other edges *an odd number of times*. Clearly, we have $m_k^{\text{odd}}(n) \geq m_k''(n) \geq m_k(n)$. It follows from the Weak Hanani-Tutte theorem that $m_0^{\text{odd}}(n) = 3n - 6$. Karl and Tóth [41] proved that $m_1^{\text{odd}}(n) \leq 5n - 9$. They conjecture that this bound is far from the truth, probably $m_1^{\mathrm{odd}}(n) = 4n - 8$. It follows that

$$\mathrm{OCR}_+(G) \geq \mathrm{OCR}(G) \geq 2e(G) - 8n(G). \tag{12}$$

and then we obtain

Theorem 25 (Improved Crossing Lemma for the odd-crossing number, [41]) *For any simple graph G, if* $e(G) \geq 6n(G)$ *then* $\mathrm{OCR}_+(G) \geq \frac{1}{54}\frac{e(G)^3}{n(G)^2}$.

6 Open Problems

1. According to Theorem 6 at the end of Sect. 2 if $n \ll e \ll n^2$, then

$$\lim_{n\to\infty} \kappa(n, e)\frac{n^2}{e^3} = c > 0.$$

The constant c is called the *midrange crossing constant*, see also the end of Sect. 2. It is not known whether the condition $n \ll e \ll n^2$ can be replaced by a weaker condition, say, $C_1 n \leq e \leq C_2 n^2$. It is not hard to see that $(4+\varepsilon)n < e$ and $(1/2-\varepsilon)n^2 > e$ are necessary conditions. Find the weakest conditions on e in terms of n such that the statement of Theorem 6 holds.

We have $0.036 \leq c \leq 0.09$, improve these bounds. In particular, are the upper bound constructions in [18, 56] optimal, apart from lower order terms?

2. In Sect. 4 we have seen that there is a generalization of the Crossing Lemma for non-homotopic multigraphs [35]. Is there any other natural condition on the drawing that guarantees a Crossing Lemma-type statement for multigraphs?

3. According to the Strong Hanani-Tutte theorem, Theorem 23 if $\mathrm{OCR}_-(G) = 0$, then it is planar so all other crossing numbers are also 0. The statement also holds on the projective plane [54] and on the torus [26] but does not hold on orientable surfaces of genus at least 4 [25]. Does it hold on other surfaces without boundary?

4. Is it true that $\mathrm{CR}(G) = \mathrm{PCR}(G)$ for every graph G?

Or a less ambitious problem, is there a constant $c > 0$ such that $\mathrm{CR}(G) \leq c\mathrm{PCR}(G)$ for every graph G? The best bound we know is that $\mathrm{CR}(G) = O(\mathrm{PCR}(G)^{3/2} \log \mathrm{PCR}(G)$ [62].

5. Similar problem for the odd-crossing number, is there a constant $c > 0$ such that $\mathrm{CR}(G) \leq c\mathrm{OCR}(G)$ for every graph G? There are examples where OCR is different from PCR and CR, [53, 67] and from the other direction we only have $\mathrm{PCR}(G) < 2\mathrm{OCR}(G)^2$ [57].

6. In Sect. 6 we have seen that Theorem 1, that is, the simplest version of the Crossing Lemma, with constant 1/64, can be generalized for all versions of the crossing number, pair-crossing number, and odd-crossing number in Table 1. However, the improvements so far work only for the "+" versions.

Improve the constant in the Crossing Lemma for the other versions of the crossing numbers in Table 1.

Acknowledgements We are very grateful to Dömötör Pálvölgyi for his helpful comments, suggestions.

References

1. Aougab, T., Biringer, I., Gaster, J.: Packing curves on surfaces with few intersections. Int. Math. Res. Not. **16**, 5205–5217 (2019)
2. Angelini, P., Bekos, M., Kaufmann, M., Pfister, M., Ueckerdt, T.: Beyond-planarity: density results for bipartite graphs (2017)
3. Abrego, B., Cetina, M., Fernández-Merchant, S., Leaños, J., Salazar, G.: On $\leq k$-edges, crossings, and halving lines of geometric drawings of k_n. Discret. Comput. Geom. **48**(1), 192–215 (2012)
4. Ackerman, E.: On topological graphs with at most four crossings per edge. Comput. Geom. **85**, 101574 (2019)
5. Ajtai, M., Chvátal, V., Newborn, M., Szemerédi, E.: Crossing-free subgraphs. In: Theory and Practice of Combinatorics, volume 60 of North-Holland Mathematical Studies, pp. 9–12. North-Holland, Amsterdam-New York (1982)
6. Aichholzer, O., Duque, F., Fabila-Monroy, R., García-Quintero, O., Hidalgo-Toscano, C.: An ongoing project to improve the rectilinear and the pseudolinear crossing constants. J. Graph Algorithms Appl. **24**(3), 421–432 (2020)
7. Ágoston, P., Pálvölgyi, D.: An improved constant factor for the unit distance problem. Stud. Sci. Math. Hung. **59**(1), 40–57 (2022)
8. Ackerman, E., Schaefer, M.: A crossing lemma for the pair-crossing number. In: International Symposium on Graph Drawing, pp. 222–233. Springer, Berlin, Heidelberg (2014)
9. Aougab, T., Souto, J.: Counting curve types. Am. J. Math. **140**(6), 1423–1441 (2018)
10. Aigner, M., Ziegler, G.: Proofs from the Book. Springer, Berlin (1999)
11. Bienstock, D., Dean, N.: Bounds for rectilinear crossing numbers. J. Graph Theory **17**, 333–348 (1993)
12. Bienstock, D.: Some provably hard crossing number problems. Discret. Comput. Geom. **6**, 443–459 (1991)
13. Büngener, A., Kaufmann, M.: Improving the crossing lemma by characterizing dense 2-planar and 3-planar graphs (2024)
14. Bhatt, S.N., Leighton, F.T.: A framework for solving vlsi graph layout problems. J. Comput. Syst. Sci. **28**, 300–343 (1984)
15. Balogh, J., Lidicky, B., Salazar, G.: Closing in on hill's conjecture. SIAM J. Discret. Math. **33**(3), 1261–1276 (2019)
16. Chojnacki, C.: Über wesentlich unplättbare kurven im dreidimensionalen raume. Fundam. Math. **23**, 135–142 (1934)
17. Chen, K., Ma, J.: On a conjecture of pach-spencer-tóth for graph crossing numbers (2025)
18. Czabarka, É., Singgih, I., Székely, L., Wang, Z.: Some remarks on the midrange crossing constant. Stud. Sci. Math. Hung. **57**(2), 187–192 (2020)
19. Dey, T.K.: Improved bounds for planar k-sets and related problems. Discret. Comput. Geom. **19**, 373–382 (1998)
20. Didimo, W., Liotta, G., Montecchiani, F.: A survey on graph drawing beyond planarity. ACM Comput. Surv. **52**(1), 1–37 (2019)
21. Douba, S.: 2-systems of arcs on spheres with prescribed endpoints. Mich. Math. J. **71**, 321–346 (2022)

22. Erdős, P., Guy, R.K.: Crossing number problems. Am. Math. Monthly **80**, 52–58 (1973)
23. Elekes, G.: On the number of sums and products. Acta Arith **81**(4), 365–367 (1997)
24. Erdős, P., Rényi, A., Sós, V.T.: On a problem of graph theory. Stud. Sci. Math. Hung. **1**, 215–235 (1966)
25. Fulek, R., Kynčl, J.: Counterexample to an extension of the hanani-tutte theorem on the surface of genus 4. Combinatorica **39**(6), 1267–1279 (2019)
26. Fulek, R., Pelsmajer, M.J., Schaefer, M.: Strong Hanani-Tutte for the Torus (2020)
27. Fox, J., Pach, J., Suk, A.: On the number of edges of separated multigraphs. In: Graph Drawing and Network Visualization, pp. 223–227. Springer, Cham (2021)
28. Fulek, R., Pelsmajer, M., Schaefer, M., Štefankovič, D.: Adjacent crossings do matter. In: Graph Drawing, volume 7034 of Lecture Notes in Computer Science, pp. 343–354. Springer, Berlin, Heidelberg (2011)
29. Girão, A., Illingworth, F., Scott, A., Wood, D.R.: Non-homotopic Drawings of Multigraphs (2024)
30. Garey, M.R., Johnson, D.S.: Crossing number is np-complete. SIAM J. Algeb. Disc. Methods **4**, 312–316 (1983)
31. Greene, J.E.: On loops intersecting at most once. Geom. Funct. Anal. **29**, 1828–1843 (2019)
32. Guy, R.K.: The decline and fall of zarankiewicz's theorem. In: Proof Techniques in Graph Theory, pp. 63–69. Academic Press, New York (1969)
33. Harary, F., Hill, A.: On the number of crossings in a complete graph. Proc. Edinb. Math. Soc. **13**(2), 333–338 (1962)
34. Hubard, A., Parlier, H.: Crossing lemmas for k-systems of arcs (2024)
35. Hubard, A., Parlier, H.: Crossing number inequalities for curves on surfaces (2025)
36. Hong, S.-H., Tokuyama, T.: Beyond planar graphs, volume 1. Communications of NII Shonan Meetings (2020)
37. Jensen, H.F.: An upper bound for the rectilinear crossing number of the complete graph. J. Combin. Theory Ser. B **10**, 212–216 (1971)
38. Juvan, M., Malnič, A., Mohar, B.: Systems of curves on surfaces. J. Combin. Theory Ser. B **68**(1), 7–22 (1996)
39. Kaufmann, M.: Personal communication at the workshop "beyond-planar graphs: Algorithmics and combinatorics", pp. 6–11. Schloss Dagstuhl, Germany, November (2016)
40. Kaufmann, M., Pach, J., Tóth, G., Ueckerdt, T.: The number of crossings in multigraphs with no empty lens. In: Graph Drawing, volume 11282 of Lecture Notes in Computer Science, pp. 242–254. Springer, Cham (2018)
41. Karl, J., Tóth, G.: Crossing lemma for the odd-crossing number. Comput. Geom. **108**, 101901 (2023)
42. Leighton, F.T.: Complexity Issues in VLSI. MIT Press, Cambridge (1983)
43. Lyndon, R.C., Schupp, P.E.: Combinatorial Group Theory. Ergebnisse der Mathematik und ihrer Grenzgebiete, vol. 89. Springer, Berlin, New York (1977)
44. Lipton, R.J., Tarjan, R.E.: A planar separator theorem. SIAM J. Appl. Math. **36**, 177–189 (1979)
45. Matoušek, J.: Lectures on discrete geometry. Graduate Texts in Mathematics, vol. 212. Springer, New York (2002)
46. Mirzakhani, M.: Growth of the number of simple closed geodesics on hyperbolic surfaces. Ann. Math. **168**(1), 97–125 (2008)
47. Malestein, J., Rivin, I., Theran, L.: Topological designs. Geom. Dedicata. **168**, 221–233 (2014)
48. Pach, J., Radoicic, R., Tardos, G., Tóth, G.: Improving the crossing lemma by finding more crossings in sparse graphs. Discret. Comput. Geom. **36**(4), 527–552 (2006)
49. Przytycki, P.: Arcs intersecting at most once. Geom. Funct. Anal. **25**, 658–670 (2015)
50. Przytycki, P., Smith, C.: Arcs on spheres intersecting at most twice. Indiana Univ. Math. J. **68**, 157–178 (2019)
51. Pach, J., Shahrokhi, F., Szegedy, M.: Applications of the crossing number. Algorithmica **16**(1), 111–117 (1996)

52. Pelsmajer, M., Schaefer, M., Štefankovič, D.: Removing even crossings. J. Combin. Theory, Ser. B **97**, 489–500 (2007)
53. Pelsmajer, M., Schaefer, M., Štefankovič, D.: Odd crossing number and crossing number are not the same. Discrete and Computational Geometry, pp. 1–13 (2009)
54. Pelsmajer, M.J., Schaefer, M., Stasi, D.: Strong hanani-tutte on the projective plane. SIAM J. Discret. Math. **23**, 1317–1323 (2009)
55. Pach, J., Spencer, J., Tóth, G.: New bounds on crossing numbers. In: Proceedings of the fifteenth annual symposium on Computational Geometry, pp. 124–133 (1999)
56. Pach, J., Tóth, G.: Graphs drawn with few crossings per edge. Combinatorica **17**(3), 427–439 (1997)
57. Pach, J., Tóth, G.: Thirteen problems on crossing numbers. Geombinatorics **9**(4), 194–207 (2000)
58. Pach, J., Tóth, G.: Which crossing number is it anyway? J. Combin. Theory Ser. B **80**(2), 225–246 (2000)
59. Pach, J., Tóth, G.: A crossing lemma for multigraphs. Discret. Comput. Geom. **63**, 918–933 (2020)
60. Pach, J., Tardos, G., Tóth, G.: Crossings between non-homotopic edges. J. Combin. Theory Ser. B **156**, 389–404 (2022)
61. Schaefer, M.: The graph crossing number and its variants: a survey. Electron. J. Combin. (2012). DS21-May
62. Schaefer, M.: Crossing Numbers of Graphs. CRC Press (2018)
63. Schaefer, M., Štefankovič, D.: Removing independently even crossings. SIAM J. Discret. Math. **24**, 379–393 (2010)
64. Shahrokhi, F., Sýkora, O., Székely, L.A., Vrťo, I.: Crossing numbers: bounds and applications. In: Intuitive Geometry. volume 6 of Bolyai Society Mathematical Studies, pp. 179–206. János Bolyai Mathematical Society, Budapest (1997)
65. Solymosi, J., Tóth, C.D.: Distinct distances in the plane. Discret. Comput. Geom. **25**, 629–634 (2001)
66. Székely, L.: Crossing numbers and hard erdős problems in discrete geometry. Comb. Probab. Comput. **6**(3), 353–358 (1997)
67. Tóth, G.: Note on the pair-crossing number and the odd-crossing number. Discret. Comput. Geom. **39**(4), 791–799 (2008)
68. Turán, P.: On an extremal problem in graph theory. Matematikai és Fizikai Lapok **48**, 436–452 (1941)
69. Tutte, W.T.: Toward a theory of crossing numbers. J. Combin. Theory **8**, 45–53 (1970)
70. White, A.T., Beineke, L.W.: Topological graph theory. In: Selected Topics in Graph Theory, pp. 15–49. Academic Press, London-New York (1983)

Recent Progress in Ramsey Theory

Jacques Verstraete

Abstract The classical Ramsey numbers $r(s,t)$ denote the minimum n such that every red-blue coloring of the edges of the complete graph K_n contains either a red clique of order s or a blue clique of order t. These quantities are the centerpiece of graph Ramsey Theory, and have been studied for almost a century. The Erdős-Szekeres Theorem (1935) shows that for each $s \geq 2$, $r(s,t) = O(t^{s-1})$ as $t \to \infty$. The celebrated work of Kim (1995) together with the work of Ajtai et al. (1980) and Shearer (1983) shows $r(3,t) = \Theta(t^2/\log t)$ as $t \to \infty$. We introduce a new approach using pseudorandom graphs, which shows $r(4,t) = \Omega(t^3/(\log t)^4)$ as $t \to \infty$, answering an old conjecture of Erdős, and we illustrate how to apply this approach to many other Ramsey and related combinatorial problems.

1 Introduction

For integers $s,t \geq 2$, the centerpiece of graph Ramsey Theorey is the classical *Ramsey number* $r(s,t)$, which denotes the minimum n such that every n-vertex graph contains either a clique of order s or an independent set of size t. These quantities are the cornerstone of Ramsey Theory, and have been studied for many decades; their existence is proved by Ramsey's Theorem [74]—see also the book by Graham et al. [44]. The quantities $r(t,t)$ are sometimes referred to as *diagonal Ramsey Numbers*, whereas $r(s,t)$ for fixed s and $t \to \infty$ are referred to as *off-diagonal Ramsey numbers*. In a major recent breakthrough, Campos et al. [18] proved that there exists $\epsilon > 0$ such that for large enough t, $r(t,t) \leq (4-\epsilon)^t$. Explicit constructions giving lower bounds for Ramsey numbers have also been studied extensively, with many new results since the influential paper of Frankl and Wilson [39]. However, no explicit construction showing $r(t,t)$ is exponential in t is known.

Research supported by NSF awards DMS-1952786 and DMS-2347832.

J. Verstraete (✉)
Department of Mathematics, University of California, San Diego, San Diego, CA, USA
e-mail: jverstraete@ucsd.edu

G. O. H. Katona et al. (eds.), *Sum(m)it280*, Bolyai Society Mathematical Studies 32,
https://doi.org/10.1007/978-3-032-18810-6_17

The focus of this paper is to outline a new approach to Ramsey theory via pseudorandom graphs, based on the works [26, 64, 67]. These and related techniques have been used effectively for not only estimating Ramsey numbers [26], but also to *hypergraph Ramsey numbers* [53], *multicolor Ramsey numbers* [6, 47], *Ramsey minimal graphs* [7], *Erdős-Rogers functions* [43, 48, 69] and *coloring hypergraphs* [68]. A highlight of this approach is that it determines $r(4,t)$ up to a factor of order $\log^2 t$, as proved by Mattheus and the author [64]:

Theorem 1.1 *There exist constants $c_1, c_2 > 0$ such that for all $t \geq 3$,*

$$c_1 \frac{t^3}{\log^4 t} \;\leq\; r(4,t) \;\leq\; c_2 \frac{t^3}{\log^2 t}. \tag{1}$$

This solves a conjecture of Erdős [21], and also gives almost tight bounds on some multicolor Ramsey numbers.

1.1 Ramsey Numbers

Since the original bounds of Erdős and Szekeres [36] in 1935, estimates for $r(s,t)$ have been the subject of extensive research; for instance the conjecture $r(s,t) = t^{s-1}/\text{polylog}(t)$ as $t \to \infty$ dates back to at least 1947–see Erdős [29] and Chung and Graham [21]. The only off-diagonal Ramsey numbers $r(s,t)$ for $s \geq 3$ whose order of magnitude is known is $r(3,t)$, as it was shown by Kim [50] that $r(3,t) = \Omega(t^2/\log t)$ as $t \to \infty$, matching previous upper bounds by Ajtai et al. [1] and Shearer [77, 78], and improving earlier bounds of Spencer [79, 80]. The current state of the art is due to Campos et al. [19], namely as $t \to \infty$:

$$(1 - o(1))\frac{t^2}{3\log t} \;\leq\; r(3,t) \;\leq\; (1 + o(1))\frac{t^2}{\log t}. \tag{2}$$

This supersedes by a constant factor the earlier lower bound $(1 - o(1))t^2/4\log t$ as $t \to \infty$, due to Fiz Pontiveros, Griffiths and Morris [37] and Bohman and Keevash [13]. Erdős [21] proposed the following conjecture:

Conjecture 1.2 For all $s \geq 4$, there exists $c = c(s) > 0$ such that

$$r(s,t) \;\geq\; \frac{t^{s-1}}{(\log t)^c}. \tag{3}$$

The previous best bounds for $r(s,t)$ with fixed $s \geq 4$ are

$$a\frac{t^{\frac{s+1}{2}}}{(\log t)^{\frac{s+1}{2} - \frac{1}{s-2}}} \;\leq\; r(s,t) \;\leq\; (1 + o(1))\frac{t^{s-1}}{(\log t)^{s-2}} \tag{4}$$

where $a > 0$ is a constant depending only on s. The lower bound comes from the so-called *K_s-free random process*, studied by Bohman and Keevash [14], improving earlier bounds of Spencer [80], while the upper bound comes from results of Li et al. [58]. In light of Theorem 1.1, which solves Conjecture 1.2 for $s = 4$, we propose the following conjecture:

Conjecture 1.3 The order of magnitude of $r(4,t)$ is $t^3/(\log t)^2$ as $t \to \infty$.

In the next section, we present general theorems connecting pseudorandom graphs, spectral graph theory and Ramsey theory, which lead to Theorem 1.1 as well as more general results on related Ramsey problems.

1.2 Spectrally Extremal Graphs

An (n,d,λ)-*graph* is an n-vertex d-regular graph whose adjacency matrix has eigenvalues $\lambda_1 \geq \lambda_2 \geq \cdots \geq \lambda_n$ such that $\max\{|\lambda_2|, |\lambda_n|\} = \lambda$. A survey of these graphs and related notions of pseudorandom graphs is given by Krivelevich and Sudakov [56]. In the present context, the following theorem due to Mubayi and the author [67] connects (n,d,λ)-graphs with Ramsey Theory:

Theorem 1.4 *Let F be a graph. If there exists an F-free (n,d,λ)-graph with $4e^2\lambda \geq (\log n)^2$, then*

$$r(F,t) \quad > \quad \frac{n(\log n)^2}{8e^2\lambda}. \tag{5}$$

The key to effectively applying this theorem is to find an F-free (n,d,λ)-graph where λ is as small as possible relative to d, while d is as large as possible relative to n. In general, the Alon-Boppana Theorem [72] shows that $\lambda = \Omega(\sqrt{d})$ when d is not too close to n, so we refer to an (n,d,λ)-graph with $\lambda = O(\sqrt{d})$ as *spectrally extremal*. Using the approach of Sudakov et al. [81], one can in theory then determine an upper bound on the value of d relative to n in an F-free (n,d,λ)-graph; for example if $F = K_s$ then $d = O(n^{1-1/(2s-3)})$. Remarkably, directly using Theorem 1.4, this leads to the following theorem giving almost tight asymtotic bounds on $r(s,t)$ together with (4):

Theorem 1.5 *If there exists a spectrally extremal K_s-free (n,d,λ)-graph with $d = \Omega(n^{1-1/(2s-3)})$, then*

$$r(s,t) \quad = \quad \Omega\Big(\frac{t^{s-1}}{(\log t)^{2s-4}}\Big). \tag{6}$$

This would prove Conjecture 1.2. Such spectrally extremal triangle-free (n,d,λ)-graphs were constructed by Alon [3] and later by Kopparty [52], but it is not known for any $s \geq 4$ whether there exists a spectrally extremal K_s-free graph with $d = n^{1-1/(2s-3)-o(1)}$. The current record is due to Bishnoi et al. [11], with

$d = \Omega(n^{1-1/(s-1)})$ for $s \geq 4$. Since there exist spectrally extremal triangle-free n-vertex graphs with $d = \Theta(n^{2/3})$, as constructed in [2, 52], an application of Theorem 1.5 gives $r(3,t) = \Omega(t^2/(\log t)^2)$, which matches the lower bound obtained by direct application of the Lovász local lemma [79]. The best explicit constructions giving lower bounds on $r(s,t)$ for $s \in \{4, 5, 6\}$ are described by Kostochka et al. [54].

More generally, for a graph F and an integer $t \geq 2$, let $r(F,t)$ denote the minimum n such that every n-vertex F-free graph contains an independent set of size t. The notorious *cycle complete-graph Ramsey numbers* are concerned with the case $F = C_k$, that is, F is a cycle of length k. Perhaps one of the most challenging open conjectures in Ramsey Theory is the following conjecture of Erdős [21]:

Conjecture 1.6 There exists $\varepsilon > 0$ such that if t is large enough, then

$$r(C_4, t) \;\; \leq \;\; t^{2-\varepsilon}. \tag{7}$$

There are currently no techniques available for proving such an upper bound, and the author believes that determining the order of magnitude of $r(C_4,t)$ as $t \to \infty$ may be one of the most difficult open problems in Ramsey theory. The current best bounds for $r(C_4,t)$ are $r(C_4,t) = O(t^2/(\log t)^2)$ due to Szemerédi (see [31]) and $r(C_4,t) = \Omega(t^{3/2}/\log t)$, the latter coming from the random C_4-free process. A *polarity graph* of a projective plane of order q may be viewed as an (n,d,λ)-graph where $n = q^2+q+1$, $d = q+1$ and $\lambda = \sqrt{q}$, and these graphs do not contain cycles of length four. Applying Theorem 1.4 to these graphs, one immediately obtains $r(C_4,t) = \Omega(t^{3/2}/\log t)$. Using finite geometry, we obtain in a similar way the following bounds on $r(C_\ell,t)$, which sometimes exceed by polynomial factors in t the previous best known bounds from the random C_ℓ-free process:

Theorem 1.7 *As* $t \to \infty$,

$$r(C_6,t) \;=\; \Omega\left(\frac{t^{5/4}}{\log^{1/2} t}\right) \quad \text{and} \quad r(C_{10},t) \;=\; \Omega\left(\frac{t^{9/8}}{\log^{1/4} t}\right). \tag{8}$$

These results are obtained by applying Theorem 1.4 to the polarity graphs [57] of *generalized quadrangles* and *generalized hexagons*. The best upper bounds on cycle-complete graph Ramsey numbers are due to Caro et al. [20] for even cycles and Sudakov [82] for odd cycles. For a survey on extremal problems for cycles in graphs, see [84].

1.3 Transference from Extremal to Ramsey

The lack of dense spectrally extremal F-free graphs, and in particular K_4-free graphs, precludes an application of Theorem 1.4 to obtain Theorem 1.1. Instead, when F is a non-bipartite graph, we use a weaker notion of pseudorandom graphs combined with the following approach in the language of hypergraphs [26]. Let $r \geq 3$ and let H be an r-uniform hypergraph. We say H is a *linear hypergraph* if no two edges of H intersect in at least two vertices. In all subsequent material, we restrict our attention to linear hypergraphs. Let ∂H denote the graph whose vertex set is $V(H)$ and whose edge set consists of pairs of vertices in a common edge of H–this is the *shadow graph* of H. We say that H is *strongly F-free* if for every isomorphic copy $F' \subseteq \partial H$ of F, there exists an edge $e \in E(H)$ such that the subgraph of F' induced by $e \cap V(F')$ is non-bipartite. For instance, if F is a triangle then H does not contain three edges e_1, e_2, e_3 such that $|e_i \cap e_j| = 1$ for $i \neq j$ and $e_1 \cap e_2 \cap e_3 = \emptyset$. The following theorem [26] connects the problem of finding extremal strongly F-free hypergraphs to bounds on Ramsey numbers:

Theorem 1.8 *Let F be a non-bipartite graph, and suppose there exists an r-uniform n-vertex d-regular strongly F-free hypergraph, where $r \geq 2048(\log n)^3$. If $t = \lceil 256n(\log n)^2/rd \rceil$, then*

$$r(F,t) = \Omega\Big(\frac{rt}{\log n}\Big). \tag{9}$$

We shall show how Theorem 1.1 may be derived from Theorem 1.8. The approach used to prove Theorem 1.1 gives the current best bounds [26, 67] on cycle-complete graph Ramsey numbers $r(C_k, t)$ for all odd values of k and $k \in \{4, 6, 10\}$. For instance, in the case of the pentagon, we have the following result [26]—the upper bound in the following theorem is due to Sudakov [82]:

Theorem 1.9 *There exist constants $c_1, c_2 > 0$ such that for all $t \geq 3$,*

$$c_1 \frac{t^{10/7}}{(\log t)^{13/7}} \leq r(C_5, t) \leq \frac{t^{3/2}}{(\log t)^{1/2}}. \tag{10}$$

The proof of Theorem 1.9 uses a finite geometric object known as a twisted triality hexagon [26], which in the language of hypergraphs allows an application of Theorem 1.8. This theorem suggests that it is plausible that $r(C_5, t)$ has order of magnitude close to $t^{3/2}$:

Conjecture 1.10 $r(C_5, t) = t^{3/2 - o(1)}$ as $t \to \infty$.

1.4 Polynomial Hypergraph Ramsey Numbers

For a k-uniform hypergraph F, let $r_k(F, t)$ denote the minimum n such that every red-blue coloring of the subsets of $[n]$ of size k contains either a red copy of F or a set of t elements of $[n]$ all of whose subsets are colored blue—the k-uniform clique on t vertices, which is denoted K_t^k. We write $r_k(K_s^k, t) = r_k(s, t)$. In the case F is a complete graph on s vertices, $F = K_s^2$, we recover the Ramsey numbers

$r(s,t)$. The behavior of hypergraph Ramsey numbers $r_k(F,t)$ when $k \geq 3$ is well-known to be quite different to the case of graphs. The *polynomial to exponential transition* conjecture by Erdős and Hajnal [32] is a prime example. This conjecture was completely solved by Mubayi and Razborov [66] for k-uniform hypergraphs when $k \geq 4$. Concerning off-diagonal hypergraph Ramsey numbers for triples, it is known [25] that there exist constants $c_1, c_2 > 0$ such that

$$2^{c_1 t \log t} \leq r_3(4,t) \leq 2^{c_2 t^2 \log t}. \tag{11}$$

So $r_3(4,t)$ is at least super-exponential in t. It remains an open problem to determine the order of magnitude of $r_3(t,t)$. Erdős et al. [33] showed that for some constants $b_1, b_2 > 0$,

$$2^{b_1 t^2} \leq r_3(t,t) \leq 2^{2^{b_2 t}} \tag{12}$$

and conjectured $r_3(t,t)$ is double exponential in t. The study of $r_3(s,t)$ turns out to be important for studying $r_k(s,t)$, due to a remarkable construction of Erdős and Hajnal [32] called the *stepping up lemma*. In this space we are not able to do justice to the rich theory of hypergraph Ramsey numbers. For a survey on hypergraph Ramsey numbers, we refer the reader to Conlon et al. [25].

In this section, we consider hypergraph Ramsey numbers $r_3(F,t)$ which are polynomial in t. In recent work [23, 24], many triple systems F for which $r_3(F,t)$ is polynomial in t were characterized. An *iterated tripartite hypergraph* (ITH) is any triple system F obtained by starting with a tripartite hypergraph and iteratively placing a tripartite hypergraph in each of the parts. Erdős and Hajnal [32] proved that if F is ITH, then $r_3(F,t)$ is polynomial in t. The main open conjecture [24] is as follows:

Conjecture 1.11 Let F be a triple system. Then the hypergraph Ramsey number $r_3(F,t)$ is polynomial in t if and only if F is an ITH.

This conjecture is proved in [24] when F is *tightly connected*—that is to say for any two edges $e, f \in E(F)$, there exists a sequence of edges $e = e_0, e_1, \dots, e_r = f$ in F such that $|e_i \cap e_{i+1}| = 2$ for $0 \leq i < r$. In particular, if F is tightly connected and not ITH, then $r_3(F,t) = 2^{\Omega(n^{2/3})}$. It is shown further that if F is the union of two tightly connected hypergraphs which are not ITH, then $r_3(F,t) = 2^{\Omega(\log^2 n)}$. A salient open case of the conjecture is when F is the Fano plane— this is a linear triple system with seven edges and seven vertices, and is not ITH.

The order of magnitude of $r_3(F,t)$ is in general difficult to determine. In the work of Kostochka, Mubayi and the author [53], the question of the order of magnitude of $r(F,t)$ is raised when F is a *loose ℓ-cycle* C_ℓ–this consists of the edges $\{2i, 2i+1, 2i+2\}$ for $0 \leq i < \ell$ and arithmetic modulo 2ℓ. In fact, one may obtain good lower bounds along the lines of the proof of Theorem 1.8, and a special case is the lower bound in the following theorem, proved in [53]:

Theorem 1.12 *There exist constants $c_1, c_2 > 0$ such that*

$$\frac{c_1 t^{3/2}}{(\log t)^{3/4}} \leq r_3(C_3, t) \leq c_2 t^{3/2}. \tag{13}$$

It is conjectured that for $\ell \geq 4$, $r_3(C_\ell, t) = t^{\ell/(\ell-1)-o(1)}$ as $t \to \infty$. The following is also conjectured in [53]:

Conjecture 1.13 As $t \to \infty$, $r_3(C_3, t) = o(t^{3/2})$.

We refer the reader to [23, 63] for further examples where $r_3(F, t)$ is determined up to $t^{o(1)}$ factors.

1.5 Erdős-Rogers Functions

The construction in [64] has been used to derive bounds on quantities known as *Erdős-Rogers functions*. For an integer $t > s \geq 2$, the *Erdős-Rogers function* $f_{s,t}(n)$ is the maximum integer m such that every n-vertex K_t-free graph has a K_s-free induced subgraph with m vertices. These quantities were introduced by Erdős and Rogers [34] more than sixty years ago. These quantities are generalizations of Ramsey numbers, since $r(s, t)$ and $f_{2,t}(n)$ determine one another for each $t \geq 2$. New upper bounds on $f_{s,s+2}(n)$ were recently obtained by Janzer and Sudakov [48] using the construction in [64] and the method of containers [9, 76], as well as new results by Gowers and Janzer [43] on $f_{s,t}(n)$. However, there is no pair (s, t) with $t \geq s + 2 \geq 4$ for which it is known that $f_{s,t}(n) = n^{\alpha+o(1)}$ for some $\alpha = \alpha(s, t)$. We focus here on $f_{s,s+1}(n)$. Since the quantities $r(3, t)$ are known to within a constant factor, one deduces $f_{2,3}(n)$ has order of magnitude $\sqrt{n \log n}$ as $n \to \infty$. Building on earlier works of Dudek and Rödl [27], Dudek et al. [28] and Wolfovits [86], we significantly improve the upper bounds on $f_{s,s+1}(n)$ in [69] as follows:

Theorem 1.14 *For each fixed $s \geq 3$,*

$$f_{s,s+1}(n) \ \leq \ 2^{100s}\sqrt{n}\, \log n. \tag{14}$$

A lower bound $f_{s,s+1}(n) = \Omega(\sqrt{n \log n}/ \log\log n)$ can be obtained from results of Shearer [78]. Therefore this theorem incidentally shows $f_{s,s+1}(n) = n^{1/2+o(1)}$ for $s = o(\log n)$. Dudek et al. [28] asked whether for $s \geq 3$,

$$\lim_{n\to\infty} \frac{f_{s,s+1}(n)}{f_{s-1,s}(n)} = \infty \tag{15}$$

as $n \to \infty$, which would not be the case for $s \geq 4$ if our upper bound is tight up to a constant factor. We tentatively make the following conjecture:

Conjecture 1.15 For all $A > 0$ and $s \geq 3$, there exists $n_0(A, s)$ such that if $n \geq n_0(A, s)$, then

$$f_{s,s+1}(n) \geq A\sqrt{n \log n}. \tag{16}$$

It may be that $f_{s,s+1}(n) = \Theta(\sqrt{n} \log n)$ for all $s \geq 3$, matching our upper bound in Theorem 1.14. It seems likely that the K_{s+1}-free construction in [69] contains a K_s-free induced subgraph with $\Theta(\sqrt{n} \log n)$ vertices, but this was not verified in [69]. Conjecture 1.15 is in contrast to $f_{2,3}(n)$, which has order of magnitude $\sqrt{n \log n}$ as $n \to \infty$. The key barrier in proving lower bounds on say $f_{3,4}(n)$ is showing that in a K_4-free graph one can typically find induced triangle-free subgraphs with substantially more vertices than the independence number of the graph. We remark that it is not known whether every K_4-free n-vertex graph of maximum degree d has an independent set of size at least $n(\log d)/d$; the current best bound $n(\log d)/d \log \log d$ is due to Shearer [78]. It was shown in [65] that for infinitely many d, there exists an n-vertex K_4-free graph of maximum degree d in which every induced triangle-free subgraph has $O(n \log d/d)$ vertices.

One may consider more generally the function $f_{F,G}(n)$, which is defined to be the maximum size of an induced F-free subgraph that every G-free n-vertex graph must have. In the case $F = K_s$ and $G = K_{s+1}$, we recover the Erdős-Rogers functions defined above. A somewhat surprising phenomenon occurs when F is a triangle-free graph, proved in [70]:

Theorem 1.16 *For any non-empty triangle-free graph F and any graph G containing a triangle, there exist constants $a_1, a_2 > 0$ such that*

$$a_1\sqrt{\frac{n \log n}{\log \log n}} \leq f_{F,G}(n) \leq \sqrt{n} \cdot e^{a_2\sqrt{\log n \log \log n}}. \tag{17}$$

Strikingly, if F is a K_4-free graph and G contains K_4, the function $f_{F,G}(n)$ can vary substantially, in the sense that $f_{F,G}(n)$ can be as low as $n^{1/3+o(1)}$—namely when F is a tree using Theorem 1.1—and as high as $n^{1/2+o(1)}$ from the results of [8, 42] when F is sufficiently dense. The study of $f_{F,G}(n)$ in general is undertaken in the work of Balogh et al. [8], Gishboliner et al. [42] and Mubayi and the author [70].

2 Pseudorandom Graphs

If A is a square symmetric matrix, then the eigenvalues of A are real. When A is an n by n matrix, we denote them $\lambda_1 \geq \lambda_2 \geq \cdots \geq \lambda_n$. If the corresponding eigenvectors forming an orthonormal basis are $e_1, e_2, \ldots, e_n$, respectively, then for any $x \in \mathbb{R}^n$, we may write $x = \sum x_i e_i$ and

$$\langle Ax, x\rangle = \sum_{i=1}^{n} \lambda_i x_i^2. \tag{18}$$

When A is the adjacency matrix of a graph G, we define $\lambda_i(G) = \lambda_i$ and

$$\lambda(G) = \max\{|\lambda_i| : 2 \leq i \leq n\} \tag{19}$$

and we refer to the eigenvalues of the graph rather than the eigenvalues of A. For $X \subseteq V(G)$, let $e(X)$ denote the number of edges $\{u, v\} \in E(G)$ with $u, v \in X$. If x is the characteristic vector of X and $x = \sum x_i e_i$, then

$$2e(X) = \langle Ax, x\rangle . \tag{20}$$

If A is doubly stochastic, then e_1 is the constant unit vector with eigenvalue equal to the common row sum. In particular, if A is the adjacency matrix of a d-regular graph G, then $\lambda_1(G) = d$ and if $\lambda(G) = \lambda$, then G is an (n, d, λ)-graph. We shall see shortly that if all the eigenvalues of G except $\lambda_1(G)$ are small in absolute value, then large subsets of the vertices induce close to the expected number of edges—this is the expander mixing lemma—and hence such graphs could be referred to collectively as pseudorandom graphs [56].

2.1 *The Alon-Boppana Theorem*

The infinite d-regular tree T_d is the universal cover of d-regular graphs, so for any d-regular graph G with n vertices, the number of closed walks of length $2k$ from $x \in V(G)$ to x is at least the number of closed walks of length $2k$ from the root of T_d to the root of T_d. This is at least

$$\frac{1}{k}\binom{2k-2}{k-1} d(d-1)^{k-1}. \tag{21}$$

The total number of walks of length $2k$ in G is at most $d^{2k} + (n-1)\lambda^{2k}$ by the trace formula, so

$$d^{2k} + (n-1)\lambda^{2k} \geq \frac{n}{k}\binom{2k-2}{k-1} d(d-1)^{k-1}. \tag{22}$$

For fixed d, by selecting k to depend appropriately on n in (22), we obtain the Alon-Boppana Theorem [72]:

Theorem 2.1 *Let $d \geq 1$. If G_n is a d-regular n-vertex graph then*

$$\liminf_{n\to\infty} \lambda(G_n) \geq 2\sqrt{d-1}. \tag{23}$$

A very short proof of the Alon-Boppana Theorem was found by Alon [72, 73]. In light of this theorem, we refer to an (n, d, λ) graph with $\lambda = O(\sqrt{d})$ as a *spectrally extremal graph*. For instance, *Ramanujan graphs* are spectrally extremal d-regular graphs with $\lambda \leq 2\sqrt{d-1}$, and were constructed by Lubotzky et al. [59] and Margulis [61] as Cayley graphs for certain prime powers d (bipartite constructions with $\max\{|\lambda_2|, |\lambda_{n-1}|\} \leq 2\sqrt{d-1}$ were given for all values of d via polynomial interlacing by Marcus et al. [62]). It turns out that random d-regular graphs are also "asymptotically" Ramanujan graphs with high probability, as proved in a major work by Friedman [40].

2.2 The Expander Mixing Lemma

Let G be an (n, d, λ)-graph. If $X \subseteq V(G)$ has characteristic vector x and $x = \sum_{i=1}^{n} x_i e_i$ with $\{e_1, e_2, \ldots, e_n\}$ an orthonormal basis of eigenvectors, then

$$\langle Ax, x\rangle = \sum_{i=1}^{n} \lambda_i x_i^2. \tag{24}$$

Since e_1 is the constant vector, $x_1 = \langle x, e_1\rangle = |X|/\sqrt{n}$. Estimating the remaining terms in (24) using $|\lambda_i| \leq \lambda$, and applying (20), we obtain the *expander mixing lemma* of Alon [3]:

Theorem 2.2 *Let G be an (n, d, λ)-graph and let $X \subseteq V(G)$. Then*

$$\lambda_n|X| \leq 2e(X) - \tfrac{d}{n}|X|^2 \leq \lambda|X| \tag{25}$$

In particular, the smaller the value of λ, the more tightly concentrated the number of edges in subsets X is around $d|X|^2/2n$.

2.3 Spectrally Extremal Graphs

Mantel's Theorem [60] states that a triangle-free graph with $2n$ vertices has at most n^2 edges, with equality only for complete n by n bipartite graphs, which are triangle-free $(2n, n, n)$-graphs. In general, one may ask for triangle-free (n, d, λ)-graphs, where d, n and λ must satisfy (22), and in particular, $\lambda = \Omega(\sqrt{d})$. This leads to an extremal problem: when $\lambda = O(\sqrt{d})$, what is the largest d for which there is a triangle-free (n, d, λ)-graph, or more generally an F-free (n, d, λ)-graph? If G is any triangle-free (n, d, λ)-graph with adjacency matrix A, then

$$0 = \mathrm{tr}(A^3) \geq d^3 - \lambda^3(n-1) \tag{26}$$

and so $d \leq \lambda(n-1)^{1/3}$. If $\lambda = O(\sqrt{d})$, then this gives $d = O(n^{2/3})$. Alon [2] gave an ingenious construction of a triangle-free Cayley (n, d, λ)-graph with $d = \Omega(n^{2/3})$ and $\lambda = O(n^{1/3})$, showing the the above bounds are tight (see also Conlon [22] and Kopparty [52]):

Theorem 2.3 *There exist triangle-free (n, d, λ)-graphs with $\lambda = O(n^{1/3})$ and $d = \Omega(n^{2/3})$ as $n \to \infty$.*

Similar arguments show that if k is odd and G is a C_k-free graph then $d = O(\lambda n^{1/k})$ and when $\lambda = O(\sqrt{d})$ this gives $d = O(n^{2/k})$, and such spectrally extremal graphs exist [4]. If F is a bipartite graph, then in many cases, the densest known n-vertex F-free graphs are graphs whose degrees are all close to some number d and with $\lambda = O(\sqrt{d})$—see Füredi and Simonovits [41] for a survey of bipartite extremal problems. A particular rich source of such graphs is the random polynomial model, due to Bukh and Conlon [16], as well as *strongly regular graphs*—see Brouwer and van Maldeghem [15] for a survey. Another fruitful source of examples is *Cayley graphs*, since the eigenvalues are determined by character sums.

2.4 Spectrally Extremal Clique-Free Graphs

More generally, Sudakov et al. [81] showed

$$d = O(\lambda^{\frac{1}{s-1}} n^{1-\frac{1}{s-1}}). \tag{27}$$

when G is a K_s-free (n, d, λ)-graph. When $\lambda = O(\sqrt{d})$ this gives

$$d = O(n^{1-\frac{1}{2s-3}}). \tag{28}$$

Constructing such spectrally extremal K_s-free graphs for $s \geq 4$ is considered to be one of the major open problems in the area. Examples of constructions were given by Alon and Krivelevich [5] and Krivelevich [55], and the current best constructions were given by Bishnoi et al. [11], who define a graph $\Gamma[q, s]$ using the bilinear form $Q(x, y) = \xi x_1 y_1 + \sum_{i=2}^{s+1} x_i y_i$ with ξ a quadratic non-residue in $\mathbb{F}_q$. Let χ be the quadratic character of $\mathbb{F}_q$. The vertex set of $\Gamma[q, s]$ is a set of one-dimensional subspaces of $\mathbb{F}_q^{s+1}$, defined as follows:

$$\{\langle x \rangle < \mathbb{F}_q^{s+1} : \chi(Q(x, x)) = 1\} \tag{29}$$

and two one-dimensional subspaces $\langle x \rangle$ and $\langle y \rangle$ are adjacent if $\chi(Q(x, y)) = 1$. A key fact is that the neighborhood of a vertex in $\Gamma[q, s]$ induces $\Gamma[q, s-1]$, and $\Gamma[q, 1]$ is empty with $q/2$ vertices. Then an induction shows $\Gamma[q, s]$ is K_{s+1}-free, and $\Gamma[q, s]$ is an (n, d, λ)-graph with $d = \Theta(n^{1-\frac{1}{s}})$ and $\lambda = O(\sqrt{d})$.

2.5 Independent Sets

An *independent set* in a graph G is simply a set X of vertices such that no edge of G is contained in X. From the expander mixing lemma, one may obtain the *Hoffman-Delsarte bound*, which states that if X is an independent set in a d-regular graph G, then $|X| \leq -n\lambda_n/(d - \lambda_n)$ (see Haemers [46]). It turns out that counting independent sets in *pseudoramdom graphs* is key to Ramsey theoretic arguments. For our purposes, an n-vertex graph G is (α, m)*-pseudorandom* if for every set $X \subseteq V(G)$ of size at least m,

$$e(X) \geq \alpha\binom{|X|}{2}. \tag{30}$$

According to the expander mixing lemma, an (n, d, λ)-graph is indeed (α, m)-pseudorandom for some $m = \Theta(n\lambda/d)$ and $\alpha = \Theta(d/n)$ as $d \to \infty$. By Turán's Theorem [83], every graph of maximum degree d has an independent set of size at least $t = n/(d + 1)$, and the number of such independent sets in any n-vertex d-regular graph when $d + 1$ divides n is at least

$$\frac{1}{t!}n(n - d - 1)(n - 2d - 2)\cdots = (d + 1)^t. \tag{31}$$

A disjoint union of cliques K_{d+1} shows this is tight. In pseudorandom graphs, it turns out there are few independent sets that are much larger than this. The following theorem due to Kohayakawa, Lee et al. [51], which is an element of the *method of containers* [9, 76], implies a similar earlier theorem of Alon and Rödl [6] for (n, d, λ)-graphs:

Theorem 2.4 *Let G be an n-vertex (α, m)-pseudorandom graph, and suppose $s \geq 1$ satisfies*

$$\exp(-\alpha s)n \leq m. \tag{32}$$

Then the number of independent sets of size $t \geq s$ in G is at most

$$\binom{n}{s}\binom{m}{t - s}. \tag{33}$$

In particular, if G is an (n, d, λ)-graph, then the number of independent sets of size $t \geq 2n(\log n)^2/d$ is at most

$$\left(\frac{4e^2\lambda}{\log^2 n}\right)^t. \tag{34}$$

The statement (34) follows from (33) using the fact that an (n, d, λ)-graph is (α, m)-pseudorandom for some $m = \Theta(n\lambda/d)$ and $\alpha = \Theta(d/n)$ as $d \to \infty$, and choosing roughly $r = \Theta(t/\log n)$. Alon and Rödl [6] used their theorem to obtain remarkably good bounds on multicolor Ramsey numbers, by simply randomly superimposing certain (n, d, λ)-graphs on top of each other. It is not clear whether the above theorem is sharp up to constant factors, for instance we may ask the following:

Conjecture 2.5 Let G be an (n, d, λ)-graph where $\lambda \geq \log n \geq 1$. Does there exist a constant $C > 0$ such that the number of independent sets of size $t \geq C(n \log n)/d$ in G is at most $(C\lambda/\log n)^t$?

A positive answer to this question would for instance give $r(3, t) = \Omega(t^2/\log t)$, and under the spectral conditions of Theorem 1.5, one would obtain $r(s, t) = \Theta(t^{s-1}/(\log t)^{s-2})$. A survey on counting independent sets is given by Samotij [75].

3 Proofs of Theorems

3.1 Proof of Theorem 1.4

We are to show that if there exists an F-free (n, d, λ)-graph G, then when $t \geq 2n(\log n)^2/d$,

$$r(F, t) > \frac{n(\log n)^2}{8e^2\lambda}. \tag{35}$$

By Theorem 2.4, the number of independent sets of size $t \geq 2n(\log d)^2/d$ in G is at most

$$\left(\frac{4e^2\lambda}{\log^2 n}\right)^t. \tag{36}$$

Sample vertices of G independently with probability $p = (\log^2 n)/(4e^2\lambda)$ and let X be the set of sampled vertices. Note that since $\lambda \geq (\log^2 n)/4e^2$, this is valid. Then $\mathrm{E}(X) = pn$ and the expected number of independent sets of size t in X is 1. Removing one vertex from each independent set of size t in X gives a set X' of $pn - 1$ vertices with no independent set of size t and inducing no copy of F. Therefore

$$r(F, t) > pn - 1 > \frac{n(\log n)^2}{8e^2\lambda}. \tag{37}$$

This proves Theorem 1.4.

3.2 Proof of Theorems 1.5 and 1.7

To derive Theorem 1.5, simply apply Theorem 1.4 to a spectrally extremal n-vertex d-regular K_s-free graph, where $d = \Theta(n^{1-1/(2s-3)})$, and to obtain Theorem 1.7 for cycles of length 6, apply Theorem 1.4 to the spectrally extremal polarity graphs of generalized quadrangles [57]: for prime powers q, these are (n, d, λ)-(pseudo)graphs with $d = q + 1$ and $n = q^3 + q^2 + q + 1$ and no cycles of length 3,4 or 6. To obtain Theorem 1.7 for cycles of length 10, apply Theorem 1.4 to the spectrally extremal polarity graphs of generalized quadrangles [57]: for prime powers q, these are (n, d, λ)-(pseudo)graphs with $d = q + 1$ and $n = q^5 + q^4 + q^3 + q^2 + q + 1$ and no cycles of length 3,4,5,6,8 or 10.

3.3 Proof of Theorem 1.8

The proof of Theorem 1.8 involves constructing a graph based on the shadow graph of a strongly F-free hypergraph H as stated in the theorem. We observe that for any set $X \subseteq V(G)$,

$$\sum_{e \in E(H)} |e \cap X| = \sum_{x \in X} d(x) \tag{38}$$

where $d(x)$ is the number of edges in H containing x. Since $d(x) \geq d$ for all $x \in V(H)$, we find that on average, for each edge of H, $|e \cap X| \geq r|X|/n$. If $|X| \geq 2n/r$, this implies that the number of edges of ∂H contained in X is

$$\sum_{e \in E(H)} \binom{|e \cap X|}{2} \geq \frac{dn}{r} \binom{\frac{r|X|}{n}}{2} \geq \frac{dr}{2n} \binom{|X|}{2}. \tag{39}$$

In particular, the shadow graph is (α, m)-pseudorandom with $\alpha = dr/2n$ and $m = 2n/r$. The next step is to define a random F-free subgraph of the shadow graph. Independently for each edge of H, randomly color the vertices of that edge with two colors, and let G be obtained from the shadow graph ∂H by taking all edges which have ends of different colors. Since H is strongly F-free, we conclude that G is F-free. The main task is to show that G inherits the pseudorandomness properties of ∂H. Indeed, we expect about half the edges of ∂H to be edges of G in any subset X. The details, which are given in [26], involve the use of Hoeffding's martingale inequality [45], and the conclusion is that with high probability, G is (α, m)-pseudorandom with $\alpha = \Theta(dr/n)$ and $m = \Theta(n/r)$ as $n \to \infty$. According to Theorem 2.4, specifically (33), there are at most $\binom{n}{s}\binom{m}{t-s}$ independent sets of size $t \geq s$ in G, when $\exp(-\alpha s)n \leq m$. We choose t as given in the theorem and $s = \Theta(t/\log n)$ so that the number of independent sets of size t is at most $\Theta(n \log n/rt)^t$. Sampling vertices independently with suitable probability $\Theta(rt/n \log n)$, we destroy all these independent sets to obtain an F-free induced subgraph with $\Omega(rt/\log n)$ vertices and no independent set of size t, thereby showing $r(F, t) = \Omega(rt/\log n)$.

3.4 *Proof of Theorem 1.1*

Theorem 1.1 is proved using Theorem 1.8. The hypergraph to which we apply Theorem 1.8 to obtain the claimed bounds on $r(4, t)$ has vertex set equal to the set of lines of the *Hermitian unital* and edges consisting of all lines through a point of the unital: this is a K_4-free n-vertex r-uniform d-regular hypergraph with n vertices, where for some prime power q, we have $r = q^2$, $n = q^2(q^2 - q + 1)$, $d = q + 1$. For background on unitals, see Barwick and Ebert [10]. In particular, with $t = \lceil 256n(\log n)^2/rd\rceil = \Theta(q(\log q)^2)$, we have

$$r(4, t) = \Omega\Big(\frac{rt}{\log^2 n}\Big) = \Omega\Big(\frac{t^3}{(\log t)^4}\Big). \tag{40}$$

3.5 *Proof of Theorem 1.9*

We apply Theorem 1.8 similarly to obtain Theorem 1.9: here H is the strongly C_5-free hypergraph whose vertices are the lines of the *twisted triality hexagon* and whose edges are the set of lines through each point of the hexagon. The parameters for prime powers q are $n = (q^3 + 1)(q^8 + q^4 + 1)$, $r = q^3 + 1$ and $d = q + 1$. Then H is C_5-free and so Theorem 1.8 with $t = \Theta(q(\log q)^2)$ gives

$$r(C_5, t) = \Omega\Big(\frac{(q^3 + 1)t}{\log^2 n}\Big) = \Omega\Big(\frac{t^{10/7}}{(\log t)^{13/7}}\Big). \tag{41}$$

This supersedes the C_5-free process by a polynomial in t.

3.6 *Proof of Theorem 1.12*

Let H be the strongly triangle-free hypergraph whose vertices are the $n = q^3 + q^2 + q + 1$ points in a generalized quadrangle and whose edges are the lines, each of size $q + 1$. Let G be the tripartite triple system on $[q + 1]$ obtained by partitioning $[q + 1]$ into sets $V = \{v_{ij} : 1 \le i, j \le \tau = \lceil 8\log q\rceil\}$ and W_i and X_j where $1 \le i, j \le 8\log q$ and $|W_i|$ and $|X_j|$ all differ in size by at most 1. Then the edges of G are all the triples $\{v_{ij}, w, x\}$ where $w \in W_i$ and $x \in X_j$, for $1 \le i, j \le \tau$. Importantly, it is straightforward to check that G does not contain a loose triangle. Now we let H^* be the triple system obtained by independently and randomly placing a copy of G in each edge of H. In other words, for each edge $e \in E(H)$, we uniformly select a bijection $\phi_e : V(G) \to e$ and add all edges $\{\phi(x), \phi(y), \phi(z)\}$ such that $\{x, y, z\} \in E(G)$. This random image of G is denoted $\phi_e(G)$. The resulting H^* has no loose triangles, and it suffices to show that with positive probability, H^* contains no independent set of size $2\tau n/q$. Suppose I is an independent set in H^*. We have

$$\sum_{e \in E(H)} |e \cap I| = (q+1)|I|. \tag{42}$$

If I is independent, then for all $e \in E(H)$, $e \cap I$ is an independent set in $\phi_e(G)$. A counting argument in G shows the probability of this event is at most $1 - |e \cap I|^3/16q$ if q is sufficiently large. Therefore the probability that I is an independent set in H^* is at most

$$\prod_{e \in E(H)} \Big(1 - \frac{|e \cap I|^3}{16q}\Big) \le \exp\Big(-\sum_{e \in E(H)} \frac{|e \cap I|^3}{16q}\Big). \tag{43}$$

The exponent is a maximum when for all $e \in E(H)$, $|e \cap I| = (q+1)|I|/n = 2(q+1)\tau/q$, in which case the probability I is independent is at most

$$\exp\Big(-\frac{n\tau^3}{2q}\Big). \tag{44}$$

The expected number of independent sets I of size $2\tau n/q$ is therefore at most

$$\exp\Big(-\frac{n\tau^3}{2q}\Big)\binom{n}{2\tau n/q} \le \exp\Big(-\frac{n\tau^3}{2q} + \frac{2\tau n \log n}{q}\Big). \tag{45}$$

Since $\tau = \lceil 8 \log q \rceil$, we easily have $\tau^3 > 4\tau \log n$, and so the expected number of independent sets I of size $2\tau n/q$ is less than 1.

3.7 Proof of Theorem 1.16

For any non-empty graph F and every graph G containing a triangle, if $s = |V(G)|$ then the results of Shearer [78] give

$$f_{F,G}(n) \ge f_{K_2,K_s}(n) \ge a_1 \sqrt{\frac{n \log n}{\log \log n}} \tag{46}$$

for some $a_1 > 0$ depending only on s. This gives the lower bound in Theorem 1.16. For the upper bound, we slightly modify a construction of Erdős et al. [30] of a linear triangle-free R-uniform N-vertex hypergraph:

Proposition 3.1 *For any $R, N \geq 3$ and $N \geq R \geq \log N$, there exists an N-vertex R-uniform hypergraph H with the following properties:*

(i) $|E(H)| \geq N^2/R^{8\sqrt{\log_R N}}$
(ii) *H is linear, that is, for any distinct edges $e, f \in H$, $|e \cap f| \leq 1$.*
(iii) *H is triangle-free, that is, for any three distinct edges $e, f, g \in H$, if $|e \cap f| = |f \cap g| = |g \cap e| = 1$ then $|e \cap f \cap g| = 1$.*

Proof of Theorem **1.16.** We are to prove

$$f_{F,K_3}(n) = n^{\frac{1}{2}+O(\sqrt{\frac{\log\log n}{\log n}})}. \tag{47}$$

Let $t = |V(F)|$. We apply Proposition 3.1 with $R = \lceil 3t \log t \log N \rceil$, where $t = |V(F)|$. Then (i) yields

$$|E(H)| \geq \frac{N^2}{R^{8\sqrt{\log_R N}}} = N^{2-O(\sqrt{\frac{\log\log N}{\log N}})}. \tag{48}$$

Let J be the graph whose vertex set is $E(H)$ and where $E(J) = \{e, f \in E(H) : e \cap f \neq \emptyset\}$. For each vertex $v \in V(H)$, the set $K_v = \{e \in E(H) : v \in e\}$ induces a clique in J. If for some distinct $v, w \in V(H)$ there exist distinct $e, f \in K_v \cap K_w$, then by definition $v, w \in e \cap f$, which contradicts that H is linear. Therefore $|V(K_v) \cap |V(K_w)| \leq 1$ for all distinct $v, w \in V(H)$, and the cliques K_v are edge-disjoint in J. Similarly, since H is triangle-free, every triangle in J is contained in a clique K_v for some $v \in V(H)$.

Independently for $v \in V(H)$, let $\chi_v : V(K_v) \to V(F)$ be a random coloring of K_v. Next, we remove all edges $\{x, y\}$ of $J[K_v]$ such that $\chi_v(x) = \chi_v(y)$ or $\chi_v(x)\chi_v(y) \notin E(F)$. In other words, we have placed a blowup of a copy of F in each set K_v. Since F contains no triangle, the resulting graph J^* is triangle-free. We now prove that J^* has no F-free induced subgraph with at least N vertices. To see this, fix a set Z of N vertices of J^*. The probability that Z is an F-free set of J^* is

$$\Pr(Z) \leq \prod_{v \in V(H)} t \cdot \left(1 - \frac{1}{t}\right)^{|K_v \cap Z|}. \tag{49}$$

Since H is R-uniform,

$$\sum_{v \in V(H)} |K_v \cap Z| = \sum_{e \in Z} |e| = R|Z| = RN. \tag{50}$$

Using $(1-x)^y \leq e^{-xy}$ for $0 \leq x \leq 1$ and $y \geq 1$,

$$\Pr(Z) \leq t^N \left(1 - \frac{1}{t}\right)^{RN} \leq e^{N \log t - RN/t} < N^{-2N}. \tag{51}$$

The number of sets of size N in J^* is no more than

$$\binom{N^2}{N} \leq \frac{N^{2N}}{N!}. \tag{52}$$

Therefore the expected number of F-free sets Z of size N in J^* is less than $1/N!$. We may therefore select J^* so as to contain no F-free subgraph with at least N vertices. Since J^* is triangle-free, and $n := |V(J^*)| = |E(H)|$, the bound (48) gives

$$f_{F,K_3}(n) < N = n^{\frac{1}{2}+O(\sqrt{\frac{\log\log n}{\log n}})}. \tag{53}$$

This proves the theorem. □

4 Concluding Remarks

The techniques presented in this paper have been used effectively for not only estimating Ramsey numbers [26], but also to *hypergraph Ramsey numbers* [53], multicolor Ramsey numbers [6, 47], *Erdős-Rogers functions* [8, 42, 43, 48, 67, 70], *degree-minimal* Ramsey graphs [7] and *coloring hypergraphs* [69]. In this section, we give a brief description of new results.

• Using Theorem 1.1, one can obtain almost tight lower bounds on k-color Ramsey numbers: let $r_k(4, t)$ denote the minimum n such that in every k-coloring of the edges of K_n, one has either a monochromatic K_4 in one of the first $k-1$ colors or a monochromatic K_t in the last color. Note that $r_2(4, t) = r(4, t)$. Then Theorem 1.1 and the upper bounds of He and Wigderson [47] give the following theorem:

Theorem 4.1 *There exist absolute constants $c_1, c_2 > 0$ such that for all $t \geq 3$,*

$$c_1 \frac{t^{2k-1}}{(\log t)^{4k-4}} \leq r_k(4, t) \leq c_2 \frac{t^{2k-1}}{(\log t)^{2k-2}}. \tag{54}$$

• A graph G is *r-Ramsey minimal* for a graph H if every r-coloring of $E(G)$ contains a monochromatic copy of H, but no proper subgraph of G possesses this property. The smallest minimum degree $s_r(H)$ over all r-Ramsey minimal graphs G was first studied by Burr et al. [17], who showed $s_2(K_k) = (k-1)^2$. Using the construction in [64], we improve the most recent upper bounds on $s_r(K_k)$ due to Fox et al. [38] and Bishnoi and LesGourges [12] as follows:

Theorem 4.2 *There exists an absolute constant $c > 0$ such that for $k, r \geq 3$, $s_r(k) \leq 2^{100k} r^2 \log r$ and*

$$s_r(k) \leq (rk)^{2+\frac{c}{k}}[(\log k)^2 + (\log k)^{1+\frac{c}{k}}(\log r)^{1+\frac{c}{k}}]. \tag{55}$$

On the other hand, for all $k, r \geq 3$,

$$s_r(k) = \Omega((k-1)r^2). \tag{56}$$

When $k \geq \log r$, this theorem gives $s_r(k) = O(r^2k^2(\log k)^2 \log r)$, which improves earlier bounds $s_r(k) = O(r^3k^3(\log k)^3)$ of Bishnoi and LesGourges [12]. The main open problem is to find better lower bounds on $s_r(k)$, especially when r and k are both large–the order of magnitude is at most $(rk \log k)^2 \log r$ from the above theorem when $k \geq \log r$, whereas the lower bound is of order $(k-1)r^2$.

• Let $q \geq k \geq 2$. Consider a hypergraph H which comes from taking all k-element subsets of sets in a family F of e sets of size q which pairwise intersect in at most one vertices. Such hypergraphs arise naturally in finite geometry, for instance, if F is the set of lines of a projective plane then $e = q^2 + q + 1$, and for $k = 3$, then the hypergraph H is the family of all collinear triples. In the case $k = 2$, the Erdős-Faber-Lovász conjecture states that if $|F| = q$ then the vertex chromatic number of the graph H is q. In other words, the edge-disjoint union of q cliques of size q has chromatic number q. This was proved by Kang et al. [49]. In [68], we consider the analog of this for $k > 2$: let $\chi(e, k, q)$ denote the maximum possible chromatic number of the hypergraph H over all families F of e sets of size q which pairwise intersect in at most one vertex. In [68], we determine the order of magnitude of $\chi(e, k, q)$ up to a factor $q^{1-2/(k-1)}$ as $e \to \infty$:

Theorem 4.3 *Let* $k, e \geq 3$ *and* $q \geq 1$. *Then there exist absolute positive constants* c_1, c_2 *such that*

$$c_1 \left(\frac{e^{1/2}}{\log e}\right)^{\frac{1}{k-1}} q^{1-\frac{2}{k-1}} < \chi(e, k, q) < c_2 \left(\frac{e^{1/2}}{\log e}\right)^{\frac{1}{k-1}} q$$

where the lower bound holds for $e > q^{2k+3}$.

It seems likely that $\chi(e, k, q) = \Theta(e^{1/2}/\log e) \cdot q$ as $e \to \infty$.

• In [53], it is conjectured that if C_ℓ denotes the loose ℓ-cycle of triples, then $r(C_\ell, t) = t^{\ell/(\ell-1)-o(1)}$. This is true for $\ell = 3$ from Theorem 1.12, but open for all $\ell \geq 4$. The lower bound $r(C_\ell, t) \geq t^{\ell/(\ell-1)-o(1)}$ for odd values of ℓ can be obtained in the same way as the proof of Theorem 1.12 if there exists an n by n bipartite graph of girth larger than 2ℓ with $n^{1+1/\ell-o(1)}$ edges–this is an outstanding conjecture of Erdős and Simonovits [35]. The existence of generalized hexagons provide such bipartite graphs when $\ell = 5$, and therefore $r(C_5, t) \geq t^{5/4-o(1)}$. Even in this case, the upper bound $r(C_5, t) = O(t^{5/4})$ seems challenging to prove. Partial results are obtained by Nie and the author [71].

References

1. Ajtai, M., Komlós, J., Szemerédi, E.: A note on Ramsey numbers. J. Comb. Theory Ser. A **29**(3), 354–360 (1980)
2. Alon, N.: Explicit Ramsey graphs and orthonormal labelings. Electron. J. Combin. **1** (1994), Research Paper 12, approx. 8 pp
3. Alon, N., Chung, F.R.K.: Explicit construction of linear sized tolerant networks. Discret. Math. **72**(1–3), 15–19 (1988)
4. Alon, N., Kahale, N.: Approximating the independence number via the θ-function. Math. Program. **80**(3), 253–264 (1998)
5. Alon, N., Krivelevich, M.: Constructive bounds for a Ramsey-type problem. Graphs Comb. **13**(3), 217–225 (1997)
6. Alon, N., Rödl, V.: Sharp bounds for some multicolor Ramsey numbers. Combinatorica **25**(2), 125–141 (2005)
7. Attwa, Y., Mattheus, S., Szabó, T., Verstraete, J.A.: On the Minimum Degree of Minimal Ramsey Graphs, Preprint (2025)
8. Balogh, J., CHen, C., Luo, H.: On the maximum F-free induced subgraphs in K_t-free graphs, Preprint at arXiv:2406.13780
9. Balogh, J., Morris, R., Samotij, W.: The method of hypergraph containers. In: Proceedings of the International Congress of Mathematicians (2018) Rio de Janeiro
10. Barwick, S.G., Ebert, G.L.: Unitals in projective planes. In: Springer Monographs in Mathematics, pp. xii+193. Springer, New York (2008)
11. Bishnoi, A., Ihringer, F., Pepe, V.: A construction for clique-free pseudorandom graphs, Preprint at https://arxiv.org/abs/1905.04677
12. Bishnoi, A., LesGourges, T.: A new upper bound on the minimum degree of minimal Ramsey graphs, Preprint at https://arxiv.org/html/2209.05147v2
13. Bohman, T., Keevash, P.: Dynamic concentration of the triangle-free process. Random Struct. Alg. **58**, 221–293 (2021)
14. Bohman, T., Keevash, P.: The early evolution of the H-free process. Invent. Math. **181**, 291–336 (2010)
15. Brouwer, A.E., van Maldeghem, H.: Strongly Regular Graphs. Cambridge University Press (2022)
16. Bukh, B., Conlon, D.: Rational exponents in extremal graph theory. J. Eur. Math. Soc. (JEMS) **20**(7), 1747–1757 (2018)
17. Burr, S.A., Erdős, P., Lovász, L.: On graphs of Ramsey type. Ars Comb. **1**(1), 167–190 (1976)
18. Campos, M., Griffiths, S., Morris, R., Sahasrabudhe, J.: An exponential improvement for diagonal Ramsey, Preprint at https://arxiv.org/abs/2303.09521
19. Campos, M., Jenssen, M., Michelen, M., Sahasrabudhe, J.: A new lower bound for the Ramsey numbers $R(3, k)$, Preprint at https://arxiv.org/abs/2505.13371
20. Caro, Y., Li, Y., Rousseau, C., Zhang, Y.: Asymptotic bounds for some bipartite graph-complete graph Ramsey numbers. Discret. Math. **220**, 51–56 (2000)
21. Chung, F.R.K., Graham, R.L.: Erdős on Graphs: His Legacy of Unsolved Problems. Fan Chung, Ron Graham, p. 142. Taylor & Francis (1998)
22. Conlon, D.: A sequence of triangle-free pseudorandom graphs. Comb. Probab. Comput. **26**(2), 195–200 (2017)
23. Conlon, D., Fox, J., Gunby, B., He, X., Suk, A., Verstraete, J.A.: On off-diagonal hypergraph Ramsey numbers. Int. Math. Res. Notices (IMRN) **11**, 1–22 (2025)
24. Conlon, D., Fox, J., Gunby, B., He, X., Mubayi, D., Suk, A., Verstraete, J.A., Yu, H.: When are off-diagonal hypergraph Ramsey numbers polynomial? To appear in Proc. Amer. Math. Soc. (12 pages), Preprint at https://arxiv.org/abs/2411.13812
25. Conlon, D., Fox, J., Sudakov, B.: Hypergraph Ramsey numbers. J. Am. Math. Soc. **23**, 247–266 (2010)
26. Conlon, D., Mattheus, S., Mubayi, D., Verstraete, J.: Ramsey numbers and the Zarankiewicz problem. Bull. Lond. Math. Soc. **56**(6), 2014–2023 (2024)

27. Dudek, A., Rödl, V.: On K_s-free subgraphs in K_{s+k}-free graphs and vertex Folkman numbers. Combinatorica **31**(1), 39–53 (2011)
28. Dudek, A., Retter, T., Rödl, V.: On generalized Ramsey numbers of Erdős and Rogers. J. Comb. Theory Ser. B **109**, 213–227 (2014)
29. Erdős, P.: Some remarks on the theory of graphs. Bull. Am. Math. Soc. **53**, 292–294 (1947)
30. Erdős, P., Frankl, P., Rödl, V.: The asymptotic number of graphs not containing a fixed subgraph and a problem for hypergraphs having no exponent. Graphs Comb **2**, 113–121 (1986)
31. Erdős, P., Faudree, R.J., Rousseau, C.C., Schelp, R.H.: On cycle-complete graph Ramsey numbers. J. Graph Theory **2**, 53–64 (1978)
32. Erdős, P., Hajnal, A., Ramsey, O.: Like theorems. Problems and results. In: Combinatorics (Proc. Conf. Combinatorial Math., Math. Inst., Oxford,: 123–140, p. 1972. Inst. Math. Appl, Southend-on-Sea (1972)
33. Erdős, P., Hajnal, A., Rado, R.: Partition relations for cardinal numbers. Acta Math. Acad. Sci. Hungar. **16**, 93–196 (1965)
34. Erdős, P., Rogers, C.A.: The construction of certain graphs. Canad. J. Math **14**, 702–707 (1962)
35. Erdős, P., Simonovits, M.: Compactness results in extremal graph theory. Combinatorica **2**(3), 275–288 (1982)
36. Erdős, P., Szekeres, G.: A combinatorial problem in geometry. Compos. Math. **2**, 463–470 (1935)
37. Fiz Pontiveros, G., Griffiths, S., Morris, R.: The triangle-free process and the Ramsey numbers. Mem. Am. Math. Soc., **263**, 125 (2020)
38. Fox, J., Grinshpun, A., Liebenau, A., Person, Y., Szabó, T.: On the minimum degree of minimal Ramsey graphs for multiple colours. J. Comb. Theory Ser. B **120**, 64–82 (2016)
39. Frankl, P., Wilson, R.M.: Intersection theorems with geometric consequences. Combinatorica **1**(4), 357–368 (1981)
40. Friedman, J.: A proof of Alon's second eigenvalue conjecture and related problems. Mem. Amer. Math. Soc. **195**(910), viii+100 (2008)
41. Füredi, Z., Simonovits, M.: The history of degenerate (bipartite) extremal graph problems. János Bolyai Math.l Soc. Budapest **25**, 169–264 (2013)
42. Gishboliner, L., Janzer, O., Sudakov, B.: Induced subgraphs of K_r-free graphs and the Erdős-Rogers problem, Preprint at https://arxiv.org/abs/2409.06650
43. Gowers, W.T., Janzer, O.: Improved bounds for the Erdős-Rogers functio. Adv. Comb. Paper No. 3, 27 (2020)
44. Graham, R.L., Rothschild, B., Solymosi, J., Spencer, J.: Ramsey Theory, 3rd edn. Wiley, New York (2015)
45. Hoeffding, W.: Probability inequalities for sums of bounded random variables. J. Am. Stat. Assoc. **58**(301), 13–30 (1963)
46. Haemers, W.H.: Interlacing eigenvalues and graphs. Linear Algebra Appl. **226**(228), 593–616 (1995)
47. He, X., Wigderson, Y.: Multicolor Ramsey numbers via pseudorandom graphs. Electron. J. Combin. **27**(1), Paper No. 1. 32, 8 (2020)
48. Janzer, O., Sudakov, B.: Improved bounds for the Erdős-Rogers $(s, s+2)$-problem. Adv. Comb. (3), 27 (2020)
49. Kang, D.Y., Kelly, T., Kühn, D., Methuku, A., Osthus, D.: A proof of the Erdős-Faber-Lovász conjecture: algorithmic aspects. In: 2021 IEEE 62nd Annual Symposium on Foundations of Computer Science–FOCS, pp. 1080–1089 (2021)
50. Kim, J.H.: The Ramsey number $R(3, t)$ has order of magnitude $t^2/\log t$. Random Struct. Algorithms **7**, 173–207 (1995)
51. Kohayakawa, Y., Lee, S.J., Rödl, V., Samotij, W.: The number of Sidon sets and the maximum size of Sidon sets contained in a sparse random set of integers. Random Struct. Algorithms **46**(1), 1–25 (2015)
52. Kopparty, S.: A constructive lower bound on $R(3, k)$, Course notes available at https://www.math.toronto.edu/swastik/courses/rutgers/graphtheory-F11/cayley.pdf

53. Kostochka, A.V., Mubayi, D., Verstraete, J.A.: Hypergraph Ramsey numbers: triangles versus cliques. J. Comb. Theory Ser. A **120**(7), 1491–1507 (2013)
54. Kostochka, A.V., Alexandr, Pudlák, P., Rödl, V.: Some constructive bounds on Ramsey numbers. J. Comb. Theory Ser. B **100**(5), 439–445 (2010)
55. Krivelevich, M.: K_s-free graphs without large K_r-free subgraphs. Comb. Probab. Comput. **3**(3), 349–354 (1994)
56. Krivelevich, M., Sudakov, B.: Pseudo-random graphs. More sets, graphs and numbers. In: Bolyai Society Mathematical Studies, vol. 15, pp. 199–262. Springer, Berlin (2006)
57. Lazebnik, F., Ustimenko, V.A., Woldar, A.J.: New upper bounds on the order of cages. The Wilf Festschrift (Philadelphia, PA, 1996). Electron. J. Comb. **4**(2), Research Paper 13, approx. 11 (1997)
58. Li, Y., Rousseau, C.C., Zang, W.: An upper bound for Ramsey numbers. Appl. Math. Lett. **17**(6), 663–665 (2004)
59. Lubotzky, A., Phillips, R., Sarnak, P.: Ramanujan graphs. Combinatorica **8**(3), 261–277 (1988)
60. Mantel, W.: Problem 28 (Solution by H. Gouwentak, W. Mantel, J. Teixeira de Mattes, F. Schuh and W. A. Wythoff). Wiskundige Opgaven **10**, 60–61 (1907)
61. Margulis, G.A.: Explicit constructions of graphs without short cycles and low density codes. Combinatorica **2**(1), 71–78 (1982)
62. Marcus, A., Spielman, D., Srivastava, N.: Interlacing families IV: Bipartite Ramanujan graphs of all sizes. SIAM J. Comput. **47**(6), 2488–2509 (2018)
63. Mattheus, S., Mubayi, D., Nie, J., Verstraete, J.A.: Off-diagonal Ramsey numbers for slowly growing hypergraphs. Random Struct. Algorithms **66**(1), Paper No. e21284 (2025)
64. Mattheus, S., Verstraete, J.A.: The asymptotics of $r(4, t)$. Ann. Math. **199**(2), 919–941 (2024)
65. Morawski, P., Verstraete, J.A.: Induced subgraphs of G-free graphs, Preprint (2025)
66. Mubayi, D., Razborov, A.: Polynomial to exponential transition in Ramsey theory. Proc. Lond. Math. Soc. **122**, 69–92 (2021)
67. Mubayi, D., Verstraete, J.A.: A note on pseudorandom Ramsey graphs. J. Eur. Math. Soc. (JEMS) **26**(1), 153–161 (2024)
68. Mubayi, D., Verstraete, J.A.: Coloring unions of nearly disjoint hypergraph cliques. Mathematika **70**(1), Paper No. e12234, 12 (2024)
69. Mubayi, D., Verstraete, J.A.: On the order of Erdős-Rogers functions, Preprint at https://arxiv.org/abs/2401.02548v2
70. Mubayi, D., Verstraete, J.A.: Erdős-Rogers functions for arbitrary pairs of graphs. https://arxiv.org/abs/2407.03121
71. Nie, J., Verstraete, J.A.: Ramsey Numbers for Non-trivial Berge Cycles. https://arxiv.org/abs/2102.03720
72. Nilli, A.: On the second eigenvalue of a graph. Discrete Math. **91**(2), 207–210 (1991)
73. Nilli, A.: Tight estimates for eigenvalues of regular graphs. Electron. J. Combin. **11**(1), Note 9, 4 (2004)
74. Ramsey, F.P.: On a problem of formal logic. Proc. Lond. Math. Soc. **30**, 264–286 (1930)
75. Samotij, W.: Counting independent sets in graphs. Eur. J. Comb. **48**, 5–18 (2015)
76. Saxton, D., Thomason, A.: Hypergraph containers. Invent. Math. **201**(3), 925–992 (2015)
77. Shearer, J.B.: A note on the independence number of triangle-free graphs. Discrete Math. **46**, 83–87 (1983)
78. Shearer, J.B.: On the independence number of sparse graphs. Random Struct Algorithms **7**(3), 269–271 (1995)
79. Spencer, J.: Ramsey's theorem—a new lower bound. J. Combin. Theory Ser. A **18**, 108–115 (1975)
80. Spencer, J.: Asymptotic lower bounds for Ramsey functions. Discrete Math. **20**(1), 69–76 (1977/1978)
81. Sudakov, B., Szabó, T., Vu, V.H.: A generalization of Turán's theorem. J. Graph Theory **49**(3), 187–195 (2005)
82. Sudakov, B.: A note on odd cycle-complete graph Ramsey numbers. Electron. J. Combin. **9**(1), Note 1, 4 (2002)

83. Turán, P.: On the theory of graphs. Colloq. Math. **3**, 19–30 (1954)
84. Verstraete, J.A.: Extremal problems for cycles in graphs. Recent trends in combinatorics. IMA Math. Appl. **159**, 83–116. Springer, Cham (2016)
85. Warnke, L.: Personal Communication (2023)
86. Wolfovitz, G.: K_4-free graphs without large induced triangle-free subgraphs. Combinatorica **33**(5), 623–631 (2013)

Developments of the Shifting Method in Extremal Set Theory

Jian Wang

Abstract The shifting method is one of the most powerful tools in extremal set theory, invented by Erdős, Ko and Rado, and further developed by Frankl, Ahlswede and Khachatrian. In this paper, we survey some developments of the shifting method including the random walk method, the generating set method and the recently developed shifting ad extremis by Frankl. To illustrate the power of the shifting method, we review some proofs and proof ideas.

1 Introduction

Let $[n]$ denote the standard n-element set $\{1, 2, \ldots, n\}$. Let $2^{[n]}$ denote the collection of all subsets of $[n]$ and let $\binom{[n]}{k}$ denote the collection of all k-element subsets of $[n]$. A subfamily of $2^{[n]}$ is called a *family* and a subfamily of $\binom{[n]}{k}$ is called a *k-uniform family*.

The central topic in extremal set theory is to determine the maximum or minimum size of a family that satisfies certain conditions involving set operations such as union, intersection and symmetric difference etc.

We say that $\mathcal{F} \subset \binom{[n]}{k}$ is an *intersecting family* if $F \cap F' \neq \emptyset$ for any $F, F' \in \mathcal{F}$. The Erdős-Ko-Rado Theorem is one of the most important results in extremal set theory.

Theorem 1.1 (The Erdős-Ko-Rado Theorem [11]) *Let $n \geq 2k$. If $\mathcal{F} \subset \binom{[n]}{k}$ is an intersecting family, then*

$$|\mathcal{F}| \leq \binom{n-1}{k-1}. \tag{1}$$

For $n > 2k$, the *full star* $\mathcal{S}_x = \left\{F \in \binom{[n]}{k} : x \in F\right\}$ is the unique family attaining equality in (1). A subfamily of $\mathcal{S}_x$ is called a *star*.

J. Wang (✉)
Department of Mathematics, Sichuan University, Chengdu, People's Republic of China
e-mail: wangjian01@tyut.edu.cn

G. O. H. Katona et al. (eds.), *Sum(m)it280*, Bolyai Society Mathematical Studies 32,
https://doi.org/10.1007/978-3-032-18810-6_18

By now there are many different proofs for the Erdős-Ko-Rado Theorem. Katona [47] gave a very short and beautiful proof, using the so-called Katona's cycle. Daykin [8] gave a proof using the Kruskal-Katona theorem. Lovász [53] determined all the eigenvalues of the Kneser graph. Then the Erdős-Ko-Rado Theorem follows from Hoffman's bound [43]. Füredi, Hwang and Weichsel [38] have another proof that uses linearly independent polynomials.

In their original paper, Erdős, Ko and Rado invented the shifting method, which proved to be one of the most powerful tools in extremal set theory. For $\mathcal{F} \subset \binom{[n]}{k}$ and $1 \le i < j \le n$, define the shift

$$S_{ij}(\mathcal{F}) = \left\{S_{ij}(F) \colon F \in \mathcal{F}\right\},$$

where

$$S_{ij}(F) = \begin{cases} F' := (F \setminus \{j\}) \cup \{i\}, & \text{if } j \in F, i \notin F \text{ and } F' \notin \mathcal{F}; \\ F, & \text{otherwise.} \end{cases}$$

The definition of the shifting operator is natural. The conditions $j \in F$ and $i \notin F$ guarantee $|F'| = k$. The condition $F' \notin \mathcal{F}$ guarantees that the shifting operator S_{ij} preserves the size of the family.

The important thing is that the shifting operator S_{ij} preserves the intersecting property.

Lemma 1.2 ([11]) *If $\mathcal{F} \subset \binom{[n]}{k}$ is intersecting, then $S_{ij}(\mathcal{F})$ is also intersecting.*

Proof We have to show that $S_{ij}(F_1) \cap S_{ij}(F_2) \neq \emptyset$ for any $F_1, F_2 \in \mathcal{F}$. Clearly if $F_1 \cap F_2 \neq \{j\}$ then $S_{ij}(F_1) \cap S_{ij}(F_2) \neq \emptyset$. Thus we may assume that $F_1 \cap F_2 = \{j\}$. If $i \in F_1 \setminus F_2$ then $S_{ij}(F_1) = F_1$ and $S_{ij}(F_1) \cap S_{ij}(F_2) = \{j\}$ or $S_{ij}(F_1) \cap S_{ij}(F_2) = \{i\}$. Thus we assume that $i \notin F_1 \cup F_2$. Since $F_1' \cap F_2 = \emptyset = F_1 \cap F_2'$ and $\mathcal{F}$ is intersecting, we infer that $S_{ij}(F_1) = F_1'$ and $S_{ij}(F_2) = F_2'$. It follows that $S_{ij}(F_1) \cap S_{ij}(F_2) = \{i\}$. □

Let $\mathcal{F} \subset \binom{[n]}{k}$ be a family. We call $\mathcal{F}$ *t-intersecting* if $|F \cap F'| \ge t$ for every $F, F' \in \mathcal{F}$. The *matching number* of $\mathcal{F}$, denoted by $\nu(\mathcal{F})$, is defined as the maximum number of pairwise disjoint members in $\mathcal{F}$. A set $E \in \binom{[n]}{k-1}$ is called a *shadow* of $\mathcal{F}$ if there exists $F \in \mathcal{F}$ such that $E \subset F$.

Let us mention that the shifting operator also has the following properties.

Lemma 1.3 ([17])

(i) The shifting operator preserves the t-intersecting property;
(ii) The shifting operator does not increase the matching number;
(iii) The shifting operator does not increase the number of shadows.

A family $\mathcal{F} \subset \binom{[n]}{k}$ is called *shifted* (or *initial*) if $S_{ij}(\mathcal{F}) = \mathcal{F}$ holds for all $1 \le i < j \le n$.

Proposition 1.4 ([17]) *By repeated shifting, one can transform an arbitrary k-graph into a shifted k-graph with the same number of edges.*

Proof Let $\mathcal{F} \subset \binom{[n]}{k}$. Whenever there is a pair (i, j) with $1 \le i < j \le n$ such that $S_{ij}(\mathcal{F}) \ne \mathcal{F}$, we apply S_{ij} to $\mathcal{F}$. Eventually, we shall arrive at a shifted family. Indeed, in each step the function $w(\mathcal{F}) = \sum_{F \in \mathcal{F}} \sum_{x \in F} x$ decreases strictly. The procedure will terminate in finite steps. □

Let us define the *shifting partial order* $\prec$. For two k-sets A and B where $A = \{a_1, \ldots, a_k\}$, $a_1 < \ldots < a_k$ and $B = \{b_1, \ldots, b_k\}$, $b_1 < \ldots < b_k$ we say that *A precedes B* and denote it by $A \prec B$ if $a_i \le b_i$ for all $1 \le i \le k$.

Proposition 1.5 ([17]) *If $\mathcal{F} \subset \binom{[n]}{k}$ is shifted, then $A \prec B$ and $B \in \mathcal{F}$ imply $A \in \mathcal{F}$.*

Proof Let $A \setminus B = \{a_1, \ldots, a_\ell\}$, $a_1 < \ldots < a_\ell$ and $B \setminus A = \{b_1, \ldots, b_\ell\}$, $b_1 < \ldots < b_\ell$. Note that $A \prec B$ implies $a_i < b_i$ for all $1 \le i \le \ell$. Since $S_{a_1,b_1}(\mathcal{F}) = \mathcal{F}$, we infer that $(B \setminus \{b_1\}) \cup \{a_1\} \in \mathcal{F}$. Since $S_{a_2,b_2}(\mathcal{F}) = \mathcal{F}$, we infer that $(B \setminus \{b_1, b_2\}) \cup \{a_1, a_2\} \in \mathcal{F}$. Repeating the same argument, we conclude that $A = (B \setminus \{b_1, \ldots, b_\ell\}) \cup \{a_1, \ldots, a_\ell\} \in \mathcal{F}$. □

1.1 Proof of the Erdős-Ko-Rado Theorem Via shifting

Now to prove the Erdős-Ko-Rado Theorem, one can assume that $\mathcal{F}$ is a shifted family.

For $i \in [n]$, define

$$\mathcal{F}(i) = \{F \setminus \{i\} \colon i \in F \in \mathcal{F}\}, \ \mathcal{F}(\bar{i}) = \{F : i \notin F \in \mathcal{F}\}.$$

For $A \subset B \subset [n]$, define

$$\mathcal{F}(A, B) = \{F \setminus B \colon F \cap B = A\}.$$

We also use $\mathcal{F}(i, \bar{j})$ to denote $\mathcal{F}(\{i\}, \{i, j\})$, $\mathcal{F}(i, j)$ to denote $\mathcal{F}(\{i, j\}, \{i, j\})$ and $\mathcal{F}(\bar{i}, \bar{j})$ to denote $\mathcal{F}(\emptyset, \{i, j\})$.

Fact 1.6 *Let $n \ge 2k$. If $\mathcal{F} \subset \binom{[n]}{k}$ is a shifted intersecting family, then $\mathcal{F}(n)$ is intersecting.*

Proof Suppose that there are $E_1, E_2 \in \mathcal{F}(n)$ such that $E_1 \cap E_2 = \emptyset$. Note that $|E_1| + |E_2| = 2k - 2$. There exists $x \in [n-1] \setminus (E_1 \cup E_2)$. Since $\mathcal{F}$ is shifted and $E_1 \cup \{x\} \prec E_1 \cup \{n\} \in \mathcal{F}$, we infer $E_1 \cup \{x\} \in \mathcal{F}$. But then $(E_1 \cup \{x\}) \cap (E_2 \cup \{n\}) = \emptyset$, contradicting the fact that $\mathcal{F}$ is intersecting. □

Fact 1.6 allows us to prove the Erdős-Ko-Rado Theorem by induction on (n, k). The case $k = 1$ is trivial. For $n = 2k$, since at most one of F, $F^c := [n] \setminus F$ is in $\mathcal{F}$, we infer that

$$|\mathcal{F}| \le \frac{1}{2}\binom{2k}{k} = \binom{2k-1}{k-1}.$$

Now assume that the theorem holds for $(n-1, k-1)$ and $(n-1, k)$ and we prove it for (n, k). Clearly $\mathcal{F}(\bar{n}) \subset \binom{[n-1]}{k}$ is intersecting. By Fact 1.6, $\mathcal{F}(n) \subset \binom{[n-1]}{k-1}$ is also intersecting. Then by the induction hypothesis

$$|\mathcal{F}| = |\mathcal{F}(n)| + |\mathcal{F}(\bar{n})| \le \binom{n-2}{k-2} + \binom{n-2}{k-1} = \binom{n-1}{k-1}.$$

This completes the proof of the Erdős-Ko-Rado Theorem.

1.2 The Hilton-Milner Theorem

Hilton and Milner [42] proved a strong stability result for the Erdős-Ko-Rado Theorem.

Theorem 1.7 (The Hilton-Milner Theorem [42]) *Let $n \ge 2k$. If $\mathcal{F} \subset \binom{[n]}{k}$ is an intersecting family that is not a star, then*

$$|\mathcal{F}| \le \binom{n-1}{k-1} - \binom{n-k-1}{k-1} + 1. \tag{2}$$

Let us define the Hilton-Milner Family

$$\mathcal{H}(n,k) = \left\{F \in \binom{[n]}{k} : 1 \in F,\ F \cap [2, k+1] \ne \emptyset\right\} \cup \{[2, k+1]\},$$

showing that (2) is best possible.

By now there are many different proofs of the Hilton-Milner Theorem [3, 5, 20, 25, 26, 40, 51, 55, etc.]. Here let us recall a slightly modified version of the proof given by Frankl and Füredi [25] using the shifting method.

We call an intersecting family *non-trivial* if it is not a star. For $\mathcal{F} \subset \binom{[n]}{k}$, a set T is called a *cover* of $\mathcal{F}$ if $T \cap F \ne \emptyset$ for all $F \in \mathcal{F}$.

Let $\mathcal{F} \subset \binom{[n]}{k}$ be a non-trivial intersecting family of the maximum size. Let us apply the shift operator S_{ij} to $\mathcal{F}$ for all $1 \le i < j \le n$ repeatedly. The only problem is that S_{ij} may turn $\mathcal{F}$ into a star. If there is a pair (i, j) such that $S_{ij}(\mathcal{F})$ is a star, then it has to be a star with center i. It follows that $\{i, j\}$ is a cover of $\mathcal{F}$. By the maximality of $\mathcal{F}$, we infer that $\{i, j\}$ is full in $\mathcal{F}$, that is, $\mathcal{F}(i, j) = \binom{[n]\setminus\{i,j\}}{k-2}$.

Without loss of generality, assume that $S_{12}(\mathcal{F})$ is a star. It follows that $\{1, 2\}$ is full in $\mathcal{F}$. Then we apply the S_{ij} to $\mathcal{F}$ for all $3 \le i < j \le n$ repeatedly. If there is another pair, say $(3, 4)$, such that $S_{34}(\mathcal{F})$ is a star, then $\{3, 4\}$ is also full in $\mathcal{F}$, contradicting the fact that $\mathcal{F}$ is intersecting. Thus we may assume that $\mathcal{F}$ is shifted on $[3, n]$. Then $\{3, 4, \ldots, k+1\} \in \mathcal{F}(1, \bar{2}) \cap \mathcal{F}(\bar{1}, 2)$. Now $\{2, 3, \ldots, k+1\} \in S_{12}(\mathcal{F})$ and

$S_{12}(\mathcal{F})$ is also non-trivial. Now we can apply the shifting operator S_{12} to $\mathcal{F}$. Thus, by applying the shifting operator to $\mathcal{F}$ repeatedly, eventually we shall arrive at a shifted non-trivial intersecting family. Therefore one can assume that $\mathcal{F}$ is shifted when proving the Hilton-Milner Theorem.

We prove the Hilton-Milner Theorem by induction on (n, k). For $n = 2k$ and $k = 2$, the theorem holds trivially. Let $n > 2k$ and $k \geq 3$. We prove the theorem for (n, k) by assuming it for $(n-1, k)$ and $(n-1, k-1)$. Since $\mathcal{F}$ is non-trivial, by shiftedness we have $\{2, 3, \ldots, k+1\} \in \mathcal{F}$. For $n > 2k$, it follows that $\mathcal{F}(\bar{n})$ is non-trivial intersecting. Thus, by the induction hypothesis

$$|\mathcal{F}(\bar{n})| \leq \binom{n-2}{k-1} - \binom{n-k-2}{k-1} + 1. \tag{3}$$

By Fact 1.6, $\mathcal{F}(n)$ is intersecting. If $\mathcal{F}(n)$ is a star, then by shiftedness $\mathcal{F}(n)$ has to be a star of center 1. Since $\{2, 3, \ldots, k+1\} \in \mathcal{F}$,

$$|\mathcal{F}(n)| \leq \binom{n-2}{k-2} - \binom{n-k-2}{k-2}. \tag{4}$$

Adding (3) and (4), we obtain that

$$|\mathcal{F}| = |\mathcal{F}(n)| + |\mathcal{F}(\bar{n})| \leq \binom{n-1}{k-1} - \binom{n-k-1}{k-1} + 1.$$

If $\mathcal{F}(n)$ is non-trivial intersecting, then by the induction hypothesis

$$|\mathcal{F}(n)| \leq \binom{n-2}{k-2} - \binom{n-k-1}{k-2} + 1. \tag{5}$$

Adding (3) and (5), we conclude that for $k \geq 3$,

$$\begin{aligned} |\mathcal{F}| = |\mathcal{F}(n)| + |\mathcal{F}(\bar{n})| &\leq \binom{n-1}{k-1} - \binom{n-k-1}{k-1} + 1 + \left(1 - \binom{n-k-2}{k-3}\right) \\ &\leq \binom{n-1}{k-1} - \binom{n-k-1}{k-1} + 1. \end{aligned}$$

This completes the proof of the Hilton-Milner Theorem.

1.3 The Kruskal-Katona Theorem

Another important result is the Kruskal-Katona Theorem, which has many applications in various problems in extremal set theory.

For $\mathcal{F} \subset \binom{[n]}{k}$, let us define the *$\ell$th shadow of $\mathcal{F}$* as

$$\partial^{(\ell)}\mathcal{F} = \left\{E \in \binom{n}{k-\ell} : \text{there exists } F \in \mathcal{F} \text{ such that } E \subset F\right\}.$$

For $\ell = 1$, we simply write $\partial\mathcal{F}$ and call it the *shadow* of $\mathcal{F}$. The celebrated Kruskal-Katona Theorem [46, 48] gives the best possible lower bounds on $|\partial\mathcal{F}|$ for given size of $\mathcal{F}$.

Recall the *co-lexicographic order* $A <_C B$ for $A, B \in \binom{[n]}{k}$ defined by, $A <_C B$ if and only if $\max\{i : i \in A \setminus B\} < \max\{i : i \in B \setminus A\}$. Let $C(n, k, m)$ be the collection of the first m sets $A \in \binom{[n]}{k}$ in the co-lexicographic order.

Theorem 1.8 (The Kruskal-Katona Theorem [46, 48]) *If $\mathcal{F} \subset \binom{[n]}{k}$, $|\mathcal{F}| = m$, then*

$$|\partial\mathcal{F}| \geq |\partial C(n, k, m)|. \tag{6}$$

For every positive integer m, one can write m in k-cascade form uniquely:

$$m = \binom{a_k}{k} + \binom{a_{k-1}}{k-1} + \ldots + \binom{a_s}{s}$$

with $a_k > a_{k-1} > \ldots > a_s \geq 1$. Let $b_i = a_i + 1$, $i = k, k-1, \ldots, s+1$. If $m = \binom{a_k}{k} + \binom{a_{k-1}}{k-1} + \ldots + \binom{a_s}{s}$, then

$$C(n, k, m) = \bigcup_{s \leq i \leq k} \left\{\{b_k, b_{k-1}, \ldots, b_{i+1}\} \cup E : E \in \binom{[a_i]}{i}\right\},$$

where $\{b_k, b_{k-1}, \ldots, b_{i+1}\} = \emptyset$ for $i = k$. Then it is easy to see that

$$\partial C(n, k, m) = \bigcup_{s \leq i \leq k} \left\{\{b_k, b_{k-1}, \ldots, b_{i+1}\} \cup E : E \in \binom{[a_i]}{i-1}\right\}$$

and

$$|\partial C(n, k, m)| = \binom{a_k}{k-1} + \binom{a_{k-1}}{k-2} + \ldots + \binom{a_s}{s-1}.$$

Thus the Kruskal-Katona Theorem can also be rewritten in the following form.

Theorem 1.9 (The Kruskal-Katona Theorem) *If $\mathcal{F} \subset \binom{[n]}{k}$, $|\mathcal{F}| \geq m = \binom{a_k}{k} + \binom{a_{k-1}}{k-1} + \ldots + \binom{a_s}{s}$, then*

$$|\partial\mathcal{F}| \geq \binom{a_k}{k-1} + \binom{a_{k-1}}{k-2} + \ldots + \binom{a_s}{s-1}. \tag{7}$$

For real $x > k - 1$, define

$$\binom{x}{k} := \frac{x(x-1)\cdots(x-k+1)}{k!}.$$

Lovász [54] gave the following version of the Kruskal-Katona Theorem.

Theorem 1.10 (Lovász version of the Kruskal-Katona Theorem [54]) *Let* $\mathcal{F} \subset \binom{[n]}{k}$. *If* $|\mathcal{F}| \geq \binom{x}{k}$ *for some real* x *with* $k \leq x \leq n$, *then*

$$|\partial\mathcal{F}| \geq \binom{x}{k-1}. \tag{8}$$

Let us recall a very short proof of Theorem 1.10 via shifting given by Frankl [16]. Let us mention that recently Chao and Yu [6] gave an entropy method proof of Theorem 1.10.

By Lemma 1.3 (iii) we may assume that $\mathcal{F}$ is shifted. The key to the proof is the following fact.

Fact 1.11 *([16])*

$$\partial\mathcal{F}(\bar{1}) \subset \mathcal{F}(1). \tag{9}$$

Proof Let $F \in \mathcal{F}(\bar{1})$ and $E = F \setminus \{x\} \in \partial\mathcal{F}(\bar{1})$ for some $x \in F$. By shiftedness we have $(F \setminus \{x\}) \cup \{1\} \in \mathcal{F}$. Thus $E \in \mathcal{F}(1)$ and the fact follows. □

By Fact 1.11 we infer that

$$\partial\mathcal{F} = \bigcup_{1 \in F \in \mathcal{F}} \binom{F}{k-1}.$$

Then the number of shadows that do not contain 1 is exactly $|\mathcal{F}(1)|$. The number of shadows containing 1 is exactly $|\partial\mathcal{F}(1)|$. Thus,

$$|\partial\mathcal{F}| = |\mathcal{F}(1)| + |\partial\mathcal{F}(1)|.$$

Now we apply induction on (n, k). For $k = 1$ and $n = k$, Theorem 1.10 holds trivially. If $|\mathcal{F}(1)| \geq \binom{x-1}{k-1}$ then by the induction hypothesis $|\partial\mathcal{F}(1)| \geq \binom{x-1}{k-1}$. It follows that

$$|\partial\mathcal{F}| = |\mathcal{F}(1)| + |\partial\mathcal{F}(1)| \geq \binom{x-1}{k-1} + \binom{x-1}{k-2} = \binom{x}{k-1}.$$

If $|\mathcal{F}(1)| < \binom{x-1}{k-1}$, then $|\mathcal{F}(\bar{1})| = |\mathcal{F}| - |\mathcal{F}(1)| \geq \binom{x-1}{k}$. By (9) and the induction hypothesis,

$$|\mathcal{F}(1)| \geq |\partial\mathcal{F}(\bar{1})| \geq \binom{x-1}{k-1},$$

contradicting the assumption. Thus Theorem 1.10 is proven.

Recall the *lexicographic order* $A <_L B$ for $A, B \in \binom{[n]}{k}$ defined by, $A <_L B$ if and only if $\min\{i : i \in A \setminus B\} < \min\{i : i \in B \setminus A\}$. For $n > k > 0$ and $\binom{n}{k} \geq m > 0$, let $\mathcal{L}(n, k, m)$ denote the collection of the first m sets $A \in \binom{[n]}{k}$ in the lexicographic order and let $\overline{\mathcal{L}}(n, k, m)$ denote the collection of the last m sets $A \in \binom{[n]}{k}$ in the lexicographic order.

There are two natural maps on families. Let us define $\phi\colon \binom{[n]}{k} \mapsto \binom{[n]}{n-k}$ by $\phi(A) = [n] \setminus A$ for every $A \in \binom{[n]}{k}$ and define $\psi : \binom{[n]}{k} \mapsto \binom{[n]}{k}$ by $\psi(A) = \{n + 1 - a : a \in A\}$ for every $A \in \binom{[n]}{k}$. For $\mathcal{F} \subset \binom{[n]}{k}$, define $\phi(\mathcal{F}) = \{\phi(F) : F \in \mathcal{F}\}$ and $\psi(\mathcal{F}) = \{\psi(F) : F \in \mathcal{F}\}$.

The following two equalities are easy to check.

$$\phi(\mathcal{L}(n, k, m)) = \overline{\mathcal{L}}(n, n - k, m), \tag{10}$$

$$\psi(\overline{\mathcal{L}}(n, k, m)) = C(n, k, m). \tag{11}$$

By (11), we see that the Kruskal-Katona Theorem can also be stated as follows.

Theorem 1.12 (The Krusakl-Katona Theorem) *If* $\mathcal{F} \subset \binom{[n]}{k}$, $|\mathcal{F}| = m$, *then*

$$|\partial\mathcal{F}| \geq |\partial\overline{\mathcal{L}}(n, k, m)|. \tag{12}$$

Let us recall a reformulation of the Kruskal–Katona Theorem, called Hilton's lemma [41].

Hilton's Lemma ([41])**.** Let n, a, b be positive integers, $n \geq a + b$. Suppose that $\mathcal{A} \subset \binom{[n]}{a}$ and $\mathcal{B} \subset \binom{[n]}{b}$ are cross-intersecting. Then $\mathcal{L}(n, a, |\mathcal{A}|)$ and $\mathcal{L}(n, b, |\mathcal{B}|)$ are cross-intersecting as well.

Proof Note that $\mathcal{A}$, $\mathcal{B}$ are cross-intersecting if and only if

$$\partial^{(n-a-b)}\phi(\mathcal{A}) \cap \mathcal{B} = \emptyset. \tag{13}$$

As $\partial^{(n-a-b)}\phi(\mathcal{A})$ and $\mathcal{B}$ are both subfamilies of $\binom{[n]}{b}$, $|\partial^{(n-a-b)}\phi(\mathcal{A})| + |\mathcal{B}| \leq \binom{n}{b}$. We are left to check that

$$\partial^{(n-a-b)}\phi(\mathcal{L}(n, a, |\mathcal{A}|)) \cap \mathcal{L}(n, b, |\mathcal{B}|) = \emptyset. \tag{14}$$

By (10), $\phi(\mathcal{L}(n, a, |\mathcal{A}|) = \overline{\mathcal{L}}(n, n - a, |\mathcal{A}|)$. Then to show (14) it suffices to show that

$$\partial^{(n-a-b)}\overline{\mathcal{L}}(n, n - a, |\mathcal{A}|) \cap \mathcal{L}(n, b, |\mathcal{B}|) = \emptyset. \tag{15}$$

Let $m = |\partial^{(n-a-b)}\overline{\mathcal{L}}(n, n - a, |\mathcal{A}|)|$. Then $\partial^{(n-a-b)}\overline{\mathcal{L}}(n, n - a, |\mathcal{A}|) = \overline{\mathcal{L}}(n, b, m)$ and (15) can be rewritten as

$$\overline{\mathcal{L}}(n, b, m) \cap \mathcal{L}(n, b, |\mathcal{B}|) = \emptyset,$$

which is equivalent to $m + |\mathcal{B}| \leq \binom{n}{b}$.

By (12) it follows that

$$|\partial^{(n-a-b)}\phi(\mathcal{A})| \geq |\partial^{(n-a-b)}\overline{\mathcal{L}}(n, n-a, |\phi(\mathcal{A})|)| = |\partial^{(n-a-b)}\overline{\mathcal{L}}(n, n-a, |\mathcal{A}|)| = m.$$

Then by (13) we get

$$m + |\mathcal{B}| \leq |\partial^{(n-a-b)}\phi(\mathcal{A})| + |\mathcal{B}| \leq \binom{n}{b},$$

and the lemma follows. □

Katona [45] also gives a very useful inequality concerning the sizes of a t-intersecting family and its ℓth shadow.

Theorem 1.13 (Katona's Intersecting Shadow Theorem [45]) *Let $n \geq 2k - t$ and $t \geq \ell \geq 1$. Suppose that $\emptyset \neq \mathcal{F} \subset \binom{[n]}{k}$ is t-intersecting. Then*

$$|\partial^{(\ell)}\mathcal{F}| \geq |\mathcal{F}|\frac{\binom{2k-t}{k-\ell}}{\binom{2k-t}{k}}, \tag{16}$$

with equality holding if and only if $\mathcal{F}$ is isomorphic to $\binom{[2k-t]}{k}$.

There are also many different proofs of Katona's Intersecting Shadow Theorem, see [17, 18, 39].

2 The Random Walk Method

In their original paper [11], Erdős, Ko and Rado also prove the following result.

Theorem 2.1 ([11]) *Let $n > k > t > 0$ and suppose that $\mathcal{F} \subset \binom{[n]}{k}$ is t-intersecting. Then for $n \geq n_0(k, t)$,*

$$|\mathcal{F}| \leq \binom{n-t}{k-t}. \tag{17}$$

Problem 1 What is the best possible value of $n_0(k, t)$?

For $t = 1$, it was determined already by Erdős, Ko and Rado, it is $2k$.

The problem was answered by Frankl [13] and the best possible value of $n_0(k, t)$ is $(t + 1)(k - t + 1)$ for $t \geq 15$. Later, Wilson [58] closed the gap $2 \leq t \leq 14$ by a completely different proof valid for all t.

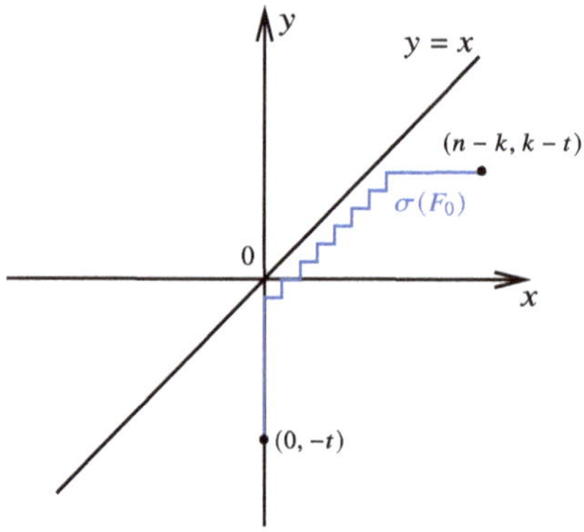

Fig. 1 The lattice path $\sigma(F_0)$ with $F_0 = \{1, 2, \dots, t-1, t+1, t+3, \dots, 2k-t+1\}$

In [13], Frankl established a very nice injection σ from a shifted family $\mathcal{F}$ to a family of lattice paths, which is called the random walk method. The random walk method extends the power of the shifting method.

Let $n \geq 2k - t$. For $F \in \binom{[n]}{k}$, define $\sigma(F)$ as a lattice path in the two-dimensional integer grid $\mathbb{Z}^2$ from $(0, -t)$ to $(n-k, k-t)$ as follows. The lattice path $\sigma(F)$ starts at point $(0, -t)$ and change the current position by $(+1, 0)$ (goes right) or by $(0, +1)$ (goes up) in each step. For $i = 1, 2, \dots, n$, $\sigma(F)$ goes up if $i \in F$ and goes right if $i \notin F$. Since $|F| = k$, there are exactly k steps going up and $n - k$ steps going right. Thus $\sigma(F)$ ends at point $(n-k, k-t)$ (see Fig. 1).

The following observation provides a geometric view of the shifting partial order.

Observation 2.2 *([13]) Let $A, B \in \binom{[n]}{k}$ with $A \prec B$. Then $\sigma(A)$ goes above $\sigma(B)$.*

Let $\mathcal{F} \subset \binom{[n]}{k}$ be a shifted family. If we know $A \notin \mathcal{F}$ then by shiftedness every $B \in \binom{[n]}{k}$ with $A \prec B$ is not in $\mathcal{F}$ either. By Observation 2.2, $\sigma(F)$ can not be below $\sigma(A)$. Thus to get an upper bound on $|\mathcal{F}|$, we only need to count the number of lattice paths from $(0, -t)$ to $(n-k, k-t)$ that do not go below $\sigma(A)$.

Fact 2.3 *([13]) Let $\mathcal{F} \subset \binom{[n]}{k}$ be a shifted t-intersecting family with $n > 2k - t$. Then*

$$F_0 := \{1, 2, \dots, t-1, t+1, t+3, \dots, 2k-t+1\} \notin \mathcal{F}.$$

Proof Indeed, if $F_0 \in \mathcal{F}$ then by shiftedness

$$F_0' = \{1, 2, \dots, t-1, t, t+2, \dots, 2k-t\} \in \mathcal{F}.$$

However, $|F_0 \cap F_0'| = t - 1$, contradicting the fact that $\mathcal{F}$ is t-intersecting. □

Frankl noticed that if a lattice path $\sigma(F)$ from $(0, -t)$ to $(n-k, k-t)$ does not go below $\sigma(F_0)$, then $\sigma(F)$ has to hit the line $y = x$ (as shown in Fig. 2). Thus the number of lattice paths from $(0, -t)$ to $(n-k, k-t)$ that do not go below $\sigma(A)$ is exactly the number of lattice paths from $(0, -t)$ to $(n-k, k-t)$ that hits the line

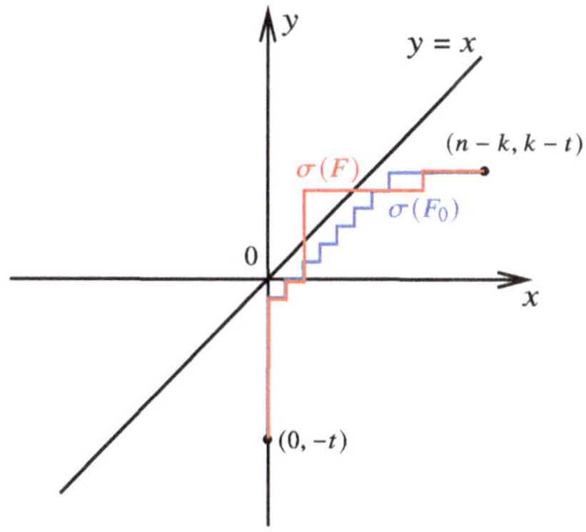

Fig. 2 The lattice path $\sigma(F)$ goes cross $\sigma(F_0)$ and hits the line $y = x$

$y = x$. By the reflection principle, it equals the number of lattice paths from $(-t, 0)$ to $(n - k, k - t)$, which is $\binom{n}{k-t}$. By this argument, Frankl obtained a universal bound on the size of a t-intersecting family for $n > 2k - t$.

Theorem 2.4 ([13]) *Let $\mathcal{F} \subset \binom{[n]}{k}$ be a t-intersecting family with $n > 2k - t$. Then*

$$|\mathcal{F}| \leq \binom{n}{k-t}.$$

Let $\mathcal{F} \subset \binom{[n]}{k}$ be a t-intersecting family and let $F \in \mathcal{F}$. Recall that $\sigma(F)$ hits the line $y = x$. Define $i(F)$ to be the integer i such that (i, i) is the first point that $\sigma(F)$ hits. Then we have $|F \cap [t + 2i(F)]| = t + i(F)$ and $|F \cap [t + 2j]| < t + j$ for all $1 \leq j < i(F)$. Clearly $0 \leq i(F) \leq k - t$. For $0 \leq i \leq k - t$, define

$$\mathcal{F}_i = \{F \in \mathcal{F} : i(F) = i\}.$$

Then $(\mathcal{F}_0, \mathcal{F}_1, \ldots, \mathcal{F}_{k-t})$ forms a partition of $\mathcal{F}$. Let $\mathcal{F}_{\geq 2} = \mathcal{F}_2 \cup \ldots \cup \mathcal{F}_{k-t}$. Note that the number of lattice paths that go through point $(0, 0)$ is $\binom{n-t}{k-t}$ and the number of lattice paths that go through point $(1, 1)$ but not $(0, 0)$ is $t\binom{n-t-2}{k-t-1}$. Thus,

$$|\mathcal{F}_{\geq 2}| \leq \binom{n}{k-t} - \binom{n-t}{k-t} - t\binom{n-t-2}{k-t-1}.$$

By a more detailed argument, Frankl gave an upper bound on $|\mathcal{F}_0| + |\mathcal{F}_1|$ and showed that $|\mathcal{F}| \leq \binom{n-t}{k-t}$ for $n \geq (t+1)(k-t+1)$ and $t \geq 15$.

Two families $\mathcal{A}, \mathcal{B} \subset \binom{[n]}{k}$ are called *cross t-intersecting* if $|A \cap B| \geq t$ for any $A \in \mathcal{A}$ and $B \in \mathcal{B}$.

By a similar arugment, Frankl also proved a cross t-intersecting version of Theorem 2.4.

Theorem 2.5 ([13]) *Let $n \geq 2k - t$. Suppose that $\mathcal{A}, \mathcal{B} \subset \binom{[n]}{k}$ are cross t-intersecting, $|\mathcal{A}| \leq |\mathcal{B}|$. Then either*

$$|\mathcal{B}| \leq \binom{n}{k-t} \text{ or } |\mathcal{A}| \leq \binom{n}{k-t-1}.$$

3 The Generating Set Method

In [13], Frankl also proposed the problem to determine the maximum size of a t-intersecting family for $n < (t+1)(k-t+1)$. Define the *Frankl family* as

$$\mathcal{A}_i(n,k,t) = \left\{ F \in \binom{[n]}{k} : |F \cap [t+2i]| \geq t+i \right\}.$$

Conjecture 3.1 ([13]) If $\mathcal{F} \subset \binom{[n]}{k}$ is a t-intersecting family, then

$$|\mathcal{F}| \leq \max_{0 \leq i \leq k-t} |\mathcal{A}_i(n,k,t)|.$$

In 1997, Ahlswede and Khachatrian [1] proved Frankl's conjecture by introducing the *generating set method.*

Let $\binom{[n]}{\leq k}$ denote the collection of all subsets of $[n]$ with size at most k. For $\mathcal{G} \subset \binom{[n]}{\leq k}$, define

$$\langle \mathcal{G} \rangle = \left\{ F \in \binom{[n]}{k} : \text{ there exists } G \in \mathcal{G} \text{ such that } G \subset F \right\}.$$

For $\mathcal{F} \subset \binom{[n]}{k}$ and $\mathcal{G} \subset \binom{[n]}{\leq k}$, we call $\mathcal{G}$ a *generating family* of $\mathcal{F}$ if $\langle \mathcal{G} \rangle = \mathcal{F}$. Members of $\mathcal{G}$ are called *generating sets.*

A family $\mathcal{F} \subset \binom{[n]}{k}$ is called *saturated* t-intersecting if $\mathcal{F}$ is t-intersecting and any addition of an extra k-set to $\mathcal{F}$ would destroy the t-intersecting property.

Let $F_1, F_2, \ldots, F_r$ be k-sets. If $F_i \cap F_j = F_1 \cap F_2 \cap \ldots \cap F_r$ holds for all $1 \leq i < j \leq r$, then we say that $F_1, F_2, \ldots, F_r$ form a sunflower of r petals. In 1960, Erdős and Rado proved the so-called sunflower lemma.

Lemma 3.2 *([10]) For $r \geq 2$, if $\mathcal{F}$ is k-graph that contains no sunflower of r petals, then*

$$|\mathcal{F}| < k!r^k.$$

Deza (cf. [9]) was the first to realize that the sunflower lemma can be used to prove results related to the Erdős-Ko-Rado Theorem ([11]). This idea was further developed by Füredi ([37]). Frankl [14] used the sunflower lemma to define the *basis* (or *kernel*) of a t-intersecting family, which is very powerful in resolving problems related to t-intersecting families for n sufficiently large.

For saturated t-intersecting families, Frankl [19] found that one can define the basis by minimal transversals.

Let $\mathcal{F} \subset \binom{[n]}{k}$ be a saturated t-intersecting family. Define the *transversal family*

$$\mathcal{T}_t(\mathcal{F}) := \{T \subset [n]\colon |T| \leq k, |T \cap F| \geq t \text{ for all } F \in \mathcal{F}\}.$$

Define the *basis* $\mathcal{B} = \mathcal{B}_t(\mathcal{F})$ as the family of all minimal members of $\mathcal{T}_t(\mathcal{F})$, that is , the collection of all *minimal transversals*. We call the basis of $\mathcal{F}$ *the standard generating family* of $\mathcal{F}$. For $t \leq \ell \leq k$, we use $\mathcal{B}^{(\ell)}$ to denote the set $\{B \in \mathcal{B}\colon |B| = \ell\}$.

The next lemma shows that the basis is indeed a generating family.

Lemma 3.3 ([30]) *Suppose that* $\mathcal{F} \subset \binom{[n]}{k}$ *is a saturated* t*-intersecting family and* $\mathcal{B} = \mathcal{B}_t(\mathcal{F})$. *Then (i) and (ii) hold.*

(i) $\mathcal{B}$ *is* t*-intersecting and* $\langle \mathcal{B} \rangle = \mathcal{F}$, *that is,* $\mathcal{B}$ *is a generating family of* $\mathcal{F}$.
(ii) For $t \leq \ell \leq k$,

$$|\mathcal{B}^{(\ell)}| \leq \binom{\ell}{t} k^{\ell - t}.$$

Since $\mathcal{B}_t(\mathcal{F})$ consists of the minimal members in $\mathcal{T}_t(\mathcal{F})$, by Lemma 3.3 (i) $\mathcal{T}_t(\mathcal{F})$ is also t-intersecting.

Fact 3.4 *Let* $\mathcal{F}$ *be a shifted and saturated* t*-intersecting family. Let* $\mathcal{B} = \mathcal{B}_t(\mathcal{F})$ *be the standard generating family. Then for any* $B \in \mathcal{B}$ *with* $j \in B$ *and* $i \notin B$, $i < j$ *there exists* $B' \in \mathcal{B}$ *such that* $B' \subset (B \setminus \{j\}) \cup \{i\}$.

For any $A \subset [n]$, let $s^+(A) = \max\{i : i \in A\}$. Let $s^+(\mathcal{B}) = \max\{s^+(B)\colon B \in \mathcal{B}\}$. Let $\ell = s^+(\mathcal{B})$. Then $\mathcal{B} \subset 2^{[\ell]}$. If $\ell \in B \in \mathcal{B}$, then we call B a *boundary generating set*. Let $\mathcal{B}_*$ be the collection of all boundary generating sets. That is,

$$\mathcal{B}_* = \{B \in \mathcal{B}\colon \ell \in B\}.$$

Lemma 3.5 *Let* $\mathcal{F}$ *be a shifted and saturated* t*-intersecting family. Let* $\mathcal{B} = \mathcal{B}_t(\mathcal{F})$ *be the standard generating set. Then for any boundary generating set* $B \in \mathcal{B}_*$, *there exists another boundary generating set* $B' \in \mathcal{B}_*$ *such that* $|B \cap B'| = t$ *and* $B \cup B' = [\ell]$.

Remark. The lemma implies that $\ell + t = |B| + |B'| \leq 2k$. Thus $\ell \leq 2k - t$. That is, the standard generating family $\mathcal{B}$ is a subfamily of $2^{[2k-t]}$.

Proof Since B is a minimal transversal of $\mathcal{F}$ and $\ell \in B$, we infer that $E := B \setminus \{\ell\}$ is no longer a transversal. It follows that there exists $F \in \mathcal{F}$ such that $|E \cap F| < t$. Then $|B \cap F| \geq t$ implies that $|B \cap F| = t$. Since $\langle \mathcal{B} \rangle = \mathcal{F}$, there exists $B' \in \mathcal{B}$ such that $B' \subset F$ (as shown in Fig. 3 (a)). Note that $\mathcal{B}$ is t-intersecting. It follows that $|B \cap B'| = t$.

If $B \cup B' \neq [\ell]$, then there exists $x \in [\ell]$ such that $x \notin B \cup B'$ (as shown in Fig. 3 (b)). Since $x < \ell$, by Fact 3.4 there exists $B'' \in \mathcal{B}$ such that $B'' \subset (B' \setminus \{\ell\}) \cup \{x\}$. It follows that $|B'' \cap B| < t$, contradicting the fact that $\mathcal{B}$ is t-intersecting. □

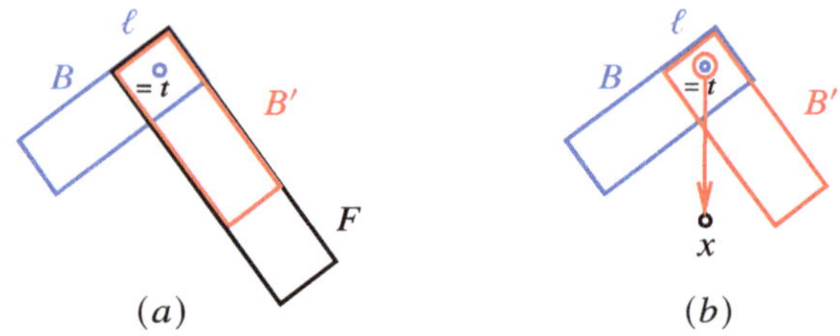

Fig. 3 The paired boundary generating sets

Ahlswede and Khachatrian [1] found a very nice way to analyze the structure of the generating families of a saturated and shifted t-intersecting family. In the rest of this seciton, we review their proof and extract the part for the exact Erdős-Ko-Rado theorem.

For $B \in \mathcal{B}$, define

$$\mathscr{D}(B) = \left\{ F \in \binom{[n]}{k} : F \cap [s^+(B)] = B \right\}.$$

Fact 3.6 *For any distinct $B_1, B_2 \in \mathcal{B}$, $\mathscr{D}(B_1) \cap \mathscr{D}(B_2) = \emptyset$.*

Proof Indeed, otherwise let $F \in \mathscr{D}(B_1) \cap \mathscr{D}(B_2)$, $\ell_i = s^+(B_i)$, $i = 1, 2$. Without loss of generality assume that $\ell_1 \le \ell_2$. Then $B_1 = F \cap [\ell_1] \subset F \cap [\ell_2] = B_2$, contradicting the fact that $\mathcal{B}$ is the collection of all minimal transversals. □

Lemma 3.7 ([1]) *Let $\mathcal{F} \subset \binom{[n]}{k}$ be a saturated, shifted t-intersecting family and let $\mathcal{B}$ be its standard generating family. Then $\mathcal{F} = \cup_{B \in \mathcal{B}} \mathscr{D}(B)$.*

Proof By the definition of the generating family, we have $\mathcal{F} \supset \cup_{B \in \mathcal{B}} \mathscr{D}(B)$. We are left to show that for any $F \in \mathcal{F}$, there exists $B \in \mathcal{B}$ such that $F \in \mathscr{D}(B)$. Since $\mathcal{B}$ is a generating family, there exists $B \in \mathcal{B}$ such that $B \subset F$. Choose such a B with $s^+(B)$ minimal. If $F \cap [s^+(B)] \ne B$ then there is some $x \in F \cap [s^+(B)]$ such that $x \notin B$. It follows that there exists $B' \in \mathcal{B}$ such that $B' \subset (B \setminus \{s^+(B)\}) \cup \{x\}$. Clearly $B' \subset F$ and $s^+(B') < s^+(B)$, contradicting the choice of B. Thus $F \cap [s^+(B)] = B$ and $F \in \mathscr{D}(B)$ follows. □

By Fact 3.6 and Lemma 3.7, we obtain the following corollary.

Corollary 3.8 *Let $\mathcal{F} \subset \binom{[n]}{k}$ be a saturated, shifted t-intersecting family and let $\mathcal{B}$ be its standard generating family. Then*

$$|\mathcal{F}| = \sum_{B \in \mathcal{B}} \binom{n - s^+(B)}{k - |B|}. \tag{18}$$

Lemma 3.9 *Let $\mathcal{F} \subset \binom{[n]}{k}$ be a shifted t-intersecting family of the maximum size. Let $\mathcal{B}$ be its standard generating family with $s^+(\mathcal{B}) = \ell$. If $\mathcal{B}_*^{(p)} \ne \emptyset$ for some $p \ne \frac{\ell+t}{2}$ and $n > 2k$, then $\mathcal{F}$ is a t-star.*

Proof Suppose that $\mathcal{F}$ is not a t-star. Then $\ell \geq t+1$. If $\mathcal{B}_*^{(p)} \neq \emptyset$ then by Lemma 3.5 we have $\mathcal{B}_*^{(\ell+t-p)} \neq \emptyset$. Let $q = \ell + t - p$, $\mathcal{P} = \mathcal{B}_*^{(p)}$ and $\mathcal{Q} = \mathcal{B}_*^{(q)}$. Define

$$\mathcal{B}_1 = (\mathcal{B} \setminus \mathcal{P}) \cup \mathcal{Q}' \text{ and } \mathcal{B}_2 = (\mathcal{B} \setminus \mathcal{Q}) \cup \mathcal{P}',$$

where $\mathcal{P}' = \{P \setminus \{\ell\} \colon P \in \mathcal{P}\}$ and $\mathcal{Q}' = \{Q \setminus \{\ell\} \colon Q \in \mathcal{Q}\}$. Then it is easy to check that $\mathcal{B}_1, \mathcal{B}_2$ are both t-intersecting. Let $\mathcal{F}_i = \cup_{B \in \mathcal{B}_i} \mathcal{D}(B)$, $i = 1, 2$. Then

$$|\mathcal{F}_1| = |\mathcal{F}| - |\mathcal{P}|\binom{n-\ell}{k-p} + |\mathcal{Q}|\binom{n-\ell}{k-q+1}$$

and

$$|\mathcal{F}_2| = |\mathcal{F}| - |\mathcal{Q}|\binom{n-\ell}{k-q} + |\mathcal{P}|\binom{n-\ell}{k-p+1}.$$

By the maximality of $|\mathcal{F}|$, we infer that

$$|\mathcal{Q}|\binom{n-\ell}{k-q+1} \leq |\mathcal{P}|\binom{n-\ell}{k-p}, \quad |\mathcal{P}|\binom{n-\ell}{k-p+1} \leq |\mathcal{Q}|\binom{n-\ell}{k-q}.$$

Thus,

$$\binom{n-\ell}{k-q+1}\binom{n-\ell}{k-p+1} \leq \binom{n-\ell}{k-p}\binom{n-\ell}{k-q}.$$

It follows that

$$\frac{n-\ell-k+q}{k-q+1} \cdot \frac{n-\ell-k+p}{k-p+1} \leq 1.$$

Since $n > 2k$ then $\frac{n-\ell-k+q}{k-p+1} > 1$ and $\frac{n-\ell-k+p}{k-q+1} > 1$, a contradiction. Thus the lemma holds. □

Lemma 3.10 *Let $\mathcal{F} \subset \binom{[n]}{k}$ be a shifted t-intersecting family of the maximum size. Let $\mathcal{B}$ be its standard generating set with $s^+(\mathcal{B}) = \ell$. If $\mathcal{B}_*^{(p)} \neq \emptyset$ with $p = \frac{\ell+t}{2}$ and $n > (t+1)(k-t+1)$, then $\mathcal{F}$ is a t-star.*

Proof For $1 \leq i \leq \ell - 1$, let

$$\mathcal{P}_i = \left\{P \in \mathcal{B}_*^{(p)} \colon i \notin P\right\}.$$

and let

$$\mathcal{B}_i = (\mathcal{B} \setminus \mathcal{B}_*^{(p)}) \cup \mathcal{P}_i', \text{ where } \mathcal{P}_i' = \{P \setminus \{\ell\} \colon P \in \mathcal{P}_i\}.$$

Then $\mathcal{B}_i$ is t-intersecting, $i = 1, 2, \ldots, \ell - 1$. Let $\mathcal{F}_i = \cup_{B \in \mathcal{B}_i} \mathcal{D}(B)$, $i = 1, 2, \ldots, \ell - 1$. Since

$$\sum_{i=1}^{\ell-1} |\mathcal{P}_i| = |\mathcal{B}_*^{(p)}|(\ell - p),$$

there exists $i \in [\ell - 1]$ such that

$$|\mathcal{P}_i| \geq \frac{\ell - p}{\ell - 1}|\mathcal{B}_*^{(p)}|.$$

Then

$$\begin{aligned}|\mathcal{F}_i| &= |\mathcal{F}| - |\mathcal{B}_*^{(p)}|\binom{n-\ell}{k-p} + |\mathcal{P}_i|\binom{n-\ell+1}{k-p+1} \\ &\geq |\mathcal{F}| - |\mathcal{B}_*^{(p)}|\left(\binom{n-\ell}{k-p} - \frac{\ell-p}{\ell-1}\binom{n-\ell+1}{k-p+1}\right).\end{aligned}$$

Thus,

$$\binom{n-\ell}{k-p} \geq \frac{\ell-p}{\ell-1}\binom{n-\ell+1}{k-p+1}.$$

It follows that

$$1 \geq \frac{\ell-p}{\ell-1}\frac{n-\ell+1}{k-p+1}.$$

Equivalently,

$$n \leq \frac{\ell-1}{\ell-p}(k-t+1).$$

Since $\ell \geq t+1$ and $\ell + t$ is even, we infer that $\ell \geq t+2$. It follows that

$$\frac{\ell-1}{\ell-p} = \frac{2\ell-2}{\ell-t} = 2 + \frac{2t-2}{\ell-t} \leq t+1.$$

But we have $n > (t+1)(k-t+1)$, a contradiction. Thus the lemma holds. □

Combining Lemma 3.9 and Lemma 3.10, concludes the proof of the Exact Erdős-Ko-Rado Theorem.

Theorem 3.11 (The Exact Erdős-Ko-Rado Theorem [13, 58]) *Let* $\mathcal{F} \subset \binom{[n]}{k}$ *be a* t*-intersecting family. If* $n > (t+1)(k-t+1)$ *then*

$$|\mathcal{F}| \leq \binom{n-t}{k-t}.$$

By more involved arguments, Ahlswede and Khachatrian proved the Complete Intersecting Theorem and the exact Hilton-Milner-Frankl Theorem.

Theorem 3.12 (The Complete Intersecting Theorem [1]) *If* $\mathcal{F} \subset \binom{[n]}{k}$ *is a* t*-intersecting family, then*

$$|\mathcal{F}| \leq \max_{0 \leq i \leq k-t} |\mathcal{A}_i|.$$

Define

$$\mathcal{H}(n,k,t) = \left\{H \in \binom{[n]}{k} : [t] \subset H,\ H \cap [t+1, k+1] \neq \emptyset\right\} \cup \{[k+1] \setminus \{j\} : 1 \leq j \leq t\}.$$

Theorem 3.13 (The Exact Hilton-Milner-Frankl Theorem [2, 14]) *If $\mathcal{F} \subset \binom{[n]}{k}$ is a t-intersecting family that is not a t-star, then for $n \geq (t+1)(k-t+1)$,*

$$|\mathcal{F}| \leq \max\{|\mathcal{A}_1(n,k,t)|, |\mathcal{H}(n,k,t)|\}.$$

Let us also mention that very recently by using the generating method, Zhang and Wu confirmed a conjecture of Tokushige [56] for $t \geq 3$.

Theorem 3.14 ([60]) *Let $n \geq k \geq t \geq 3$ and $n \geq (t+1)(k-t+1)$. If $\mathcal{A}, \mathcal{B} \subset \binom{[n]}{k}$ are cross t-intersecting, then*

$$|\mathcal{A}||\mathcal{B}| \leq \binom{n-t}{k-t}^2.$$

4 Shifting Ad Extremis

The shifting method is very powerful because of the random walk method and the generating set method. Let us mention that

- the random walk method can be used when the family is *shifted*;
- the generating set method can only be used when the family is *both shifted and saturated.*

However, if the family has certain properties that might be destroyed by shifting, then the shifting method cannot be used. For example, shifting may turn a non-trivial intersecting family to a trivial one. The shifting operator may decrease the covering number.

Recently, Frankl [24] introduced a modified version of the shifting method, which is called *shifting ad extremis*. Let us illustrate it by a two-families version. Let $\mathcal{F} \subset \binom{[n]}{k}$, $\mathcal{G} \subset \binom{[n]}{\ell}$ be families having certain properties (e.g., intersecting, cross t-intersecting) that are maintained by simultaneous shifting and certain properties (e.g., $\tau(\mathcal{F}) \geq s$) that might be destroyed by shifting. Let $\mathcal{P}$ be the collection of the latter properties that we want to maintain.

Definition 4.1 Suppose that $\mathcal{F} \subset \binom{[n]}{k}$, $\mathcal{G} \subset \binom{[n]}{\ell}$ are families having property $\mathcal{P}$. We say that $\mathcal{F}$ and $\mathcal{G}$ have been *shifted ad extremis* with respect to $\mathcal{P}$ if $S_{ij}(\mathcal{F}) = \mathcal{F}$ and $S_{ij}(\mathcal{G}) = \mathcal{G}$ for every pair $1 \leq i < j \leq n$ whenever $S_{ij}(\mathcal{F})$ and $S_{ij}(\mathcal{G})$ also have property $\mathcal{P}$.

Lemma 4.2 ([24]) *Let* $\mathcal{F} \subset \binom{[n]}{k}$, $\mathcal{G} \subset \binom{[n]}{\ell}$ *be families having property* $\mathcal{P}$. *By repeated shifting, one can obtain families* $\mathcal{F}'$ *and* $\mathcal{G}'$ *that are shifted ad extremis with respect to* $\mathcal{P}$.

4.1 Intersecting Families with Covering Number s

Define its *covering number* $\tau(\mathcal{F})$ as

$$\tau(\mathcal{F}) = \min\{|T| : T \cap F \neq \emptyset \text{ for all } F \in \mathcal{F}\}.$$

Let us define the function

$$f(n, k, s) = \max\left\{|\mathcal{F}| : \mathcal{F} \subset \binom{[n]}{k} \text{ is intersecting and } \tau(\mathcal{F}) \geq s\right\}.$$

With this terminology the Erdős-Ko-Rado Theorem and the Hilton-Milner Theorem can be stated as for $n \geq 2k$,

$$f(n, k, 1) = \binom{n-1}{k-1}, \; f(n, k, 2) = \binom{n-1}{k-1} - \binom{n-k-1}{k-1} + 1.$$

Erdős and Lovász [12] proved that $\lfloor k!(e-1)\rfloor \leq f(n, k, k) \leq k^k$ and $f(n, 3, 3) = 10$. Lovász [52] conjectured that $\lfloor k!(e-1)\rfloor$ is the exact bound. Later, Lovász's conjecture was disproved by Frankl, Ota and Tokushige [29] for $k \geq 4$ and the lower bound of $f(n, k, k)$ was improved to $(1 + o(1)) \left(\frac{k}{2}\right)^k$. The upper bound of $f(n, k, k)$ was further improved in [4, 7, 21, 57, 59].

In 1980, Frankl [15] determined $f(n, k, 3)$ for $k \geq 4$, $n > n_0(k)$.

Theorem 4.3 ([33]) *For* $k \geq 7$ *and* $n \geq 2k$,

$$f(n, k, 3) = \binom{n-1}{k-1} - \binom{n-k}{k-1} - \binom{n-k-1}{k-1} + \binom{n-2k}{k-1} + \binom{n-k-2}{k-3} + 3. \quad (19)$$

Let us mention that Kupavskii [50] proved the same result for $k \geq 100$ and $n \geq 2k$ using a different method.

Define

$$\mathcal{B} = \{[2, k+1], \{2\} \cup [k+2, 2k], \{3\} \cup [k+2, 2k]\}$$

and

$$\mathcal{A} = \left\{A \in \binom{[n]}{k} : 1 \in A \text{ and } A \cap B \neq \emptyset \text{ for each } B \in \mathcal{B}\right\}.$$

Set $\mathcal{G}(n, k) = \mathcal{A} \cup \mathcal{B}$ (as shown in Fig. 4).

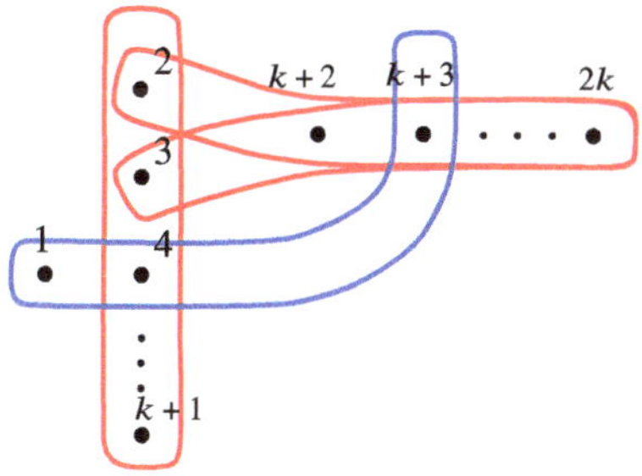

Fig. 4 The extremal construction $\mathcal{G}(n, k)$

It is easy to check that for $n \geq 2k$, $\mathcal{G}(n, k)$ is an intersecting k-graph with $\tau(\mathcal{G}(n, k)) = 3$ and

$$|\mathcal{G}(n,k)| = \binom{n-1}{k-1} - \binom{n-k}{k-1} - \binom{n-k-1}{k-1} + \binom{n-2k}{k-1} + \binom{n-k-2}{k-3} + 3.$$

The 3-element covers are $\{1, 2, 3\}$ and $\{1, u, v\}$ with $u \in [2, k+1]$, $v \in [k+2, 2k]$.

Let $\mathcal{F} \subset \binom{[n]}{k}$ be an intersecting family satisfying property $\mathcal{P} = \{\tau(\mathcal{F}) \geq 3\}$. Suppose that $\mathcal{F}$ is shifted ad extremis with respect to $\mathcal{P}$.

A pair (i, j), $1 \leq i < j \leq n$, is called a *shift-resistant pair* if $\tau(S_{ij}(\mathcal{F})) = 2$. Define the *shift-resistant graph* $\mathbb{H} = \mathbb{H}_{\mathcal{P}}(\mathcal{F})$ as the collection of all pairs (i, j) such that

$$\tau(S_{ij}(\mathcal{F})) = 2 \text{ for } (i, j) \in \mathbb{H} \text{ and } S_{ij}(\mathcal{F}) = \mathcal{F} \text{ for } (i, j) \notin \mathbb{H}.$$

Clearly $\mathcal{F}$ is shifted if and only if $\mathbb{H}$ is empty.

Let us mention that $\mathcal{G}(n, k)$ is *not shifted but shifted ad extremis* with respect to $\tau(\mathcal{G}(n, k)) \geq 3$. The shift-resistant graph is

$$\mathbb{H} = [2, k+1] \times [k+2, 2k] \cup \{(1, 2), (1, 3)\}.$$

For example, $\{1, 2\}$ is a *cover* of $S_{2,k+2}(\mathcal{G}(n, k))$.

To illustrate the use of shifting ad extremis, let us give a proof sketch of Theorem 4.3.

Let $\mathcal{F} \subset \binom{[n]}{k}$ be an intersecting family of the maximal size with covering number at least 3. Assume that $\mathcal{F}$ is shifted ad extremis with respect to $\tau(\mathcal{F}) \geq 3$ and let $\mathbb{H}$ be the correponding shift-resistant graph. Define the family of *3-covers* as

$$\mathcal{T}^{(3)}(\mathcal{F}) = \left\{T \in \binom{[n]}{3} : T \text{ is a cover of } \mathcal{F}\right\}$$

By the maximality of $|\mathcal{F}|$, $\mathcal{T}^{(3)}(\mathcal{F})$ is intersecting. Then the proof splits into three cases:

- The shifted case: $\mathcal{F}$ is shifted.
- The shift-resistant case I: $\mathcal{T}^{(3)}(\mathcal{F})$ is not a star.
- The shift-resistant case II: $\mathcal{T}^{(3)}(\mathcal{F})$ is a star.

Case 1. The shifted case.
The *clique number* $\omega(\mathcal{F})$ for $\mathcal{F} \subset \binom{[n]}{k}$ is defined as

$$\omega(\mathcal{F}) = \max\left\{q : \exists Q \in \binom{[n]}{q}, \binom{Q}{k} \subset \mathcal{F}\right\}.$$

Frankl determined the maximum size of an intersecting family with clique number at least $k + s - 1$.

Theorem 4.4 ([22]) *Let $n > 2k \geq 2s$. Suppose that $\mathcal{F} \subset \binom{[n]}{k}$ is intersecting and $\omega(\mathcal{F}) \geq k + s - 1$. Then*

$$|\mathcal{F}| \leq |\mathcal{K}(n, k, s)|, \tag{20}$$

where

$$\mathcal{K}(n, k, s) = \left\{K \in \binom{[n]}{k} : 1 \in K, |K \cap [2, k+s-1]| \geq s-1\right\} \bigcup \binom{[2, k+s-1]}{k}.$$

If $\mathcal{F}$ is a shifted family with $\tau(\mathcal{F}) \geq 3$, then we have $\omega(\mathcal{F}) \geq k + 2$. Indeed, since $\tau(\mathcal{F}) \geq 3$ and shiftedness imply $\{3, 4, \ldots, k+2\} \in \mathcal{F}$, by shiftedness again $[k+2]$ forms a clique in $\mathcal{F}$. By Theorem 4.4, we conclude that $|\mathcal{F}| \leq |\mathcal{K}(n, k, 3)| \leq |\mathcal{G}(n, k)|$.

Case 2. The shift-resistant case I: $\mathcal{T}^{(3)}(\mathcal{F})$ is not a star.
Note that $\mathcal{F}$ is not shifted but shifted ad extremis. It follows that $\mathbb{H} \neq \emptyset$. Let $(i, j) \in \mathbb{H}$. Then $\tau(S_{ij}(\mathcal{F})) = 2$. Assume that $\{i, x_{ij}\}$ is a cover of $S_{ij}(\mathcal{F})$. Then $\{i, j, x_{ij}\}$ is a cover of $\mathcal{F}$. Thus $\mathcal{T}^{(3)}(\mathcal{F}) \neq \emptyset$.

We claim that $\mathbb{H}$ has matching number at least two. Indeed, otherwise assume that $(1, 2) \in \mathbb{H}$. Then $\mathcal{F}$ is shifted on $\{3, 4, \ldots, n\}$. Without loss of generality further assume that $x_{12} = 3$, that is, $\{1, 2, 3\}$ is a cover of $\mathcal{F}$. Note that $(1, 2) \in \mathbb{H}$ implies $\{1, 3\}$ is a cover of $S_{12}(\mathcal{F})$. Since $\{1, 3\}$ is not a cover of $\mathcal{F}$, by shiftedness we infer that $\{4, 5, \ldots, k+2\} \in \mathcal{F}(2)$. Since $\{2, 3\}$ is not a cover of $\mathcal{F}$, by shiftedness we have $\{4, 5, \ldots, k+2\} \in \mathcal{F}(1)$. Then $\{2, 4, 5, \ldots, k+2\} \in S_{12}(\mathcal{F})$, contradicting the fact that $\{1, 3\}$ is a cover of $S_{12}(\mathcal{F})$. Thus $\mathbb{H}$ has matching number at least two.

Let $(1, 2), (3, 4) \in \mathbb{H}$. Since $\mathcal{T}^{(3)}$ is intersecting, we may assume that $x_{12} = x_{34} = 5$. Let $T_1 = \{1, 2, 5\}$, $T_2 = \{3, 4, 5\}$. Since $\mathcal{T}^{(3)}(\mathcal{F})$ is non-trivially intersecting, one can find $T_3 \in \mathcal{T}^{(3)}(\mathcal{F})$ such that T_1, T_2, T_3 form a generalized triangle. By symmetry either $T_3 = \{2, 4, 6\}$ or $T_3 = \{2, 3, 4\}$ (as shown in Fig. 5).

Note that $\mathcal{T}^{(3)}(\mathcal{F})$ is intersecting and $\mathcal{T}^{(3)}(\mathcal{F}), \mathcal{F}$ are cross-intersecting. For any $\{x, y\} \in \binom{[n]}{2}$, we have $|\mathcal{T}^{(3)}(\mathcal{F})(x, y)| \leq k$. Indeed, for otherwise $F \cap \{x, y\} \neq \emptyset$ for any $F \in \mathcal{F}$, contradicting $\tau(\mathcal{F}) \geq 3$. Let $\mathcal{P} = \{T_1, T_2, T_3\}$. One can check that $|\mathcal{T}^{(2)}(\mathcal{P})| \leq 7$. It follows that $|\mathcal{T}^{(3)}(\mathcal{F})| \leq 7k < k^2 - k + 1 = |\mathcal{T}^{(3)}(\mathcal{G}(n, k))|$ for

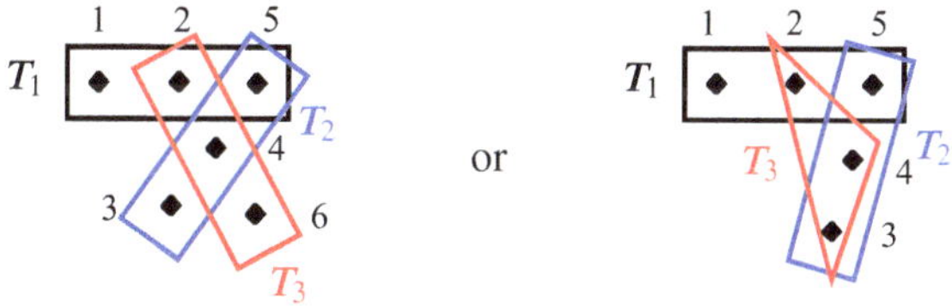

Fig. 5 The generalized triangle (T_1, T_2, T_3) in $\mathcal{T}^{(3)}(\mathcal{F})$

$k \geq 8$. Thus $|\mathcal{F}| < |\mathcal{G}(n, k)|$ for $k \geq 8$ and n sufficiently large with respect to k. By some involved computations, we can show that $|\mathcal{F}| < |\mathcal{G}(n, k)|$ for $n > 2k$ and $k \geq 7$. For more details, please see [33].

Case 3. The shift-resistant case II: $\mathcal{T}^{(3)}(\mathcal{F})$ is a star.
Suppose that $\mathcal{T}^{(3)}(\mathcal{F})$ is a star with *center 1*. Define

$$\mathcal{A} := \{F \setminus \{1\} \colon 1 \in F \in \mathcal{F}\}, \ \mathcal{B} := \{F \colon 1 \notin F \in \mathcal{F}\}.$$

We say that $\mathcal{A}$, $\mathcal{B}$ are *saturated cross-intersecting with respect to* $\mathcal{A}$ if any addition of an extra $(k-1)$-set to $\mathcal{A}$ would destroy the cross-intersecting property. Since $\mathcal{F}$ is of the maximum size, we infer that $\mathcal{A}$, $\mathcal{B}$ are saturated cross-intersecting with respect to $\mathcal{A}$. Since $\tau(\mathcal{F}) \geq 3$, $\mathcal{B}$ is non-trivially intersecting.

Lemma 4.5 ([33]) *If $\mathcal{F}$ is shifted ad extremis with respect to $\tau(\mathcal{F}) \geq 3$, then $\mathcal{A}$, $\mathcal{B}$ are shifted ad extremis with respect to $\tau(\mathcal{B}) \geq 2$. Moreover, the corresponding shift-resistant graph is $\mathbb{H}' = \mathbb{H} \cap \binom{[2,n]}{2}$.*

By the same argument as in Case 2, $\mathbb{H}$ has matching number at least two. It follows that $\mathbb{H}' = \mathbb{H} \cap \binom{[2,n]}{2} \neq \emptyset$. Define *2-cover graph* $\widehat{\mathbb{H}}$ of $\mathcal{B}$ to be the graph of all covers of $\mathcal{B}$ of size 2. It is easy to see that the shift-resistant graph $\mathbb{H}'$ is a subgraph of the 2-cover graph $\widehat{\mathbb{H}}$. Moreover, if $(i, j) \in \widehat{\mathbb{H}}$ then $\{1, i, j\}$ is a cover of $\mathcal{F}$ and (i, j) is *full* in $\mathcal{A}$, that is, any $(k-1)$-set containing $\{i, j\}$ is in $\mathcal{A}$.

Let us mention that if $\mathcal{F} = \mathcal{G}(n, k)$ then

$$\mathbb{H}' = \mathbb{H} = [2, k+1] \times [k+2, 2k] \text{ and } \widehat{\mathbb{H}} = \mathbb{H}' \cup \{(2, 3)\}.$$

The key Lemma to the proof is the following.

Lemma 4.6 ([33]) *The 2-cover graph $\widehat{\mathbb{H}}$ is partially shifted. That is, $(i, j) \notin \widehat{\mathbb{H}}$ and $(j, \ell) \in \widehat{\mathbb{H}}$ imply $(i, \ell) \in \widehat{\mathbb{H}}$.*

By the partial shiftedness, it is showed in [33] that $\widehat{\mathbb{H}}$ either contains a triangle or is a complete bipartite graph.

Theorem 4.7 ([33]) *$\widehat{\mathbb{H}}$ either contains a triangle or is a complete bipartite graph on partite sets X, Y with $2 \leq |X|, |Y| \leq k$. Moreover, $X \cup Y$ is the set of the first $|X| + |Y|$ elements in $[2, n]$.*

Theorem 4.8 (Frankl's non-trivial cross-intersecting inequality [24]) *Let $2 \le a \le b$, $n \ge a+b$. Suppose that $\mathcal{A} \subset \binom{[n]}{a}$ and $\mathcal{B} \subset \binom{[n]}{b}$ are non-trivial and cross-intersecting. Then*

$$|\mathcal{A}| + |\mathcal{B}| \le \binom{n}{b} - 2\binom{n-a}{b} + \binom{n-2a}{b} + 2.$$

Finally, by using Frankl's non-trivial cross-intersecting inequality, one can complete the proof of Theorem 4.3 for the triangle case and the complete bipartite case, separately.

Let $\mathcal{F} \subset \binom{[n]}{k}$ be an intersecting family. Recall the definition of $\mathcal{B}(\mathcal{F})$, the family of covers of $\mathcal{F}$ that are minimal for containment.

Example 1 Let $X = \{x_1, \ldots, x_{k-1}\}$, $Y = \{y_1, \ldots, y_{k-1}\}$, $Z = \{z_1, \ldots, z_{k-1}\}$ be three pairwise disjoint subsets of $[2, n]$. Let

$$\mathcal{H}^* = \{X \cup \{y_i\} : i = 1, 2\} \cup \{Y \cup \{z_i\} : i = 1, 2\} \cup \{Z \cup \{x_i\} : i = 1, 2\}.$$

Define

$$\mathcal{F}^* = \left\{ F \in \binom{[n]}{k} : 1 \in F \text{ and there exists } B \in \mathcal{B}(\mathcal{H}^*) \text{ such that } B \subset F \right\} \cup \mathcal{H}^*.$$

The value of $f(n, k, 4)$ was determined for $k \ge 9$ and $n > n_0(k)$ in [28].

Theorem 4.9 ([28]) *For $k \ge 9$ and $n > n_0(k)$, $f(n, k, 4) = |\mathcal{F}^*|$.*

Example 2 Let Y, V and X, Z be $(k-1)$-subsets of $[2, n]$ with $(Y \cup V) \cap (X \cup Z) = \emptyset$, $Y \cap V = \{\tilde{y}\}$, $X \cap Z = \{\tilde{x}\}$. In each of the four $(k-2)$-sets $Y \setminus \{\tilde{y}\}$, $V \setminus \{\tilde{y}\}$, $X \setminus \{\tilde{x}\}$, $Z \setminus \{\tilde{x}\}$, we fix 2 elements $y_1, y_2, v_1, v_2, x_1, x_2$ and z_1, z_2. Let

$$\begin{aligned} \mathcal{G} :=& \{X \cup \{v_i\} : i = 1, 2\} \cup \{Y \cup \{x_i\} : i = 1, 2\} \cup \{Z \cup \{y_i\} : i = 1, 2\} \\ & \cup \{V \cup \{z_i\} : i = 1, 2\} \cup \{(Z \setminus \{\tilde{x}\}) \cup \{x_1, y_i\} : i = 1, 2\}. \end{aligned}$$

Define

$$\mathcal{G}_0 = \mathcal{G} \cup \{(V \setminus \{\tilde{y}\}) \cup \{y_1, y_2\}\} \text{ and } \mathcal{G}_1 = \mathcal{G} \cup \{(Y \setminus \{\tilde{y}\}) \cup \{v_1, v_2\}\}.$$

For $j = 0, 1$, define

$$\mathcal{F}_j = \left\{ F \in \binom{[n]}{k} : 1 \in F \text{ and there exists } B \in \mathcal{B}(\mathcal{G}_j) \text{ such that } B \subset F \right\} \cup \mathcal{G}_j.$$

Recently, Frankl and Wang determined the value of $f(n, k, 5)$ for $k \ge 67$ and $n \ge 5k^6$.

Theorem 4.10 ([36]) *Suppose that* $k \geq 69$ *and* $n \geq 5k^6$, *then* $f(n, k, 5) = \max\{|\mathcal{F}_0|, |\mathcal{F}_1|\}$.

4.2 Diversity and the Maximum Degree Ratio of Intersecting Families

Let $\mathcal{F} \subset \binom{[n]}{k}$ be an intersecting family. Recall that

$$\mathcal{F}(i) = \{F \setminus \{i\} : i \in F \in \mathcal{F}\},\ \mathcal{F}(\bar{i}) = \{F : i \notin F \in \mathcal{F}\}$$

and note that $|\mathcal{F}| = |\mathcal{F}(i)| + |\mathcal{F}(\bar{i})|$. It should be pointed out that $\mathcal{F}(1) \subset \binom{[2,n]}{k-1}$ and $\mathcal{F}(\bar{1}) \subset \binom{[2,n]}{k}$ are cross-intersecting.

There are two ways of measuring how far an intersecting family is from a star. Define

$$\varrho(\mathcal{F}) = \max_{i \in [n]} \frac{|\mathcal{F}(i)|}{|\mathcal{F}|},\ \gamma(\mathcal{F}) = \min_{i \in [n]} |\mathcal{F}(\bar{i})|,$$

where $\varrho(\mathcal{F})$ is called the *maximum degree ratio* of $\mathcal{F}$ and $\gamma(\mathcal{F})$ is called the *diversity* of $\mathcal{F}$. If $\mathcal{F}$ is a star then $\varrho(\mathcal{F}) = 1$ and $\gamma(\mathcal{F}) = 0$. That is, the larger $\varrho(\mathcal{F})$ is and the smaller $\gamma(\mathcal{F})$ is, the closer $\mathcal{F}$ is to a star.

By using the shifting ad extremis, we obtained several results for the maximum degree ratio of an intersecting family.

Theorem 4.11 ([32]) *Let* n, k, d *be integers,* $k > d \geq 2$, $n \geq 4(d-1)dk$. *If* $\mathcal{F} \subset \binom{[n]}{k}$ *is an intersecting family with* $|\mathcal{F}| \geq 2^d d^{2d+1}\binom{n-d-1}{k-d-1}$, *then*

$$\varrho(\mathcal{F}) > \frac{1}{d}.$$

Theorem 4.12 ([31]) *Suppose that* $\mathcal{F} \subset \binom{[n]}{k}$ *is an intersecting family with* $n \geq 2k$ *and* $|\mathcal{F}| \geq 36\binom{n-3}{k-3}$. *Then*

$$\varrho(\mathcal{F}) > \frac{1}{2}.$$

Theorem 4.13 ([31]) *Suppose that* $\mathcal{F} \subset \binom{[n]}{k}$ *is an intersecting family with* $|\mathcal{F}| \geq 36\binom{n-3}{k-3}$ *and* $n \geq 24k$. *Then*

$$\varrho(\mathcal{F}) > \frac{2}{3} - \frac{k}{n}.$$

Define the *triangle family* as

$$\mathcal{T}(n, k) = \left\{F \in \binom{[n]}{k} : |F \cap [3]| \geq 2\right\}.$$

Theorem 4.14 ([35]) *Let $\mathcal{F} \subset \binom{[n]}{k}$ be an intersecting family with $n > 36k$. If $|\mathcal{F}| \geq |\mathcal{T}(n,k)| + 1$, then*

$$\varrho(\mathcal{F}) > 1 - \max\{k+4, 40\} \cdot \frac{k}{3n}.$$

If $|\mathcal{F}| \geq |\mathcal{T}(n,k)| + \frac{p}{2}\binom{n-p+3}{k-p+3}$ for some $p \geq 40$, then

$$\varrho(\mathcal{F}) > 1 - \frac{pk}{3n}.$$

Let us mention that all these results were already proved in [14] for n sufficiently large with respect to k.

In [23], Frankl proved the following.

Theorem 4.15 ([23]) *Suppose that $n > 72k$, $\mathcal{F} \subset \binom{[n]}{k}$ is intersecting. Then*

$$\gamma(\mathcal{F}) \leq \binom{n-3}{k-2}. \tag{21}$$

By using Theorem 4.12 and repeating the proof of Theorem 4.15 in [23], Theorem 4.15 holds for $n > 36k$. Frankl conjectured in [19] that (21) holds for $n > 3k$. Counter-examples by Huang [44] and Kupavskii [49] showed that one needs to assume at least $n \geq (2+\sqrt{3})k$.

To illustrate the use of shifting ad extremis, let us prove a weaker version of Theorem 4.12. We should mention that almost all the proof ideas were given by Frankl in [23].

Theorem 4.16 (A weaker version of Theorem 4.12) *Suppose that $\mathcal{F} \subset \binom{[n]}{k}$ is an intersecting family with $n \geq 2k$ and $|\mathcal{F}| \geq 48\binom{n-3}{k-3}$. Then*

$$\varrho(\mathcal{F}) > \frac{1}{2}.$$

Let $\Delta_2(\mathcal{F})$ denote the maximum 2-degree of $\mathcal{F}$, that is, the maximum of $|\mathcal{F}(x,y)|$ over all $x, y \in [n]$. Let us prove two useful lemmas.

Lemma 4.17 *Let $\mathcal{F} \subset \binom{[n]}{k}$ be an intersecting family. If $\rho(\mathcal{F}) \leq \frac{1}{2}$ then $\Delta_2(\mathcal{F}) \leq 2\binom{n-3}{k-3}$.*

Proof Let $x, y \in [n]$. Then $\mathcal{F}(x,y), \mathcal{F}(\bar{x},\bar{y})$ are cross-intersecting. If $|\mathcal{F}(x,y)| \geq 2\binom{n-3}{k-3} > \binom{n-3}{k-3} + \binom{n-4}{k-3} + \binom{n-6}{k-4}$, then by Hilton's lemma we infer that

$$|\mathcal{F}(\bar{x},\bar{y})| \leq \binom{n-5}{k-3} + \binom{n-6}{k-3} < |\mathcal{F}(x,y)|.$$

By symmetry we may assume that $|\mathcal{F}(x,\bar{y})| \leq |\mathcal{F}(\bar{x},y)|$. It follows that $|\mathcal{F}(\bar{y})| < |\mathcal{F}(y)|$. Thus $\varrho(\mathcal{F}) > \frac{1}{2}$, contradicting $\rho(\mathcal{F}) \leq \frac{1}{2}$. □

Lemma 4.18 *Let $\mathcal{F} \subset \binom{[n]}{k}$ be an intersecting family. If $\mathcal{F}$ is shifted then $\rho(\mathcal{F}) > \frac{1}{2}$.*

Proof By shiftedness,

$$\partial\mathcal{F}(\bar{1}) \subset \mathcal{F}(1).$$

By Theorem 1.13, we have

$$|\mathcal{F}(\bar{1})| < |\partial\mathcal{F}(\bar{1})| \leq |\mathcal{F}(1)|.$$

Then $\varrho(\mathcal{F}) > \frac{1}{2}$ follows. □

Proof of Theorem 4.16 Suppose for contradiction that $\mathcal{F} \subset \binom{[n]}{k}$ is an intersecting family with $|\mathcal{F}| > 48\binom{n-3}{k-3}$ and $\varrho(\mathcal{F}) \leq \frac{1}{2}$. Without loss of generality assume that $\mathcal{F}$ is shifted ad extremis with respect to $\varrho(\mathcal{F}) \leq \frac{1}{2}$. Let $\mathbb{H}$ be the corresponding shift resistant graph. By Lemma 4.18 we may assume $\mathbb{H} \neq \emptyset$. By Lemma 4.17, $\Delta_2(\mathcal{F}) < 2\binom{n-3}{k-3}$.

Claim $\mathbb{H}$ has matching number at most two.

Proof Suppose for contradiction that $(i_1, j_1), (i_2, j_2), (i_3, j_3)$ are three pairwise disjoint pairs in $\mathbb{H}$. Let

$$\mathcal{G}_r = \{F \in \mathcal{F} : F \cap \{i_r, j_r\} \neq \emptyset\},\ r = 1, 2, 3.$$

Since $(i_r, j_r) \in \mathbb{H}$, we have $|\mathcal{G}_r| \geq \frac{1}{2}|\mathcal{F}|$. Note also that for $1 \leq r < r' \leq 3$

$$|\mathcal{G}_r \cap \mathcal{G}_{r'}| \leq \sum_{x\in\{i_r, j_r\}, y\in\{i_{r'}, j_{r'}\}} |\mathcal{F}(x, y)| < 4 \times \Delta_2(\mathcal{F}) < 8\binom{n-3}{k-3} \leq \frac{1}{6}|\mathcal{F}|.$$

It follows that

$$|\mathcal{F}| \geq |\mathcal{G}_1 \cup \mathcal{G}_2 \cup \mathcal{G}_3| = \sum_{1\leq r\leq 3} |\mathcal{G}_r| - \sum_{1\leq r<r'\leq 3} |\mathcal{G}_r \cap \mathcal{G}_{r'}| > \frac{3}{2}|\mathcal{F}| - \frac{1}{2}|\mathcal{F}| = |\mathcal{F}|,$$

a contradiction. Thus the claim holds. □

If $\nu(\mathbb{H}) = 1$ then we may assume that $(n-1, n) \in \mathbb{H}$. If $\nu(\mathbb{H}) = 2$ then we may assume that $(n-1, n), (n-3, n-2) \in \mathbb{H}$. In either case $\mathcal{F}$ is shifted on $[n-4]$. Let

$$\mathcal{H} = \{F \in \mathcal{F} : F \cap \{n-3, n-2, n-1, n\} = \{n-1\} \text{ or } F \cap \{n-3, n-2, n-1, n\} = \{n\}\}$$

and let

$$\mathcal{G} = \{F \in \mathcal{F} : F \subset [n-4]\}.$$

Then $\mathcal{G}$ is a shifted family. By $\nu(\mathbb{H}) \leq 2$, $S_{ij}(\mathcal{H}) = \mathcal{H}$ for all $1 \leq i < j \leq n-5$. Since $(n-1, n) \in \mathbb{H}$, we infer that

$$|\mathcal{H}| \geq \frac{1}{2}|\mathcal{F}| - |\mathcal{F}(n-1,n)| - \sum_{x\in\{n-1,n\}, y\in\{n-3,n-2\}} |\mathcal{F}(x,y)| \geq \frac{1}{2}|\mathcal{F}| - 5\Delta_2(\mathcal{F}).$$

Then

$$|\mathcal{H}(\bar{1},\bar{2})| = |\mathcal{H}| - \sum_{x\in\{n-1,n\}, y\in\{1,2\}} |\mathcal{F}(x,y)| \geq \frac{1}{2}|\mathcal{F}| - 9\Delta_2(\mathcal{F}) \geq 6\binom{n-3}{k-3}.$$

Claim $|\mathcal{G}| < 4\binom{n-3}{k-3}$.

Proof By Lemma 4.17 we have $|\mathcal{G}(1,2)| \leq \Delta_2(\mathcal{F}) < 2\binom{n-3}{k-3}$. By shiftedness, $\mathcal{G}(\bar{1},\bar{2})$ are 3-intersecting. By Theorem 2.4, we have $|\mathcal{G}(\bar{1},\bar{2})| \leq \binom{n-6}{k-3}$. By shiftedness of $\mathcal{G}$, $\mathcal{H}(\bar{1},\bar{2}), \mathcal{G}(1,\bar{2})$ are cross 2-intersecting. Since $|\mathcal{H}(\bar{1},\bar{2})| > \binom{n-6}{k-3}$, by shiftedness and Theorem 2.5 we infer that $|\mathcal{G}(\bar{1},2)| \leq |\mathcal{G}(1,\bar{2})| \leq \binom{n-6}{k-4}$. Thus,

$$|\mathcal{G}| = |\mathcal{G}(1,2)| + |\mathcal{G}(1,\bar{2})| + |\mathcal{G}(\bar{1},2)| + |\mathcal{G}(\bar{1},\bar{2})| \leq 2\binom{n-3}{k-3} + 2\binom{n-6}{k-4} + \binom{n-6}{k-3} < 4\binom{n-3}{k-3}.$$

□

Since $\varrho(\mathcal{F}) \leq \frac{1}{2}$, we infer that $|\mathcal{F}(\bar{n})| \geq \frac{1}{2}|\mathcal{F}| \geq 24\binom{n-3}{k-3}$. Let $D = \{n-3, n-2, n-1, n\}$. For $i \in D$, recall that

$$\mathcal{F}(\{i\}, D) = \{F \setminus D \colon F \in \mathcal{F}, F \cap D = \{i\}\}.$$

Then

$$\sum_{n-3\leq i\leq n-1} |\mathcal{F}(\{i\}, D)| \geq |\mathcal{F}(\bar{n})| - \sum_{n-3\leq x<y\leq n-1} |\mathcal{F}(x,y)| - |\mathcal{G}| \geq 14\binom{n-3}{k-3}.$$

It follows that there exists $u \in \{n-3, n-2, n-1\}$ such that $|\mathcal{F}(\{u\}, D)| > \frac{14}{3}\binom{n-3}{k-3}$.

Now let $V = D \setminus \{u\}$. Since $|\mathcal{F}(\bar{u})| \geq \frac{1}{2}|\mathcal{F}| \geq 24\binom{n-3}{k-3}$. It follows that

$$\sum_{i\in V} |\mathcal{F}(\{i\}, D)| \geq |\mathcal{F}(\bar{u})| - \sum_{x,y\in V,\ x<y} |\mathcal{F}(x,y)| - |\mathcal{G}| \geq 14\binom{n-3}{k-3}.$$

Then there exists $v \in V$ such that $|\mathcal{F}(\{v\}, D)| > \frac{14}{3}\binom{n-3}{k-3}$. Moreover,

$$|\mathcal{F}(\{u\}, D \cup \{1\})| \geq |\mathcal{F}(\{u\}, D)| - |\mathcal{F}(u,1)| > \binom{n-5}{k-3}.$$

Similarly, $|\mathcal{F}(\{v\}, D \cup \{1\})| > \binom{n-5}{k-3}$. However, by shiftedness $\mathcal{F}(\{u\}, D \cup \{1\})$, $\mathcal{F}(\{v\}, D \cup \{1\})$ are cross 2-intersecting, contradicting Theorem 2.5. □

5 Open Problems

Although the shifted ad extremis is powerful, it seems still difficult to attack the following conjectures.

Conjecture 5.1 ([29]) For $k > k_0(s)$, there exist constants $c_0(k)$ and $c_1(k)$ depending on k such that

$$c_0(k)k^{s-3} \le \frac{f(n,k,s)}{\binom{n-s}{k-s}} - k^{s-1} + \binom{s-1}{2}k^{s-2} \le c_1(k)k^{s-3}. \tag{22}$$

holds for $n > n_0(k)$.

The conjecture was based on the following construction. Let $X_0, X_1, \ldots, X_{s-2}$ be disjoint subsets of $[2, n]$ with $|X_i| = k - i$, $i = 0, 1, \ldots, s-2$. Define

$$\mathcal{E}_i = \{X_i \cup \{x_0, \ldots, x_{i-1}\} : x_j \in X_j \text{ for } 0 \le j \le i-1\}.$$

Set $\mathcal{E}(s) = \mathcal{E}_0 \cup \mathcal{E}_1 \cup \ldots \cup \mathcal{E}_{s-2}$. One can check that $\mathcal{E}(s)$ is an intersecting family with covering number $s-1$. Define

$$\mathcal{F}(n,k,s) = \left\{F \in \binom{[n]}{k} : 1 \in F \text{ and there exists } B \in \mathcal{B}(\mathcal{E}(s)) \text{ such that } B \subset F\right\} \cup \mathcal{E}(s).$$

Then $\mathcal{F}(n,k,s)$ is an intersecting family with covering number s. Note that for k sufficiently large with respect to s, the number of covers of $\mathcal{E}(s)$ is exactly $k(k-1)\cdots(k-s+2) = k^{s-1} - \binom{s-1}{2}k^{s-2} + O(k^{s-3})$. It follows that

$$|\mathcal{F}(n,k,s)| = \left(k^{s-1} - \binom{s-1}{2}k^{s-2} + O(k^{s-3})\right)\binom{n-s}{k-s}.$$

Conjecture 5.2 ([34]) There exists an $\epsilon > 0$ such that for $n > n_0(k)$ and $k > k_0$,

$$f(n,k,k) \le ((1-\epsilon)k)^k.$$

Conjecture 5.3 ([31]) Suppose that $\mathcal{F} \subset \binom{[n]}{k}$ is intersecting, $n > 100k$, $|\mathcal{F}| > \binom{n-3}{k-3}$. Then

$$\varrho(\mathcal{F}) \ge \frac{3}{7}. \tag{23}$$

Note that the following construction attains the equality in (23). Let $L_1, L_2, \ldots, L_7$ be the 7 sets of size 3 corresponding to the seven lines of the Fano plane. Without loss of generality, we may assume that

$$L_1 = (1, 2, 3),\ L_2 = (1, 4, 5),\ L_3 = (1, 6, 7),\ L_4 = (2, 4, 6),$$
$$L_5 = (2, 5, 7),\ L_6 = (3, 5, 6),\ L_7 = (3, 4, 7).$$

For $n \geq 10$ and $k \geq 3$, let

$$\mathcal{F} = \left\{ F \in \binom{[n]}{k} : F \cap [7] = L_i \text{ for some } i \in [7] \right\}.$$

Then it is easy to see that $\varrho(\mathcal{F}) = \frac{3}{7}$.

Conjecture 5.4 ([35]) Suppose that $\mathcal{F} \subset \binom{[n]}{k}$ is intersecting. If $n \geq 4k$, then

$$\gamma(\mathcal{F}) \leq \binom{n-3}{k-2}.$$

In [31] it was showed that the Conjecture holds for $n > 36k$. Thus the conjecture is open for the range $4k < n \leq 36k$.

Acknowledgements I would like to thank Peter Frankl for bringing me into the field of extremal set theory and sharing me many nice ideas. I also would like to thank Gyula O.H. Katona for inviting me to give a talk on the SUMMIT 280 conference, which was the initial motivation of writing this survey. The work was supported by National Natural Science Foundation of China Grant no. 12471316.

References

1. Ahlswede, R., Khachatrian, L.H.: The complete intersection theorem for systems of finite sets. Eur. J. Combin. **18**, 125–136 (1997)
2. Ahlswede, R., Khachatrian, L.H.: The complete nontrivial-intersection theorem for systems of finite sets. J. Combin. Theory Ser. A **76**, 121–138 (1996)
3. Alon, N.: Ph.D. Thesis, Hebrew University (1983)
4. Arman, A., Troy, R.: An upper bound for the size of a k-uniform intersecting family with covering number k. J. Comb. Theory, Ser. A **147**, 18–26 (2017)
5. Borg, P.: Intersecting and cross-intersecting families of labeled sets. Electron. J. Combin. **15**(1), N9 (2008)
6. Chao, T., Yu, H.H.: Kruskal-Katona-type problems via the entropy method. J. Comb. Theory, Ser. B **169**, 480–506 (2024)
7. Cherkashin, D.: About maximal number of edges in hypergraph-clique with chromatic number 3 (2011). arXiv:1107.1869
8. Daykin, D.E.: Erdős-Ko-Rado from Kruskal-Katona. J. Comb. Theory Ser. A **17**, 254–255 (1974)

9. Deza, M., Erdős, P., Frankl, P.: Intersection properties of systems of finite sets. Proc. London Math. Soc. (3) **36**(2), 369–384 (1978)
10. Erdős, P., Rado, R.: Intersection theorems for systems of sets. J. Lond. Math. Soc. **35**, 85–90 (1960)
11. Erdős, P., Ko, C., Rado, R.: Intersection theorems for systems of finite sets. Quart. J. Math. Oxford Ser. **12**, 313–320 (1961)
12. Erdős, P., Lovász, L.: Problems and results on 3-chromatic hypergraphs and some related questions. In: Hajnal, A. et al. (ed.) Infinite and Finite Sets, Proceedings of Colloquium Mathematical Society Janos Bolyai, vol. 10, pp. 609–627, North-Holland, Amsterdam (1975)
13. Frankl, P.: The Erdős-Ko-Rado theorem is true for $n = ckt$. Coll. Math. Soc. J. Bolyai **18**, 365–375 (1978)
14. Frankl, P.: On intersecting families of finite sets. J. Combin. Theory Ser. A **24**, 146–161 (1978)
15. Frankl, P.: On intersecting families of finite sets. Bull. Austral. Math. Soc. **21**(3), 363–372 (1980)
16. Frankl, P.: A new short proof for the Kruskal-Katona theorem. Discret. Math. **48**, 327–329 (1984)
17. Frankl, P.: The shifting technique in extremal set theory. Surv. Combi. **123**, 81–110 (1987)
18. Frankl, P.: Linear independence, a unifying approach to shadow theorems **57**, 36–39 (2016)
19. Frankl, P.: Antichains of fixed diameter. Mosc. J. Comb. Number Theory **7**, 189–219 (2017)
20. Frankl, P.: A simple proof of the Hilton-Milner theorem. Moscow J. Comb. Number Theory **8**, 97–101 (2019)
21. Frankl, P.: A near-exponential improvement of a bound of Erdős and Lovász on maximal intersecting families. Comb. Prob. Comput. **28**(5), 733–739 (2019)
22. Frankl, P.: Erdős-Ko-Rado theorem for a restricted universe. Electron. J. Combin. (2) #18 (2020)
23. Frankl, P.: Maximum degree and diversity in intersecting hypergraphs. J. Combin. Theory, Ser. B **144**, 81–94 (2020)
24. Frankl, P.: On the maximum of the sum of the sizes of non-trivial cross-intersecting families. Combinatorica **44**, 15–35 (2024)
25. Frankl, P., Füredi, Z.: Non-trivial intersecting families. J. Combin. Theory Ser. A **41**, 150–153 (1986)
26. Frankl, P., Tokushige, N.: Some best possible inequalities concerning cross-intersecting families. J. Combin. Theory Ser. A **61**, 87–97 (1992)
27. Frankl, P., Tokushige, N.: Extremal problems for finite sets. Am. Math. Soc. (2018)
28. Frankl, P., Ota, K., Tokushige, N.: Uniform intersecting families with covering number four. J. Comb. Theory, Ser. A **71** (1), 127–145 (1995)
29. Frankl, P., Ota, K., Tokushige, N. Covers in uniform intersecting families and a counterexample to a conjecture of Lovász. J. Comb. Theory, Ser. A **74**(1), 33–42 (1996)
30. Frankl, P., Wang, J.: Intersections and distinct intersections in cross-intersecting families. Eur. J. Combin. **110**, 103665 (2022)
31. Frankl, P., Wang, J.: Improved bounds on the maximum diversity of intersecting families. Eur. J. Combin. **118**, 103885 (2024)
32. Frankl, P., Wang, J.: Improved bounds concerning the maximum Degree of intersecting hypergraphs. Electron. J. Comb. **31**(2), #P2.33 (2024)
33. Frankl, P., Wang, J.: Intersecting families with covering number three. J. Comb. Theory, Ser. B **171**, 96–139 (2025)
34. Frankl, P., Wang, J.: Best possible bounds on the double-diversity of intersecting hypergraphs. Comb. Number Theory **14**, 91–112 (2025)
35. Frankl, P., Wang, J.: On the C-diversity of intersecting hypergraphs. Eur. J. Combin. **130**, 104199 (2025)
36. Frankl, P., Wang, J.: Intersecting families with covering number five. Discret. Math. **348**, 114546 (2025)
37. Füredi, Z.: On finite set-systems whose every intersection is a kernel of a star. Discrete Math. **47**, 129–132 (1983)

38. Füredi, Z., Hwang, K-W., Weichsel, P.M.: A proof and generalizations of the Erdős-Ko-Rado theorem using the method of linearly independent polynomials. In: Topics in Discrete Mathematics, vol. 26, pp. 215–224. Springer, Berlin (2006)
39. Gao, J., Liu, H., Xu, Z.: Stability through non-shadows. Combinatorica **43**, 1125–1137 (2023)
40. Hurlbert, G., Kamat, V.: New injective proofs of the Erdős-Ko-Rado and Hilton-Milner theorems. Discrete Math. **341**, 1749–1754 (2018)
41. Hilton, A.J.W.: The Erdős-Ko-Rado Theorem with valency conditions, unpublished manuscript (1976)
42. Hilton, A.J.W., Milner, E.C.: Some intersection theorems for systems of finite sets. Quart. J. Math. Oxford Ser. **18**, 369–384 (1967)
43. Hoffman, A.: On eigenvalues and colourings of graphs. In: Graph Theory and its Applications, pp. 79–91. Academic, New York (1970)
44. Huang, H.: Two extremal problems on intersecting families. Eur. J. Comb. **76**, 1–9 (2019)
45. Katona, G.O.H.: Intersection theorems for systems of finite sets. Acta Math. Acad. Sci. Hung. **15**, 329–337 (1964)
46. Katona, G.O.H.: A theorem of finite sets, Theory of Graphs. Proc. Colloq. Tihany, Akad. Kiadó, 187–207 (1966)
47. Katona, G.O.H.: A simple proof of the Erdős-Ko-Rado theorem. J. Combin. Theory, Ser. B **13**, 183–184 (1972)
48. Kruskal, J.B.: The number of simplices in a complex, Mathematical Optimization. Techniques **251**, 251–278 (1963)
49. Kupavskii, A.: Diversity of uniform intersecting families. Eur. J. Comb. **74**, 39–47 (2018)
50. Kupavskii, A.: Intersecting families with covering number 3. J. Combin. Theory, Ser. B **177**, 216–233 (2026)
51. Kupavskii, A., Zakharov, D.: Regular bipartite graphs and intersecting families. J. Comb. Theory Ser. A **155**, 180–189 (2018)
52. Lovász, L.: On minimax theorems of combinatorics. Math. Lapok **26**, 209–264 (1975)
53. Lovász, L.: On the shannon capacity of a graph. IEEE Trans. Inf. Theory **25**(1), 1–7 (1979)
54. Lovász, L.: Combinatorial Problems and Exercises, 13.31, North-Holland, Amsterdam (1979)
55. Mörs, M.: A generalization of a theorem of Kruskal. Graphs Combin. **1**, 167–183 (1985)
56. Tokushige, N.: The eigenvalue method for cross-t-intersecting families. J. Algebraic Combin. **38**(3), 653–662 (2013)
57. Tuza, Z.: Inequalities for minimal covering sets in set systems of given rank. Discret. Appl. Math. **51**, 187–195 (1994)
58. Wilson, R.M.: The exact bound in the Erdős-Ko-Rado theorem. Combinatorica **4**, 247–257 (1984)
59. Zakharov, D.: On the size of maximal intersecting families. Comb. Prob. Comput. **33**, 32–49 (2024)
60. Zhang, H., Wu, B.: On a conjecture of Tokushige for cross-t-intersecting families. J. Comb. Theory, Ser. B **171**, 49–70 (2025)

GPSR Compliance
The European Union's (EU) General Product Safety Regulation (GPSR) is a set of rules that requires consumer products to be safe and our obligations to ensure this.

If you have any concerns about our products, you can contact us on

ProductSafety@springernature.com

In case Publisher is established outside the EU, the EU authorized representative is:

Springer Nature Customer Service Center GmbH
Europaplatz 3
69115 Heidelberg, Germany

www.ingramcontent.com/pod-product-compliance
Ingram Content Group UK Ltd.
Pitfield, Milton Keynes, MK11 3LW, UK
UKHW020933180926
475692UK00003B/19
* 9 7 8 3 0 3 2 1 8 8 0 9 0 *